Geometric Analysis

IAS/PARK CITY
MATHEMATICS SERIES
Volume 22

Geometric Analysis

Hubert L. Bray
Greg Galloway
Rafe Mazzeo
Natasa Sesum
Editors

American Mathematical Society
Institute for Advanced Study

Hubert Lewis Bray, Gregory J. Galloway, Rafe Mazzeo,
and Natasa Sesum, Volume Editors

IAS/Park City Mathematics Institute runs mathematics education programs that bring together high school mathematics teachers, researchers in mathematics and mathematics education, undergraduate mathematics faculty, graduate students, and undergraduates to participate in distinct but overlapping programs of research and education. This volume contains the lecture notes from the Graduate Summer School program

2010 *Mathematics Subject Classification.* Primary 53-06, 35-06, 83-06.

Library of Congress Cataloging-in-Publication Data

Geometric analysis / Hubert L. Bray, editor [and three others].
pages cm. — (IAS/Park City mathematics series ; volume 22)
Includes bibliographical references.
ISBN 978-1-4704-2313-1 (alk. paper)
1. Geometric analysis. 2. Mathematical analysis. I. Bray, Hubert L., editor.

QA360.G455 2015
515′.1—dc23 2015031562

♾ The paper used in this book is acid-free and falls within the guidelines
established to ensure permanence and durability.
Visit the AMS home page at `http://www.ams.org/`

10 9 8 7 6 5 4 3 2 1 22 21 20 19 18 17 16

Contents

Preface

The IAS/Park City Mathematics Institute (PCMI) was founded in 1991 as part of the "Regional Geometry Institute" initiative of the National Science Foundation. In mid 1993 the program found an institutional home at the Institute for Advanced Study (IAS) in Princeton, New Jersey.

The IAS/Park City Mathematics Institute encourages both research and education in mathematics and fosters interaction between the two. The three-week summer institute offers programs for researchers and postdoctoral scholars, graduate students, undergraduate students, high school teachers, undergraduate faculty, and researchers in mathematics education. One of PCMI's main goals is to make all of the participants aware of the total spectrum of activities that occur in mathematics education and research. We wish to involve professional mathematicians in education and to bring modern concepts in mathematics to the attention of educators. To that end, the summer institute features general sessions designed to encourage interaction among the various groups. In-year activities at the sites around the country form an integral part of the High School Teachers Program.

Each summer a different topic is chosen as the focus of the Research Program and Graduate Summer School. Activities in the Undergraduate Summer School deal with this topic as well. Lecture notes from the Graduate Summer School are being published each year in this series. The first twenty one volumes are:

- Volume 1: *Geometry and Quantum Field Theory* (1991)
- Volume 2: *Nonlinear Partial Differential Equations in Differential Geometry* (1992)
- Volume 3: *Complex Algebraic Geometry* (1993)
- Volume 4: *Gauge Theory and the Topology of Four-Manifolds* (1994)
- Volume 5: *Hyperbolic Equations and Frequency Interactions* (1995)
- Volume 6: *Probability Theory and Applications* (1996)
- Volume 7: *Symplectic Geometry and Topology* (1997)
- Volume 8: *Representation Theory of Lie Groups* (1998)
- Volume 9: *Arithmetic Algebraic Geometry* (1999)
- Volume 10: *Computational Complexity Theory* (2000)
- Volume 11: *Quantum Field Theory, Supersymmetry, and Enumerative Geometry* (2001)
- Volume 12: *Automorphic Forms and their Applications* (2002)
- Volume 13: *Geometric Combinatorics* (2004)
- Volume 14: *Mathematical Biology* (2005)
- Volume 15: *Low Dimensional Topology* (2006)
- Volume 16: *Statistical Mechanics* (2007)

- Volume 17: *Analytic and Algebraic Geometry: Common Problems, Different Methods* (2008)
- Volume 18: *Arithmetic of L-functions* (2009)
- Volume 19: *Mathematics in Image Processing* (2010)
- Volume 20: *Moduli Spaces of Riemann Surfaces* (2011)
- Volume 21: *Geometric Group Theory* (2012)
- Volume 22: *Geometric Analysis* (2013)

Volumes are in preparation for subsequent years.

Some material from the Undergraduate Summer School is published as part of the Student Mathematical Library series of the American Mathematical Society. We hope to publish material from other parts of the IAS/PCMI in the future. This will include material from the High School Teachers Program and publications documenting the interactive activities that are a primary focus of the PCMI. At the summer institute late afternoons are devoted to seminars of common interest to all participants. Many deal with current issues in education: others treat mathematical topics at a level which encourages broad participation. The PCMI has also spawned interactions between universities and high schools at a local level. We hope to share these activities with a wider audience in future volumes.

Rafe Mazzeo
Director, PCMI
January, 2016

Introduction

Amongst the many great advances in mathematics in the last part of the twentieth century, the successes in geometric analysis rank very highly. The most spectacular are the resolutions of the Poincaré conjecture and Thurston's geometrization conjecture, through the work of Perelman and the many people who clarified and extended his ideas. These were proved using Hamilton's Ricci flow, a powerful tool in the subject later used by Brendle and Schoen to prove the differentiable sphere theorem in higher dimensions. Beyond these, we mention the many dramatic advances in the study of minimal submanifolds, harmonic mappings and related variational problems, the deeper understanding of general relativity using tools from PDE, Riemannian and Lorentzian geometry, the use of gauge theory to detect subtle new topological invariants, and the relationship between the spectral behavior of the Laplace-Beltrami operator on a Riemannian manifold and the dynamical properties of the underlying geodesic flow.

It is not easy to give a comprehensive definition of this subject, and the name 'geometric analysis' has only been in common currency for the last 25 years or so. Loosely speaking, this field involves the many interlocking relationships between geometry and partial differential equations. These interconnections go both ways. For example, a 'purely geometric' problem, such as finding the optimal shape of a manifold, can be translated into an equivalent problem which involves solving a PDE. If a solution of that equation can be found, this can then be translated back into a solution of the original geometric problem. A classic instance is the uniformization theorem, where one seeks optimal (constant curvature) metrics on surfaces. There are several different analytic approaches, the earliest involving complex analysis and later ones involving semilinear elliptic PDE's. In the other direction, new perspectives in the field of PDE and many new techniques to solve various classes of equations have been inspired by the geometry underlying these equations. Among the many examples here, deep advances in fully nonlinear elliptic equations originated in the fundamental breakthroughs by Yau and others on Monge-Ampere equations arising in geometry. In a different direction, the entire modern theory of linear partial differential using microlocal analysis, pioneered by Hörmander, Kohn, Nirenberg and others, relies on a new way of viewing linear PDE through a geometric lens and exploiting the deep connections with symplectic geometry.

The research area highlighted in the 2013 session of the Park City Mathematics Institute was geometric analysis. The program of this summer school included lectures by: Michael Eichmair, Fernando Coda Marques, Tristan Riviere, Igor Rodnianski, Peter Topping, Jeff Viaclovsky, Ben Weinkove, Brian White, Steve Zelditch, and the Clay Scholars Gerhard Huisken and Richard Schoen. All were chosen both for the excellence of their mathematical work as well as their expository talents.

This volume collects papers contributed by Huisken, Marques, Riviere, Topping, Viaclovsky, Weinkove, White and Zelditch. The topics covered include general relativity, the proof of the long-standing Willmore conjecture as well as analytic aspects of the Willmore equation, Ricci and Kähler-Ricci flow, critical metrics, minimal surfaces and spectral theory. The papers here represent the lectures for the Graduate Summer School at PCMI, presented to an audience of 80 graduate students and 60 researchers. We also include a paper by the Clay Senior Scholar Gerhard Huisken, loosely based on his public lecture during the summer session. The organizers are grateful to the Clay Mathematics Institute for their sponsorship and support of two Clay Senior Scholars during the 2013 PCMI session: Gerhard Huisken and Richard Schoen.

As with every PCMI volume, this collection of papers is meant to serve as a high level introduction to many of the most important topics in geometric analysis by some of the great experts in the field, and is intended for graduate students or anyone else wishing an entrée into the field. The 2013 session was marked by recollections by the more seasoned researchers of their participation in the previous PCMI session on geometric analysis in 1992, and the lasting influence that workshop had on their careers. We can only hope that the 2013 session of PCMI will have a similarly long-lasting and far-reaching effect in this wonderful field.

Hugh Bray, Greg Galloway, Rafe Mazzeo, Natasa Sesum

Heat Diffusion in Geometry

Gerhard Huisken

IAS/Park City Mathematics Series
Volume 22, 2013

Heat Diffusion in Geometry

Gerhard Huisken

This article is an expanded version of a lecture the author held in July 2013 at the PCMI summer school at Park City. The author had the position of Clay Senior Scholar during this summer school, and gratefully acknowledges this support from the Clay Mathematics Institute. The article starts with a description of linear heat diffusion in a simple one dimensional setting that should be accessible to advanced high school students and is just based on the fundamental theorem of calculus. Later sections explain how properties of heat diffusion already present in the basic one dimensional setting extend to geometric deformations like curve shortening flow, mean curvature flow of hypersurfaces and the Ricci flow of Riemannian metrics. The discussion focuses on selected examples and results to demonstrate characteristic features of geometric evolution equations and does not claim to be complete in any way as many beautiful results had to be left out.

1. Heat diffusion

Consider a long, thin bar of a metal such as copper. In such material heat loss to the outside through contact with the air or through radiation is slow compared to the heat flux inside the metal, so it is a good approximation to describe *heat diffusion* in this case by a temperature distribution depending on just one space and one time variable. Suppose the metal bar has length L and

$$u : [0, L] \times [0, T) \to \mathbb{R}$$

describes the temperature distribution in the bar on some time interval $[0, T)$, i.e. $u(x, t)$ is the temperature at length x at time t. We assume the *heat capacity* of the material to be constant ($c_h = 1$) and define the *heat content* in the interval $[a, b] \subset (0, L)$ at time t as

$$Q_{a,b}(t) := \int_a^b u(x, t)\, dx.$$

To explore how the temperature distribution u changes in time, we have to use a crucial input from Physics: The *heat flux* $\Phi(x, t)$ inside the material in direction of the $x - axis$ is proportional to the (negative) spatial derivative of u at (x, t), i.e. heat flows from hot to cold at a rate proportional to the infinitesimal temperature differences. If we assume the *thermal conductivity* again to be constant ($k = 1$) this heat transfer law reads

$$\Phi(x, t) = -\frac{\partial}{\partial x} u(x, t).$$

Fachbereich Mathematik, Universität Tübingen Auf der Morgenstelle 10, 72076 Tübingen, Germany
E-mail address: gerhard.huisken@uni-tuebingen.de

If there is no heating source in the metal such that heat can only enter and leave through the end points of an interval $(a,b) \subset [0,L]$, we then get from the fundamental theorem of calculus

$$\begin{aligned}\frac{d}{dt}Q_{a,b}(t) &= \Phi(a,t) - \Phi(b,t)\\ &= -\frac{d}{dx}u(a,t) + \frac{d}{dx}u(b,t)\\ &= \int_a^b \frac{d^2}{dx^2}u(x,t)\,dx\end{aligned}$$

On the other hand we can compute the change of heat from the time derivative of the temperature

$$\frac{d}{dt}Q_{a,b}(t) = \int_a^b \frac{d}{dt}u(x,t)\,dx$$

such that we have the identity

$$\int_a^b \frac{d}{dt}u(x,t) - \frac{d^2}{dx^2}u(x,t)\,dx = 0$$

for *any* interval $(a,b) \subset [0,L]$ and at each time $t \in [0,T)$. Since we assume here all functions involved to be continuous, it must be true that

$$\frac{d}{dt}u(x,t) = \frac{d^2}{dx^2}u(x,t) \qquad (*)$$

holds everywhere on $[0,L] \times [0,T)$. (Otherwise we could find at least a tiny interval somewhere where the previous integral identity was wrong). The equation $(*)$ is called the *heat equation.*

To avoid the discussion of boundary data let us assume that $L = 2\pi$ and the functions $u(\cdot,t)$ are periodic, $u(x,t) = u(x+2\pi,t)$. Equivalently,

$$u : S^1 \times [0,T) \to \mathbb{R}$$

describes the temperature distribution in a thin ring of metal. Using separation of variables $u(x,t) = \psi(x)\varphi(t)$ as an ansatz, explicit solutions to the heat equation can be easily found in the form

$$u(x,t) = a_m \sin(mx)\exp(-m^2t) + b_m \cos(mx)\exp(-m^2t)$$

for $m = 0,1,2,\cdots$. These special solutions already exhibit many of the essential features of solutions to the heat equation. Indeed, since the heat equation is linear, any superposition of such solutions is again a solution and by choosing the coefficients a_m, b_m appropriately (as Fourier coefficients of the initial temperature $u(\cdot,0)$) we can solve the heat equation for very general initial data. For this lecture the following properties of all these solutions are crucial:

i) *The total heat content is constant:* Using integration by parts we get

$$\frac{d}{dt}Q = \int_{S^1} \frac{d^2}{dt^2}u\,dx = 0.$$

So in the special case where $u > 0$ and $Q = 1$ we may view the heat equation as a flow on the space of probability measures.

ii) *The comparison principle:* If there are two solutions u_1 and u_2 such that $u_1(x,0) \le u_2(x,0)$ for all $x \in S^1$ then $u_1(x,t) \le u_2(x,t)$ for all $x \in S^1, t \ge 0$.
Proof Let $\epsilon > 0$ be arbitrary and assume that there is a first time $t_0 > 0$ where

at some point x_0 the function u_1 reaches the function $v(x,t) := u_2(x,t) + \epsilon \exp t$, i.e. $u_1(x_0,t_0) = u_2(x_0,t_0) + \exp t_0$. At (x_0,t_0) we must have $\frac{d}{dt}(v-u_1)(x_0,t_0) \leq 0$ and $\frac{d^2}{dx^2}(v-u_1)(x_0,t_0) \geq 0$ since at (x_0,t_0) the new minimum value 0 is assumed. But by the heat equation $(\frac{d}{dt} - \frac{d^2}{dx^2})(v-u_1)(x_0,t_0) = \epsilon \exp t > 0$, a contradiction. Hence we must have $u_1 < v$ for all (x,t). Since this is true for all $\epsilon > 0$, the result follows.

iii) *Smoothing property:* If the solution u satisfies a bound $\sup |u| \leq M$ on some time interval $[0,T]$ then at the end of the time interval we get a derivative bound:

$$\sup_{S^1} \left| \frac{\mathrm{d}}{dx} u(x,T) \right|^2 \leq \frac{M^2}{2T},$$

with similar, scaling invariant estimates holding for all higher derivatives.
Proof Compute from the heat equation

$$(\frac{d}{dt} - \frac{d^2}{dx^2})\left(u^2 + 2t|\frac{d}{dx}u|^2\right) \leq 0$$

and the conclusion follows from the comparison principle.

iv) *Asymptotic convergence:* As $t \to \infty$ all solutions converge exponentially fast to the constant function.
Proof Note that all the special solutions exhibited above converge to a constant exponentially fast, hence any superposition of them.

v) *Entropy monotonicity:* Suppose $u > 0$ and define the entropy E by setting

$$E(u) := \int_{S^1} (\log(u^{-1}))u\,dx = -\int_{S^1} (\log u)u\,dx.$$

Then

$$\begin{aligned} \frac{d}{dt}E(u(\cdot,t)) &= -\int_{S^1} \frac{d^2}{dx^2}u\,dx - \int_{S^1} (\log u)\frac{d^2}{dx^2}u\,dx \\ &= \int_{S^1} u^{-1}\left|\frac{d}{dx}u\right|^2 dx \geq 0. \end{aligned}$$

This monotonicity formula describes the rate at which information is lost as u approaches a constant function for $t \to \infty$.

All the properties of the heat equation on the circle above carry over to analogous results for the linear heat equation on arbitrary closed n-dimensional Riemannian manifolds (M^n, g), where we replace the second derivative $(d^2/dx^2)u$ by the Laplace-Beltrami operator $\Delta_g u$ and solve

$$\frac{d}{dt}u = \Delta_g u.$$

The nonlinear geometric evolution equations considered in the following sections are all of parabolic type and hence inherit versions of the basic properties of the heat equation discussed above while exhibiting additional nonlinear features arising from the geometric situation under consideration.

2. Curve shortening

The image of the ordinary heat equation smoothing out the graph of a function over a circle until eventually the graph of the solution turns into a round circle itself inspires the question whether we can find a more geometric version of the heat equation which can deform arbitrary closed, embedded curves into circles. This seems to be not entirely hopeless since the well-known Jordan curve theorem tells us that this can certainly be done if we are allowed to use arbitrary continuous deformations of the embedded closed curve.

Let $\gamma_0 : S^1 \to \mathbb{R}^2$ be an initial curve in the plane with outer unit normal ν which is parametrized by arc length s such that the *curvature vector* $\vec{\kappa} = -\kappa\nu$ is given by

$$\vec{\kappa} = \frac{d^2}{ds^2}\gamma.$$

In particular, the curvature κ of a round circle of radius r equals $1/r$. We then look for a one-parameter family of curves $\gamma : S^1 \times [0,T) \to \mathbb{R}^2$ satisfying the system of partial differential equations

$$\frac{d}{dt}\gamma = \vec{\kappa} = \frac{d^2}{ds^2}\gamma, \qquad \gamma(\cdot,0) = \gamma_0.$$

Here $s = s(t)$ is the arc length of the evolving curve, such that this system of equations is non-linear. It is not hard to compute the following basic facts about curve shortening flow:

i) *Example: A round circle shrinks to a point in finite time.* In view of symmetry, the evolution equation reduces to an ODE for the radius of the circle:

$$\frac{d}{dt}r(t) = -\frac{1}{r(t)}$$

such that for an initial radius r_0 the solution is given by $r(t) = \sqrt{r_0^2 - 2t}$ on the time interval $[0, T_{max}), T_{max} = r_0^2/2$.

ii) *Length and area are decreasing:*

$$\frac{d}{dt}L(t) = -\int_{S^1} \kappa^2(x,t)\,dx, \quad \frac{d}{dt}A(t) = -\int_{S^1} \kappa(x,t)\,dx = -2\pi.$$

iii) *The "isoperimetric defect" decreases:*

$$\frac{d}{dt}\Big(L^2(t) - 4\pi A(t)\Big) = -2L\int_{S^1} \kappa^2\,dx + 8\pi^2 \le -2\Big(\int_{S^1} \kappa\,dx\Big)^2 + 8\pi^2 = 0.$$

iv) *A reaction diffusion equation for the curvature:* A short calculation shows that the curvature of the evolving curve satisfies a nonlinear reaction diffusion equation:

$$\frac{d}{dt}\kappa = \frac{d^2}{ds^2}\kappa + \kappa^3.$$

The following theorem due to Grayson [**16**] was first established in the convex case by Gage [**14**] and Gage-Hamilton [**15**]. It can be extended to curve shortening flow in general Riemann surfaces, see [**17**].

Theorem (M. Grayson) *Any closed, embedded initial curve remains embedded and contracts to a point in finite time. It turns first convex and approaches a round shrinking circle near the finite time.*

As immediate corollaries we obtain the *Jordan curve theorem* stating that any closed embedded curve in the plane bounds a region that is diffeomorphic to a disc as well as the isoperimetric inequality:

Corollary *Any planar region of area A bounded by a curve of length L satisfies the inequality*

$$L^2 \ge 4\pi A.$$

Proof Evolve the boundary curve by curve shortening flow. In view of Grayson's theorem it will smoothly contract to a point as $t \to T_{max} = A/2\pi$ when both area $A(t)$ and length $L(t)$ tend to zero. But as we saw in (iii) the isoperimetric defect is decreasing along the flow, proving the desired inequality at $t = 0$.

The original proof of the theorem due to Grayson controls the shape of the curve along its evolution with the help of a number of delicate estimates exploiting the fact that the number of inflection points is decreasing. Here we sketch the steps of a different approach [**25**]which only examines the first singularity of the flow: If it can be shown that the curve remains embedded and that near the first singular time the curve approaches a shrinking circle, the theorem is proven. These are the crucial steps:

a) One can use the comparison principle for parabolic equations similar as in section 1 to show that disjoint curves remain disjoint and embedded curves remain embedded under the flow.

b) Since the area is decreasing at a constant rate in view of property (ii) above, a singularity must occur in finite time.

c) In view of the scaling invariant smoothing properties of curve shortening flow resembling that of the heat equation in section 1, it is not hard to see that a singularity can only occur when the curvature becomes unbounded.

d) Using parabolic rescaling near the singularity together with the scaling invariant singularity estimates used already in the previous step it is possible to obtain a nontrivial *tangent curve shortening flow* existing for $-\infty < t < T \leq \infty$ which is approximating the original flow near the singularity. Certain nonlinear monotonicity properties of the curve shortening flow imply that any such singularity model must be strictly convex and either a homothetically shrinking or a translating solution of the curve shortening flow. See [**26**] and [**18**] for details.

e) Since self similar solutions as obtained from d) satisfy a simple ODE, it is not hard to see that the only embeddded convex homothetically shrinking solution is the shrinking round circle as desired in the theorem. The only possible translating convex solution is (up to scaling) given as the graph of $y(x,t) = -\log\cos x + t$, which looks like a *hairpin curve* approaching two parallel lines near infinity. If we can rule out this hairpin curve which is also called the *grim reaper curve* in the literature since every other solution of curve shortening flow in its path has to become distinct before it passes through, then Grayson's theorem is proved.

f) The hairpin curve can be ruled out as a singularity model for embedded curves by proving a quantitative estimate on the quality of the embedding that compares the intrinsic arc length $l(x,y,t)$ between two points $\gamma(x,t), \gamma(y,t)$ on the curve to their extrinsic distance $d(x,y,t) = |\gamma(x,t) - \gamma(y,t)|$ across the plane. In [**25**] the author proved that

$$\inf_{x \neq y} \frac{d(x,y,t)}{L(t)} \left(\sin \left(\frac{l(x,y,t)}{L(t)} \pi \right) \right)^{-1}$$

is non-decreasing in time; it is strictly monotonically increasing unless the curve is a round circle. Since the estimate is scaling invariant, it survives the rescaling process near a singularity and rules out the arbitrarily small ratio between extrinsic and intrinsic distances encountered on the hairpin curve.

The strategy of focusing on the singularities of the flow turns out to be crucial for the study of geometric evolution equations in higher dimensions.

3. Mean curvature flow

Consider a family of smooth, closed hypersurface immersions $F : M^n \times [0,T) \to \mathbb{R}^{n+1}$ with outer unit normal ν and mean curvature $H = \lambda_1 + \cdots + \lambda_n$ being the sum of the principal curvatures λ_i. We say that $M_t^n = F(\cdot,t)(M^n)$ is a solution of mean curvature flow if it satisfies the evolution system

$$\frac{d}{dt}F = \vec{H} = -H\nu = \Delta_g F.$$

Here $\vec{H}$ is the mean curvature vector and Δ_g is the Laplace-Beltrami operator with respect to the evolving induced metric $g = g(t)$ on the hypersurface M_t^n. This generalisation of the curve shortening flow in section 2 is again a quasilinear parabolic system for F, it can be viewed as the L^2- gradient flow of area, we have

$$\frac{d}{dt}|M_t^n| = -\int_{M^n} H^2\, d\mu$$

where $d\mu$ denotes the surface measure on the evolving surfaces. Many properties of the curve shortening flow remain true in this higher dimensional flow, such as the *comparison principle* and the *scaling invariant regularity estimates*, see [**13**], [**12**]. There is a rich collection of well-known examples:

(i) *The shrinking sphere:* If the initial surface is sphere of radius R_0, mean curvature flow reduces to an ODE for the radius of shrinking concentric spheres: d/dt r(t) = -n/r(t) with solution $r(t) = \sqrt{r_0^2 - 2nt}$ on the finite time interval $[0, T_{max}), T_{max} = r_0^2/2n$.

(ii) *Shrinking cylinders:* If $\Sigma_t^m \subset \mathbb{R}^{m+1}$ is a solution of mean curvature flow, the extension $M_t^n = \Sigma_t^m \times \mathbb{R}^{n-m} \subset \mathbb{R}^{n+1}$ is also a solution of mean curvature flow, in particular we have the shrinking cylinders $S^m_{r(t)} \times \mathbb{R}^{n-m}$ with radius $r(t) = \sqrt{r_0^2 - 2mt}$.

(iii) *Convex surfaces:* It was shown in [**27**] that convex surfaces contract to a point in finite time in such a way that rescalings approach the shrinking round sphere solution of example (i).

(iii) *Shrinking donut:* Angenent proved the existence of an embedded, axially symmetric torus in $\mathbb{R}^3$ shrinking to a point under homotheties, see [**2**].

(iv) *Neck pinches:* There are axially symmetric solutions of dumbbell shape that develop a neck pinch in the middle: The existence of such singularities can be easily seen from the comparison principle: If we make the two bubbles at the ends of the dumbbell sufficiently large that two spheres of large radii can serve as inner barriers and at the same time make the part between the bubbles sufficiently small that a small *Angenent torus* can surround it as an outer barrier, then it is clear from the comparison principle that some singularity in the middle has to occur before the bubbles have a chance to shrink off. A rescaling of this singularity approaches the shrinking cylinder $S^{n-1}_{r(t)} \times \mathbb{R}$ from the second example above.

(v) *Degenerate neck pinches:* It has been shown that a degenerate neck pinch can occur if the initial surface is a delicately unbalanced dumbbell surface, where one bubble shrinks into a cusp singularity while the other bubble still contains a smooth region with finite area. In this case a rescaling near the singularity can produce both shrinking cylinders (when rescaling near the waist of the shrinking bubble) and translating surfaces shaped like a parabola (when rescaling at the tip of the developing cusp) as asymptotic singularity model.

(vi) *Non-collapsing:* In mean curvature flow of embedded mean convex hypersurfaces the quality of the embedding can be quantitatively controlled through a non-collapsing estimate similar as in the case of curve shortening in section 2: In work of White [**38**], Sheng-Wang [**34**] and Andrews [**1**] it is shown that at each point of the surface the inscribed radius, i.e. the radius of the largest ball still enclosed by the surface, is bounded from below by a uniform fraction of the inverse of the mean curvature at the same point. The non-collapsing property provides immediate regularity properties of the flow and prevents the development of singularities where several sheets of the solution come together, compare also the recent work of Haslhofer-Kleiner in [**24**].

To understand the behaviour near singularities and to extend the flow beyond singularities in a controlled way, one has to prove certain a priori estimates for the evolving surfaces. These estimates depend on evolution equations for the induced metric on the surface and evolution equations for the curvature of the evolving surfaces that can be computed from the basic mean curvature flow equation $d/dtF = -H\nu$. In particular, the selfadjoint *Weingarten map* $W = \{h^i_j\}_{1\le i,j\le n}$ encapsulating the curvature properties of the surface with eigenvalues equal to the *principal curvatures* $\lambda_1 \cdots, \lambda_n$ is found to satisfy the following system of reaction diffusion equations:

$$\frac{d}{dt}h^i_j = \Delta_g h^i_j + (\lambda_1^2 + \cdots + \lambda_n^2)h^i_j$$

or, in more abstract, covariant notation

$$\frac{d}{dt}W = \Delta_g W + |W|^2 W.$$

Note that the reaction part of this system, $|W|^2W$, is always pointing in radial direction in the space of Weingarten maps, such that the system of ODEs

$$\frac{d}{dt}W = |W|^2 W.$$

preserves all *cones* in the space of Weingarten maps as invariant sets. Since the diffusion part of the system, $\Delta_g W$, is isotropic with the same diffusion rate in all directions and has the effect of averaging solutions, a comparison principle for the full system of reaction diffusion equations established by Hamilton [**20**] can be applied, extending the basic comparison principle for the heat equation proven in section 1 in a natural way:

Comparison principle for mean curvature flow: *If there is a convex cone C in the space of Weingarten maps, such that for each point on the initial hypersurface $p \in M^n_0$ we have $W(p,0) \in C$, then $W(p,t) \in C$ for all $p \in M^n_t$ and all $t \ge 0$.*

If we take C to be the half space $\{W|H > 0\}$ or the positive cone $\{W|\lambda_i > 0, 1 \le i \le n\}$, we see that mean convex surfaces remain mean convex and convex surfaces remain convex along mean curvature flow. Other classes of surfaces invariant under mean curvature flow include k-convex surfaces where the sum of the smallest k principal curvatures is positive, here $C = \{W|\lambda_{i_1} + \cdots + \lambda_{i_k} > 0\}$, as well as surfaces where one of the elementary symmetric functions σ_k of the principal curvatures is everywhere positive, i.e. $C = \{W|\sigma_k(\lambda_1, \cdots, \lambda_n) > 0\}$.

While the comparison principle is powerful in showing that certain assumptions on the initial hypersurface are preserved during mean curvature flow, the radially directed reaction terms in the evolution system for the curvature cannot on their own help to show that the shape of the surfaces improves in any way. It is therefore necessary to exploit the diffusion part of the flow in an essential way. This can be achieved with the help of integral estimates where integration by parts brings in global properties of the evolving surfaces and a well known Sobolev inequality on submanifolds due to Michael and Simon allows the use of iteration techniques due to DeGiorgi and Stampacchia that eventually lead to estimates describing the non-linear uniformisation properties of the flow. Brakke [**4**], Huisken [**26**], White [**37**], [**38**], Ecker-Huisken [**13**], Huisken-Sinestrari [**28**], [**29**] amongst others have developed a regularity theory in this spirit, see also [**12**] and [**36**]. The following convexity result summarises some of these results and provides an important restraint on the possible shape of singularities in the mean convex case.

Theorem *Let M_0^n be a smooth closed initial surface with $H > 0$. Then*
(i) For any $\eta > 0$ there is a constant C_η depending only on the initial data such that the smallest principal curvature satisfies the estimate

$$\lambda_1 \geq -\eta H - C_\eta .$$

(ii) Any singularity model obtained from M_t^n by rescaling is weakly convex. If it is not strictly convex, it splits into a Euclidean factor and a lower-dimensional strictly convex solution of mean curvature flow.

The convexity estimates are a first step towards a closer quantitative control of singularity profiles that is necessary to extend the flow by surgery. We leave this discussion to the last section.

4. Ricci flow

We have seen that dumbbell shaped 2-dimensional surfaces will develop neck pinch singularities in finite time under mean curvature flow. To avoid such singularities one might think of using the Gauss curvature as deformation speed instead of mean curvature, since the Gauss curvature along a neck is small or negative, intuitively preventing a neck in the surface from shrinking further. Unfortunately, the motion of a hypersurface in direction of its Gauss curvature is not parabolic unless the surface is convex, preventing us from going down that route. Instead, it turns out to be fruitful to use the Gauss curvature and Riemann curvature tensor to define an intrinsic parabolic flow of Riemannian metrics: In a seminal paper Hamilton [**19**] proposed the quasilinear parabolic system of evolution equations

$$\frac{d}{dt} g_{ij} = -2R_{ij}$$

where $R_{ij} = g^{kl} R_{ikjl}$ are the components of the Ricci tensor, which is the trace of the Riemann curvature tensor Riem $= \{R_{ijkl}\}$ of the metric g on a Riemannian manifold $M^n, g)$. This flow exhibits many features that we are already familiar with from mean curvature flow in section 3. This lecture cannot begin to cover all the fascinating results on Ricci flow, we just try to point out some of the similarities with mean curvature flow:

(i) *The shrinking sphere:* If the initial manifold is a sphere of radius r_0, Ricci flow reduces to an ODE for the radius $r(t)$ of shrinking spheres: $(d/dt)r^2(t) = -2(n-1)$ with solution $r^(t) = \sqrt{r_0^2 - 2(n-1)t}$ on the finite time interval $[0, T_{max}), T_{max} = r_0^2/2(n-1)$.

(ii) *Shrinking cylinders:* If $(M^m, g(t))$ is a solution of Ricci flow, the product manifold $(N^n, \tilde{g}(t)) = (M^m \times \mathbb{R}^{n-m}, \tilde{g}(t))$, where $\tilde{g}(t)$ is the product metric of $g(t)$

with the Euclidean metric, is also a solution of Ricci flow, in particular we have the shrinking cylinders $S^m_{r(t)} \times \mathbb{R}^{n-m}$ with radius $r(t) = \sqrt{r_0^2 - 2(m-1)t}$.

(iii) *The reaction diffusion system for the curvature:* While the evolution system for the Riemannian curvature tensor derived from the basic equation $(d/dt)g = -2Ric(g)$,

$$\frac{d}{dt}\text{Riem} = \Delta_g \text{Riem} + \text{Riem} \times \text{Riem} + \text{Riem} \,\#\, \text{Riem},$$

features the same scaling and the same diffusion law as in mean curvature flow, the quadratic reaction terms in this system have a much more complicated structure which turns out to allow the proof of many important estimates via the parabolic comparison principle.

(iv) *The two dimensional case:* In the case of 2-dimensional Riemann surfaces, the evolution equation for the metric describes a conformal change of metric that is equivalent to a quasi-linear second order parabolic equation for the conformal factor with respect to some background metric:

$$\frac{d}{dt}g_{ij} = -Rg_{ij}.$$

In this case the scalar curvature R is just twice the Gauss curvature of the surface and satisfies the evolution equation

$$\frac{d}{dt}R = \Delta_g R + R^2.$$

In this case the behaviour of solutions turns out to be just as nice as in the 1-dimensional mean curvature flow, the curve shortening flow: Given any smooth initial metric on a closed Riemann surface Σ^2, the solution of the Ricci flow equation converges (after a rescaling that keeps the area constant) to a smooth metric of constant Gauss curvature in the same conformal class. This new proof of the uniformisation theorem was shown by Cao [**9**] in the case of non-positive curvature and by Hamilton [**21**] and Chow [**11**] in the case of the sphere, see also Struwe [**35**] for an approach based on concentration compactness. A crucial ingredient of Hamiltons proof in the positive case was a non-linear entropy monotonicity similar to the monotonicity for the 1-dimensional heat equation discussed in section 1.

(v) *Manifolds with positive curvature:* Beginning with the original result of Hamilton [**19**] showing that Riemannian 3-manifolds of positive Ricci curvature contract to zero volume in finite time while approaching a shrinking metric of constant positive curvature, many conditions for the initial manifold were established that ensure convergence to a round shrinking metric of positive constant curvature, just like convexity ensures this result in the case of mean curvature flow: Positive curvature operator was shown to suffice in the case of 4-manifolds by Hamilton, [**20**], before Böhm and Wilking discovered a new class of preserved curvature conditions in higher dimensions and showed that all Riemannian manifolds of positive curvature operator are space forms, see [**3**]. Following this work Brendle and Schoen [**8**] established the diffeomorphic 1/4 pinching theorem, see also the work of Nguyen [**31**] and the account of this part of the theory by Brendle [**5**].

(vi) *Positivity estimates near singularities:* If the initial manifold does not satisfy one of the curvature conditions discussed in (v), a wide range of singularities can occur in Ricci flow in finite time without the volume of the metric becoming extinct and only few of them are understood at this stage. So far progress was only possible in configurations where the rescaling of a singularity leads to solutions of Ricci flow that have non-negative sectional curvature. Positivity estimates for the curvature of Ricci flow solutions that are analogous to the convexity estimates in mean curvature flow discussed at the end of the last section are so far only known in two cases: a) Hamilton-Ivey proved that the rescaling of any finite time singularity of 3-dimensional Ricci flow leads to a solution of non-negative sectional curvature, see [**22**]. b) In the 4-dimensional case Hamilton [**23**] proved that the rescaling of any finite time singularity must have non-negative curvature operator if the initial metric had positive isotropic curvature. In both these cases a) and b) the singularity model obtained from rescaling either has strictly positive curvature or can be split into a Euclidean factor and a lower dimensional Ricci flow of strictly positive curvature.

As in the case of mean curvature flow the positivity result just described is the starting point for a further classification of singularities that eventually allows the extension of mean curvature flow and Ricci flow via surgery.

5. Towards surgery

Both in mean curvature flow and in Ricci flow the reaction diffusion systems for the curvature play a dual role: On the one hand the reaction terms are causing singularities in finite time, but on the other hand the interplay between reaction and diffusion terms ensures that the singularities themselves exhibit features of uniformisation and self similarity. This leads to the uniformisation theorems for embedded curve shortening flow, for 2-dimensional Ricci flow and for initial data of sufficiently positive curvature in both mean curvature flow and Ricci flow as discussed in previous sections.

If singularities occur before uniformisation of the geometry on the whole manifold is achieved, the natural question arises whether the flow can be extended beyond such singularities while keeping complete control of topological changes that occur in the singularity. So far such a controlled extension has only been achieved in configurations where the singularities encountered are distinctly one-dimensional, such that a cylindrical singularity model of the flow with exactly one flat direction can be identified near any singularity that is not uniformly positively curved. Only in these situations a *surgery* can be performed, where at a singular time all cylindrical regions of high curvature are cut out along codimension-one spheres and replaced by spherical caps, reducing the maximum curvature substantially such that the flow can be restarted and extended for a further time interval. To implement a surgery procedure that has only to be repeated finitely often, delicate estimates on the cylindrical regions are needed in both mean curvature flow and Ricci flow that build on but also go beyond the convexity and positivity estimates for the curvature mentioned at the end of sections 3 and 4. A precisely controlled cylindrical structure with only one flat direction of the non-positively curved singular set and a surgery algorithm has been established so far in the following cases:

Ricci flow surgery: (i) A precise structure and normal form for cylindrical regions near singularities of Ricci flow was first established by Hamilton in the case of 4-manifolds of positive isotropic curvature, [**23**]. In this paper Hamilton also describes the general construction for the attachment of spherical caps to cylindrical regions that preserves the positivity estimates valid on smooth Ricci flow both in three and four dimensions. See also [**8**] for a proof based on Perelmanns work in

[**33**] that the number of surgeries required is finite for four-manifolds of positive isotropic curvature.
(ii) On general 3-manifolds Hamilton [**22**] and Perelman [**32**] establish that near singularities that are not strictly positively curved cylindrical rescalings can be found. A crucial step in this classification of singularities is an entropy monotonicity formula of Perelmann in [**32**] ruling out regions of collapsed volume in the evolving 3-manifold at finite times and pointing to fascinating new connections between geometry, optimal transport and renormalisation group flows. Perelmann then implements the surgery construction of [**23**] on cylindrical regions into a delicate algorithm for Ricci flow with surgery in [**33**] and proves that only finitely many surgeries can occur on finite time intervals. This leads to a proof of the famous *Poincare conjecture* stating that any closed, simply connected 3-manifold is diffeomorphic to the sphere as well as to a proof of the geometrisation conjecture of Thurston establishing a complete topological classification of 3-manifolds.

Mean curvature flow surgery: (i) For 2-convex closed immersed hypersurfaces $M^n \to \mathbb{R}^{n+1}$ for $n \geq 3$ Huisken and Sinestrari [**30**] build on the convexity estimates from section 3 to prove the cylindrical structure of non-convex singularities and proceeded to construct a surgery algorithm for such surfaces that terminates after finitely many steps. This leads to a complete classification of such surfaces and their enclosed regions.
(ii) Note that since the conference in Park City Brendle and Huisken have extended in [**7**] the surgery construction from (i) to the case of 2-dimensional embedded mean convex surfaces in $\mathbb{R}^3$, making use of a sharpened version of the non-collapsing estimate of Andrews [**1**] due to Brendle [**6**].

Bibliography

[1] B. Andrews, *Non-collapsing in mean convex mean curvature flow*, preprint (2011), arXiv:1108.0247. MR2967056

[2] S. Angenent, *Shrinking doughnuts, Nonlinear diffusion equations and their equilibrium states*, 3 (Gregynog, 1989), Progr. Nonlinear Differential Equations Appl., vol. 7, Birkhäuser Boston, Boston, MA, 1992, pp. 21–38. MR1167827 (93d:58032)

[3] C. Böhm and B. Wilking, *Manifolds with positive curvature operators are space forms*, Annals of Mathematics **167** (2008), 1079–1097. MR2415394 (2009h:53146)

[4] K. Brakke, *The motion of a surface by its mean curvature*, Princeton Univ. Press , (1978). MR485012 (82c:49035)

[5] S. Brendle, *Ricci flow and the sphere theorem*, Grad. Studies in Math., American Math. Society **111** (2010). MR2583938 (2011e:53035)

[6] S. Brendle, *A sharp bound for the inscribed radius under mean curvature flow*, preprint, (2013), arxiv:13091459.

[7] S. Brendle and G. Huisken, *Mean curvature flow with surgery of mean convex surfaces in* $\mathbb{R}^3$, preprint (2013) arxiv:1309.1461.

[8] S. Brendle and R. Schoen, *Manifolds with 1/4 pinched curvature are space forms*, J. Amer. Math. Soc. **22** (2009), 287–307. MR2449060 (2010a:53045)

[9] H.D. Cao, *Deformation of Kähler metric to Kähler-Einstein metrics on compact Kähler manifolds*, Invent. Math. **81** (1985), 359–372. MR799272 (87d:58051)

[10] B.L. Chen and X.P. Zhu, *Ricci flow with surgery on four-manifolds with positive isotropic curvature*, arXiv:math.DG/0504478, April 2005. MR2409626 (2009h:53147)

[11] B. Chow, *The Ricci flow on the 2-sphere*, J. Differential Geom. **33**, no.2, (1991), pp. 325–334. MR1094458 (92d:53036)

[12] K. Ecker, *Regularity theory for mean curvature flow*, Birkhäuser, Boston (2004). MR2024995 (2005b:53108)

[13] K. Ecker and G. Huisken, *Interior estimates for hypersurfaces moving by mean curvature*, Invent. Math. **105** (1991), 547–569. MR1117150 (92i:53010)

[14] M.E. GAGE, *Curve shortening makes convex curves circular*, Invent. Math. **76**, (1984) no. 2, 357–364. MR742856 (85i:52004)
[15] M.E. GAGE AND R.S. HAMILTON, *The heat equation shrinking convex plane curves*, J. Diff. Geom. **23**, (1996), 69–96. MR840401 (87m:53003)
[16] M. GRAYSON, *The heat equation shrinks embedded plane curves to round points*, J. Diff. Geom. **26**, (1987) no. 2, 285–314. MR906392 (89b:53005)
[17] M. GRAYSON, *Shortening embedded curves*, Annals of Math. **129**, (1989), 71–111. MR979601 (90a:53050)
[18] R.S. HAMILTON, *The Harnack estimate for the mean curvature flow*, J. Differential Geom. **41** (1995), 215–226. MR1316556 (95m:53055)
[19] R.S. HAMILTON, *Three-manifolds with positive Ricci curvature*, J. Differential Geom. **17** (1982), 255–306. MR664497 (84a:53050)
[20] R.S. HAMILTON, *Four-manifolds with positive curvature operator*, J. Differential Geom. **24** (1986), 153–179. MR862046 (87m:53055)
[21] R.S. HAMILTON, *The Ricci flow on surfaces*, Contemporary Mathematics **71** (1988), pp.237–261. MR954419 (89i:53029)
[22] R.S. HAMILTON, *The formation of singularities in the Ricci flow*, Surveys in Differential Geometry **2** International Press (1995), pp.7–136. MR1375255 (97e:53075)
[23] R.S. HAMILTON, *Four-manifolds with positive isotropic curvature*, Comm. in Anal. and Geom. **5** (1997), pp.1–92. MR1456308 (99e:53049)
[24] R. HASLHOFER AND B. KLEINER, *Mean curvature flow of mean convex surfaces*, preprint (2013), arxiv:1304.0926.
[25] G. HUISKEN, *A distance comparison principle for evolving curves*, Asian J. of Math. **2**, (1998), 127–133. MR1656553 (99m:58052)
[26] G. HUISKEN, *Asymptotic behaviour for singularities of the mean curvature flow*, J. Diff. Geometry **31** (1990), 285–299. MR1030675 (90m:53016)
[27] G. HUISKEN, *Flow by mean curvature of convex surfaces into spheres*, J. Diff. Geometry **20** (1984), 237–266. MR772132 (86j:53097)
[28] G. HUISKEN AND C. SINESTRARI, *Mean curvature flow singularities for mean convex surfaces*, Calc. Variations, **8** (1999) 1–14. MR1666878 (99m:58057)
[29] G. HUISKEN AND C. SINESTRARI, *Convexity estimates for mean curvature flow and singularities of mean convex surfaces*, Acta Math., **183** (1999), 45–70. MR1719551 (2001c:53094)
[30] G. HUISKEN AND C. SINESTRARI, *Mean curvature flow with surgeries of two–convex hypersurfaces*, Invent. Math., **175** (2009) 137–221. MR2461428 (2010a:53138)
[31] H.T. NGUYEN, *Isotropic curvature and the Ricci flow*, IMRN. **2010**, no.3, (2010), 536–558. MR2587576 (2011e:53108)
[32] G. PERELMAN, *The entropy formula for the Ricci flow and its geometric applications*, arXiv:math. DG/0211159 v1 November 11,(2002).
[33] G. PERELMAN, *Ricci flow with surgery on three manifolds*, arXiv:math.DG/0303109 v1 March 10, (2003).
[34] W. SHENG AND X.-J. WANG, *Singularity profile in the mean curvature flow*, Methods Appl. Anal. **16** no. 2 (2009), pp.139–155. MR2563745 (2011c:53163)
[35] M. STRUWE, *Curvature flows on surfaces*, Ann. Scuola Norm. Sup. Pisa Cl. Sci. **5** no. 1 (2002), pp.247–274. MR1991140 (2004f:53083)
[36] X.-J. WANG, *Convex solutions to mean curvature flow*, Annals of Math. **173** Nr 3 (2011), 1185–1239. MR2800714
[37] B. WHITE, *The size of the singular set in mean curvature flow of mean convex sets*, J. Amer. Math. Soc. **13** (2000), 665–695. MR1758759 (2001j:53098)
[38] B. WHITE, *The nature of singularities in mean curvature flow of mean convex sets*, J. Amer. Math. Soc. **16** (2003), 123–138. MR1937202 (2003g:53121)

Applications of Hamilton's Compactness Theorem for Ricci flow

Peter Topping

IAS/Park City Mathematics Series
Volume 22, 2013

Applications of Hamilton's Compactness Theorem for Ricci flow

Peter Topping

Overview

In these lectures, I will try to give an introduction to two separate aspects of Ricci flow, namely Hamilton's compactness theorem and the very neat theory of Ricci flow in 2D. The target audience consists of graduate students with some background in differential geometry and PDE theory.

Hamilton's compactness theorem is an absolutely fundamental tool in the modern theory of Ricci flow. I will spend the early part of the course explaining what this result says – roughly that given an appropriate sequence of Ricci flows, one can pass to a subsequence and get smooth convergence to a limit Ricci flow. In order to make sense of that, we will first look at the details of what it means for a sequence of Ricci flows, or simply of Riemannian manifolds, to converge in the Cheeger-Gromov sense. We will not assume any prior knowledge of this notion, although it will be almost essential to have some basic prior knowledge of Riemannian geometry, including the basic idea of the Riemannian curvature tensor.

I will then go on to illustrate the most basic application of the compactness theorem, as envisaged by Hamilton and realised fully by Perelman, by blowing up a singularity to obtain a limit ancient Ricci flow modelling the singularity. To do this we will take a brief detour to mention Perelman's 'no local collapsing theorem".

The rough idea of how this looks in practice in 3D will be explained. But to fully illustrate this application, and also some other key ideas – particularly Perelman's notion of κ-solutions and their most basic theory – we will focus on the 2D situation, and give a Perelman-style proof of the beautiful results of Hamilton and Chow explaining what Ricci flow does to an arbitrary compact surface.

From there we will consider the problem of starting a Ricci flow with a completely general metric, typically on a noncompact surface. This raises some interesting well-posedness issues as one struggles to find the right way of posing the problem to obtain both existence and uniqueness of solutions. We will resolve this problem with the notion of *instantaneously complete* Ricci flows.

The lectures will then complete a full circle by applying this 2D theory in order to understand better Hamilton's compactness theorem and the various extensions that one needs (or desires) to take the theory further. More precisely, we will use the 2D theory (including some additional examples of 'contracting cusp' Ricci flows) to construct some visual examples which violate some extended forms of Hamilton's result that were previously widely believed and used.

Mathematics Institute, University of Warwick, Coventry, CV4 7AL, UK.
E-mail address: p.m.topping@warwick.ac.uk

These lecture notes were written for a mini-course at the Park City Mathematics Institue, Utah (July 2013) based on lectures I gave the year before at ICTP, Trieste (June 2012) and at the 6th KIAS Winter School in South Korea (Feb. 2012). I would like to thank Hugh Bray, Greg Galloway, Rafe Mazzeo, Natasa Sesum, Claudio Arezzo and Jaigyoung Choe for the invitations to these events. Thanks to Neil Course and Gregor Giesen for preparing some of the pictures, which were originally used for [**28**] and [**11**], and to Michael Coffey for helping with the problem sheets that accompany the lectures.

Background reading

Hopefully you already know some Riemannian geometry. If not, you will need to read some of the basics quickly – try Fran Burstall's three lecture introduction [**1**] and the references therein.

You will also need some basic intuition about parabolic equations, or simply just for the linear heat equation, which you will get from many basic introductions to PDE theory.

For the fundamentals of Ricci flow, try my earlier lecture notes [**28**]. In particular, we will follow Chapter 7 of those, and some other parts here and there.

For the remaining parts of Perelman's theory that we will require, you will be fine with Perelman's first paper [**25**], although if you end up wanting to see that material with more detail, then there are some very useful resources such as the notes of Kleiner and Lott [**23**].

For the two-dimensional theory of Ricci flow from the viewpoint of these lectures, one can see particularly [**13, 31, 15, 33**], and we will also draw on elements of [**12**] and [**14**] in order to explain some subtleties of Hamilton's compactness theorem [**32**].

LECTURE 1

Ricci flow basics – existence and singularities

1.1. Initial PDE remarks

A simple but (as it turns out) very natural nonlinear version of the classical linear heat equation is the *logarithmic fast diffusion equation*

$$\frac{\partial u}{\partial t} = e^{-2u}\Delta u, \tag{1.1}$$

where $u : \Omega \times [0,T] \to \mathbb{R}$, for some domain $\Omega \subset \mathbb{R}^2$. This is different from the normal heat equation because the speed of diffusion is scaled by e^{-2u}, which depends on u. Historically this equation has been used extensively for modelling the evolution of the thickness of a thin colloidal film, but it also turns out to have a very geometric flavour.

This equation, being parabolic, is well-posed (in particular one has existence and uniqueness of solutions) if we specify the initial function $u(\cdot, 0)$, and also u on $\partial\Omega \times [0,T]$, just as for the ordinary heat equation. There is also an extensive literature dealing with the case $\Omega = \mathbb{R}^2$ (see, for example, [**7**] as a starting point) although one has extreme nonuniqueness in this case in general in the absence of any growth conditions for u as $x \to \infty$.

Given a solution u to (1.1), we can define a t-dependent Riemannian metric $g(t) = e^{2u}(dx^2 + dy^2)$, and calculate

$$\frac{\partial g}{\partial t} = 2\frac{\partial u}{\partial t} g(t) = 2e^{-2u}\Delta u\, g(t) = -2K g(t),$$

where $K = -e^{-2u}\Delta u$ is the *Gauss curvature* of $g(t)$. In particular, we have

$$\frac{\partial g}{\partial t} = -2\mathrm{Ric}_{g(t)},$$

which is the *Ricci flow equation* of Hamilton [**18**].

This simple observation suggests that the right way of posing (1.1) is to take a Riemannian surface $(\mathcal{M}^2, g_0)$ as initial data, and try to find a t-dependent family $g(t)$, $t \in [0,T]$, such that $g(0) = g_0$ and

$$\frac{\partial g}{\partial t} = -2K g(t),$$

(which forces the flow $g(t)$ to retain the same conformal structure). With respect to any local isothermal coordinate chart, such a solution induces a solution to (1.1), but different charts give different solutions and it does not make sense to discuss regions of slow diffusion (u large) or fast diffusion ($u \ll -1$). Taking this geometric viewpoint puts a collection of powerful techniques at our disposal to study (1.1), and also leads us to the right well-posedness class in full generality, as we will see in Lecture 5.

1.2. Basic Ricci flow theory

In general dimension $n \in \mathbb{N}$, a Ricci flow is a one-parameter family $g(t)$ of Riemannian metrics on an n-dimensional manifold $\mathcal{M}$, satisfying the nonlinear PDE

$$\frac{\partial g}{\partial t} = -2\mathrm{Ric}_{g(t)}.$$

Aside from trivial examples such as the real line (which has no curvature and cannot move) the simplest example is possibly the round shrinking sphere $(S^n, g^n(t))$ where $g^n(t) = (1 - 2(n-1)t)g_0^n$, and g_0^n is the metric of the round sphere of constant sectional curvature 1. Taking products of Ricci flows yields further Ricci flows, and thus another example is the shrinking cylinder $\mathbb{R} \times (S^2, g^2(t))$, or for a similar, compact example one could take $S^1 \times (S^2, g^2(t))$. More generally, we have:

THEOREM 1.1 (Hamilton [**18**], Shi [**27**], Chen-Zhu [**3**], Kotschwar [**24**]). *Given a complete, bounded curvature Riemannian manifold $(\mathcal{M}^n, g_0)$, there exists $T \in (0, \infty]$ and a unique complete, bounded curvature Ricci flow $g(t)$ for $t \in [0, T)$ on $\mathcal{M}^n$ such that $g(0) = g_0$.*

To clarify, such a Ricci flow is said to be *complete* if $(\mathcal{M}^n, g(t))$ is complete for all $t \in [0, T)$, and *here* the Ricci flow is said to have *bounded curvature* if for all $t_0 \in [0, T)$, we have[1]

$$\sup_{\mathcal{M}^n \times [0, t_0]} |\mathrm{Rm}| < \infty.$$

In particular we allow the curvature to blow up as $t \uparrow T$, which is possibly a little unconventional. Indeed, in Theorem 1.1 we may assume that either $T = \infty$ or $\sup_{\mathcal{M}^n} |\mathrm{Rm}|(\cdot, t) \to \infty$ as $t \uparrow T$, as was explained by Hamilton (see for example [**28**, §5.3]).

Clearly the shrinking sphere flow given above is an example where $T < \infty$ and the curvature blows up, but this can occur also without the whole manifold disappearing. For example, if $\mathcal{M} = S^3$, we can have a neck pinch singularity:

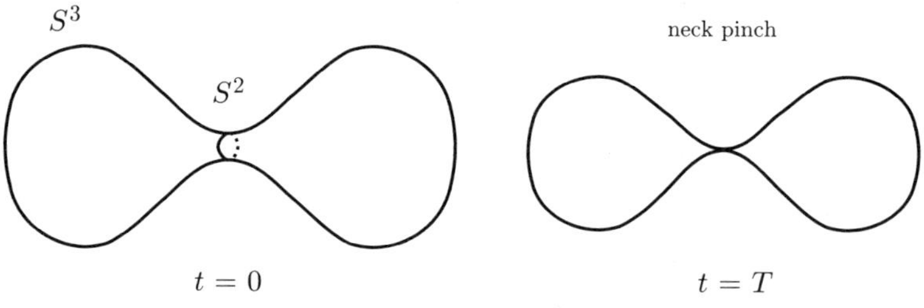

We want to analyse such a singularity by blowing up, i.e. rescaling the flow parabolically (see below) and trying to extract a Ricci flow that models the singularity (i.e. captures some of its information). For the neck pinch example above, the rescaling procedure should produce a shrinking cylinder Ricci flow in the limit:

[1] Rm denotes the full Riemannian curvature tensor. This condition is equivalent to upper and lower bounds (depending on t_0) on all sectional curvatures.

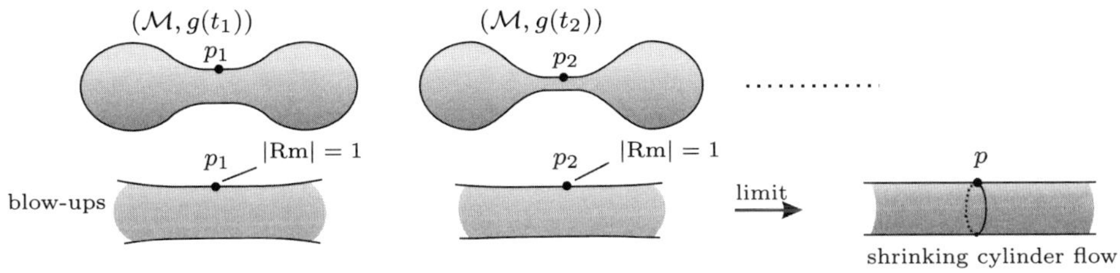

The rescaling alluded to above works by scaling space and time appropriately: If $g(t)$ is a Ricci flow for $t \in [a, b]$ and $\lambda > 0$, then

$$g_\lambda(t) := \lambda g(t/\lambda)$$

is a new Ricci flow for $t \in [\lambda a, \lambda b]$ as described in [**28**, §1.2.3]. Under this rescaling, lengths increase by a factor of $\lambda^{\frac{1}{2}}$, and sectional curvatures are scaled by a factor λ^{-1}. Therefore, we can scale up Ricci flows with large curvatures to Ricci flows with curvatures of order 1 by suitably choosing λ.

In order to make this procedure work, we will at least need a notion of convergence of Ricci flows, and before that a notion of convergence of Riemannian manifolds, and we address these points in the next lecture.

LECTURE 2

Cheeger-Gromov convergence and Hamilton's compactness theorem

The basic reference for this lecture is [**28**, Sections 7.1 and 7.2], although we need some slight modifications here.

Motivated by the discussion in the previous lecture, we need to find a good notion of convergence for a sequence of flows so it will make sense to pass to a limit of a sequence of scaled-up Ricci flows. We will then need Hamilton's compactness theorem to tell us that a reasonable sequence of rescaled flows will in fact converge (after passing to a subsequence). Before that we need to consider the slightly simpler problem of finding a good notion of convergence of a sequence of Riemannian manifolds.

2.1. Convergence and compactness of manifolds

It is reasonable to suggest that a sequence $\{g_i\}$ of Riemannian metrics on a manifold $\mathcal{M}$ should converge to a metric h on $\mathcal{M}$ when $g_i \to h$ as tensors. However, we would like a notion of convergence of Riemannian manifolds that is diffeomorphism invariant: it should not be affected if we modify each metric g_i by an i-dependent diffeomorphism. Once we have asked for such invariance, it is necessary to discuss convergence with respect to a point of reference on each manifold, for reasons that we will see in a moment.

DEFINITION 2.1 (Smooth, pointed "Cheeger-Gromov" convergence of manifolds). A sequence $(\mathcal{M}_i, g_i, p_i)$ of (smooth) complete, *pointed* Riemannian manifolds (that is, Riemannian manifolds $(\mathcal{M}_i, g_i)$ and points $p_i \in \mathcal{M}_i$) is said to *converge* (smoothly) to the smooth pointed Riemannian manifold $(\mathcal{N}, h, q)$ as $i \to \infty$ if there exist

(i) a sequence of domains $\Omega_i \subset\subset \mathcal{N}$, exhausting $\mathcal{N}$ (that is, so that any compact set $K \subset \mathcal{N}$ satisfies $K \subset \Omega_i$ for sufficiently large i) with $q \in \Omega_i$ for each i, and
(ii) a sequence of smooth maps $\phi_i : \Omega_i \to \mathcal{M}_i$ that are diffeomorphic onto their image and satisfy $\phi_i(q) = p_i$ for all i,

such that

$$\phi_i^* g_i \to h$$

smoothly locally on $\mathcal{N}$ as $i \to \infty$.

REMARK 2.2. With this notion, limits will be unique in the sense that if $(\mathcal{N}_1, h_1, q_1)$ and $(\mathcal{N}_2, h_2, q_2)$ are two such limits and both are complete, then there exists an isometry $I : (\mathcal{N}_1, h_1) \to (\mathcal{N}_2, h_2)$ that sends q_1 to q_2.

REMARK 2.3. To demonstrate why we need to include the points p_i in the above definition, consider the following example.

$(\mathcal{N}, h) =$ 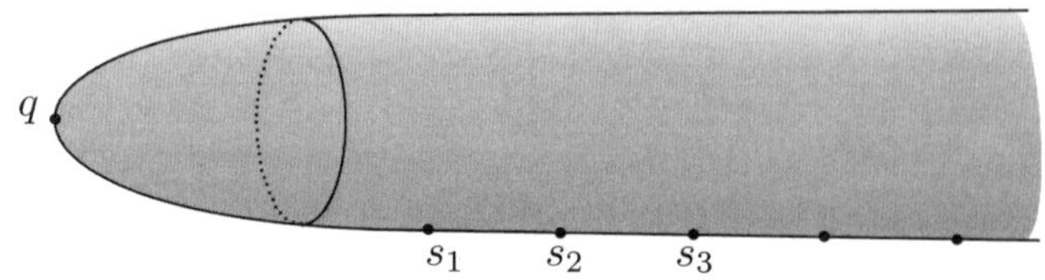

We will take the same sequence of Riemannian manifolds, with different points p_i, and get two different limits. Suppose first that for every i, $(\mathcal{M}_i, g_i)$ is equal to the $(\mathcal{N}, h)$ as shown above. Then $(\mathcal{M}_i, g_i, q) \to (\mathcal{N}, h, q)$, but $(\mathcal{M}_i, g_i, s_i)$ converges to a cylinder.

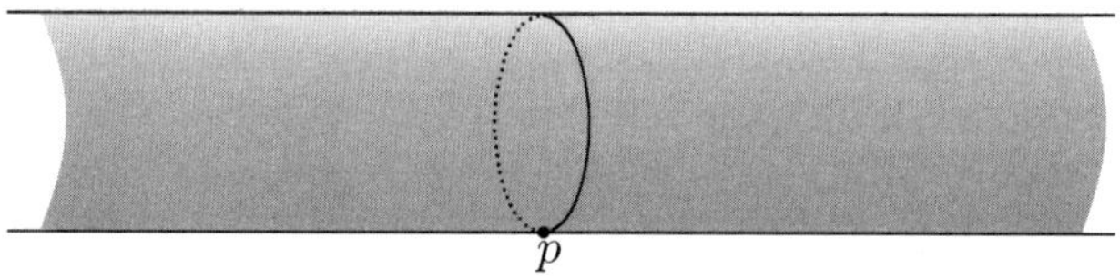

REMARK 2.4. It is possible to have $\mathcal{M}_i$ compact for all i, but have the limit $\mathcal{N}$ non-compact. For example:

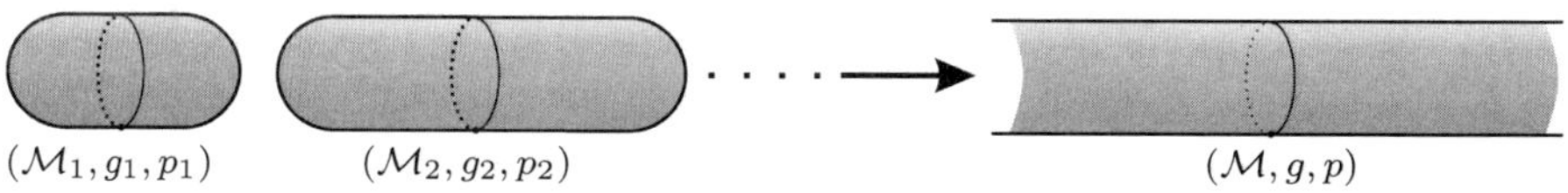

However, if $\mathcal{N}$ is compact, then in the Definition 2.1 of convergence, we must have $\Omega_i = \mathcal{N}$ for sufficiently large i, and the maps ϕ_i will then serve as diffeomorphisms $\mathcal{N} \to \mathcal{M}_i$ (i.e. all the $\mathcal{M}_i$ are the same as the limit) for sufficiently large i.

Two consequences of the convergence $(\mathcal{M}_i, g_i, p_i) \to (\mathcal{N}, h, q)$ are that

(i) for all $s > 0$ and $k \in \{0\} \cup \mathbb{N}$,

$$\sup_{i \in \mathbb{N}} \sup_{B_{g_i}(p_i, s)} \left|\nabla^k \mathrm{Rm}(g_i)\right| < \infty, \tag{2.1}$$

(ii)

$$\inf_i \ \mathrm{inj}(\mathcal{M}_i, g_i, p_i) > 0, \tag{2.2}$$

where $\mathrm{inj}(\mathcal{M}_i, g_i, p_i)$ denotes the injectivity radius of $(\mathcal{M}_i, g_i)$ at p_i.

In fact, conditions (i) and (ii) are *sufficient* for subconvergence. Various incarnations of the following result appear in, or can be derived from, papers of (for example) Greene and Wu [**16**], Fukaya [**10**] and Hamilton [**20**], all of which can be traced back to original ideas of Gromov [**17**] and Cheeger.

THEOREM 2.5 (Compactness – manifolds). *Suppose that $(\mathcal{M}_i, g_i, p_i)$ is a sequence of (smooth) complete, pointed Riemannian manifolds (all of dimension n) satisfying* (2.1) *and* (2.2). *Then there exists a (smooth) complete, pointed Riemannian manifold $(\mathcal{N}, h, q)$ (of dimension n) such that after passing to some subsequence* *in i, we have*

$$(\mathcal{M}_i, g_i, p_i) \to (\mathcal{N}, h, q),$$

as $i \to \infty$.

REMARK 2.6. Although we have explained that (i) and (ii) are necessary to have any hope of convergence, it may be instructive to consider what might go wrong if either condition were dropped entirely.

We clearly need some curvature control, otherwise we may have:

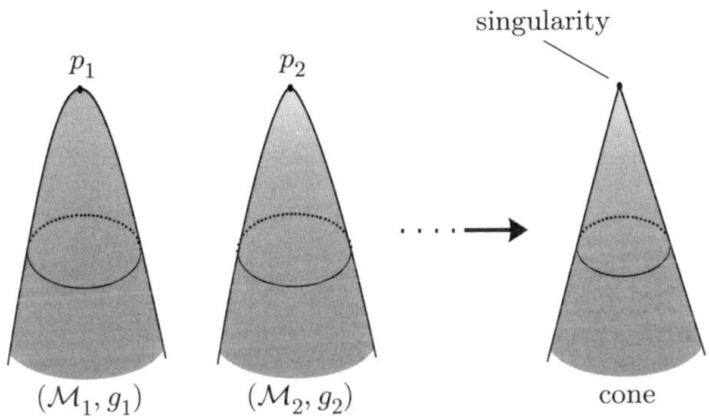

There are other notions of convergence which can handle this type of limit. For example, Gromov-Hausdorff convergence allows us to take limits of metric spaces. (See, for example [**17**].)

The uniform positive lower bound on the injectivity radius is also necessary, since otherwise we could have, for example, degeneration of two-dimensional cylinders:

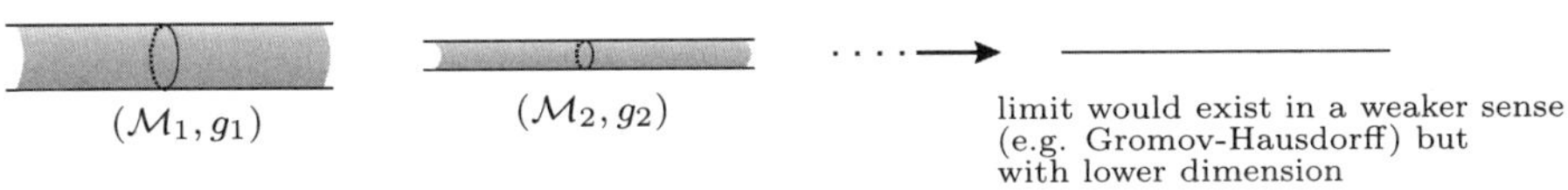

REMARK 2.7. Given curvature bounds as in (2.1), the injectivity radius lower bound at p_i implies a positive lower bound for the injectivity radius at other points $q \in \mathcal{M}_i$ in terms of the distance from p_i to q, and the curvature bounds, as discussed in [**20**], say.

2.2. Convergence and compactness of flows

One can derive Hamilton's compactness theorem for Ricci flow from the compactness theorem for manifolds (Theorem 2.5). In order to do this, we need to make sense of what it means for a sequence of flows to converge.

DEFINITION 2.8 (Convergence of flows). Let $h(t)$ be a (smooth) one-parameter family of Riemannian metrics on a fixed underlying manifold $\mathcal{N}$, for t within some fixed time interval $I \subset \mathbb{R}$. We will call such a $(\mathcal{N}, h(t))$ a *flow* on I. Let $q \in \mathcal{N}$. Let $(\mathcal{M}_i, g_i(t))$ be a sequence of flows on I, and let $p_i \in \mathcal{M}_i$ for each i. We say that

$$(\mathcal{M}_i, g_i(t), p_i) \to (\mathcal{N}, h(t), q)$$

as $i \to \infty$ if there exist

(i) a sequence of domains $\Omega_i \subset\subset \mathcal{N}$ exhausting $\mathcal{N}$ and satisfying $q \in \Omega_i$ for each i, and

(ii) a sequence of smooth maps $\phi_i : \Omega_i \to \mathcal{M}_i$, diffeomorphic onto their image, and with $\phi_i(q) = p_i$,

such that

$$\phi_i^* g_i(t) \to h(t)$$

smoothly locally on $\mathcal{N} \times I$ as $i \to \infty$.

REMARK 2.9. It also makes sense to talk about convergence on (for example) the time interval $(-\infty, 0)$, even when flows are defined only for $(-T_i, 0)$ with $T_i \to \infty$.

We now wish to specialise to the case of *Ricci* flows – i.e. flows as above which also satisfy the Ricci flow equation. Hamilton [**20**] showed how to combine parabolic regularity theory (via Shi's local derivative estimates, which we will not be discussing in detail) and the Cheeger-Gromov compactness of Theorem 2.5 applied only at one time $t = 0$, to prove the following result at the heart of these lectures:

THEOREM 2.10 (Hamilton's compactness theorem for Ricci flows). *Let $\mathcal{M}_i$ be a sequence of manifolds of dimension n, and let $p_i \in \mathcal{M}_i$ for each i. Suppose that $g_i(t)$ is a sequence of complete Ricci flows on $\mathcal{M}_i$ for $t \in (a, b)$, where $-\infty \leq a < 0 < b \leq \infty$. Suppose that*

$$(i) \qquad \sup_i \sup_{x \in \mathcal{M}_i,\ t \in (a,b)} \left|\mathrm{Rm}_{g_i(t)}\right|(x) < \infty, \qquad \textit{and}$$

$$(ii) \qquad \inf_i \ \mathrm{inj}(\mathcal{M}_i, g_i(0), p_i) > 0.$$

Then there exist a manifold $\mathcal{N}$ of dimension n, a complete Ricci flow $h(t)$ on $\mathcal{N}$ for $t \in (a, b)$, and a point $q \in \mathcal{N}$ such that, after passing to a subsequence in i, we have

$$(\mathcal{M}_i, g_i(t), p_i) \to (\mathcal{N}, h(t), q),$$

as $i \to \infty$.

REMARK 2.11. In this theorem, we could equally well work on a time interval $(a, 0]$, with $a < 0$, but it is not reasonable to work on a time interval $[0, b)$ with $b > 0$, because there would be a problem obtaining such strong compactness at time $t = 0$, before the parabolic nature of Ricci flow has had a chance to assert itself by smoothing out the flow.

REMARK 2.12. In the presence of condition (i), there is an alternative way of phrasing condition (ii), namely

$$(ii)^{alt} \qquad \inf_i \mathrm{Vol}_{g_i(0)}[B_{g_i(0)}(p_i, 1)] > 0.$$

The intuition here is that once the curvature is controlled, the volume of a unit ball should be roughly comparable to what it is in Euclidean space, unless the injectivity radius is small. For example, in a very thin (flat) torus $S^1_\varepsilon \times S^1$, where $0 < \varepsilon \ll 1$, the volume of a unit ball is of order ε, not of order 1, because the injectivity radius is of order ε. (See [**2**, Theorem 4.7].)

LECTURE 3

Applications to Singularity Analysis

3.1. The rescaled flows

Now that we have discussed the convergence and compactness of Ricci flows, we are in a better position to do the rescaling near a singularity leading to a limit Ricci flow modelling the singularity, as discussed earlier. If this material is very new to you, it would be an option to fast-forward to Theorem 3.2 on the first reading.

Let $(\mathcal{M}, g(t))$ be a Ricci flow with $\mathcal{M}$ closed, on a maximal time interval $[0, T)$ – as in Section 1.2 – and assume that $T < \infty$, so that

$$\sup_{\mathcal{M}} |\mathrm{Rm}_{g(t)}| \to \infty$$

as $t \uparrow T$. Choose points $p_i \in \mathcal{M}$ and times $t_i \uparrow T$ such that

$$Q_i := |\mathrm{Rm}_{g(t_i)}|(p_i) = \sup_{x \in \mathcal{M},\ t \in [0, t_i]} |\mathrm{Rm}_{g(t)}|(x),$$

by, for example, picking (p_i, t_i) to maximise $|\mathrm{Rm}_{g(t)}|$ over $\mathcal{M} \times [0, T - \frac{1}{i}]$. Notice in particular that $Q_i \to \infty$ as $i \to \infty$. Define rescaled (and translated) flows $g_i(t)$ by

$$g_i(t) = Q_i\, g\left(t_i + \frac{t}{Q_i}\right),$$

which as discussed in Section 1.2, will be Ricci flows on the time intervals $[-t_i Q_i, (T - t_i) Q_i)$.

Moreover, for each i, we have $|\mathrm{Rm}_{g_i(0)}|(p_i) = 1$ and for $t \in (-t_i Q_i, 0]$, we have

$$\sup_{\mathcal{M}} |\mathrm{Rm}_{g_i(t)}| \leq 1.$$

Therefore, for all $a < 0$, $g_i(t)$ is defined for $t \in (a, 0]$, for sufficiently large i, and

$$\sup_i \sup_{\mathcal{M} \times (a, 0]} |\mathrm{Rm}_{g_i(t)}| < \infty.$$

By Hamilton's compactness theorem 2.10 (and Remarks 2.11 and 2.9), we can pass to a subsequence in i, and get convergence $(\mathcal{M}, g_i(t), p_i) \to (\mathcal{N}, h(t), q)$ to a "singularity model" Ricci flow $(\mathcal{N}, h(t))$, defined for $t \leq 0$, *provided* that we can establish the injectivity radius estimate

$$\inf_i \mathrm{inj}(\mathcal{M}, g_i(0), p_i) > 0, \tag{3.1}$$

or (as discussed in Remark 2.12) equivalently that

$$\inf_i \mathrm{Vol}_{g_i(0)}[B_{g_i(0)}(p_i, 1)] > 0. \tag{3.2}$$

This missing step was historically a major difficulty, except in some special cases. However, as we will now see, this issue was resolved by Perelman.

3.2. Perelman's no local collapsing theorem

The core result which solves the issue we have just discussed is:

THEOREM 3.1 (Perelman [**25**]). *Suppose that $\mathcal{M}$ is a closed manifold and $g(t)$ is a Ricci flow on $\mathcal{M}$ for $t \in [0,T)$ with $T > 0$. Then for all $p \in \mathcal{M}$ and $t \in [0,T)$, if $r \in (0,1]$ is sufficiently small so that $|\mathrm{Rm}_{g(t)}| \leq r^{-2}$ on $B_{g(t)}(p,r)$ (we call the largest such r the* curvature scale*) then*

$$\frac{\mathrm{Vol}_{g(t)}[B_{g(t)}(p,r)]}{r^n} \geq \kappa(n, g(0), T) > 0.$$

The proof of this theorem is an appealing argument using a monotonic entropy that was discovered by Perelman, which allows one to deduce the existence of a certain log-Sobolev inequality. The details may be found in [**28**, §8.3].

It may be worth pausing to visualise what Theorem 3.1 is ruling out. It says that in finite time, the flow cannot collapse by shrinking a circle (which might happen without making the curvature scale shrink to nothing) although this *can* happen if we drop the hypothesis that $\mathcal{M}$ is closed, or if we consider the flow as $t \to \infty$. The flow can collapse by shrinking a 2-sphere, as we saw in the shrinking cylinder example in Section 1.2.

Let us try to apply Theorem 3.1 to the Ricci flow $g(t)$ from Section 3.1. Rephrased in terms of the rescaled Ricci flows $g_i(t)$, that result tells us that for all $p \in \mathcal{M}$, $t \in [-t_i Q_i, (T - t_i)Q_i)$ and $r \in (0, Q_i^{\frac{1}{2}}]$ such that $|\mathrm{Rm}_{g_i(t)}| \leq r^{-2}$ on $B_{g_i(t)}(p,r)$, we have

$$\frac{\mathrm{Vol}_{g_i(t)}[B_{g_i(t)}(p,r)]}{r^n} \geq \kappa > 0. \tag{3.3}$$

This statement will have two applications – first it will shortly allow us to establish (3.2) in order to be able to apply Hamilton's compactness theorem to give a limit Ricci flow $h(t)$, but it will also allow us to deduce a 'non-collapsing' property of this limit.

For the first application, note that by construction, the curvature is controlled for $t \leq 0$ by $|\mathrm{Rm}_{g_i(t)}| \leq 1$ and so we are free to take $r = 1$ and $t = 0$ in (3.3) to give

$$\mathrm{Vol}_{g_i(0)}[B_{g_i(0)}(p,1)] \geq \kappa$$

for all $p \in \mathcal{M}$, which is stronger than (3.2). By the discussion in Section 3.1 this allows us to extract a limit "singularity model" Ricci flow $(\mathcal{N}, h(t))$ for $t \leq 0$. Moreover, as the second application of (3.3), after a little thought we see that we can pass to the limit $i \to \infty$ (which ultimately amounts to replacing Q_i by ∞ and $g_i(t)$ by the limit $h(t)$) to see that for all $p \in \mathcal{N}$, $t \leq 0$ and $r > 0$, if $|\mathrm{Rm}_{h(t)}| \leq r^{-2}$ on $B_{h(t)}(p,r)$, then we have

$$\frac{\mathrm{Vol}_{h(t)}[B_{h(t)}(p,r)]}{r^n} \geq \kappa > 0. \tag{3.4}$$

To summarise, we have shown the following:

THEOREM 3.2. *Suppose that $\mathcal{M}^n$ is a closed n-dimensional manifold, and $g(t)$ is a Ricci flow on $\mathcal{M}$ for $t \in [0,T)$, where $T \in (0,\infty)$ is maximal. Then there exist sequences $p_i \in \mathcal{M}$ and $t_i \uparrow T$ such that*

$$Q_i := |\mathrm{Rm}_{g(t_i)}|(p_i) = \sup_{t\in[0,t_i]} \sup_{\mathcal{M}} |\mathrm{Rm}_{g(t)}|,$$

and so that if we define rescaled Ricci flows

$$g_i(t) := Q_i \, g\left(t_i + \frac{t}{Q_i}\right),$$

then there exists a complete bounded curvature Ricci flow $(\mathcal{N}^n, h(t))$ *defined for* $t \leq 0$, *and* $q \in \mathcal{N}$ *such that*

$$(\mathcal{M}, g_i(t), p_i) \to (\mathcal{N}, h(t), q)$$

as $i \to \infty$ *Moreover, we have*

$$\sup_{t \leq 0} \sup_{\mathcal{N}} |\mathrm{Rm}_{h(t)}| \leq 1 = |\mathrm{Rm}_{h(0)}|(q),$$

and $h(t)$ *is* κ*-noncollapsed at all scales in the sense that for all* $p \in \mathcal{N}$, $t \leq 0$ *and* $r > 0$, *if* $|\mathrm{Rm}_{h(t)}| \leq r^{-2}$ *on* $B_{h(t)}(p, r)$, *then we have*

$$\frac{\mathrm{Vol}_{h(t)}[B_{h(t)}(p, r)]}{r^n} \geq \kappa > 0. \tag{3.5}$$

LECTURE 4

The case of compact surfaces – an alternative approach to the results of Hamilton and Chow

In this lecture we will apply the theory we have developed so far in the case that the underlying manifold is two-dimensional (and compact). In this dimension, and with a little more effort in three dimensions, the rescaled Ricci flows $h(t)$ we saw in the previous lecture will be so-called κ-solutions, which are special ancient solutions, that can be well understood. This dramatically restricts the types of singularity that can occur for compact $\mathcal{M}$, particularly in two dimensions when all singularities will be seen to behave asymptotically like a shrinking sphere. (This special structure does not follow in the noncompact case, as we shall see.)

The first simple ingredient in the 2D case that we will exploit is a lower bound for the Gauss curvature:

LEMMA 4.1. *Given a Ricci flow $(\mathcal{M}^2, g(t))$ for $t \in (0,T]$ on a closed surface $\mathcal{M}^2$, the Gauss curvature $K_{g(t)}$ satisfies*

$$K_{g(t)} \geq -\frac{1}{2t}.$$

PROOF. A computation reveals that under Ricci flow on a surface, the Gauss curvature evolves according to the PDE

$$\frac{\partial K}{\partial t} = \Delta K + 2K^2.$$

(Although we don't need it here, a similar equation holds for the scalar curvature in higher dimensions [**28**, Proposition 2.5.4].) The lemma is then easy to see by the maximum principle: Note that $-\frac{1}{2t}$ is a solution to the ODE $\frac{dK}{dt} = 2K^2$. (See [**28**, Section 3.2] for more details.) In fact, although we have assumed that $\mathcal{M}$ is compact, the result turns out to be true in much more general situations, even where the normal maximum principle fails [**4**], as we will see in Lemma 5.4. □

Because the curvature decreases when we scale up a metric (see the end of Section 1.2) an immediate consequence of the lemma is that the limit Ricci flows $h(t)$ arising in the previous lecture must all have weakly positive (i.e. nonnegative) curvature in this case. Combining this fact with what we already know, these limit Ricci flows $h(t)$ will then be so-called κ-solutions in the sense made precise in the following theorem.

THEOREM 4.2. *When the underlying manifold $\mathcal{M}$ is a compact surface, any limit Ricci flow $h(t)$ as constructed in the previous lecture must satisfy the following properties, which define what it means to be a κ-**solution**:*

(1) *It is **Ancient**, i.e. it is defined for $-\infty < t \leq T$, for some $T \in \mathbb{R}$. (In this particular, case, we can set $T = 0$.)*

(2) *It has* **Bounded curvature**, *but is not everywhere flat. (In this particular case we have* $|\mathrm{Rm}| \leq 1$ *everywhere, and* $|\mathrm{Rm}| = 1$ *at at least one point, by construction.)*

(3) *It is* **Complete**. *(In this case, that was a consequence of Hamilton's compactness theorem.)*

(4) *It is κ-noncollapsed at all scales, i.e. there exists $\kappa > 0$ such that for each time t at which $h(t)$ is defined, and each point $p \in \mathcal{N}$ and radius $r > 0$ for which $|\mathrm{Rm}| \leq r^{-2}$ on $B_{h(t)}(p,r)$, we have*

$$\frac{\mathrm{Vol}_{h(t)} B_{h(t)}(p,r)}{r^n} \geq \kappa.$$

(In this case, that was a consequence of Perelman's theory.)

(5) *It has weakly positive (i.e. nonnegative) curvature operator. (In this case, that simply means that the Gauss curvature is weakly positive, which was the most recent thing we established.)*

The notion of κ-solutions was introduced by Perelman [**25**, §11] principally to study Ricci flow of three-dimensional manifolds. By virtue of the so-called Hamilton-Ivey pinching result [**21**] (which we will not discuss here) it is also possible to establish that limit Ricci flows $h(t)$ have (weakly) positive curvature in that case too. The true significance of κ-solutions is that in low dimensions, they are restrictive enough that we can understand what they look like. In two-dimensions, this is particularly clean:

THEOREM 4.3 (Perelman [**25**, §11.3]). *The only oriented 2D κ-solution is the round shrinking sphere $(S^2, g^2(t))$, as found in Section* 1.2.

We may therefore deduce that for a Ricci flow on an oriented, compact surface, whenever a finite-time singularity occurs, it can be rescaled to give the Ricci flow that is a round shrinking sphere. But we already saw in Remark 2.4 that the only way a Cheeger-Gromov limit can be a compact manifold is if the original sequence of manifolds is (eventually) diffeomorphic to this limit. That means that we can only have a finite-time singularity in a Ricci flow on an oriented, compact surface if that surface is S^2. Note that we then have $\mathrm{Vol}_{g(t_i)} \mathcal{M} \to 0$ as $i \to \infty$. On the other hand, if the underlying manifold is S^2, then we can make a short calculation (see [**28**, (2.5.7) and (2.5.8)]) to see that the volume measure evolves according to

$$\frac{\partial}{\partial t} d\mu_{g(t)} = -2K d\mu_{g(t)},$$

and hence by the Gauss-Bonnet theorem we have

$$\frac{d}{dt} \mathrm{Vol}_{g(t)} \mathcal{M} = -2 \int_{\mathcal{M}} K d\mu_{g(t)} = -8\pi,$$

and the Ricci flow must shrink to nothing in a specific finite time $T = \frac{\mathrm{Vol}_{g_0} \mathcal{M}}{8\pi}$. Note that at the times t_i as which we are blowing up, the flow must look more and more like a round sphere of volume $8\pi(T - t_i)$, i.e. of curvature $1/2(T - t_i)$. We have proved the following result, with very different techniques to those originally used.

THEOREM 4.4 (Hamilton [**19**], Chow [**5**]). *Let $\mathcal{M}^2$ be a closed, oriented surface and g_0 any smooth metric. Then there exists a unique Ricci flow $g(t)$ on $\mathcal{M}$, for $t \in [0, T)$ so that $g(0) = g_0$. We may assume that $T = \infty$ unless $\mathcal{M} = S^2$, in* *which case we have* $T = \frac{\mathrm{Vol}_{g_0} \mathcal{M}}{8\pi}$, *and there exist sequences of times* $t_i \uparrow T$ *and diffeomorphisms* $\varphi_i : \mathcal{M} \to \mathcal{M}$ *such that*

$$\frac{\varphi_i^*(g(t_i))}{2(T - t_i)} \to g_{+1}$$

as $i \to \infty$, *where* g_{+1} *is a round metric of constant curvature* $+1$.

REMARK 4.5. With a little more work [**19**, **5**, **11**], one can show in fact that

$$\frac{g(t)}{2(T - t)} \to g_{+1}$$

uniformly as $t \uparrow T$ in Theorem 4.4.

The convergence in the higher genus case is possibly a little easier to deal with, and one can prove:

THEOREM 4.6 (Hamilton [**19**]). *Suppose the surface in Theorem 4.4 is a torus* T^2. *Then the Ricci flow* $g(t)$ *satisfies* $g(t) \to g_f$ *smoothly as* $t \to \infty$, *where* g_f *is a flat metric.*

If instead the surface in Theorem 4.4 is orientable, of genus $\gamma > 1$, *then* $\frac{g(t)}{2t} \to g_{-1}$ *smoothly, where* g_{-1} *is the unique hyperbolic metric that shares a conformal class with* $g(t)$.

The easiest way of deriving the 2D compact Ricci flow theory appeals to the Uniformisation theorem, which tells us that any Riemann surface admits a compatible Riemannian metric (i.e. inducing the same conformal structure) which has constant curvature $+1$, 0 to -1, depending on its genus. More recently, Chen, Lu and Tian [**6**] have shown that in this case of compact underlying manifold, one can develop the Ricci flow theory independently of the Uniformisation theorem, thus yielding a new proof of that fundamental result in the special case of compact underlying manifold.

To motivate the material of the next lecture, one might ask the question of whether Ricci flow will try to perform the same task of finding a constant curvature metric in the case that the underlying manifold is noncompact. Now, even posing the Ricci flow is an issue, as we shall see.

LECTURE 5

The 2D case in general – Instantaneously Complete Ricci flows

5.1. How to pose the Ricci flow in general

We have just seen the complete theory, including asymptotics, in the compact 2D case. We also have short-time well-posedness for Ricci flows starting with complete, bounded-curvature metrics by virtue of Theorem 1.1. In this lecture we wish to understand the case that the underlying manifold $\mathcal{M}$ is noncompact, but the starting metric g_0 could have *unbounded curvature* or even be *incomplete*.

To illustrate the difficulties involved here, we consider the following simple example:

> *How should we Ricci flow the flat disc* (D^2, g_0), *where* $g_0 = dx^2 + dy^2$ *is the Euclidean metric?*

To get a feel for this, let's return to the discussion of the logarithmic fast diffusion equation at the beginning of these lectures, and write the Ricci flow as $g(t) = e^{2u}(dx^2 + dy^2)$ where u must solve

$$\begin{cases} \dfrac{\partial u}{\partial t} = e^{-2u}\Delta u \\ u(\cdot, 0) \equiv 0. \end{cases} \tag{5.1}$$

Of course, there is an obvious solution, given that the initial metric is flat, which is the flow that does not move, and is the flat disc for all time. That is, the solution $u \equiv 0$ throughout D and for all times. However, this sort of trivial solution would not exist even for a small perturbation of the initial data above, and now that we have written out the PDE (5.1) we see that even amongst solutions u that are continuous up to the boundary of D, we are free to *prescribe* $u(\cdot, t)$ on ∂D for each positive t. We thus need an extra 'boundary'-type condition to kill this extreme nonuniqueness. The central idea of the theory we are about to see [**29**] is that in full generality for Ricci flow on surfaces, a geometric condition which leads to uniqueness (whilst preserving existence of solutions!) is that of *instantaneous completeness*, that is:

> *We consider Ricci flows* $g(t)$ *for* $t \in [0, T)$ *that are complete for all* $t \in (0, T)$.

In particular, in the case of Ricci flow starting with a disc, we have the following result which is a variant of a very special case of [**29**].

THEOREM 5.1. *There exists a Ricci flow* $g(t)$ *on the disc* D^2 *for* $t \in [0, \infty)$ *such that*

(1) $g(0) = g_0$, *the standard flat metric,*
(2) $g(t)$ *is complete for all* $t > 0$.

Any other Ricci flow $\tilde{g}(t)$ on D^2 over any time interval $[0,T)$ (with $T \in (0,\infty]$) that satisfies both (1) and (2) above must agree with $g(t)$ on $[0,T)$ where they are both defined.

Given that the unit disc is not complete, something unusual must be happening! This is easiest to see by considering the conformal factor u: For arbitrarily small $t > 0$, the conformal factor u must blow up near the boundary at a fast enough rate to be sure that the metric is complete. That means that the integral of e^u along any curve escaping to 'infinity' in the disc must be infinite. The graph of the conformal factor looks something like:[1]

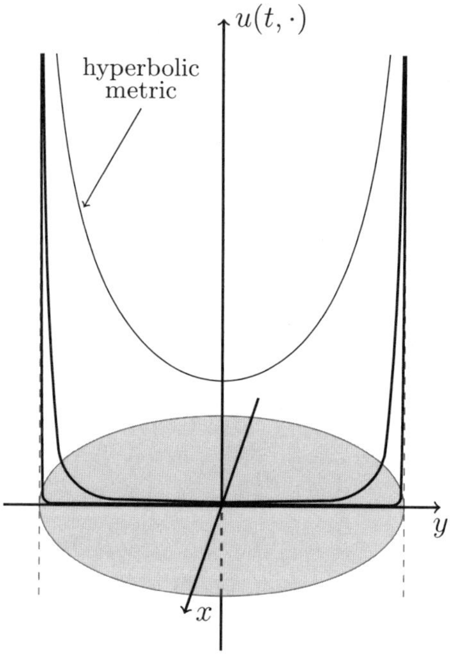

Initially the flow will look more or less like the flat disc on the interior, but in an expanding layer around the boundary, the flow will have curvature approximately its minimum possible value $-\frac{1}{2t}$ (recall Lemma 4.1). Indeed, if we write $g_{-1} := \frac{4g_0}{(1-x^2-y^2)^2}$ for the (unique) complete hyperbolic metric on the disc D that is a conformal deformation of g_0, then the Ricci flow $g(t)$ will resemble $2t\,g_{-1}$ on the expanding layer around the boundary, and eventually on the whole disc.

In fact, there is a much more general existence and uniqueness theory, which we outline in the following section.

5.2. The existence and uniqueness theory

The original existence theory for instantaneously complete Ricci flows [**29**] was the first to allow the initial curvature to be unbounded (below) and also allowed the initial metric to be incomplete. However, more recently, the full existence and uniqueness theory has been completed, which handles a completely general surface, and gives the optimal existence time. Moreover, the flow $g(t)$ we find has the

[1]Thanks to Gregor Giesen for this figure.

additional property of being 'maximally stretched,' by which we mean that any other (conformal) Ricci flow $\tilde{g}(t)$ with $\tilde{g}(0) \leq g(0)$ must satisfy $\tilde{g}(t) \leq g(t)$ for all t in any time interval $[0, a)$ during which both flows are defined (irrespective of whether or not $\tilde{g}(t)$ is complete or has bounded curvature).

THEOREM 5.2 (Existence: joint with Giesen [**13**]; uniqueness from [**33**]). *Let $(\mathcal{M}, g_0)$ be a smooth Riemannian surface which need not be complete, and could have curvature unbounded below and/or above. Depending on the conformal type, we define $T \in (0, \infty]$ by*

$$T := \begin{cases} \frac{1}{4\pi\chi(\mathcal{M})} \operatorname{Vol}_{g_0} \mathcal{M} & \text{if } (\mathcal{M}, g_0) \cong \mathcal{S}^2, \mathbb{C} \text{ or } \mathbb{R}P^2, \\ \infty & \text{otherwise.}^2 \end{cases}$$

Then there exists a unique smooth Ricci flow $(\mathcal{M}, g(t))$ for $t \in [0, T)$ such that

(1) *$g(0) = g_0$;*
(2) *$g(t)$ is instantaneously complete;*
(3) *$g(t)$ is maximally stretched.*

This flow is unique in the sense that if $\tilde{g}(t)$ is any Ricci flow on $\mathcal{M}$, for $t \in [0, \tilde{T})$, satisfying (1) and (2), then $\tilde{T} \leq T$ and $\tilde{g}(t) = g(t)$ for all $t \in [0, \tilde{T})$.

If $T < \infty$, then we have

$$\operatorname{Vol}_{g(t)} \mathcal{M} = 4\pi\chi(\mathcal{M})(T - t) \to 0 \text{ as } t \nearrow T,$$

and in particular, T is the maximal existence time.

Several aspects of this theorem resemble the results we have seen of Hamilton and Chow in the compact case (which have been absorbed into this result). However, there are some significant departures from that theory, and some shocking examples of flows fitting into this theorem, as we survey in a moment.

Our Ricci flow can start with a completely general initial surface, but in the special case that this initial surface is both complete and of curvature bounded above and below, there is a competing flow, namely that of Shi from Theorem 1.1, existing on some time interval $[0, \tilde{T})$, and by the uniqueness assertion of Theorem 5.2, these flows agree while Shi's exists.

As we discussed in Section 1.2, if $\tilde{T} < \infty$, then we can assume that the curvature of $\tilde{g}(t)$ blows up as $t \uparrow \tilde{T}$. It may be then hard to imagine how one could have $T > \tilde{T}$ since this would imply that the curvature could blow up in our flow, but that we could then keep on flowing. However, this is exactly what can happen, for example:

THEOREM 5.3 (Proved with Giesen [**15**]). *There exist a complete immortal Ricci flow $\big(g(t)\big)_{t\in[0,\infty)}$ on $\mathbb{C}$ arising from Theorem 5.2, and a time $t_1 \in [1, 3)$ such that*

$$\sup_{\mathcal{M}} \left|K_{g(t)}\right| \begin{cases} < \infty & \text{for all } t \in [0, 1) \\ = \infty & \text{for all } t \in \left(t_1, t_1 + \frac{1}{100}\right) \\ < \infty & \text{for all } t \in [4, \infty). \end{cases}$$

In fact, using the techniques we develop to prove this theorem, we can prescribe with great generality the set of times at which the curvature is unbounded. Note that this feature now means that the term 'maximal existence time' needs to be defined carefully in each noncompact situation. Our maximal existence time is

[2]Note that also $T = \infty$ if $\operatorname{Vol}_{g_0} \mathbb{C} = \infty$.

greater than, and in general *strictly* greater than the traditional maximal existence time.

When we have unbounded curvature in these examples, it is unbounded above. It can never be unbounded below for positive times, and indeed we have the following curvature estimates. Formally they follow via the maximum principle, although that does not apply directly in this general situation. In practice, they follow from the actual *construction* made in the proof of Theorem 5.2, or via the theory in [**29, 4**].

LEMMA 5.4. *In Theorem 5.2, the flow satisfies the lower curvature bound*

$$K_{g(t)} \geq -\frac{1}{2t}$$

for all $t \in (0, T)$. *If in addition we have an initial upper curvature bound* $K_{g_0} \leq K_0 \geq 0$ *for the initial metric, then the flow* $g(t)$ *has the upper curvature bound*

$$K_{g(t)} \leq \frac{K_0}{1 - 2K_0 t}$$

for all $t \geq 0$ *if* $K_0 = 0$, *and for all* $t \in [0, \frac{1}{2K_0})$ *otherwise. (Lower bounds for the existence time* T *are implicit here.) Irrespective of any initial upper curvature bound, if we have a lower curvature bound* $K_{g_0} \geq k_0 \leq 0$ *for the initial metric, and the initial metric is* **complete**, *then the flow* $g(t)$ *has the lower curvature bound*

$$K_{g(t)} \geq \frac{k_0}{1 - 2k_0 t}$$

for all $t \in [0, T)$.

It may be clear from the discussion above that while our flow from Theorem 5.2 takes a general metric and makes it immediately complete and of curvature bounded below, it need not immediately transform the metric to one of curvature bounded above. That is, we do not immediately find ourselves in the classical situation. This is illustrated in generality by:

THEOREM 5.5 (Proved with Giesen [**14**]). *For all noncompact Riemann surfaces* $\mathcal{M}^2$, *there exists a Ricci flow* $g(t)$ *on* $\mathcal{M}^2$ *for* $t \in [0, \infty)$ *(respecting the complex structure of the surface) such that*

$$\sup_{\mathcal{M}} K_{g(t)} = \infty$$

for all $t \in [0, \infty)$.

5.3. Asymptotics

One of the features of the Ricci flow theory on compact surfaces that we saw in Lecture 4, was that it 'geometrised' a surface – i.e. up to scaling, it converged to a constant curvature metric. In the noncompact case, so far we only fully understand the asymptotics in the hyperbolic case:

THEOREM 5.6 (Proved with Giesen [**13**]). *Suppose we have a Ricci flow in Theorem 5.2 on a surface which supports a complete hyperbolic metric* H *conformally equivalent to the Ricci flow (in which case* $T = \infty$ *automatically). Then we have convergence of the rescaled solution*

$$\frac{1}{2t} g(t) \longrightarrow H \quad \textit{smoothly locally as } t \to \infty.$$

If additionally there exists a constant $M > 0$ such that $g_0 \le MH$ then the convergence is global: For any $k \in \mathbb{N}_0 := \mathbb{N} \cup \{0\}$ and $\eta \in (0,1)$ there exists a constant $C = C(k, \eta, M) > 0$ such that for all $t \ge 1$ we have

$$\left\| \frac{1}{2t} g(t) - H \right\|_{C^k(\mathcal{M},H)} \le \frac{C}{t^{1-\eta}} \quad \stackrel{t\to\infty}{\longrightarrow} \quad 0.$$

In fact, in this latter case, for all $t > 0$ we have $\left\| \frac{1}{2t} g(t) - H \right\|_{C^0(\mathcal{M},H)} \le \frac{C}{t}$, and even

$$0 \le \frac{1}{2t} g(t) - H \le \frac{M}{2t} H.$$

Related results were proved by Ji-Mazzeo-Sesum [**22**] under the additional hypotheses that the initial data was complete, and of bounded curvature, and conformally equivalent to a compact surface with finitely many punctures, and with all ends asymptotic to hyperbolic cusps.

This theorem has to be reconciled with Theorem 5.5, which told us that there exists a Ricci flow with unbounded curvature for all time on any noncompact Riemann surface. In particular on any noncompact Riemann surface that supports a hyperbolic metric, for example the disc D, there exists such an unbounded curvature Ricci flow. On the other hand, in such a case Theorem 5.6 tells us that the curvature is converging to -1, which is bounded! Ricci flow resolves this paradox by sending the bad regions of high curvature out to spatial infinity, the convergence of Theorem 5.6 being only smooth local convergence in general.

5.4. Singularities not modelled on shrinking spheres

We take the opportunity to give an additional example of Theorem 5.2 in action that not only illustrates a distinction compared with Lecture 4 where all singularities were modelled on shrinking spheres, but will also be useful in Lecture 7.

The starting metric will be conformally the complex plane, and look geometrically like a spherical bulb with a hyperbolic cusp attached. Such a metric could be constructed by bare hands, or would naturally arise by taking initial data in Theorem 5.2 consisting of a punctured sphere, and flowing for a moment.

By scaling up or down, we assume our bulb metric g_{bulb} has area exactly equal to 8π, and that way Theorem 5.2 will make it flow for exactly time 2.

According to the theory developed in [**8**], [**9**], and the references therein, the end of this flow will always look like a hyperbolic cusp. On the other hand, the bulb part will gradually shrink under the flow, and just before the surface disappears (along with all its area) it will look like a 'cigar' (see [**28**, §1.2.2]) tapering into a hyperbolic cusp.

g_{bulb}:

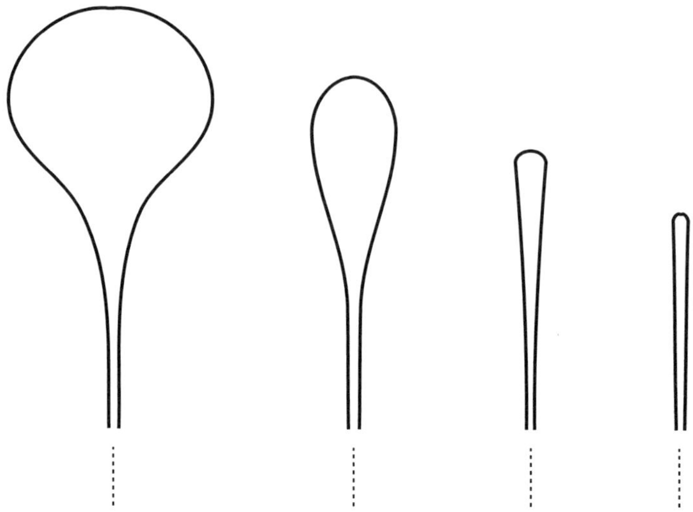

$t = 0$

Even though one can blow up this finite-time singularity similarly to as in Lecture 3, this situation does not fit into the discussion of Lecture 4 because we are not on a compact manifold, and Perelman's no local collapsing theorem 3.1 fails, and we cannot deduce that the blow-up is a κ-solution and thus a shrinking sphere as in Theorem 4.3. Indeed, as indicated above, the blow-ups in this case would converge to a cigar metric [**28**, §1.2.2].

LECTURE 6

Contracting Cusp Ricci flows

There is one further type of well-posedness issue to discuss, and a new type of Ricci flow in 2D fitting into this framework that will be needed in the next lecture. Up until now, we have posed a Ricci flow by fixing an underlying manifold and specifying an initial metric. Thus the initial metric is achieved as a limit of the tensors $g(t)$ as $t \downarrow 0$. However, given the discussion of Cheeger-Gromov convergence that we saw in Lecture 2, it could be considered more natural to ask for the initial manifold to be achieved as a Cheeger-Gromov-type limit of $(\mathcal{M}, g(t))$ as $t \downarrow 0$.

DEFINITION 6.1. ([**31**].) We say that a complete Ricci flow $(\mathcal{M}^n, g(t))$ for $t \in (0, T]$ has a complete Riemannian manifold $(\mathcal{N}^n, g_0)$ as initial condition if there exists a smooth map $\varphi : \mathcal{N} \to \mathcal{M}$, diffeomorphic onto its image, such that

$$\varphi^*(g(t)) \to g_0$$

smoothly locally on $\mathcal{N}$ as $t \downarrow 0$.

The usual way of posing Ricci flow fits into this framework by setting $\mathcal{N} = \mathcal{M}$ and taking φ to be the identity.

In practice, we will be interested in the case that $(\mathcal{N}, g_0)$ has bounded curvature but $g(t)$ is allowed to have curvature with no uniform upper bound. In this way, $\mathcal{M}$ and $\mathcal{N}$ may not be diffeomorphic since parts of $\mathcal{M}$ may be shot out to infinity as $t \downarrow 0$ resulting in a change of topology in the limit.

This generalised notion of initial condition permits some new types of solution which do not fit into the classical framework. In particular, we show that a bounded-curvature Riemannian surface with a hyperbolic cusp need not be obliged to flow forwards in time retaining the cusp (as it would in the solution of Shi or of Theorem 5.2) but can add in a point at infinity, removing the puncture in the surface, and let the cusp contract in a controlled way. More generally we have:

THEOREM 6.2 ([**31**]). *Suppose $\mathcal{M}$ is a compact Riemann surface and $\{p_1, \ldots, p_n\} \subset \mathcal{M}$ is a finite set of distinct points. If g_0 is a complete, bounded-curvature, smooth, conformal metric on $\mathcal{N} := \mathcal{M} \backslash \{p_1, \ldots, p_n\}$ with uniformly strictly negative curvature in a neighbourhood of each point p_i, then there exists a Ricci flow $g(t)$ on $\mathcal{M}$ for $t \in (0, T]$ (for some $T > 0$) having $(\mathcal{N}, g_0)$ as initial condition in the sense of Definition* 6.1. *We can take the map φ there to be the natural inclusion of $\mathcal{N}$ in $\mathcal{M}$.*

Moreover, the cusps contract logarithmically in the sense that for some $C < \infty$ and all $t \in (0, T]$ sufficiently small, we have

$$\frac{1}{C}(-\ln t) \leq diam\,(\mathcal{M}, g(t)) \leq C(-\ln t). \tag{6.1}$$

Furthermore, the curvature of $g(t)$ is bounded below uniformly as $t \downarrow 0$.

Let us consider one specific example where this theorem applies, that will be required in the next lecture. Consider a punctured torus $\mathcal{N} := T^2\backslash\{p\}$ (the torus having any complex structure) and let g_0 be g_{hyp}, the unique complete conformal hyperbolic metric on $\mathcal{N}$. The metric will have a cusp at the puncture. One Ricci flow continuation would be the homothetically expanding flow (which coincides with the solution constructed by Shi and that of Theorem 5.2) but another continuation would see the cusp contract with the subsequent Ricci flow living on the whole torus $\mathcal{M}$.

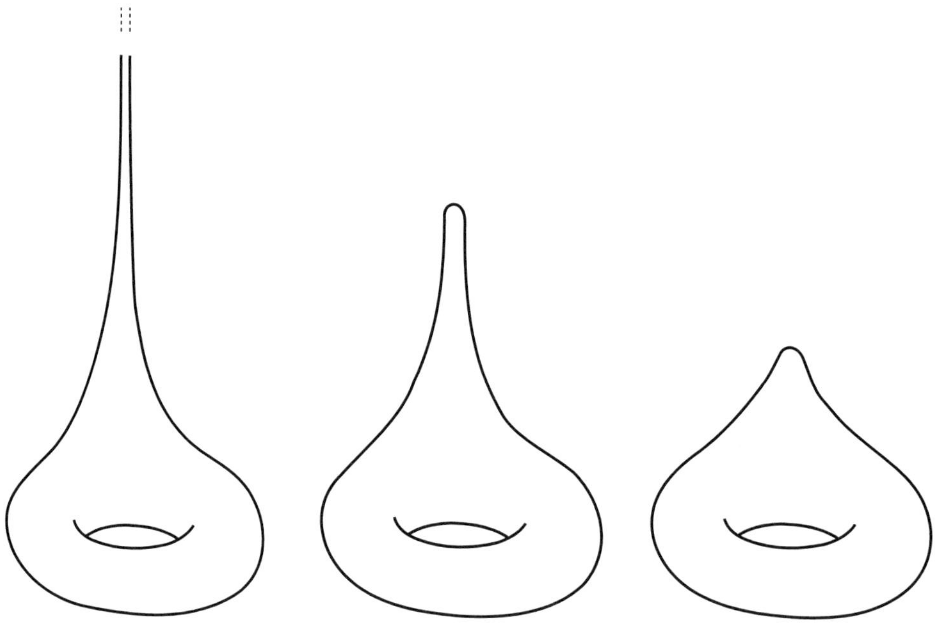

One characteristic of these nonuniqueness examples is that the initial condition $(\mathcal{N}, g_0)$ does not have a lower bound for its injectivity radius, or equivalently that one can find unit balls of arbitrarily small area. In fact, this is a necessary condition for nonuniqueness.

THEOREM 6.3 ([**31**]). *Suppose that $(\mathcal{N}, g_0)$ is a complete Riemannian surface with bounded curvature which is noncollapsed in the sense that for some $r_0 > 0$ we have*

$$\mathrm{Vol}_{g_0}(B_{g_0}(x, r_0)) \geq \varepsilon > 0 \tag{6.2}$$

for all $x \in \mathcal{N}$. If for $i = 1, 2$ we have complete Ricci flows $(\mathcal{M}_i, g_i(t))$ for $t \in (0, T_i]$ (some $T_i > 0$) with $(\mathcal{N}, g_0)$ as initial condition in the sense of Definition 6.1, then these two Ricci flows must agree over some nonempty time interval $t \in (0, \delta]$ in the sense that there exists a diffeomorphism $\psi : \mathcal{M}_1 \to \mathcal{M}_2$ with $\psi^*(g_2(t)) = g_1(t)$ for all $t \in (0, \delta]$.

Despite the nonuniqueness implied by Theorem 6.2, that construction throws up a quite different uniqueness issue: Does there exist more than one flow that does the same job of contracting the cusps? The next result shows that there does not.

THEOREM 6.4 ([**31**]). *In the situation of Theorem 6.2 (in which φ is the natural inclusion of $\mathcal{N}$ into $\mathcal{M}$) if $\tilde{g}(t)$ is a smooth Ricci flow on $\mathcal{M}$ for some time interval $t \in (0, \delta)$ ($\delta \in (0, T]$) such that $\tilde{g}(t) \to g_0$ smoothly locally on $\mathcal{N}$ as $t \downarrow 0$ and the Gauss curvature of $\tilde{g}(t)$ is uniformly bounded below, then $\tilde{g}(t)$ agrees with the flow $g(t)$ constructed in Theorem 6.2 for $t \in (0, \delta)$.*

LECTURE 7

Subtleties of Hamilton's compactness theorem

Having started with Hamilton's compactness theorem 2.10 and applied it to 2D Ricci flow theory, we now complete the circle by applying our 2D Ricci flow theory to understand the subtleties concerning extensions of Hamilton's compactness theorem that are required in applications.

Up until now, the compactness theorem was used to blow up a Ricci flow that is becoming singular, at points of maximum curvature (see Lecture 3). However, in practice, it is essential to be able to analyse other parts of the Ricci flow where the curvature may be blowing up, even though the curvature might be much larger elsewhere. Moreover, there is an entirely different situation where one needs to apply the compactness theorem, namely when making contradiction arguments. Loosely speaking, in order to prove that a certain fact about Ricci flow is true to some degree, one assumes otherwise – i.e. that there exists a sequence of Ricci flows which violates the assertion to a greater and greater degree. One then tries to appeal to compactness to get a limit Ricci flow with contradictory features.

For all these applications, Hamilton's compactness theorem 2.10 is not quite enough, because the curvature hypothesis

$$(i) \qquad \sup_i \sup_{x\in\mathcal{M}_i,\ t\in(a,b)} \left|\mathrm{Rm}_{g_i(t)}\right|(x) < \infty$$

is too restrictive. A modification of the proof yields the following modified result where only local curvature bounds are assumed:

THEOREM 7.1 (Extended Hamilton compactness theorem). *Suppose that $(\mathcal{M}_i, g_i(t))$ is a sequence of complete Ricci flows on n-dimensional manifolds $\mathcal{M}_i$ for $t \in (a,b)$, where $-\infty \leq a < 0 < b \leq \infty$. Suppose that $p_i \in \mathcal{M}_i$ for each i, and that*

(i) for all $r > 0$, there exists $M = M(r) < \infty$ such that for all $t \in (a,b)$ and for all i, there holds

$$\sup_{B_{g_i(0)}(p_i,r)} \left|\mathrm{Rm}_{g_i(t)}\right| \leq M, \qquad \text{and}$$

(ii)

$$\inf_i \mathrm{inj}(\mathcal{M}_i, g_i(0), p_i) > 0.$$

Then there exist a manifold $\mathcal{N}$ of dimension n, a Ricci flow $h(t)$ on $\mathcal{N}$ for $t \in (a,b)$ and a point $q \in \mathcal{N}$, such that $(\mathcal{N}, h(0))$ is complete, and after passing to a subsequence in i, we have

$$(7.1) \qquad (\mathcal{M}_i, g_i(t), p_i) \to (\mathcal{N}, h(t), q),$$

as $i \to \infty$.

In the literature this result is often stated (and used!) with the additional conclusion that the limit Ricci flow $h(t)$ is *complete* – that is, $h(t)$ is complete for every $t \in (a,b)$. It was noted recently [**23**, Appendix E] following questions raised by Cabezas-Rivas and Wilking that this completeness 'does not immediately follow'. Our objective in this lecture is to demonstrate that in fact the completeness fails in general, and some modification of earlier applications is required.

7.1. Intuition behind the construction

In this section we sketch an intuitive construction that indicates that a limiting Ricci flow arising in Theorem 7.1 can be complete over an open time interval containing $t = 0$, but incomplete beyond a certain time. The precise construction we make in [**32**] is a little different to reduce the amount of technology required to make it rigorous. (That technology was later developed in [**15**].)

At the core of the construction are the 'contracting cusp' examples of Ricci flows we constructed in Lecture 6, and in particular, the alternative way of flowing the torus $T^2\backslash\{p\}$ with its complete hyperbolic manifold g_{hyp} (which has a hyperbolic cusp as its end). Recall that one way of flowing this initial manifold is simply to dilate (analogously to the shrinking sphere example from Section 1.2) giving

$$G_1(t) := (1+2t)g_{hyp},$$

for all $t \in [0,\infty)$. In addition, we have the alternative flow from Lecture 6 where one imagines capping off the hyperbolic cusp infinitely far out, and letting it contract down. Let us write $G_2(t)$ for this alternative flow, which we view either as a complete Ricci flow on T^2 for $t > 0$, or as a Ricci flow on $T^2\backslash\{p\}$ for $t \geq 0$ which is incomplete for $t > 0$ but equal to g_{hyp} for $t = 0$.

From this discussion, we see that a perfectly valid smooth Ricci flow $h(t)$ on $T^2\backslash\{p\}$ would consist of $G_1(t+1)$ for $t \in [-1,1]$, followed by an appropriate scaling of $G_2(t)$, restricted to $T^2\backslash\{p\}$. Precisely, that scaled flow would be the restriction to $T^2\backslash\{p\}$ of $5G_2((t-1)/5)$ for $t \in (1,\infty)$. Beyond $t = 1$, the flow would be incomplete because we have removed the point p.

The core principle of this lecture is:

> *Such flows $h(t)$ can arise as Hamilton-Cheeger-Gromov limits of complete Ricci flows within the extension of Hamilton's compactness theorem given in Theorem* 7.1.

A precise statement can be found in [**32**]; here we sketch how one could hope to construct a sequence of complete, bounded-curvature Ricci flows satisfying the hypotheses of Theorem 7.1, with a limit flow as given above.

The basic building blocks are the complete hyperbolic metric g_{hyp} on $T^2\backslash\{p\}$ and a complete metric g_{bulb} on the punctured 2-sphere (equivalently on the plane) whose end is asymptotic to a hyperbolic cusp, and whose area is exactly 8π, as considered in Section 5.4. In particular, the Ricci flow of Theorem 5.2 would flow g_{bulb} for exactly time 2 before all area was sucked out of the bulb part of the manifold, and at each time the flow would have a cusp-like end.

The Ricci flows $g_i(t)$ we wish to imagine putting into Theorem 7.1 will exist on the whole of T^2. They will be the Ricci flows whose initial data at $t = -1$ arises by chopping off the ends of the cusps of both the metrics g_{hyp} and g_{bulb}, and gluing them together. The larger i becomes, the further out we wish to make our truncations. If we fix a base-point $q \in T^2\backslash\{p\}$, and consider p to correspond to some fixed point in the bulb surface for each i, then the distance $d_{g_i(-1)}(p,q)$ converges to infinity as $i \to \infty$.

The idea then is that the two distinct parts of $g_i(-1)$ will evolve largely independently of each other within the time interval $[-1,1)$, but then at time $t=1$, the area within the bulb part of the flow will have been exhausted, and the cusp end of the torus part of the flow should start contracting. The effect of this is that during the initial time interval $[-1,1)$, the Ricci flows $(T^2, g_i(t), q)$ will converge to a homothetically expanding hyperbolic metric on $T^2\backslash\{p\}$ in the large i limit (with the bulb part too far away to see) but for $t>1$, the manifolds $(T^2, g_i(t), q)$ should have uniformly controlled diameter (independent of i) and will converge to a smooth metric on the whole torus in the Cheeger-Gromov sense. Conversely, an observer at p will (in the limit $i \to \infty$) be living in a separate 'bulb' universe to q until time $t \sim 1$ when a big crunch occurs and the point q flies into view as p appears in the 'torus' universe. The construction is illustrated in the following figure.

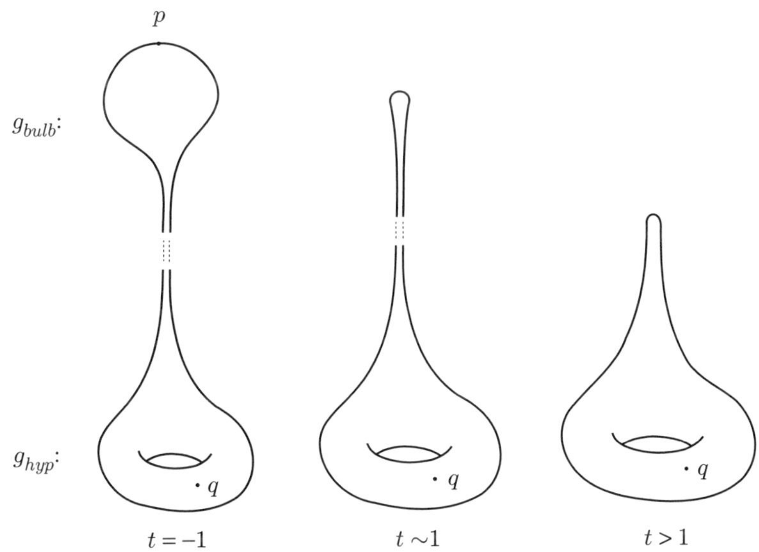

It is not too difficult to argue precisely, using pseudolocality technology (see [**25**]) that one obtains the expanding hyperbolic metric flow as the limit on an initial time interval. In order to show that the cusp will start collapsing at a uniformly controlled rate before some uniformly bounded time we require some more involved *a priori* estimates. In [**32**], we prove these estimates in a slightly different situation to the one outlined above, notably replacing the bulb metric by long thin cigars with k-dependent geometry, but with area uniformly bounded above and below.

7.2. Fixing proofs requiring completeness in the extended form of Hamilton's compactness theorem

We have seen in this lecture that under the traditional hypotheses of the extended Hamilton compactness theorem 7.1, we cannot rely on the limit being complete for all times. However, in applications, this completeness is typically essential. We would therefore like to end by commenting on some additional hypotheses that will deliver this completeness. In the major applications of this theory, such as in

Perelman's work on Ricci flow [**25**], one or other of these hypotheses can typically be seen to be available. More details can be found in [**32**].

The first fix will work in many applications where the dependency of the upper bound for the curvature on the radius is really hiding a dependence on i. One important example of this is in the proof of Perelman's pseudolocality theorem [**25**]. More precisely, the limit $h(t)$ will be complete for all $t \in (a,b)$ in Theorem 7.1 if we replace hypothesis (i) with

> (i)′ For all $r > 0$, there exist $K \in \mathbb{N}$ dependent on r, and $M < \infty$ **independent** of r, such that for all $t \in (a,b)$ and for all $i \geq K$, there holds
> $$\sup_{B_{g_i(0)}(p_i,r)} \left|\mathrm{Rm}_{g_i(t)}\right| \leq M.$$

Perhaps the greatest use of Ricci flow compactness in the work of Perelman is in his development of the theory of κ-solutions in [**25**, §11], as defined in Lecture 4, which is critical in applications. In this situation, one has *nonnegative curvature*, which prevents distances from increasing as time advances. This implies that the flow cannot turn an incomplete manifold into a complete one as t increases, and therefore that the limit flow must be complete for all negative times. The following result illustrates the idea, and can be generalised in many different directions (cf. [**23**]).

THEOREM 7.2 (Compactness of Ricci flows with a lower Ricci bound). *In the situation of Theorem* 7.1, *if each flow* $g_i(t)$ *has Ricci curvature uniformly bounded below (also uniformly in* i*) then the limit Ricci flow* $(\mathcal{N}, h(t))$ *is complete for all* $t \in (a, 0]$.

Note that the example whose construction we have sketched in Section 7.1 has a loss of completeness beyond some *positive* time. In the special case of two-dimensional Ricci flows, one can *never* have a loss of completeness for negative times in Theorem 7.1, because we know ([**4**], cf. (4.1) and (5.4)) that the Gauss curvature of each flow is bounded below by $-\frac{1}{2(t-a)}$, and one can apply Theorem 7.2 over arbitrarily smaller time intervals $(a+\varepsilon, b)$, for $\varepsilon \in (0, |a|)$.

Bibliography

[1] F.E. Burstall, *Basic Riemannian Geometry*, in 'Spectral Theory and Geometry', (eds. E.B. Davies and Y. Safarov), L.M.S. Lecture Note Series **273**, C.U.P. (1999), pages 1-29. `http://www.maths.bath.ac.uk/~feb/papers/icms/paper.pdf` MR1736864 (2001a:53001)

[2] J. Cheeger, M. Gromov and M. Taylor, *Finite propagation speed, kernel estimates for functions of the Laplace operator, and the geometry of complete Riemannian manifolds.*', J. Differential Geom. **17** (1982) 15–53. MR658471 (84b:58109)

[3] B.-L. Chen and X.-P. Zhu, *Uniqueness of the Ricci flow on complete noncompact manifolds.* J. Differential Geom. **74** (2006), 119–154. MR2260930 (2007i:53071)

[4] B.-L. Chen, *Strong uniqueness of the Ricci flow.* J. Differential Geometry, **82** (2009) 363–382. MR2520796 (2010h:53095)

[5] B. Chow, *The Ricci flow on the 2-sphere.* Journal of Differential Geometry, 33(2):325–334, 1991. MR1094458 (92d:53036)

[6] X. Chen, P. Lu, and G. Tian, *A note on uniformization of Riemann surfaces by Ricci flow.* Proc. Amer. Math. Soc. **134** (2006), 3391–3393. MR2231924 (2007d:53109)

[7] P. Daskalopoulos and M. del Pino, *On a singular diffusion equation.* Communications in Analysis and Geometry, **3** (1995) 523–542. MR1371208 (97b:35116)

[8] P. Daskalopoulos and R. S. Hamilton, *Geometric estimates for the logarithmic fast diffusion equation.* Comm. Anal. Geom. **12** (2004) 143–164. MR2074874 (2005b:53107)

[9] P. Daskalopoulos and N. Sesum, *Type II extinction profile of maximal solutions to the Ricci flow in* $\mathbb{R}^2$. J. Geometric Analysis **20** (2010) 565–591. MR2610890 (2011h:53085)

[10] K. Fukaya, *A boundary of the set of the Riemannian manifolds with bounded curvatures and diameters.* J. Differential Geom. **28** (1988) 1–21. MR950552 (89h:53090)

[11] G. Giesen, 'Instantaneously complete Ricci flows on surfaces.' PhD thesis, University of Warwick, 2012.

[12] G. Giesen and P. M. Topping, *Ricci flow of negatively curved incomplete surfaces.* Calc. Var. and PDE, **38** (2010), 357–367. MR2647124 (2011d:53155)

[13] G. Giesen and P. M. Topping, *Existence of Ricci flows of incomplete surfaces.* Comm. Partial Differential Equations, **36** (2011) 1860–1880. `http://arxiv.org/abs/1007.3146` MR2832165 (2012i:53063)

[14] G. Giesen and P. M. Topping, *Ricci flows with unbounded curvature.* Math. Zeit., **273** (2013) 449–460. `http://arxiv.org/abs/1106.2493` MR3010170

[15] G. Giesen and P. M. Topping, *Ricci flows with bursts of unbounded curvature.* To appear, Comm. Partial Differential Equations. `http://arxiv.org/abs/1302.5686` MR3010170

[16] R. Greene and H. Wu, *Lipschitz convergence of Riemannian manifolds.* Pac. J. Math. **131** (1988) 119-141. MR917868 (89g:53063)

[17] M. Gromov, Metric structures for Riemannian and non-Riemannian spaces. 'Progress in Math.' **152** Birkhäuser, 1999. MR1699320 (2000d:53065)

[18] R.S. Hamilton, *Three-manifolds with positive Ricci curvature.* J. Differential Geometry **17** (1982) 255–306. MR664497 (84a:53050)

[19] R.S. Hamilton, *The Ricci flow on surfaces. Mathematics and general relativity (Santa Cruz, CA, 1986)*, **71** *Contemporary Mathematics*, 237–262. American Mathematical Society, Providence, RI, 1988. MR954419 (89i:53029)

[20] R. S. Hamilton, *A compactness property for solutions of the Ricci flow.* Amer. J. Math. **117** (1995) 545–572. MR1333936 (96c:53056)

[21] R. S. Hamilton, *The formation of singularities in the Ricci flow.* Surveys in differential geometry, Vol. II (Cambridge, MA, 1993) 7–136, Internat. Press, Cambridge, MA, 1995. MR1375255 (97e:53075)

[22] L. JI, R. MAZZEO and N. SESUM, *Ricci flow on surfaces with cusps.* Math. Ann. **345** (2009) 819–834. MR2545867 (2011c:53161)

[23] B. KLEINER and J. LOTT, *Notes on Perelman's papers.* Geometry and Topology, **12** (2008) 2587–2855. [Revised version of February 2013, available online]. MR2460872 (2010h:53098)

[24] B. KOTSCHWAR, *An energy approach to the problem of uniqueness for the Ricci flow.* Comm. Anal. Geom. **22** (2014), no. 1, 149–176. http://arxiv.org/abs/1206.3225 MR3194377

[25] G. PERELMAN *The entropy formula for the Ricci flow and its geometric applications.* http://arXiv.org/abs/math/0211159v1 (2002).

[26] G. PERELMAN *Ricci flow with surgery on three-manifolds.* http://arxiv.org/abs/math/0303109v1 (2003).

[27] W.-X. SHI, *Deforming the metric on complete Riemannian manifolds.* J. Differential Geom. **30** (1989) 223–301.
MR1001277 (90i:58202)

[28] P. M. TOPPING, 'Lectures on the Ricci flow.' L.M.S. Lecture note series **325** C.U.P. (2006) http://www.warwick.ac.uk/~maseq/RFnotes.html

[29] P. M. TOPPING, *Ricci flow compactness via pseudolocality, and flows with incomplete initial metrics.* J. Eur. Math. Soc. (JEMS) **12** (2010) 1429–1451. MR2734348 (2011k:53089)

[30] P. M. TOPPING, *Reverse bubbling in geometric flows.* Surveys in Geometric Analysis and Relativity. Advanced Lectures in Mathematics **20**, International Press (2011) eds H. Bray and W.P. Minicozzi II. Volume dedicated to Richard Schoen on the occasion of his sixtieth birthday. MR2906061 (2012j:53003)

[31] P. M. TOPPING, *Uniqueness and nonuniqueness for Ricci flow on surfaces: Reverse cusp singularities.* I.M.R.N., **2012** (2012) 2356–2376. MR2923169

[32] P. M. TOPPING, *Remarks on Hamilton's Compactness Theorem for Ricci flow.* J. Reine Angew. Math. **692** (2014), 173–191. MR3274551

[33] P. M. TOPPING, *Uniqueness of Instantaneously Complete Ricci flows.* Geom. Topol. **19** (2015), no. 3, 1477–1492. http://arxiv.org/abs/1305.1905 MR3352241

The Kähler-Ricci flow on compact Kähler manifolds

Ben Weinkove

IAS/Park City Mathematics Series
Volume 22, 2013

The Kähler-Ricci flow on compact Kähler manifolds

Ben Weinkove

Preface

These lecture notes are based on five hours of lectures given at the Park City Math Institute in the summer of 2013. The notes are intended to be a leisurely introduction to the Kähler-Ricci flow on compact Kähler manifolds. They are aimed at graduate students who have some background in differential geometry, but do not necessarily have any knowledge of Kähler geometry or the Ricci flow. There are exercises throughout the text. The goal is that by the end, the reader will learn the basic techniques in the Kähler-Ricci flow and know enough to be able to explore the current literature.

The material covered by these notes is as follows. In the first lecture, we give a quick introduction to some of the main definitions and tools of Kähler geometry. In Lecture 2, we introduce the Kähler-Ricci flow and give some simple examples, before stating, and in Lecture 3 proving, the maximal existence time theorem for the flow. In Lecture 4 we prove long time convergence results in the cases when the manifold has negative or zero first Chern class. Finally in Lecture 5 we discuss more recent work on the behavior of the flow on Kähler surfaces. We also "go beyond" the Kähler-Ricci flow by discussing a new flow on complex manifolds called the Chern-Ricci flow.

The Kähler-Ricci flow started as a small branch of the study of Hamilton's Ricci flow, but by now is itself a vast area of research. As a consequence, we have had to omit many topics. For the interested reader seeking more complete expository sources: the chapter [**66**] by Jian Song and the author contains many of the results of these notes and much more; the works [**4**, **8**, **31**] (in the same volume as [**66**]) and the more general survey [**53**] are excellent sources of information.

The author thanks Matt Gill who was the teaching assistant for this course, for his help in writing the exercises. In addition, thanks go to the organizers, Hubert Bray, Greg Galloway, Rafe Mazzeo and Natasa Sesum, of the research program of the 2013 PCMI Summer Session for giving the author the opportunity to participate in this exciting event. Discussions with researchers and graduate students at the Park City Math Institute were invaluable in shaping the form of these notes. The author also thanks Valentino Tosatti for some helpful comments on a previous version of these notes.

Department of Mathematics, Northwestern University, 2033 Sheridan Road, Evanston IL 60208
E-mail address: weinkove@math.northwestern.edu

LECTURE 1

An Introduction to Kähler geometry

In this lecture we introduce the notion of a Kähler metric and describe the associated covariant derivatives and curvatures. We take a somewhat informal approach which emphasizes the minimal definitions and tools needed to carry out computations. The reader looking for more details may wish to consult [**38**] or [**28**], for example.

1.1. Complex manifolds

Let M be a smooth manifold of dimension $2n$. We say that M is a *complex manifold of complex dimension* n if M can be covered by charts (U, z) where U is an open subset of M and $z : U \to \mathbb{C}^n$ is a homeomorphism onto an open subset $z(U)$ of $\mathbb{C}^n$, with the following property: if $(\tilde{U}, \tilde{z})$ is another chart with $U \cap \tilde{U}$ nonempty then the *transition maps*

$$\tilde{z} \circ z^{-1} : z(U \cap \tilde{U}) \to \tilde{z}(U \cap \tilde{U})$$

and

$$z \circ \tilde{z}^{-1} : \tilde{z}(U \cap \tilde{U}) \to z(U \cap \tilde{U}),$$

are holomorphic.

We write $z = (z^1, \ldots, z^n)$ and $\tilde{z} = (\tilde{z}^1, \ldots, \tilde{z}^n)$. These are called *complex coordinates*. We also introduce the real coordinates $(x^1, \ldots, x^n, y^1, \ldots, y^n)$ defined by the usual formula

$$z^i = x^i + \sqrt{-1} y^i.$$

Note that we avoid the notation i for $\sqrt{-1}$ since i is our favorite letter for an index. We define operators $\frac{\partial}{\partial z^i}$ and $\frac{\partial}{\partial \overline{z}^i}$ on $\mathbb{C}^n$ by

$$\frac{\partial}{\partial z^i} = \frac{1}{2}\left(\frac{\partial}{\partial x^i} - \sqrt{-1}\frac{\partial}{\partial y^i}\right), \quad \frac{\partial}{\partial \overline{z}^i} = \frac{1}{2}\left(\frac{\partial}{\partial x^i} + \sqrt{-1}\frac{\partial}{\partial y^i}\right).$$

As the notation suggests, we have for all $i, j = 1, \ldots, n$,

$$\frac{\partial}{\partial z^i}(z^j) = \delta_{ij}, \quad \frac{\partial}{\partial z^i}(\overline{z}^j) = 0, \quad \frac{\partial}{\partial \overline{z}^i}(z^j) = 0, \quad \frac{\partial}{\partial \overline{z}^i}(\overline{z}^j) = \delta_{ij},$$

where δ_{ij} is the Kronecker delta symbol. In particular, a smooth function f on $\mathbb{C}^n$ is holomorphic if and only if

$$\frac{\partial f}{\partial \overline{z}^i} = 0 \quad \text{for } i = 1, \ldots, n.$$

Hence the condition that the transition maps be holomorphic can be written as

$$(1.1.1) \qquad \frac{\partial \tilde{z}^i}{\partial \overline{z}^j} = 0 \quad \text{and} \quad \frac{\partial z^i}{\partial \overline{\tilde{z}}^j} = 0, \quad \text{for all } i, j = 1, \ldots, n,$$

where defined. Here we are writing $\tilde{z}$ for the map $\tilde{z} \circ z^{-1}$ and z for the map $z \circ \tilde{z}^{-1}$.

Recall that the definition of a smooth manifold requires the existence of coordinate charts whose transition maps are smooth, and this allows for a well-defined notion of a smooth function. Similarly, on a complex manifold M we can make the following definition: a smooth function f on M is *holomorphic* if for each coordinate chart (U, z) we have

$$\frac{\partial f}{\partial \overline{z}^i} = 0, \quad \text{for } i = 1, \ldots, n,$$

on U. Of course, we are writing f for $f \circ z^{-1}$. To see that this is well-defined, suppose that $\tilde{U}$ is an overlapping coordinate chart and compute using the chain rule on $U \cap \tilde{U}$,

$$\frac{\partial f}{\partial \overline{\tilde{z}}^j} = \sum_k \frac{\partial f}{\partial \overline{z}^k} \frac{\partial \overline{z}^k}{\partial \overline{\tilde{z}}^j} + \sum_k \frac{\partial f}{\partial z^k} \frac{\partial z^k}{\partial \overline{\tilde{z}}^j} = 0,$$

as required. Note that we have used the condition (1.1.1) to see that the second term vanishes.

EXAMPLE 1.1. $\mathbb{C}^n$ is a complex manifold with a single coordinate chart $U = \mathbb{C}^n$ and $z : U \to \mathbb{C}^n$ the identity map. Taking the quotient of $\mathbb{C}^n$ by the lattice $\mathbb{Z}^{2n}$, say, gives a compact complex manifold homeomorphic to the torus T^{2n}.

EXAMPLE 1.2. Define *complex projective space* $\mathbb{P}^n$ as follows. As a topological space, $\mathbb{P}^n$ is the quotient space

$$(\mathbb{C}^{n+1} - \{0\}) / \sim .$$

where $\sim$ is the equivalence relation defined by

$$(Z_0, Z_1, \ldots, Z_n) \sim (\lambda Z_0, \lambda Z_1, \ldots, \lambda Z_n),$$

for $\lambda \in \mathbb{C}^*$. In other words, it is the space of complex lines through the origin in $\mathbb{C}^{n+1}$. Define open sets

$$U_i = \{[Z_0, \ldots, Z_n] \in \mathbb{P}^n \mid Z_i \neq 0\}, \quad \text{for } i = 0, 1, \ldots, n,$$

where we are writing $[Z_0, \ldots, Z_n]$ for the equivalence class of $(Z_0, \ldots, Z_n)$ (the Z_i are called *homogeneous coordinates*). The U_i cover $\mathbb{P}^n$. On U_0 we define complex coordinates $z^1, \ldots, z^n$ by

$$z^1 = \frac{Z_1}{Z_0}, \ldots, z^n = \frac{Z_n}{Z_0},$$

and similarly for $U_1, \ldots, U_n$. We leave it to the reader to check that the associated transition maps are holomorphic.

EXERCISE 1.1. Show that $\mathbb{P}^1$ is diffeomorphic to the sphere S^2.

1.2. Vector fields, 1-forms, Hermitian metrics and tensors

Let M be a complex manifold as above. The *complexified tangent space* $(T_pM)^{\mathbb{C}}$ at a point p is given by the span over $\mathbb{C}$ of

$$\frac{\partial}{\partial z^1}, \ldots, \frac{\partial}{\partial z^n}, \frac{\partial}{\partial \overline{z}^1}, \ldots, \frac{\partial}{\partial \overline{z}^n},$$

where we evaluate at the point p. We write $(T_pM)^{\mathbb{C}} = T_p^{1,0}M \oplus T_p^{0,1}M$, where $T_p^{1,0}M$ is given by the span of the $\frac{\partial}{\partial z^i}$ and $T_p^{0,1}M$ by the $\frac{\partial}{\partial \overline{z}^i}$. By the chain rule and the equations (1.1.1), this decomposition of $(T_pM)^{\mathbb{C}}$ is independent of choice of complex coordinate chart. Indeed,

$$\frac{\partial}{\partial z^i} = \sum_k \frac{\partial \tilde{z}^k}{\partial z^i}\frac{\partial}{\partial \tilde{z}^k} + \sum_k \frac{\partial \overline{\tilde{z}}^k}{\partial z^i}\frac{\partial}{\partial \overline{\tilde{z}}^k} = \sum_k \frac{\partial \tilde{z}^k}{\partial z^i}\frac{\partial}{\partial \tilde{z}^k} \in \text{span}\left\{\frac{\partial}{\partial \tilde{z}^1}, \ldots, \frac{\partial}{\partial \tilde{z}^n}\right\},$$

and similarly for $\frac{\partial}{\partial \overline{z}^i}$.

We define a $T^{1,0}$ *vector field* on M to be a smooth complex-valued vector field X on M with the property that $X_p \in T_p^{1,0}M$ for all $p \in M$. We write X locally as

$$X = \sum_i X^i \frac{\partial}{\partial z^i},$$

where the $X^i : U \to \mathbb{C}$ are smooth functions which satisfy the following *transformation rule*. If we write $\tilde{X}^i$ for the corresponding functions on $\tilde{U}$ then

$$X^i = \sum_j \tilde{X}^j \frac{\partial z^i}{\partial \tilde{z}^j} \quad \text{on } U \cap \tilde{U}.$$

Any collection of functions $X^i : U \to \mathbb{C}$ defined on each chart in a cover of M, which satisfy the above transformation rule, determine a globally defined vector field X. Indeed one can check that

$$X^i \frac{\partial}{\partial z^i} = \tilde{X}^i \frac{\partial}{\partial \tilde{z}^i} \quad \text{on } U \cap \tilde{U}.$$

Here and henceforth we are using the *summation convention* that we sum over repeated indices from 1 to n when one index is upper and the other is lower (we regard the index i in $\frac{\partial}{\partial z^i}$ as a lower index).

EXERCISE 1.2. We define a *holomorphic vector field* on M to be a $T^{1,0}$ vector field $X = X^i \frac{\partial}{\partial z^i}$ such that

$$\frac{\partial X^i}{\partial \overline{z^j}} = 0 \quad \text{for all } i, j = 1, \ldots, n.$$

Show that this condition is well-defined, independent of choice of coordinate chart.

We also define $T^{0,1}$ vector fields in a similar way. A $T^{0,1}$ vector field is written locally as $Y = Y^{\overline{j}} \frac{\partial}{\partial \overline{z^j}}$ where the $Y^{\overline{j}}$ transform according to the rule

$$Y^{\overline{j}} = \tilde{Y}^{\overline{\ell}} \overline{\frac{\partial z^j}{\partial \tilde{z}^\ell}} \quad \text{on } U \cap \tilde{U},$$

where of course

$$\overline{\frac{\partial z^j}{\partial \tilde{z}^\ell}} = \frac{\partial \overline{z^j}}{\partial \overline{\tilde{z}}^\ell}.$$

We can do the same for the complexified *cotangent* space $(T_p^*M)^{\mathbb{C}}$ which is spanned over $\mathbb{C}$ by the 1-forms

$$dz^1, \ldots, dz^n, d\overline{z}^1, \ldots, d\overline{z}^n.$$

Here $dz^i = dx^i + \sqrt{-1}dy^i$ and $d\overline{z^i} = dx^i - \sqrt{-1}dy^i$ are dual to $\frac{\partial}{\partial z^i}$ and $\frac{\partial}{\partial \overline{z^i}}$ respectively. We have a decomposition of the complexified cotangent space into $(1,0)$-forms and $(0,1)$-forms, spanned by the dz^i and $d\overline{z^i}$ respectively. A $(1,0)$-form a on

M is written locally as $a = a_i dz^i$, and a $(0,1)$ form b as $b = b_{\bar{j}} d\bar{z}^j$, where the a_i and $b_{\bar{j}}$ transform by

$$a_i = \tilde{a}_k \frac{\partial \tilde{z}^k}{\partial z^i}, \quad b_{\bar{j}} = \tilde{b}_{\bar{\ell}} \overline{\frac{\partial \tilde{z}^\ell}{\partial z^j}} \quad \text{on } U \cap \tilde{U}.$$

Finally, define a *Hermitian metric* g on M to be a Hermitian inner product on the n-dimensional complex vector space $T_p^{1,0}M$ for each p, which varies smoothly in p. Locally g is given by an $n \times n$ positive definite Hermitian matrix whose (i,j)th entry we denote by $g_{i\bar{j}}$, which transforms according to

$$g_{i\bar{j}} = \tilde{g}_{k\bar{\ell}} \frac{\partial \tilde{z}^k}{\partial z^i} \overline{\frac{\partial \tilde{z}^\ell}{\partial z^j}} \quad \text{on } U \cap \tilde{U}. \tag{1.2.1}$$

Given $T^{1,0}$ vector fields $X = X^i \frac{\partial}{\partial z^i}$ and $Y = Y^i \frac{\partial}{\partial z^i}$ we define their pointwise inner product by

$$\langle X, Y \rangle_g = g_{i\bar{j}} X^i \overline{Y^j},$$

and we write

$$|X|_g = \sqrt{\langle X, X \rangle_g}$$

for the *norm* of X with respect to g.

EXERCISE 1.3. Show that $\langle X, Y \rangle_g$ is well-defined, independent of choice of complex coordinates.

We can similarly use g to define an inner product on $T^{0,1}$ vectors.

REMARK 1.1. A Hermitian metric g defines a Riemannian metric g_R, which we can define locally by

$$g_R\left(\frac{\partial}{\partial x^i}, \frac{\partial}{\partial x^j}\right) = 2\mathrm{Re}(g_{i\bar{j}}) = g_R\left(\frac{\partial}{\partial y^i}, \frac{\partial}{\partial y^j}\right), \quad g_R\left(\frac{\partial}{\partial x^i}, \frac{\partial}{\partial y^j}\right) = 2\mathrm{Im}(g_{i\bar{j}}).$$

However, we won't make use of this correspondence.

Extending all of the above, we can define tensors on a complex manifold with any number of upper or lower indices, barred or unbarred. For example, the set of locally defined functions $S^{ik}_{\bar{j}} : U \to \mathbb{C}$ defines a tensor with two upper unbarred indices and one lower barred index, if it satisfies the transformation rule:

$$S^{ik}_{\bar{j}} = \tilde{S}^{ab}_{\bar{c}} \frac{\partial z^i}{\partial \tilde{z}^a} \frac{\partial z^k}{\partial \tilde{z}^b} \overline{\frac{\partial \tilde{z}^c}{\partial z^j}} \quad \text{on } U \cap \tilde{U}. \tag{1.2.2}$$

More formally, $S = S^{ik}_{\bar{j}} \frac{\partial}{\partial z^i} \otimes \frac{\partial}{\partial z^k} \otimes d\bar{z}^j$ defines a smooth section of $T^{1,0}M \otimes T^{1,0}M \otimes (T^{0,1})^*M$. However, in these lecture notes we will stick with more informal language.

The reader will notice that the transformation formulae for any kind of tensor can easily be derived by following the simple rule: match the indices according to barred/unbarred, upper/lower and z or $\tilde{z}$. For example, in (1.2.2), the indices i, k on the left hand side are unbarred upper indices with respect to z, and since they are "free indices", they match with corresponding free unbarred indices i, k with respect to z on the right. On the other hand, the upper unbarred indices a, b with respect to $\tilde{z}$ on the right are "summed indices" and so must match with *lower* unbarred a, b indices with respect to $\tilde{z}$.

EXERCISE 1.4. Let $g = (g_{k\overline{\ell}})$ be a Hermitian metric on M. Define $g^{i\overline{j}}$ to be the (i,j)th component of the inverse matrix of $(g_{k\overline{\ell}})$. Show that $g^{i\overline{j}}$ defines a tensor on M, which we call g^{-1}.

We define an pointwise inner product on $(1,0)$ forms using $g^{i\overline{j}}$, as follows. If $a = a_i dz^i$ and $b = b_i dz^i$ then

$$\langle a, b\rangle_g = g^{i\overline{j}} a_i \overline{b_j},$$

and we define the norm of a to be $|a|_g = \sqrt{\langle a, a\rangle_g}$. We can similarly define an inner product for $(0,1)$ forms.

1.3. Kähler metrics and covariant differentiation

We say that a Hermitian metric $g = (g_{i\overline{j}})$ is *Kähler* if

$$\partial_k g_{i\overline{j}} = \partial_i g_{k\overline{j}} \quad \text{for all } i,j,k = 1,\ldots,n. \tag{1.3.1}$$

Namely, $\partial_k g_{i\overline{j}}$ is unchanged when we swap the two unbarred indices k and i. Here and henceforth, to simplify notation, we are writing

$$\partial_i = \frac{\partial}{\partial z^i} \quad \text{and} \quad \partial_{\overline{j}} = \frac{\partial}{\partial \overline{z}^j}.$$

EXERCISE 1.5. Show that the condition (1.3.1) is independent of choice of complex coordinates.

EXAMPLE 1.3. If M is a complex manifold of complex dimension 1 (a Riemann surface) then every Hermitian metric is Kähler, since (1.3.1) is vacuous.

EXAMPLE 1.4. $\mathbb{C}^n$ with the Euclidean metric $g_{i\overline{j}} = \delta_{ij}$ is Kähler. More generally, we may take $g_{i\overline{j}} = A_{ij}$ where (A_{ij}) is any fixed $n\times n$ positive definite Hermitian matrix. Since $g_{i\overline{j}}$ is constant, it descends to the quotient $\mathbb{C}^n/\mathbb{Z}^{2n}$ to give a Kähler metric on the torus.

EXERCISE 1.6. Following on from Example 1.2, we can define

$$g_{i\overline{j}} = \partial_i\partial_{\overline{j}} \log(1 + |z^1|^2 + \cdots + |z^n|^2), \quad \text{on } U_0,$$

and similarly for $U_1, \ldots, U_n$. Show that $(g_{i\overline{j}})$ defines a Kähler metric on $\mathbb{P}^n$. This metric is called the *Fubini-Study metric.*

Given a Kähler metric g, we define its *Kähler form* to be

$$\omega = \sqrt{-1} g_{i\overline{j}} dz^i \wedge d\overline{z}^j.$$

Observe that since $(g_{i\overline{j}})$ is Hermitian, the form ω is real:

$$\overline{\omega} = -\sqrt{-1}\overline{g_{i\overline{j}}} d\overline{z}^i \wedge dz^j = \sqrt{-1} g_{j\overline{i}} dz^j \wedge d\overline{z}^i = \omega.$$

It is a form of type $(1,1)$ (in the span of the $dz^i \wedge d\overline{z}^j$).

Note that if g is just a Hermitian metric then one can still define an associated real $(1,1)$ form ω, which is sometimes referred to as the *fundamental 2-form* of g. Abusing notation slightly, we will often refer to the form ω as a Hermitian metric (or as a Kähler metric, if g is Kähler).

The following exercise shows that a Hermitian metric g is Kähler if and only if ω is d-closed. First we need some notation. We define an operator ∂ which takes a (p,q) form to a $(p+1,q)$ form as follows. Given a (p,q) form

$$a = a_{i_1\cdots i_p\overline{j_1}\cdots\overline{j_q}} dz^{i_1} \wedge \cdots \wedge dz^{i_p} \wedge d\overline{z}^{j_1} \wedge \cdots \wedge d\overline{z}^{j_q},$$

we define

$$\partial a = \partial_k a_{i_1\cdots i_p \overline{j_1}\cdots\overline{j_q}} dz^k \wedge dz^{i_1} \wedge \cdots \wedge dz^{i_p} \wedge d\overline{z}^{j_1} \wedge \cdots \wedge d\overline{z}^{j_q}.$$

An argument similar to the solution of Exercise 1.5 shows that ∂ is well-defined independent of choice of coordinates. The operator $\overline{\partial}$, taking (p,q) forms to $(p,q+1)$ forms is defined similarly, and we set

$$d = \partial + \overline{\partial},$$

which is the same as the usual exterior derivative on manifolds.

EXERCISE 1.7. Let g be a Hermitian metric. Show that

$$g \text{ is Kähler} \iff d\omega = 0 \iff \partial\omega = 0 \iff \overline{\partial}\omega = 0.$$

The next result gives a way to construct new Kähler manifolds from old ones.

PROPOSITION 1.1. *Let (M,g) be a Kähler manifold, and $N \subset M$ a complex submanifold. Then $g|_N$ is a Kähler metric on N.*

PROOF. N being a complex submanifold of dimension k means the following: at any point p of N we can find complex coordinates for M centered at p so that near p, N is given by $\{z^{k+1} = \cdots = z^n = 0\}$ and $z^1, \ldots, z^k$ give complex coordinates for N. It follows immediately that the metric g at p defines an inner product on the complexified tangent space of N (just restrict to the span of $\partial_1, \cdots, \partial_k$).

Let $\iota : N \to M$ be the inclusion map, so that $\omega|_N = \iota^*\omega$. Then by the standard property of the exterior derivative, we have

$$d\iota^*\omega = \iota^* d\omega = 0,$$

as required. □

It follows that:

COROLLARY 1.1. *Every smooth projective variety admits a Kähler metric.*

Indeed, a smooth projective variety can be defined to be a complex submanifold of $\mathbb{P}^N$ for some N, which by Example 1.6 admits a Kähler metric.

We now define the notion of *covariant differentiation* on a Kähler manifold (M,g). Observe that, by the argument of Exercise 1.2, if we are given a $T^{1,0}$ vector field $X = X^i\partial_i$, the object

$$\frac{\partial X^i}{\partial \overline{z}^\ell},$$

gives a well-defined tensor, meaning that it transforms according to the rule

$$\frac{\partial X^i}{\partial \overline{z}^\ell} = \frac{\partial \tilde{X}^k}{\partial \overline{\tilde{z}}^j} \overline{\frac{\partial \tilde{z}^j}{\partial z^\ell}} \frac{\partial z^i}{\partial \tilde{z}^k} \quad \text{on } U \cap \tilde{U}.$$

Indeed, this follows from the fact that the transition maps for the X^i are holomorphic and so "pass through" the operator $\partial_{\overline{\ell}}$. However, the reader can check that the object

$$\frac{\partial X^i}{\partial z^k}$$

does *not* give a well-defined tensor, and this leads us to define covariant differentiation.

We define the *Christoffel symbols* of g on the chart (U, z) to be the functions $\Gamma^i_{kp} : U \to \mathbb{C}$ defined by

$$\Gamma^i_{kp} = g^{i\overline{q}} \partial_k g_{p\overline{q}}, \tag{1.3.2}$$

where we recall that $g^{i\overline{q}}$ are the components of the inverse of g. By the Kähler condition (1.3.1),

$$\Gamma^i_{kp} = \Gamma^i_{pk}.$$

The Christoffel symbols do *not* define a tensor. However, given a $T^{1,0}$ vector field X, we define the *covariant derivative* $\nabla_k X^i$ by

$$\nabla_k X^i = \partial_p X^i + \Gamma^i_{kp} X^p,$$

and this *does* define a tensor. By the above observation, $\partial_{\overline{\ell}} X^i$ is already a tensor and we define

$$\nabla_{\overline{\ell}} X^i = \partial_{\overline{\ell}} X^i.$$

Similarly, for a $T^{0,1}$ vector field $Y = Y^{\overline{j}} \partial_{\overline{j}}$, a $(1,0)$ form $a = a_i dz^i$ and a $(0,1)$ form $b = b_{\overline{j}} d\overline{z}^j$, we define

$$\begin{array}{ll} \nabla_k Y^{\overline{j}} = \partial_k Y^{\overline{j}} & \nabla_{\overline{\ell}} Y^{\overline{j}} = \partial_{\overline{\ell}} Y^{\overline{j}} + \overline{\Gamma^j_{\ell q}} Y^{\overline{q}} \\ \nabla_k a_i = \partial_k a_i - \Gamma^p_{ki} a_p & \nabla_{\overline{\ell}} a_i = \partial_{\overline{\ell}} a_i \\ \nabla_k b_{\overline{j}} = \partial_k b_{\overline{j}} & \nabla_{\overline{\ell}} b_{\overline{j}} = \partial_{\overline{\ell}} b_{\overline{j}} - \overline{\Gamma^q_{\ell j}} b_{\overline{q}}. \end{array}$$

EXERCISE 1.8. Show that $\nabla_k X^i$, $\nabla_{\overline{\ell}} Y^{\overline{j}}$ etc. all define tensors.

Moreover, we can extend covariant differentiation naturally to any kind of tensor, such as the tensor $S^{ab}_{\overline{c}}$ described above in Section 1.2:

$$\begin{aligned} \nabla_k S^{ab}_{\overline{c}} &= \partial_k S^{ab}_{\overline{c}} + \Gamma^a_{kp} S^{pb}_{\overline{c}} + \Gamma^b_{kp} S^{ap}_{\overline{c}} \\ \nabla_{\overline{\ell}} S^{ab}_{\overline{c}} &= \partial_{\overline{\ell}} S^{ab}_{\overline{c}} - \overline{\Gamma^q_{\ell c}} S^{ab}_{\overline{q}}. \end{aligned}$$

In particular, we have $\nabla_k g_{i\overline{j}} = 0$. Indeed, this follows from the choice of the Christoffel symbols, since

$$\nabla_k g_{i\overline{j}} = \partial_k g_{i\overline{j}} - \Gamma^p_{ki} g_{p\overline{j}} = \partial_k g_{i\overline{j}} - g^{p\overline{q}} (\partial_k g_{i\overline{q}}) g_{p\overline{j}} = 0,$$

where we have used the fact that $g^{p\overline{q}} g_{p\overline{j}} = \delta_{jq}$.

REMARK 1.2. ∇ coincides with the Levi-Civita connection of the Riemannian metric g_R associated with g, extended to the complexified tangent bundle.

1.4. Curvature

We now describe the curvature associated to a Kähler metric g. Define the *curvature tensor* $R_{i\overline{j}k}{}^p$ by

$$R_{i\overline{j}k}{}^p = -\partial_{\overline{j}} \Gamma^p_{ik},$$

for Γ^p_{ik} the Christoffel symbols of g, defined by (1.3.2).

EXERCISE 1.9. Show that $R_{i\overline{j}k}{}^p$ is a tensor.

It will be convenient to define

$$R_{i\bar{j}k\bar{\ell}} = R_{i\bar{j}k}{}^{p} g_{p\bar{\ell}}.$$

That is, we lower the index p into the last slot using the metric g. We will also refer to this tensor as the curvature tensor. The tensor $R_{i\bar{j}k\bar{\ell}}$ has the following symmetries:

PROPOSITION 1.2. *The curvature tensor of a Kähler metric satisfies*

$$R_{i\bar{j}k\bar{\ell}} = R_{k\bar{j}i\bar{\ell}} = R_{i\bar{\ell}k\bar{j}} = R_{k\bar{\ell}i\bar{j}},$$

and

$$\overline{R_{i\bar{j}k\bar{\ell}}} = R_{j\bar{i}\ell\bar{k}}.$$

PROOF. From the definitions,

$$\begin{aligned} R_{i\bar{j}k\bar{\ell}} &= - g_{p\bar{\ell}} \partial_{\bar{j}} (g^{p\bar{q}} \partial_i g_{k\bar{q}}) \\ &= - g_{p\bar{\ell}} g^{p\bar{q}} \partial_{\bar{j}} \partial_i g_{k\bar{q}} + g_{p\bar{\ell}} g^{p\bar{s}} g^{r\bar{q}} \partial_{\bar{j}} g_{r\bar{s}} \partial_i g_{k\bar{q}} \\ &= - \partial_i \partial_{\bar{j}} g_{k\bar{\ell}} + g^{p\bar{q}} \partial_i g_{k\bar{q}} \partial_{\bar{j}} g_{p\bar{\ell}}, \end{aligned} \tag{1.4.1}$$

where we have used the formula for the derivative of an inverse matrix $\delta(A^{-1}) = -A^{-1}(\delta A)A^{-1}$. The proposition then follows immediately from this formula, and the Kähler condition (1.3.1). □

The curvature tensor measures the failure of covariant derivatives to commute. More precisely:

PROPOSITION 1.3. *For a $T^{1,0}$ vector field $X = X^p \partial_p$, a $T^{0,1}$ vector field $Y = Y^{\bar{q}} \partial_{\bar{q}}$, a $(1,0)$ form $a = a_p dz^p$ and a $(0,1)$ form $b = b_{\bar{q}} d\bar{z}^q$, we have the following commutation formulae:*

$$[\nabla_i, \nabla_{\bar{j}}] X^p = R_{i\bar{j}k}{}^{p} X^k, \qquad [\nabla_i, \nabla_{\bar{j}}] Y^{\bar{q}} = -R_{i\bar{j}}{}^{\bar{q}}{}_{\bar{\ell}} Y^{\bar{\ell}}$$

$$[\nabla_i, \nabla_{\bar{j}}] a_p = -R_{i\bar{j}p}{}^{q} a_q, \qquad [\nabla_i, \nabla_{\bar{j}}] b_{\bar{q}} = R_{i\bar{j}}{}^{\bar{\ell}}{}_{\bar{q}} b_{\bar{\ell}}.$$

Here, $[\nabla_i, \nabla_{\bar{j}}] = \nabla_i \nabla_{\bar{j}} - \nabla_{\bar{j}} \nabla_i$, and we are raising and lowering indices of the curvature tensor using g.

Before we prove this proposition, it is convenient to introduce the notion of a *holomorphic normal coordinate system.*

LEMMA 1.1. *Let (M, g) be a Kähler manifold. For any fixed point $x \in M$, there exists a holomorphic coordinate chart (U, z) centered at x such that, at x,*

$$g_{i\bar{j}} = \delta_{ij}, \quad \text{and} \quad \partial_k g_{i\bar{j}} = 0,$$

for all $i, j, k = 1, 2, \ldots, n$.

PROOF. By an affine linear change in coordinates, we can find coordinates $\tilde{z}^1, \ldots, \tilde{z}^n$ centered at x with $\tilde{g}_{i\bar{j}} = \delta_{ij}$ at that point. To obtain the vanishing of the first derivatives of g, define a new holomorphic coordinate system $z^1, \ldots, z^n$ by

$$\tilde{z}^i = z^i - \frac{1}{2} \tilde{\Gamma}^i_{jk}(0) z^j z^k.$$

Observe that the first derivative of $\tilde{z} = \tilde{z}(z)$ at 0 is the identity, and hence by the inverse function theorem, we can solve for z as a holomorphic function of $\tilde{z}$ in a neighborhood of zero. We leave it as an exercise to check that in the z coordinate system, we have

$$\partial_k g_{i\bar{j}} = 0$$

at x for all i, j, k. □

EXERCISE 1.10. Complete the proof of Lemma 1.1

The coordinate system we constructed in Lemma 1.1 is called a *holomorphic normal coordinate system* for g. The lemma implies in particular that we can choose coordinates for which the Christoffel symbols vanish at a point (and this implies that Γ cannot be a tensor, since if it were, it would have to vanish everywhere). Note that the Kähler condition is required for the existence of holomorphic normal coordinates. Indeed from (1.3.1) it is immediate that the existence of these coordinates for g implies that g is Kähler.

We now complete the proof of Proposition 1.3.

PROOF OF PROPOSITION 1.3. Since both sides are tensors, it is sufficient to prove the identities at a single point x, in a holomorphic normal coordinate system (if an equation of tensors holds in one coordinate system, it must hold in every coordinate system). Compute at x,

$$\begin{aligned}[\nabla_i, \nabla_{\bar{j}}]X^p &= \partial_i \nabla_{\bar{j}} X^p - \partial_{\bar{j}}(\partial_i X^p + \Gamma^p_{ik} X^k) \\ &= \partial_i \partial_{\bar{j}} X^p - \partial_{\bar{j}} \partial_i X^p - (\partial_{\bar{j}} \Gamma^p_{ik}) X^k \\ &= R_{i\bar{j}k}{}^p X^k,\end{aligned}$$

giving the first formula. The others are left as the next exercise. □

EXERCISE 1.11. Complete the proof of Proposition 1.3.

EXERCISE 1.12. Let $\omega = \sqrt{-1} g_{i\bar{j}} dz^i \wedge d\bar{z}^j$ be a Kähler metric.

(1) Show that if $\beta = \sqrt{-1}\beta_{i\bar{j}} dz^i \wedge d\bar{z}^j$ is a real $(1,1)$-form and ω a Kähler form, then

$$n\omega^{n-1} \wedge \beta = g^{i\bar{j}} \beta_{i\bar{j}} \omega^n =: (\mathrm{tr}_\omega \beta)\omega^n.$$

(2) Let f be a real-valued function. Show that

$$n\omega^{n-1} \wedge \sqrt{-1}\partial f \wedge \overline{\partial} f = |\partial f|^2_g \omega^n.$$

Hint: pick coordinates at a point for which $g_{i\bar{j}} = \delta_{ij}$.

We end this section by discussing the *Ricci curvature* of a Kähler metric, which is defined to be the tensor $R_{i\bar{j}}$ given by

$$R_{i\bar{j}} = g^{k\bar{\ell}} R_{i\bar{j}k\bar{\ell}}.$$

A key property of Kähler metrics is the following simple formula for the Ricci curvature:

PROPOSITION 1.4. *The Ricci curvature is given by*

$$R_{i\bar{j}} = -\partial_i \partial_{\bar{j}} \log \det g.$$

PROOF. The proposition follows easily from the well-known formula for the derivative of the determinant of an invertible Hermitian matrix A:

$$\delta \det A = \mathrm{trace}(A^{-1}\delta A) \det A,$$

which can be rewritten as

$$\delta \log \det A = \mathrm{trace}(A^{-1}\delta A). \tag{1.4.2}$$

We compute

$$R_{i\bar{j}} = -g^{k\bar{\ell}} g_{p\bar{\ell}} \partial_{\bar{j}} \Gamma^p_{ik} = -\partial_{\bar{j}} \Gamma^p_{ip} = -\partial_{\bar{j}}(g^{p\bar{q}} \partial_i g_{p\bar{q}}) = -\partial_{\bar{j}} \partial_i \log \det g,$$

as required. □

We define the *Ricci form* of g to be the $(1,1)$ form

$$\mathrm{Ric}(\omega) = \sqrt{-1} R_{i\bar{j}} dz^i \wedge d\overline{z^j} = -\sqrt{-1}\partial\overline{\partial} \log \det g.$$

It follows from either Proposition 1.2 or Proposition 1.4 that $(R_{i\bar{j}})$ is Hermitian, and hence $\mathrm{Ric}(\omega)$ is a real $(1,1)$ form. Proposition 1.4 implies that $\partial_k R_{i\bar{j}} = \partial_i R_{k\bar{j}}$, namely that $\mathrm{Ric}(\omega)$ is d-closed.

Note that we often write

$$\mathrm{Ric}(\omega) = -\sqrt{-1}\partial\overline{\partial} \log \omega^n.$$

We make sense of this expression as follows. If Ω is any volume form, locally written as

$$\Omega = a(z)(\sqrt{-1})^n dz^1 \wedge d\overline{z^1} \wedge \cdots \wedge dz^n \wedge d\overline{z^n},$$

then we define

$$\sqrt{-1}\partial\overline{\partial} \log \Omega = \sqrt{-1}\partial\overline{\partial} \log a.$$

EXERCISE 1.13. This definition of $\sqrt{-1}\partial\overline{\partial} \log \Omega$ is well-defined, independent of choice of local coordinates.

Then since

$$\omega^n = n!(\sqrt{-1})^n \det g \, dz^1 \wedge d\overline{z^1} \wedge \cdots \wedge dz^n \wedge d\overline{z^n},$$

we see that $-\sqrt{-1}\partial\overline{\partial} \log \omega^n = -\sqrt{-1}\partial\overline{\partial} \log \det g$.

EXERCISE 1.14. Let ω_{FS} be the Fubini-Study metric of Exercise 1.6. Show that

$$\mathrm{Ric}(\omega_{\mathrm{FS}}) = (n+1)\omega_{\mathrm{FS}}.$$

LECTURE 2

The Kähler-Ricci flow and the Kähler cone

In this lecture we introduce the Kähler-Ricci flow. We also discuss Kähler classes, the Kähler cone and the first Chern class. We describe the maximal existence time result for the Kähler-Ricci flow and give some simple examples.

2.1. The Kähler-Ricci flow and simple examples

Let (M, ω_0) be a compact Kähler manifold. If $\omega = \omega(t)$ is a smooth family of Kähler metrics on M satisfying the equation

$$\frac{\partial}{\partial t}\omega = -\mathrm{Ric}(\omega), \quad \omega|_{t=0} = \omega_0, \tag{2.1.1}$$

then we say that $\omega(t)$ is a solution of the *Kähler-Ricci flow* starting at ω_0. For the reader who is familiar with Hamilton's Ricci flow of Riemannian metrics: this is the same equation (modulo a factor of 2) starting at a Kähler metric.

We describe now some simple examples of solutions to the Kähler-Ricci flow. First, let M be a compact Riemann surface (complex dimension 1). The following theorem is known as the *Uniformization Theorem.*

THEOREM 2.1. *On any compact Riemann surface M there exists a Kähler metric ω_{KE} with*

$$\mathrm{Ric}(\omega_{\mathrm{KE}}) = \mu\,\omega_{\mathrm{KE}}, \tag{2.1.2}$$

for some constant μ.

In general, a Kähler metric satisfying (2.1.2) is called a *Kähler-Einstein* metric, which explains the notation. By multiplying ω_{KE} by a constant, we may assume that μ is equal to either 1, 0 or -1. Indeed this follows from the fact that for any Kähler metric ω and positive real number λ,

$$\mathrm{Ric}(\lambda\omega) = \mathrm{Ric}(\omega), \tag{2.1.3}$$

as can be seen immediately from the formula of Proposition 1.4.

The Gauss-Bonnet formula on a Riemann surface M can be written as

$$\int_M \mathrm{Ric}(\omega) = 2\pi(2 - 2g_M),$$

where g_M is the genus of M, and hence the sign of μ determines whether M has genus 0, 1 or greater than 1.

EXAMPLE 2.1. If $\mu = 1$ then $M = \mathbb{P}^1$. Let $\omega_0 = 2\omega_{\mathrm{FS}}$ where ω_{FS} is the Fubini-Study metric from Exercise 1.6. Then by Exercise 1.14,

$$\mathrm{Ric}(\omega_0) = \omega_0.$$

We claim that $\omega(t) = (1-t)\omega_0$ is a solution of the Kähler-Ricci flow on $[0,1)$. Indeed,

$$\frac{\partial}{\partial t}\omega(t) = -\omega_0 = -\mathrm{Ric}(\omega_0) = -\mathrm{Ric}(\omega(t)),$$

where the last equality makes use of (2.1.3). Hence there is a solution of the Kähler-Ricci flow which shrinks the Fubini-Study metric to zero in finite time, by scaling. Recall that $\mathbb{P}^1$ is diffeomorphic to S^2 (Exercise 1.1). In fact, the Fubini-Study metric is a constant multiple of the standard round metric on S^2, and so this solution of the Kähler-Ricci flow can be visualized as a shrinking round sphere (Figure 2.1).

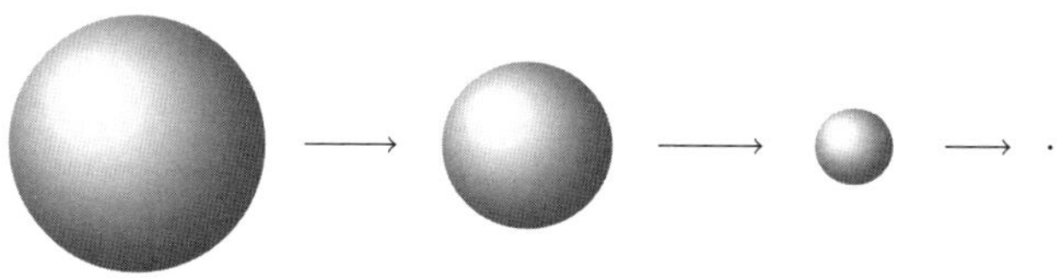

FIGURE 2.1. $\mathbb{P}^1$ shrinking along the Kähler-Ricci flow

EXAMPLE 2.2. If $\mu = 0$ then M is a torus. We have a stationary solution of the Kähler-Ricci flow starting at $\omega_0 = \omega_{\mathrm{KE}}$,

$$\omega(t) = \omega_0,$$

for $t \geqslant 0$.

EXAMPLE 2.3. If $\mu = -1$ then M is a surface of genus strictly greater than one. If ω_{KE} satisfies $\mathrm{Ric}(\omega_{\mathrm{KE}}) = -\omega_{\mathrm{KE}}$, then

$$\omega(t) = (1+t)\omega_{\mathrm{KE}}$$

solves the Kähler-Ricci flow for $t \geqslant 0$ starting at $\omega_0 = \omega_{\mathrm{KE}}$. The solution of the Kähler-Ricci flow exists for all time and expands by scaling.

2.2. The Kähler cone and the first Chern class

A Kähler metric ω is a closed real $(1,1)$ form, and hence defines an element of the cohomology group

$$H^{1,1}_{\overline{\partial}}(M,\mathbb{R}) = \frac{\{\overline{\partial}\text{-closed real } (1,1) \text{ forms}\}}{\mathrm{Im}\,\overline{\partial}}.$$

By Hodge theory, $H^{1,1}_{\overline{\partial}}(M,\mathbb{R})$ is a finite dimensional vector space over $\mathbb{R}$.

The $\partial\overline{\partial}$ *Lemma*, which holds on Kähler manifolds, and which we will not state in its full generality, implies that

$$H^{1,1}_{\overline{\partial}}(M,\mathbb{R}) = \frac{\{\overline{\partial}\text{-closed real } (1,1) \text{ forms}\}}{\mathrm{Im}\,\partial\overline{\partial}}.$$

Namely, if β and γ are two closed real $(1,1)$ forms with $\beta = \gamma + \overline{\partial}\eta$ for some $(1,0)$ form η then $\beta = \gamma + \sqrt{-1}\partial\overline{\partial}f$ for a real-valued function f.

In particular, if ω and ω' are two Kähler metrics with $[\omega] = [\omega']$ (i.e. they define the same element in $H^{1,1}_{\overline{\partial}}(M,\mathbb{R})$) then

$$\omega' = \omega + \sqrt{-1}\partial\overline{\partial}\varphi, \tag{2.2.1}$$

for some smooth real-valued function φ. Moreover, the function φ is unique up to a constant, since if $\tilde{\varphi}$ is another function satisfying (2.2.1) then $\sqrt{-1}\partial\overline{\partial}(\varphi - \tilde{\varphi}) = 0$ and by the next exercise, $\varphi - \tilde{\varphi}$ is a constant.

EXERCISE 2.1. Let (M, ω) be a compact Kähler manifold. Show that if a smooth function $f : M \to \mathbb{R}$ satisfies $\sqrt{-1}\partial\overline{\partial}f \geqslant 0$ then f is a constant on M. *Hint: integrate* $f\sqrt{-1}\partial\overline{\partial}f \wedge \omega^{n-1}$ *over* M *and use Stokes' Theorem.*

We say that a class α in $H^{1,1}_{\overline{\partial}}(M, \mathbb{R})$ is a *Kähler class* if there exists a Kähler metric ω with $[\omega] = \alpha$, and in this case we write $\alpha > 0$. If $-\alpha$ is a Kähler class then we write $\alpha < 0$.

EXERCISE 2.2. For α in $H^{1,1}_{\overline{\partial}}(M, \mathbb{R})$, show that the conditions $\alpha > 0$, $\alpha = 0$ and $\alpha < 0$ are mutually exclusive. Here $\alpha = 0$ simply means that α is the zero element of $H^{1,1}_{\overline{\partial}}(M, \mathbb{R})$.

Note that a class α in $H^{1,1}_{\overline{\partial}}(M, \mathbb{R})$ need not satisfy one of $\alpha > 0$, $\alpha = 0$ or $\alpha < 0$, as we shall see in examples later.

We define the *Kähler cone* of M to be

$$\mathrm{Ka}(M) = \{\alpha \in H^{1,1}_{\overline{\partial}}(M, \mathbb{R}) \mid \alpha > 0\}.$$

EXERCISE 2.3. Show that $\mathrm{Ka}(M)$ is an open convex cone in $H^{1,1}_{\overline{\partial}}(M, \mathbb{R})$. (Recall that being a *convex cone* means that $\alpha, \alpha' \in \mathrm{Ka}(M)$ and $s, s' \in \mathbb{R}^{>0}$ implies that $s\alpha + s'\alpha' \in \mathrm{Ka}(M)$.)

We now describe the *first Chern class* of a Kähler manifold M. This is a special element of $H^{1,1}_{\overline{\partial}}(M, \mathbb{R})$ defined by

$$c_1(M) = [\mathrm{Ric}(\omega)],$$

where ω is any Kähler metric on M. Note that, in comparison to the usual definition in the literature, we have omitted a factor of 2π.

It appears from the definition that $c_1(M)$ depends on the choice of metric ω, but in fact it does not:

PROPOSITION 2.1. *$c_1(M)$ is independent of choice of ω.*

PROOF. Let $\omega' = \sqrt{-1}g'_{i\overline{j}}dz^i \wedge d\overline{z}^j$ be any other Kähler metric. Then

$$\det g' = e^F \det g,$$

for some smooth function $F : M \to \mathbb{R}$. Then

$$\mathrm{Ric}(\omega') = -\sqrt{-1}\partial\overline{\partial}\log\det g' = -\sqrt{-1}\partial\overline{\partial}\log\left(e^F\det g\right) = \mathrm{Ric}(\omega) - \sqrt{-1}\partial\overline{\partial}F,$$

which implies that $[\mathrm{Ric}(\omega')] = [\mathrm{Ric}(\omega)]$. □

Note that the manifolds in Examples 2.1, 2.2 and 2.3 have $c_1(M) > 0$, $c_1(M) = 0$ and $c_1(M) < 0$ respectively.

EXERCISE 2.4. Let $M = M_1 \times M_2$ be a product of two Kähler manifolds (M_1, ω_1) and (M_2, ω_2), and write $\pi_1 : M \to M_1$ and $\pi_2 : M \to M_2$ for the projection maps. Let $\omega = \pi_1^*\omega_1 + \pi_2^*\omega_2$ be the product of the two metrics ω_1 and ω_2, a Kähler metric on M.

(a) Show that

$$\mathrm{Ric}(\omega) = \pi_1^*\mathrm{Ric}(\omega_1) + \pi_2^*\mathrm{Ric}(\omega_2).$$

(b) Suppose that $\omega_1(t)$ and $\omega_2(t)$ solve the Kähler-Ricci flow on M_1 and M_2 respectively, starting at ω_1 and ω_2. Then show that $\omega(t) = \pi_1^*\omega_1(t) + \pi_2^*\omega_2(t)$ solves the Kähler-Ricci flow on M.

EXERCISE 2.5. Let $M = \mathbb{P}^1 \times \mathbb{P}^1$ and write $\omega_{\mathrm{KE}} = 2\omega_{\mathrm{FS}}$ for the Kähler-Einstein metric on $\mathbb{P}^1$, which we recall has $\mathrm{Ric}(\omega_{\mathrm{KE}}) = \omega_{\mathrm{KE}}$. Let ω be the Kähler metric on M given by the product of these Kähler-Einstein metrics on $\mathbb{P}^1$. Show that

$$c_1(M) = [\omega].$$

2.3. Maximal existence time for the Kähler-Ricci flow

We now return to the Kähler-Ricci flow (2.1.1), and observe that if $\omega(t)$ is a solution of the flow then the cohomology classes $[\omega] = [\omega(t)]$ must evolve by

$$\frac{d}{dt}[\omega] = -c_1(M), \quad [\omega]|_{t=0} = [\omega_0]. \tag{2.3.1}$$

This simple ODE system has the solution

$$[\omega(t)] = [\omega_0] - tc_1(M).$$

Hence, as long as a solution to the Kähler-Ricci flow exists, we must have

$$[\omega_0] - tc_1(M) > 0,$$

since this element of $H^{1,1}_{\overline{\partial}}(M, \mathbb{R})$ contains the Kähler metric $\omega(t)$. The maximal existence time theorem for the Kähler-Ricci flow states that this necessary condition is sufficient for existence of a solution:

THEOREM 2.2. *There exists a unique maximal solution to the Kähler-Ricci flow (2.1.1) starting at ω_0 for $t \in [0, T)$, where*

$$T = \sup\{t > 0 \mid [\omega_0] - tc_1(M) > 0\}. \tag{2.3.2}$$

We say that a solution $\omega(t)$ for $t \in [0, T)$ to the Kähler-Ricci flow starting at ω_0 is *maximal* if there does not exist a solution starting at ω_0 on $[0, T')$ for any $T' > T$.

The result Theorem 2.2 is due to Cao [**7**] in the special case when $c_1(M)$ is zero, positive or negative. In this generality, it was proved by Tian-Zhang [**75**]; weaker versions of the result appeared earlier in the work of Tsuji [**81, 82**].

Theorem 2.2 says that the flow exists for as long as the straight line path $t \mapsto [\omega_0] - tc_1(M)$ remains in the Kähler cone. There are four possibilities:

(a) The path $t \mapsto [\omega_0] - tc_1(M)$ hits zero. This can only occur if $c_1(M) > 0$ and $[\omega_0] = Tc_1(M)$ with $T < \infty$.
(b) The Kähler class does not move. This occurs if and only if $c_1(M) = 0$.
(c) The path $t \mapsto [\omega_0] - tc_1(M)$ remains in the Kähler cone for all time. This could occur if $c_1(M) < 0$, for example.
(d) The path $t \mapsto [\omega_0] - tc_1(M)$ hits a non-zero element of the boundary of the Kähler cone. This kind of behavior often occurs, as we will discuss later. The behavior of the flow will depend on the kind of boundary element that the path hits (see Example 2.5 for a simple illustration of this.)

These are illustrated by Figure 2.2.

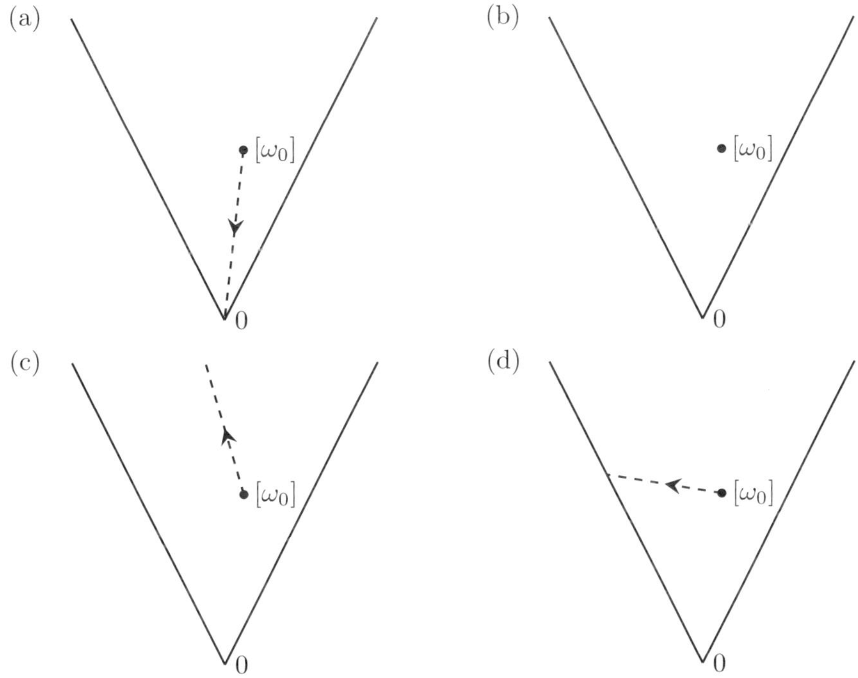

FIGURE 2.2. The Kähler-Ricci flow at the level of cohomology classes

EXAMPLE 2.4. Let M be a Riemann surface. Then $H^{1,1}_{\overline{\partial}}(M, \mathbb{R})$ is one-dimensional and the Kähler cone is the open half line. The three behaviors (a), (b) and (c) occur when $\mu = 1$, $\mu = 0$ and $\mu = -1$ respectively.

EXAMPLE 2.5. Let $M = \mathbb{P}^1 \times \mathbb{P}^1$ as in Exercise 2.5. The space $H^{1,1}_{\overline{\partial}}(M, \mathbb{R})$ is spanned by the classes $\alpha_i = [\pi_i^* \omega_{\mathrm{KE}}]$ for $i = 1, 2$, using the obvious notation. The Kähler cone is given by

$$\mathrm{Ka}(M) = \{x\alpha_1 + y\alpha_2 \mid x, y \in \mathbb{R}^{>0}\}.$$

From Exercise 2.5, the first Chern class of M is $c_1(M) = \alpha_1 + \alpha_2$. Suppose that the initial metric is a product of Kähler-Einstein metrics

$$\omega_0 = x\,\pi_1^*\omega_{\mathrm{KE}} + y\,\pi_2^*\omega_{\mathrm{KE}} \in x\alpha_1 + y\alpha_2.$$

From Exercise 2.4 we see that there are three possible behaviors of the Kähler-Ricci flow depending on the values of x, y.

(i) $[\omega_0]$ lies above the diagonal. Then the path $t \mapsto [\omega_0] - tc_1(M)$ hits a boundary element which is a multiple of $[\alpha_2]$. The first $\mathbb{P}^1$ shrinks to zero as $t \to T$ and the flow converges to a multiple of the Kähler-Einstein metric on the second $\mathbb{P}^1$.

(ii) $[\omega_0]$ lies on the diagonal $x = y$. Then the path $t \mapsto [\omega_0] - tc_1(M)$ hits zero and the two $\mathbb{P}^1$'s shrink simultaneously to a point in finite time.

(iii) $[\omega_0]$ lies below the diagonal. The same behavior as in (i) with the roles of the two $\mathbb{P}^1$'s reversed.

These are illustrated by Figure 2.3.

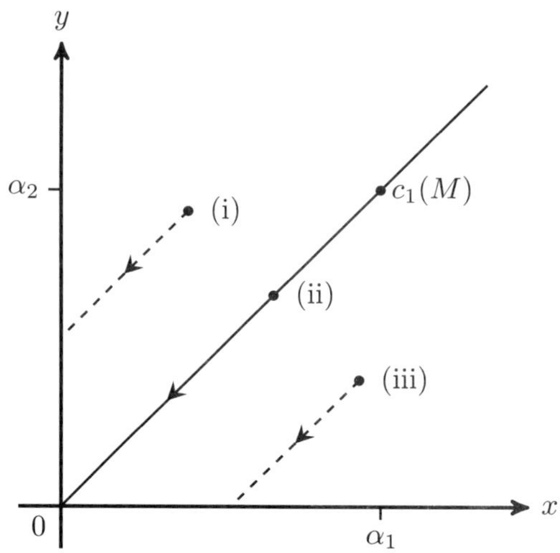

FIGURE 2.3. Three behaviors of the Kähler-Ricci flow on $\mathbb{P}^1 \times \mathbb{P}^1$

EXERCISE 2.6. Let E be a torus and S a surface of genus > 1, with Kähler metrics ω_E and ω_S respectively, satisfying

$$\mathrm{Ric}(\omega_E) = 0, \quad \mathrm{Ric}(\omega_S) = -\omega_S.$$

Let $M = E \times S$.

(a) Find $c_1(M)$ and show that $T = \infty$ for every choice of initial metric ω_0.
(b) Write down the solution $\omega(t)$ of the Kähler-Ricci flow starting with ω_0 equal to the product of ω_E and ω_S.
(c) For your solution from (b), describe geometrically what is happening to $\omega(t)/t$ as $t \to \infty$.

EXERCISE 2.7. Fix a Kähler manifold (M, ω_0). We say that a class $\alpha \in H^{1,1}_{\overline{\partial}}(M, \mathbb{R})$ is *nef* is for all $\varepsilon > 0$ there exists $\omega_\varepsilon \in \alpha$ with $\omega_\varepsilon \geqslant -\varepsilon\omega_0$.

(a) Show that a class α is nef if and only if it is in the closure of the Kähler cone of M.
(b) Show that T given by (2.3.2) can be written as

$$T = \sup\{t > 0 \mid [\omega_0] - tc_1(M) \text{ is nef}\}.$$

LECTURE 3

The parabolic complex Monge-Ampère equation

In this lecture we describe how the Kähler-Ricci flow can be reduced to a parabolic complex Monge-Ampère equation. We use this description to prove the maximal existence time result for the flow.

3.1. Reduction to the complex Monge-Ampère equation

We wish to prove Theorem 2.2 which states that there is a unique maximal solution to the Kähler-Ricci flow (2.1.1)

$$\frac{\partial}{\partial t}\omega = -\mathrm{Ric}(\omega), \quad \omega|_{t=0} = \omega_0,$$

starting at ω_0 on $[0, T)$ for

$$T = \sup\{t > 0 \mid [\omega_0] - tc_1(M) > 0\}.$$

The idea is to reduce the Kähler-Ricci flow equation to a *parabolic complex Monge-Ampère equation*. We will briefly discuss now this terminology.

On $\mathbb{C}^n$ the complex Monge-Ampère operator is the determinant of the complex Hessian:

$$\varphi \mapsto \det\left(\frac{\partial^2\varphi}{\partial z^i \partial \overline{z}^j}\right), \text{ for } \varphi \text{ with } \left(\frac{\partial^2\varphi}{\partial z^i \partial \overline{z}^j}\right) \geqslant 0.$$

However, on a compact Kähler manifold (M, g), this last condition is too strong, since it would imply that φ is constant (see Exercise 2.1). It is natural to replace the complex Hessian $(\partial_i \partial_{\overline{j}}\varphi)$ by $(g_{i\overline{j}} + \partial_i \partial_{\overline{j}}\varphi)$. We define the complex Monge-Ampère operator on M to be

$$\varphi \mapsto \det\left(g_{i\overline{j}} + \frac{\partial^2\varphi}{\partial z^i \partial \overline{z}^j}\right), \text{ for } \varphi \text{ with } \left(g_{i\overline{j}} + \frac{\partial^2\varphi}{\partial z^i \partial \overline{z}^j}\right) \geqslant 0.$$

There are many functions φ satisfying $g_{i\overline{j}} + \partial_i \partial_{\overline{j}}\varphi > 0$, and indeed these functions parametrize (modulo constants) the space of Kähler metrics in the same cohomology class as g.

By *parabolic complex Monge-Ampère equation*, we mean a nonlinear equation of the form

$$\frac{\partial}{\partial t}\varphi = \log \frac{\det\left(g_{i\overline{j}} + \frac{\partial^2\varphi}{\partial z^i \partial \overline{z}^j}\right)}{\det g} + f(\varphi, t),$$

for some function f. The metric g may in general have some dependence on t.

Before we show that the Kähler-Ricci flow (2.1.1) is equivalent to such an equation, we need the following well known theorem of Hamilton [**33**] (see [**17**] for a shorter proof, and [**14**] for a recent exposition).

THEOREM 3.1. *Given any compact Kähler manifold (M, g_0), there exists a unique solution of the Kähler-Ricci flow on a maximal time interval $[0, S)$ for some S with $0 < S \leqslant \infty$.*

In fact, Hamilton's theorem holds for the general Ricci flow on a compact Riemannian manifold. Theorem 3.1 is just the statement that a solution to the Kähler-Ricci flow exists for some short time and is unique. It then follows immediately that there must exist a solution on a maximal time interval. The theorem does not tell us anything about the quantity S. The point of Theorem 2.2 is that $S = T$, where T has the simple formulation (2.3.2) in terms of cohomology classes.

We now begin the proof of Theorem 2.2. By the discussion of Section 2.3, we must have $S \leqslant T$. Since we wish to show that $T = S$, we will assume for a contradiction that $S < T$.

First we pick a smooth family of reference Kähler metrics. Since $S < T$, the cohomology class $[\omega_0] - Sc_1(M)$ lies in the Kähler cone and hence contains a Kähler metric $\hat{\omega}_S$, say. We define $t \mapsto \hat{\omega}_t$ for $t \in [0, S]$ to be the linear path of metrics between ω_0 and $\hat{\omega}_S$ (Figure 3.1). Namely:

$$\hat{\omega}_t = \frac{1}{S}((S-t)\omega_0 + t\hat{\omega}_S) = \omega_0 + t\chi \in [\omega_0] - tc_1(M), \tag{3.1.1}$$

where we define

$$\chi = \frac{1}{S}(\hat{\omega}_S - \omega_0) \in -c_1(M). \tag{3.1.2}$$

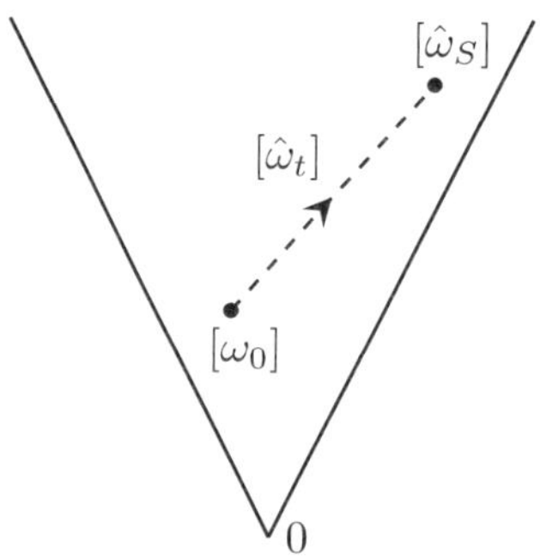

FIGURE 3.1. The path of reference metrics $\hat{\omega}_t$ in the Kähler cone

We now make the key claim: that the Kähler-Ricci flow (2.1.1) is equivalent to the parabolic complex Monge-Ampère equation

$$\frac{\partial}{\partial t}\varphi = \log \frac{(\hat{\omega}_t + \sqrt{-1}\partial\overline{\partial}\varphi)^n}{\Omega}, \quad \hat{\omega}_t + \sqrt{-1}\partial\overline{\partial}\varphi > 0, \quad \varphi|_{t=0} = 0, \tag{3.1.3}$$

where Ω is a volume form with

$$\sqrt{-1}\partial\overline{\partial}\log\Omega = \chi, \quad \int_M \Omega = \int_M \omega_0^n, \tag{3.1.4}$$

(recall Exercise 1.13).

First, why does there exist a volume form Ω satisfying (3.1.4)? Since $-\chi \in c_1(M)$ and $-\sqrt{-1}\partial\overline{\partial}\log\omega_0^n \in c_1(M)$ we have

$$\chi = \sqrt{-1}\partial\overline{\partial}\log\omega_0^n + \sqrt{-1}\partial\overline{\partial}f = \sqrt{-1}\partial\overline{\partial}\log(\omega_0^n e^f),$$

for some real valued function f. Hence we may take $\Omega = \omega_0 e^{f+c}$ where c is a constant chosen so that $\int_M \Omega = \int_M \omega_0^n$.

We now prove the claim. Suppose that φ solves (3.1.3) and set $\omega(t) = \hat{\omega}_t + \sqrt{-1}\partial\overline{\partial}\varphi$. Then

$$\frac{\partial}{\partial t}\omega = \chi + \sqrt{-1}\partial\overline{\partial}\log\frac{\omega^n}{\Omega} = -\mathrm{Ric}(\omega)$$

with $\omega|_{t=0} = \omega_0$ as required. The other direction is left as an exercise.

EXERCISE 3.1. Finish the proof of the claim that (3.1.3) is equivalent to (2.1.1).

To prove Theorem 2.2 we will establish:

PROPOSITION 3.1. *Let $\varphi = \varphi(t)$ solve (3.1.3) on $[0, S)$. Then for each $k = 0, 1, 2, \ldots$ there exists a positive constant A_k such that on $[0, S)$,*

$$\|\varphi\|_{C^k(M)} \leqslant A_k, \quad \hat{\omega}_t + \sqrt{-1}\partial\overline{\partial}\varphi \geqslant \frac{1}{A_0}\omega_0.$$

The point here is that the bounds are independent of t. Recall that the C^k norm of a function is defined by taking the sum of the sup norms of the 0th through kth derivatives of the function (with respect to some fixed Riemannian metric).

Given Proposition 3.1, we will complete the proof of Theorem 2.2. Since the bounds on φ are independent of t in $[0, S)$ we can apply the Arzelà-Ascoli Theorem to see that for a sequence of times $t_i \to S$,

$$\varphi(t_i) \to \varphi(S) \quad \text{in } C^\infty \text{ as } i \to \infty,$$

for a smooth function $\varphi(S)$ with $\omega(S) := \omega_0 + \sqrt{-1}\partial\overline{\partial}\varphi(S) \geqslant \frac{1}{C_0}\omega_0 > 0$. Indeed, the Arzelà-Ascoli Theorem says in particular that if $\{f_j\}$ is a sequence of functions bounded in $C^{k+1}(M)$ then there exists a convergence subsequence in $C^k(M)$. Using a diagonal argument, the C^∞ estimates of Proposition 3.1 give C^∞ convergence (i.e. convergence in every C^k norm) of $\varphi(t_i)$ for a sequence of times $t_i \to S$.

In fact since $\dot{\varphi}$ is bounded, $\varphi(t)$ converges to a unique $\varphi(S)$ as $t \to S$ (see Exercise 3.2 below). Now apply Hamilton's Theorem 3.1 to obtain a solution to the Kähler-Ricci flow starting at $\omega(S)$, for at least some short time $[0, \varepsilon)$. Putting the two solutions together we obtain a solution to the Kähler-Ricci flow starting at ω_0 on the interval $[0, S + \varepsilon)$. But this contradicts the maximality of S.

EXERCISE 3.2. For $t \in [0, T_0)$ (with $0 < T_0 \leqslant \infty$), let $f_t : M \to \mathbb{R}$ be a family of smooth functions which are uniformly bounded in C^∞, independent of t.

(a) Suppose that for some function $f : M \to \mathbb{R}$, we have

$$f_t \to f \text{ pointwise on } M, \text{ as } t \to T_0.$$

Then f_t converges in C^∞ to f. In particular, f is smooth.

(b) Suppose that $T_0 < \infty$ and $\partial f_t/\partial t$ is uniformly bounded. Then show that there exists a unique smooth function $f : M \to \mathbb{R}$ such that f_t converges in C^∞ to f.

It remains then to prove Proposition 3.1.

3.2. Estimates on φ and $\dot{\varphi}$

In this and the next section, we will make use of the *maximum principle* to prove Proposition 3.1. We use only a simple version of the maximum principle, a consequence of elementary calculus, which can be stated as follows:

The maximum principle. Let $f = f(x,t)$ be a smooth function on $M \times [0,a]$ for $a > 0$ and M a compact manifold. Then f achieves a global maximum at some point $(x_0, t_0) \in M \times [0,a]$. At x_0 we have

$$\sqrt{-1}\partial\overline{\partial} f(x_0, t_0) \leqslant 0.$$

The sign of $\frac{\partial}{\partial t} f(x_0, t_0)$ depends on the value of t_0. There are three cases (Figure 3.2):

(i) If $t_0 = 0$ then $\frac{\partial}{\partial t} f(x_0, t_0) \leqslant 0$.
(ii) If $t_0 \in (0,a)$ then $\frac{\partial}{\partial t} f(x_0, t_0) = 0$.
(iii) If $t_0 = a$ then $\frac{\partial}{\partial t} f(x_0, t_0) \geqslant 0$.

In particular, if $t_0 \neq 0$ then $\frac{\partial}{\partial t} f(x_0, t_0) \geqslant 0$.

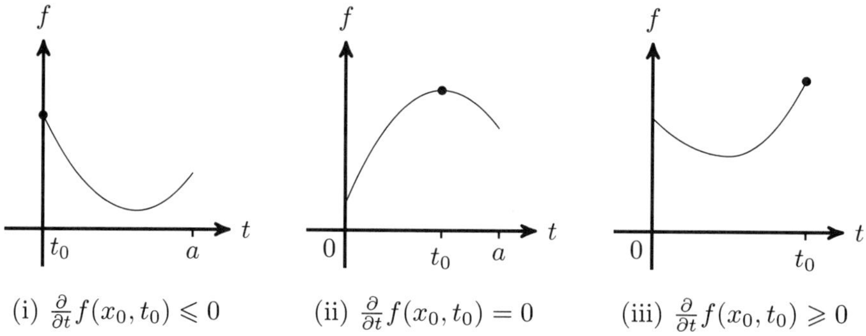

FIGURE 3.2. The time derivative at the maximum of f

Of course we can replace maximum by minimum, if we reverse all the inequalities.

We apply the maximum principle to prove estimates on $\varphi = \varphi(t)$ which we assume solves (3.1.3) on $[0, S)$. The following two lemmas are due to Tian-Zhang [**75**].

LEMMA 3.1. *There exists a uniform constant C so that on $M \times [0,S)$,*

$$|\varphi| \leqslant C.$$

PROOF. Define a function $\psi = \varphi - At$ for a constant A which we will specify later. Compute

$$\frac{\partial}{\partial t}\psi = \log \frac{(\hat{\omega}_t + \sqrt{-1}\partial\overline{\partial}\varphi)^n}{\Omega} - A = \log \frac{(\hat{\omega}_t + \sqrt{-1}\partial\overline{\partial}\psi)^n}{\Omega} - A.$$

At a point when $\sqrt{-1}\partial\overline{\partial}\psi \leqslant 0$ we have $\hat{\omega}_t + \sqrt{-1}\partial\overline{\partial}\psi \leqslant \hat{\omega}_t$ and so

$$\log \frac{(\hat{\omega}_t + \sqrt{-1}\partial\overline{\partial}\psi)^n}{\Omega} \leqslant \log \frac{\hat{\omega}_t^n}{\Omega}.$$

We now pick

$$A = \sup_{M\times[0,S]} \log \frac{\hat{\omega}_t^n}{\Omega} + 1.$$

Observe that $\hat{\omega}_t$ is a smooth family of Kähler metrics on $[0, S]$ (this uses the assumption that $S < T$) and hence A is a uniform constant. Then if $\sqrt{-1}\partial\overline{\partial}\psi \leqslant 0$, we have

$$\frac{\partial}{\partial t}\psi \leqslant -1.$$

We can now use this to deduce that the maximum of ψ must occur at $t = 0$, which gives an upper bound for ψ and hence for φ.

Let's make this last part more precise. Fix $a \in (0, S)$. Then ψ is a smooth function on $M \times [0, a]$. Suppose that ψ achieves a maximum at (x_0, t_0). If $t_0 > 0$ then $\frac{\partial}{\partial t}\psi \geqslant 0$ at (x_0, t_0), contradicting the inequality $\frac{\partial}{\partial t}\psi \leqslant -1$ above. Hence the maximum of ψ on $M \times [0, a]$ is achieved at $t = 0$. Recalling that $\varphi|_{t=0} = 0$, we obtain that $\psi \leqslant 0$ on $M \times [0, a]$ and hence

$$\varphi \leqslant At \leqslant AS \quad \text{on } M \times [0, a].$$

Since a was an arbitrary number in $(0, S)$, this gives the uniform upper bound $C = AS$ for φ on $M \times [0, S)$.

The lower bound is left as an exercise. □

EXERCISE 3.3. Complete the proof of Lemma 3.1.

Next we bound $\dot{\varphi} := \frac{\partial\varphi}{\partial t}$, which is equivalent to a bound on the volume form of the evolving metric.

LEMMA 3.2. *There exists a uniform constant $C > 0$ so that on $M \times [0, S)$,*

$$|\dot{\varphi}| \leqslant C,$$

and

$$C^{-1}\Omega \leqslant \omega^n \leqslant C\Omega.$$

PROOF. For the lower bound of $\dot{\varphi}$ define

$$Q = (S - t + \varepsilon)\dot{\varphi} + \varphi + nt,$$

where $\varepsilon > 0$ is a positive constant to be determined later. Differentiating (3.1.3) with respect to t and recalling (1.4.2), (3.1.1) and (3.1.2),

$$\frac{\partial}{\partial t}\dot{\varphi} = \mathrm{tr}_\omega\left(\frac{\partial}{\partial t}(\hat{\omega}_t + \sqrt{-1}\partial\overline{\partial}\varphi)\right) = \Delta\dot{\varphi} + \mathrm{tr}_\omega\chi,$$

where tr is defined by Exercise 1.12. Here, the Laplace operator Δ is defined by

$$\Delta f := \mathrm{tr}_\omega(\sqrt{-1}\partial\overline{\partial}f) = g^{i\overline{j}}\partial_i\partial_{\overline{j}}f,$$

for a function f. It then follows that

$$\begin{aligned}\left(\frac{\partial}{\partial t} - \Delta\right) Q &= (S - t + \varepsilon)\mathrm{tr}_\omega \chi - \dot{\varphi} + \dot{\varphi} - \Delta\varphi + n \\ &= (S - t + \varepsilon)\mathrm{tr}_\omega \chi - \mathrm{tr}_\omega(\omega - \hat{\omega}_t) + n \\ &= \mathrm{tr}_\omega((S - t + \varepsilon)\chi + \omega_0 + t\chi) \\ &= \mathrm{tr}_\omega(\hat{\omega}_S + \varepsilon\chi) > 0,\end{aligned}$$

if we choose $\varepsilon > 0$ small enough so that $\hat{\omega}_S + \varepsilon\chi$ is positive definite (recall that $\hat{\omega}_S$ is a Kähler metric). We now apply the maximum principle (strictly speaking, the minimum principle). If Q achieves a minimum on some compact time interval then at that minimum, we must have either $t_0 = 0$ or both $\frac{\partial}{\partial t} Q \leqslant 0$ and $\Delta Q \geqslant 0$. But the above inequality shows that the latter cannot occur and hence the minimum of Q must occur at $t_0 = 0$. It follows that for $t \in [0, S)$,

$$(S - t + \varepsilon)\dot{\varphi} + \varphi + nt \geqslant -C,$$

and hence

$$\dot{\varphi} \geqslant \frac{1}{\varepsilon}\left(-C - nS - \sup_{M\times[0,S)} |\varphi|\right) := C',$$

since $|\varphi|$ is bounded by Lemma 3.1. The upper bound for $\dot{\varphi}$ is left as an exercise.

The estimate on the volume form follows immediately from the equation (3.1.3) and the bound on $\dot{\varphi}$. □

EXERCISE 3.4. Complete the proof of Lemma 3.2.

A remark about constants: in what follows we will often use C, C' etc to denote a uniform constant (the uniformity should be clear from the context) which may differ from line to line.

3.3. Estimate on the metric

We next bound the evolving metric $\omega = \omega(t)$.

LEMMA 3.3. *There exists a unform constant C such that on $M \times [0, S)$,*

$$\mathrm{tr}_{\omega_0} \omega \leqslant C.$$

This result follows from the argument of Cao [**7**], and is a parabolic version of an estimate of Yau [**84**] and Aubin [**1**] (an alternative approach is to prove a parabolic Schwarz lemma [**60, 83**]). Note that once we have Lemma 3.3 together with the bound on the volume from Lemma 3.2, we have

$$C^{-1}\omega_0 \leqslant \omega \leqslant C\omega_0, \tag{3.3.1}$$

for a uniform positive constant $C > 0$. Indeed, this follows from the next exercise:

EXERCISE 3.5. Let ω and ω_0 be Kähler metrics. Suppose there exists a uniform constant $C > 0$ such that

$$\frac{1}{C}\omega_0^n \leqslant \omega^n \leqslant C\omega_0^n.$$

Then

$$\mathrm{tr}_{\omega_0}\omega \leqslant C_1 \iff \mathrm{tr}_\omega \omega_0 \leqslant C_2 \iff \frac{1}{C_3}\omega_0 \leqslant \omega \leqslant C_3\omega_0,$$

where each constant $C_i > 0$ can depend on C and the constant C_j in the bound that is being assumed. So for example, in proving the first implication $\Longrightarrow$, C_2 may depend on C and C_1. *Hint: choose coordinates so that g_0 is the identity and g is diagonal.*

PROOF OF LEMMA 3.3. We require two key calculations. The first is

$$\left(\frac{\partial}{\partial t}-\Delta\right)\log \mathrm{tr}_{\omega_0}\omega = -\frac{1}{\mathrm{tr}_{\omega_0}\omega} g^{i\bar{j}} R^{0\ \ \bar{\ell}k}_{i\bar{j}} g_{k\bar{\ell}} + (\dagger), \tag{3.3.2}$$

where

$$(\dagger) = -\frac{g_0^{k\bar{\ell}} g^{i\bar{j}} g^{p\bar{q}} \nabla^0_i g_{k\bar{q}} \nabla^0_{\bar{j}} g_{p\bar{\ell}}}{\mathrm{tr}_{\omega_0}\omega} + \frac{|\partial \mathrm{tr}_{\omega_0}\omega|^2_g}{(\mathrm{tr}_{\omega_0}\omega)^2}. \tag{3.3.3}$$

Here, we are using $R^{0\ \ \bar{\ell}k}_{i\bar{j}}$ and ∇^0 to denote the curvature tensor and covariant derivative with respect to g_0. The second calculation is the inequality:

$$(\dagger) \leqslant 0, \tag{3.3.4}$$

which we will leave as an exercise (see below).

Given these two calculations, we can easily complete the proof of the lemma. We define a quantity

$$Q = \log \mathrm{tr}_{\omega_0}\omega - A\varphi,$$

where $A \geqslant 1$ is a constant to be determined soon. We note that by an elementary local calculation,

$$|g^{i\bar{j}} R^{0\ \ \bar{\ell}k}_{i\bar{j}} g_{k\bar{\ell}}| \leqslant C_0(\mathrm{tr}_\omega\omega_0)(\mathrm{tr}_{\omega_0}\omega),$$

for a uniform constant C_0 which depends only on the curvature of g_0. Then from (3.3.2), (3.3.3), (3.3.4) we have

$$\begin{aligned}\left(\frac{\partial}{\partial t}-\Delta\right) Q &\leqslant C_0\mathrm{tr}_\omega\omega_0 - A\dot{\varphi} + A\Delta\varphi \\ &= C_0\mathrm{tr}_\omega\omega_0 - A\dot{\varphi} + A\mathrm{tr}_\omega(\omega - \hat{\omega}_t) \\ &\leqslant \mathrm{tr}_\omega(C_0\omega_0 - A\hat{\omega}_t) + C'A,\end{aligned}$$

where we have used the fact that $\dot{\varphi}$ is bounded. Now choose A large enough so that

$$C_0\omega_0 - A\hat{\omega}_t \leqslant -\omega_0.$$

Since the family of reference metrics $\hat{\omega}_t$ is uniformly bounded, the constant A is uniform. Then at a point (x_0, t_0) where Q achieves a maximum, assuming that $t_0 \neq 0$, we have by the maximum principle,

$$0 \leqslant -\mathrm{tr}_\omega\omega_0 + C'A,$$

and hence $\mathrm{tr}_\omega\omega_0 \leqslant C$ at (x_0, t_0). From Exercise 3.5 we obtain at (x_0, t_0),

$$\mathrm{tr}_{\omega_0}\omega \leqslant C.$$

But since φ is uniformly bounded, we see that Q is bounded from above at (x_0, t_0) and hence

$$\log \mathrm{tr}_{\omega_0}\omega - A\varphi \leqslant C \qquad \text{on } M \times [0, S).$$

Note that the case when $t_0 = 0$ is trivial. Again using the fact that φ is uniformly bounded we obtain

$$\mathrm{tr}_{\omega_0}\omega \leqslant C \qquad \text{on } M \times [0, S),$$

as required.

It remains to establish (3.3.2). Since it is an inequality of tensors, we are free to choose any coordinate system centered at a fixed point x, say. We choose holomorphic normal coordinates for g_0 as provided by Lemma 1.1. First compute using (2.1.1),

$$\frac{\partial}{\partial t}\log \mathrm{tr}_{\omega_0}\omega = \frac{1}{\mathrm{tr}_{\omega_0}\omega}\mathrm{tr}_{\omega_0}\left(\frac{\partial}{\partial t}\omega\right) = -\frac{1}{\mathrm{tr}_{\omega_0}\omega} g_0^{i\bar{j}} R_{i\bar{j}}. \tag{3.3.5}$$

Next, using the fact that the first derivatives of g_0 vanish at x,

$$\Delta \mathrm{tr}_{\omega_0}\omega = g^{i\bar{j}}\partial_i\partial_{\bar{j}}(g_0^{k\bar{\ell}}g_{k\bar{\ell}}) = g^{i\bar{j}}(\partial_i\partial_{\bar{j}}g_0^{k\bar{\ell}})g_{k\bar{\ell}} + g^{i\bar{j}}g_0^{k\bar{\ell}}\partial_i\partial_{\bar{j}}g_{k\bar{\ell}}.$$

But, applying the formula (1.4.1) with the metric g_0,

$$\partial_i\partial_{\bar{j}}g_0^{k\bar{\ell}} = -\partial_i(g_0^{p\bar{\ell}}g_0^{k\bar{q}}\partial_{\bar{j}}(g_0)_{p\bar{q}}) = -g_0^{p\bar{\ell}}g_0^{k\bar{q}}\partial_i\partial_{\bar{j}}(g_0)_{p\bar{q}} = R^{0\ \ \bar{\ell}k}_{i\bar{j}}.$$

Combining the above and using (1.4.1) with the metric g,

$$\begin{aligned}\Delta \mathrm{tr}_{\omega_0}\omega &= g^{i\bar{j}}R^{0\ \ \bar{\ell}k}_{i\bar{j}}g_{k\bar{\ell}} - g^{i\bar{j}}g_0^{k\bar{\ell}}R_{i\bar{j}k\bar{\ell}} + g^{i\bar{j}}g_0^{k\bar{\ell}}g^{p\bar{q}}\partial_i g_{k\bar{q}}\partial_{\bar{j}}g_{p\bar{\ell}} \\ &= g^{i\bar{j}}R^{0\ \ \bar{\ell}k}_{i\bar{j}}g_{k\bar{\ell}} - g_0^{k\bar{\ell}}R_{k\bar{\ell}} + g^{i\bar{j}}g_0^{k\bar{\ell}}g^{p\bar{q}}\nabla^0_i g_{k\bar{q}}\nabla^0_{\bar{j}}g_{p\bar{\ell}},\end{aligned}$$

where for the last equality we use $g^{i\bar{j}}R_{i\bar{j}k\bar{\ell}} = R_{k\bar{\ell}}$ and, at the point x, $\nabla^0_i = \partial_i$. Then from (3.3.5),

$$\begin{aligned}&\left(\frac{\partial}{\partial t} - \Delta\right)\log \mathrm{tr}_{\omega_0}\omega \\ &= \frac{\partial}{\partial t}\log \mathrm{tr}_{\omega_0}\omega - \frac{\Delta \mathrm{tr}_{\omega_0}\omega}{\mathrm{tr}_{\omega_0}\omega} + \frac{|\partial \mathrm{tr}_{\omega_0}\omega|^2_g}{(\mathrm{tr}_{\omega_0}\omega)^2} \\ &= -\frac{1}{\mathrm{tr}_{\omega_0}\omega}g_0^{i\bar{j}}R_{i\bar{j}} - \frac{1}{\mathrm{tr}_{\omega_0}\omega}\left(g^{i\bar{j}}R^{0\ \ \bar{\ell}k}_{i\bar{j}}g_{k\bar{\ell}} - g_0^{k\bar{\ell}}R_{k\bar{\ell}} + g_0^{k\bar{\ell}}g^{i\bar{j}}g^{p\bar{q}}\nabla^0_i g_{k\bar{q}}\nabla^0_{\bar{j}}g_{p\bar{\ell}}\right) \\ &\quad + \frac{|\partial \mathrm{tr}_{\omega_0}\omega|^2_g}{(\mathrm{tr}_{\omega_0}\omega)^2} \\ &= -\frac{1}{\mathrm{tr}_{\omega_0}\omega}g^{i\bar{j}}R^{0\ \ \bar{\ell}k}_{i\bar{j}}g_{k\bar{\ell}} + (\dagger),\end{aligned}$$

as required. □

In the proof, we made use of:

EXERCISE 3.6. Show that $(\dagger) \leqslant 0$ as follows. Define

$$B_{i\bar{j}k} = \nabla^0_i g_{k\bar{j}} - \frac{\partial_k(\mathrm{tr}_{\omega_0}\omega)}{\mathrm{tr}_{\omega_0}\omega}g_{i\bar{j}},$$

and then show that

$$0 \leqslant g_0^{i\bar{q}}g^{p\bar{j}}g^{k\bar{\ell}}B_{i\bar{j}k}\overline{B_{q\bar{p}\ell}} = g_0^{i\bar{q}}g^{p\bar{j}}g^{k\bar{\ell}}\nabla^0_k g_{i\bar{j}}\nabla^0_{\bar{\ell}}g_{p\bar{q}} - \frac{|\partial \mathrm{tr}_{\omega_0}\omega|^2_g}{\mathrm{tr}_{\omega_0}\omega}.$$

3.4. Higher order estimates

We now complete the proof of Proposition 3.1 and hence Theorem 2.2. Given Lemma 3.3 the proof of Proposition 3.1 follows from fairly standard parabolic theory, which we will quote without proof. We already have from (3.3.1) the estimate

$$\omega = \hat{\omega}_t + \sqrt{-1}\partial\bar{\partial}\varphi \geqslant \frac{1}{C_0}\omega_0,$$

and we also have immediately

$$|\sqrt{-1}\partial\overline{\partial}\varphi|_{g_0} \leqslant C.$$

We can then apply the parabolic Evans-Krylov theory [**18, 40**] (for a proof in the complex setting see [**25**]) to obtain the parabolic Schauder estimate

$$\|\varphi\|_{C^{2+\alpha,1+\alpha/2}} \leqslant C,$$

where the $1+\alpha/2$ refers to the derivative in the t direction. An alternative to using Evans-Krylov is to prove "Calabi" and curvature estimates using the maximum principle [**6, 7, 11, 50, 51, 55**]. The higher order estimates for φ follow from a simple "bootstrap" argument. Indeed, we can apply the differential operator $L = \frac{\partial}{\partial x^k}$ say to (3.1.3) to obtain, locally,

$$\frac{\partial}{\partial t}L(\varphi) = g^{i\overline{j}}\partial_i\partial_{\overline{j}}L(\varphi) + g^{i\overline{j}}L((\hat{g}_t)_{i\overline{j}}) - L\left(\log(\Omega)\right).$$

This is a linear parabolic equation in $L(\varphi)$ with coefficients in $C^{\alpha,\alpha/2}$. The standard parabolic estimates (see for example [**43**]) give a $C^{2+\alpha,1+\alpha/2}$ bound for $L(\varphi)$. Since L was any first order operator with constant coefficients, we obtain a bound for φ in $C^{3+\alpha,1+\alpha/2}$. This gives higher regularity for the coefficients of the parabolic equation above, and we repeat the argument to obtain higher regularity for φ. Thus we obtain bounds for φ in all derivatives.

LECTURE 4

Convergence results

In this lecture we describe convergence results for the Kähler-Ricci flow in the case when the first Chern class of M is negative or zero.

4.1. Negative first Chern class

Let (M, ω_0) be a compact Kähler manifold with $c_1(M) < 0$. Observe that

$$T = \sup\{t > 0 \mid [\omega_0] - tc_1(M) > 0\} = \infty$$

and hence there is a solution to the Kähler-Ricci flow for all time. Note however that the Kähler class $[\omega(t)]$ becomes unbounded as $t \to \infty$, so there is no chance that the flow can converge to a smooth Kähler metric. For this reason, we rescale the flow as follows.

Suppose that $\tilde{\omega}(s)$ solves the Kähler-Ricci flow $\frac{\partial}{\partial s}\tilde{\omega}(s) = -\mathrm{Ric}(\tilde{\omega}(s))$ on $[0, \infty)$ with $\tilde{\omega}(0) = \omega_0$. Define

$$\omega(t) = \frac{1}{s+1}\tilde{\omega}(s), \quad t = \log(s+1).$$

Then $\omega(0) = \omega_0$ and

$$\frac{\partial}{\partial t}\omega(t) = -\frac{1}{(s+1)^2}\frac{ds}{dt}\tilde{\omega}(s) - \frac{1}{s+1}\frac{ds}{dt}\frac{\partial}{\partial s}\tilde{\omega}(s) = -\omega(t) - \mathrm{Ric}(\omega(t)),$$

where we have used (2.1.3) for the last equality. We call this equation,

$$\frac{\partial}{\partial t}\omega = -\mathrm{Ric}(\omega) - \omega, \qquad \omega|_{t=0} = \omega_0, \tag{4.1.1}$$

the *normalized Kähler-Ricci flow*. Whenever $c_1(M) < 0$, we have a solution to this flow for all time. Moreover, the Kähler class $[\omega(t)]$ satisfies the ordinary differential equation

$$\frac{d}{dt}[\omega(t)] = -c_1(M) - [\omega(t)], \quad [\omega(t)] = [\omega_0],$$

whose solution is

$$[\omega(t)] = e^{-t}[\omega_0] + (1 - e^{-t})[-c_1(M)]. \tag{4.1.2}$$

Thus $[\omega(t)]$ moves in a straight line from $[\omega_0]$ to the Kähler class $-c_1(M)$.

In this section, we prove the following theorem.

THEOREM 4.1. *Suppose that $c_1(M) < 0$. Then starting at any Kähler metric ω_0 the solution to the normalized Kähler-Ricci flow (4.1.1) exists for all time and converges in C^∞ to a Kähler-Einstein metric ω_{KE}, which satisfies*

$$\mathrm{Ric}(\omega_{\mathrm{KE}}) = -\omega_{\mathrm{KE}}. \tag{4.1.3}$$

Moreover, ω_{KE} is the unique Kähler metric solving (4.1.3).

The existence of a solution to (4.1.3) was proved independently by Yau [**84**] and Aubin [**1**] in the 1970s, and the uniqueness was shown earlier by Calabi [**5**]. H.-D. Cao [**7**] proved that the Kähler-Ricci flow starting at any Kähler metric ω_0 in the class $-c_1(M)$ converges to a solution of (4.1.3), giving a parabolic proof of the result of Yau and Aubin. The theorem we state here is slightly more general than Cao's result, since we allow ω_0 to lie in an arbitrary Kähler class. Theorem 4.1 follows from the work of Tsuji [**81**] and Tian-Zhang [**75**] who dealt more generally with Kähler manifolds with $-c_1(M)$ being positive in a weaker sense (more precisely, when M has big and nef canonical bundle, also known as a smooth minimal model of general type).

As in the previous section, we reduce the flow equation (4.1.1) to a parabolic complex Monge-Ampère equation. We first need to pick reference Kähler metrics in the cohomology class $[\omega(t)]$. Since $c_1(M) < 0$ there exists a Kähler metric $\hat{\omega}_\infty$ in $-c_1(M)$, which we fix once and for all. Define

$$\hat{\omega}_t = e^{-t}\omega_0 + (1 - e^{-t})\hat{\omega}_\infty,$$

which clearly lies in $[\omega(t)]$ given by (4.1.2). Next, we fix a volume form Ω satisfying

$$\sqrt{-1}\partial\overline{\partial}\log\Omega = \hat{\omega}_\infty, \quad \int_M \Omega = \int_M \omega_0^n. \tag{4.1.4}$$

EXERCISE 4.1. Show that there exists a volume form Ω satisfying (4.1.4).

We can then rewrite the normalized Kähler-Ricci flow as the parabolic complex Monge-Ampère equation:

$$\frac{\partial}{\partial t}\varphi = \log\frac{(\hat{\omega}_t + \sqrt{-1}\partial\overline{\partial}\varphi)^n}{\Omega} - \varphi, \quad \hat{\omega}_t + \sqrt{-1}\partial\overline{\partial}\varphi > 0, \quad \varphi|_{t=0} = 0, \tag{4.1.5}$$

EXERCISE 4.2. Show that (4.1.1) is equivalent to (4.1.5).

From now on let $\varphi(t)$ solve (4.1.5). We wish to prove estimates for φ which are independent of t.

PROPOSITION 4.1. *Let $\varphi = \varphi(t)$ solve (4.1.5) for $t \in [0, \infty)$. Then for each $k = 0, 1, 2, \ldots$ there exists a positive constant A_k such that on $[0, \infty)$,*

$$\|\varphi\|_{C^k(M)} \leqslant A_k, \quad \hat{\omega}_t + \sqrt{-1}\partial\overline{\partial}\varphi \geqslant \frac{1}{A_0}\omega_0.$$

In the same way that we proved Proposition 3.1, we begin by establishing estimates on φ and $\dot{\varphi}$. In this case we can prove stronger estimates which improve as $t \to \infty$.

LEMMA 4.1. *There exists a uniform constant $C > 0$ such that on $M \times [0, \infty)$,*

(i) $|\varphi(t)| \leqslant C$.
(ii) $|\dot{\varphi}(t)| \leqslant C(t+1)e^{-t}$.
(iii) *There exists a continuous function φ_∞ on M such that*

$$|\varphi(t) - \varphi_\infty| \leqslant Ce^{-t/2}.$$

(iv) $C^{-1}\Omega \leqslant \omega^n \leqslant C\Omega$.

PROOF. Part (i) is left as an exercise. For (ii), we use an argument of Tian-Zhang [**75**]. Compute

$$\left(\frac{\partial}{\partial t} - \Delta\right)\dot{\varphi} = \mathrm{tr}_\omega\left(\frac{\partial}{\partial t}\hat{\omega}_t\right) - \dot{\varphi} = \mathrm{tr}_\omega(-e^{-t}\omega_0 + e^{-t}\hat{\omega}_\infty) - \dot{\varphi}$$
$$\left(\frac{\partial}{\partial t} - \Delta\right)\varphi = \dot{\varphi} - \mathrm{tr}_\omega(\omega - \hat{\omega}_t) = \dot{\varphi} - n + \mathrm{tr}_\omega(e^{-t}\omega_0 + (1-e^{-t})\hat{\omega}_\infty).$$

Then we have

$$\left(\frac{\partial}{\partial t} - \Delta\right)(e^t\dot{\varphi}) = \mathrm{tr}_\omega(-\omega_0 + \hat{\omega}_\infty) \tag{4.1.6}$$

and

$$\left(\frac{\partial}{\partial t} - \Delta\right)(\dot{\varphi} + \varphi + nt) = \mathrm{tr}_\omega\hat{\omega}_\infty. \tag{4.1.7}$$

Subtracting (4.1.7) from (4.1.6), we obtain

$$\left(\frac{\partial}{\partial t} - \Delta\right)\Big((e^t-1)\dot{\varphi} - \varphi - nt\Big) = -\mathrm{tr}_\omega\omega_0 < 0.$$

It then follows from the maximum principle that

$$(e^t-1)\dot{\varphi} - \varphi - nt \leqslant 0,$$

and since φ is bounded by (i), this gives the upper bound $\dot{\varphi} \leqslant C(t+1)e^{-t}$.

For the lower bound of $\dot{\varphi}$ we add a large multiple of (4.1.7) to (4.1.6),

$$\left(\frac{\partial}{\partial t} - \Delta\right)\Big((e^t+A)\dot{\varphi} + A\varphi + Ant\Big) = \mathrm{tr}_\omega(-\omega_0 + \hat{\omega}_\infty + A\hat{\omega}_\infty) > 0$$

where A is a constant chosen so that $A\hat{\omega}_\infty \geqslant \omega_0$. Then

$$\dot{\varphi} \geqslant -C(1+t)e^{-t}$$

follows from the minimum principle.

For (iii), compute for $s > t$ and $x \in M$,

$$\begin{aligned} |\varphi(x,s) - \varphi(x,t)| &= \left|\int_t^s \dot{\varphi}(x,u)du\right| \leqslant \int_t^s |\dot{\varphi}(x,u)|du \\ &\leqslant C\int_t^s e^{-u/2}du = 2C(e^{-t/2} - e^{-s/2}), \end{aligned} \tag{4.1.8}$$

since from (ii) we have $|\dot{\varphi}| \leqslant Ce^{-t/2}$. Then from (4.1.8), $\varphi(t)$ converges uniformly to a continuous function φ_∞. Taking the limit in (4.1.8) as $s \to \infty$ gives (iii).

Part (iv) follows from (i) and (ii). □

EXERCISE 4.3. Prove part (i) of Lemma 4.1.

Next we prove a bound on the metric.

LEMMA 4.2. *There exists a uniform constant C such that on $M \times [0, \infty)$,*

$$C^{-1}\omega_0 \leqslant \omega(t) \leqslant C\omega_0.$$

PROOF. The proof is similar to that of Lemma 3.3, so we will be brief. First, to bound the trace of ω with respect to ω_0, we claim that

$$\left(\frac{\partial}{\partial t} - \Delta\right) \log \mathrm{tr}_{\omega_0}\omega \leqslant C_0 \mathrm{tr}_{\omega}\omega_0 - 1, \tag{4.1.9}$$

for a uniform C_0. This calculation is left as an exercise.

Now define $Q = \log \mathrm{tr}_{\omega_0}\omega - A\varphi$, for A to be determined. Calculate

$$\left(\frac{\partial}{\partial t} - \Delta\right) Q \leqslant C_0 \mathrm{tr}_{\omega}\omega_0 - 1 - A\dot{\varphi} + A\mathrm{tr}_{\omega}(\omega - \hat{\omega}_t).$$

Choose A sufficiently large so that $A\hat{\omega}_t \geqslant (C_0+1)\omega_0$, which we can do so since the metrics $\hat{\omega}_t$ are uniformly bounded as $t \to \infty$. Hence

$$\mathrm{tr}_{\omega}\omega_0 \leqslant C$$

at the maximum of Q (if occuring at $t_0 > 0$), using the fact $\dot{\varphi}$ is bounded by part (ii) of Lemma 4.1. From Exercise 3.5 and Lemma 4.1 again, we see that $\mathrm{tr}_{\omega_0}\omega$ and hence Q is bounded from above at the maximum of Q. The lemma follows by using Exercise 3.5 once more. □

EXERCISE 4.4. Prove the inequality (4.1.9).

We now have everything we need to obtain the C^∞ estimates:

PROOF OF PROPOSITION 4.1. The argument follows in exactly the same way as in Section 3.4. The only difference is that we are dealing with the normalized Kähler-Ricci flow, but this only adds a harmless term. □

To finish the proof of Theorem 4.1, we need to prove convergence to a unique Kähler-Einstein metric. From part (iii) of Lemma 4.1:

$$|\varphi(t) - \varphi_\infty| \leqslant Ce^{-t/2},$$

we know that $\varphi(t)$ converges uniformly exponentially fast to φ_∞. But since we have C^∞ estimates from Proposition 4.1 we can apply the Exercise 3.2 to see that $\varphi(t)$ converges to φ_∞ in C^∞, and in particular φ_∞ is smooth.

Next, we apply part (ii) of Lemma 4.1:

$$|\dot{\varphi}| \leqslant C(t+1)e^{-t},$$

to see that $\dot{\varphi}$ converges in C^∞ to 0. Then taking the limit as $t \to \infty$ of (4.1.5) we obtain

$$0 = \log \frac{(\hat{\omega}_\infty + \sqrt{-1}\partial\overline{\partial}\varphi_\infty)^n}{\Omega} - \varphi_\infty.$$

Taking $\sqrt{-1}\partial\overline{\partial}$ of both sides of this equation and recalling (4.1.4), we have

$$\mathrm{Ric}(\hat{\omega}_\infty + \sqrt{-1}\partial\overline{\partial}\varphi_\infty) + \hat{\omega}_\infty + \sqrt{-1}\partial\overline{\partial}\varphi_\infty = 0,$$

or, in other words, $\hat{\omega}_\infty + \sqrt{-1}\partial\overline{\partial}\varphi$ satisfies the Kähler-Einstein equation (4.1.3).

It remains to prove the uniqueness of solutions to the Kähler-Einstein equation. Suppose that ω_{KE} and ω'_{KE} are two solutions of (4.1.3). Then $\omega_{\mathrm{KE}}, \omega'_{\mathrm{KE}}$ both lie in $-c_1(M)$ and so by the $\partial\overline{\partial}$ Lemma we can write

$$\omega'_{\mathrm{KE}} = \omega_{\mathrm{KE}} + \sqrt{-1}\partial\overline{\partial}\psi$$

for some function ψ. We have

$$\mathrm{Ric}(\omega'_{\mathrm{KE}}) = -\omega'_{\mathrm{KE}} = -\omega_{\mathrm{KE}} - \sqrt{-1}\partial\overline{\partial}\psi = \mathrm{Ric}(\omega_{\mathrm{KE}}) - \sqrt{-1}\partial\overline{\partial}\psi,$$

and hence

$$-\sqrt{-1}\partial\overline{\partial}\log\frac{\omega'^n_{\mathrm{KE}}}{\omega^n_{\mathrm{KE}}}=-\sqrt{-1}\partial\overline{\partial}\psi.$$

Applying Exercise 2.1, we have

$$\log\frac{(\omega_{\mathrm{KE}}+\sqrt{-1}\partial\overline{\partial}\psi)^n}{\omega^n_{\mathrm{KE}}}=\psi+C,$$

for some constant C. We now apply the maximum principle to ψ (a simpler version of the maximum principle, where the function has no dependence on t). At the maximum of $\psi+C$, we have $\sqrt{-1}\partial\overline{\partial}\psi\leqslant 0$ and so at this point $\psi+C\leqslant 0$. By considering similarly the minimum of $\psi+C$ we obtain $\psi+C\geqslant 0$ and hence ψ is constant and $\omega_{\mathrm{KE}}=\omega'_{\mathrm{KE}}$.

This completes the proof of Theorem 4.1.

4.2. Zero first Chern class

We now consider the case when $c_1(M)=0$. Fix any Kähler metric ω_0. We have

$$T=\sup\{t>0\mid[\omega_0]-tc_1(M)>0\}=\infty,$$

and so a solution to the Kähler-Ricci flow exists for all time. The Kähler class $[\omega(t)]$ does not move, and so unlike the case of $c_1(M)<0$ there is no need to rescale the flow.

The behavior of the Kähler-Ricci flow is given by the following theorem:

THEOREM 4.2. *Suppose $c_1(M)=0$. Then starting at any Kähler metric ω_0, the Kähler-Ricci flow* (2.1.1) *exists for all time and converges in C^∞ to a Kähler-Einstein metric ω_{KE} satisfying*

$$\mathrm{Ric}(\omega_{\mathrm{KE}})=0. \tag{4.2.1}$$

Moreover, ω_{KE} is the unique Kähler metric in $[\omega_0]$ satisfying (4.2.1).

The existence of a Kähler metric ω_{KE} in each Kähler class satisfying (4.2.1) is due to Yau, and the uniqueness part was already established by Calabi [**5**]. H.-D. Cao [**7**] proved Theorem 4.2, making use of Yau's L^∞ estimate for the complex Monge-Ampère equation [**84**]:

THEOREM 4.3. *Let (M,ω_0) be a compact Kähler manifold and let $F:M\to\mathbb{R}$ be a smooth function. Suppose that θ satisfies the complex Monge-Ampère equation*

$$(\omega_0+\sqrt{-1}\partial\overline{\partial}\theta)^n=e^F\omega_0^n,\quad \omega_0+\sqrt{-1}\partial\overline{\partial}\theta>0.$$

Then

$$\mathrm{osc}(\theta):=\sup_M\theta-\inf_M\theta\leqslant C,$$

for a constant depending only on (M,ω_0) and $\sup_M F$.

Note that if θ in this theorem is normalized by $\int_M\theta\,\omega^n=0$, say, then the conclusion of the theorem is that $\|\theta\|_{L^\infty}=\sup_M|\theta|\leqslant C$. We will omit the proof of Yau's Theorem 4.3 and proceed to prove Theorem 4.2.

We first prove the uniqueness part of Theorem 4.2. Suppose ω_{KE} and ω'_{KE} are solutions of (4.2.1) in the same Kähler class. Then we can write $\omega'_{\mathrm{KE}}=\omega_{\mathrm{KE}}+\sqrt{-1}\partial\overline{\partial}\psi$ for some smooth function ψ. Since $\mathrm{Ric}(\omega_{\mathrm{KE}})=\mathrm{Ric}(\omega'_{\mathrm{KE}})=0$ we have

$$\sqrt{-1}\partial\overline{\partial}\log\frac{\omega'^n_{\mathrm{KE}}}{\omega^n_{\mathrm{KE}}}=0,$$

and so $\log(\omega'^n_{\mathrm{KE}}/\omega^n_{\mathrm{KE}})$ is a constant. Since the integral of ω^n_{KE} is the same as the integral of ω'^n_{KE} we have $\omega'^n_{\mathrm{KE}} = \omega^n_{\mathrm{KE}}$. Then using Stokes' Theorem,

$$\begin{aligned} 0 = \int_M \psi(\omega^n_{\mathrm{KE}} - \omega'^n_{\mathrm{KE}}) &= \int_M \psi(\omega_{\mathrm{KE}} - \omega'_{\mathrm{KE}}) \wedge \sum_{i=0}^{n-1} \omega^i_{\mathrm{KE}} \wedge \omega'^{n-1-i}_{\mathrm{KE}}. \\ &= -\int_M \psi\sqrt{-1}\partial\overline{\partial}\psi \wedge \sum_{i=0}^{n-1} \omega^i_{\mathrm{KE}} \wedge \omega'^{n-1-i}_{\mathrm{KE}} \\ &= \int_M \sqrt{-1}\partial\psi \wedge \overline{\partial}\psi \wedge \sum_{i=0}^{n-1} \omega^i_{\mathrm{KE}} \wedge \omega'^{n-1-i}_{\mathrm{KE}} \\ &\geqslant \int_M \sqrt{-1}\partial\psi \wedge \overline{\partial}\psi \wedge \omega^{n-1}_{\mathrm{KE}} = \frac{1}{n}\int_M |\partial\psi|^2_{\omega_{\mathrm{KE}}} \omega^n_{\mathrm{KE}}, \end{aligned}$$

where for the last inequality we are using the fact that all the terms that we are throwing away are nonnegative. Hence ψ must be a constant and $\omega_{\mathrm{KE}} = \omega'_{\mathrm{KE}}$.

We now return to the Kähler-Ricci flow, and as usual reduce the flow equation (2.1.1) to a parabolic complex Monge-Ampère equation. The Kähler class does not change along the flow so we can choose ω_0 as a reference metric. Since $c_1(M) = 0$, there exists a volume form Ω with

$$\sqrt{-1}\partial\overline{\partial}\log\Omega = 0 \quad \text{and} \quad \int_M \Omega = \int_M \omega_0^n.$$

Then the Kähler-Ricci flow is equivalent to:

$$\frac{\partial}{\partial t}\varphi = \log\frac{(\omega_0 + \sqrt{-1}\partial\overline{\partial}\varphi)^n}{\Omega}, \quad \omega_0 + \sqrt{-1}\partial\overline{\partial}\varphi > 0, \quad \varphi|_{t=0} = 0. \tag{4.2.2}$$

From now on, let $\varphi = \varphi(t)$ solve this equation (4.2.2). We have the following lemma.

LEMMA 4.3. *There exists a uniform constant $C > 0$ such that on $M \times [0, \infty)$,*
 (i) $|\dot{\varphi}| \leqslant C$.
 (ii) $C^{-1}\Omega \leqslant \omega^n \leqslant C\Omega$.
 (iii) $\mathrm{osc}(\varphi) \leqslant C$.

PROOF. For (i), compute

$$\frac{\partial}{\partial t}\dot{\varphi} = \Delta\dot{\varphi}.$$

Then the bound on $\dot{\varphi}$ follows from the maximum principle (see the next exercise). Part (ii) follows from (i). Part (iii) also follows from (i) together with Theorem 4.3. □

We made use of:

EXERCISE 4.5. On a compact Kähler manifold M, let $f = f(x,t)$ satisfy the *heat equation*

$$\frac{\partial}{\partial t}f = \Delta f, \quad f|_{t=0} = f_0,$$

where Δ is the Laplacian associated to $g(t)$, an arbitrary family of Kähler metrics. Then show that on $M \times [0, \infty)$,

$$|f| \leqslant \sup_M |f_0|.$$

Hint: consider $f \pm \varepsilon t$.

Next we prove an estimate on the evolving metric $\omega(t)$. Note that there is a complication here which did not appear in the case of $c_1(M) < 0$ nor in the proof of Theorem 2.2: it arises because we only have a bound on the oscillation $\operatorname{osc}(\varphi)$ and not on $|\varphi|$.

LEMMA 4.4. *There exists a uniform constant C such that on $M \times [0, \infty)$,*

$$C^{-1}\omega_0 \leqslant \omega(t) \leqslant C\omega_0.$$

PROOF. We claim that there exist uniform constants C and A such that for $(x,t) \in M \times [0,\infty)$,

$$(\operatorname{tr}_{\omega_0}\omega)(x,t) \leqslant C \exp\left(A\left(\varphi(x,t) - \inf_{M\times[0,t]}\varphi\right)\right). \tag{4.2.3}$$

To prove (4.2.3), apply the maximum principle to the quantity $Q = \log \operatorname{tr}_{\omega_0}\omega - A\varphi$ for a large constant A, as in the proofs of Lemmas 3.3 and 4.2. The details are left as the next exercise.

Now define

$$\tilde{\varphi} = \varphi - \frac{1}{V}\int_M \varphi\,\Omega, \quad \text{where } V = \int_M \Omega = \int_M \omega^n.$$

Since the oscillation of φ is bounded, $|\tilde{\varphi}|$ is uniformly bounded. From (4.2.3), we have

$$\begin{aligned}(\operatorname{tr}_{\omega_0}\omega)(x,t) &\leqslant C\exp\left(A\left(\tilde{\varphi}(x,t) + \frac{1}{V}\int_M \varphi(t)\Omega - \inf_{M\times[0,t]}\tilde{\varphi} - \inf_{[0,t]}\frac{1}{V}\int_M\varphi\,\Omega\right)\right)\\ &\leqslant C'\exp\left(\frac{A}{V}\left(\int_M\varphi(t)\Omega - \inf_{[0,t]}\int_M\varphi\,\Omega\right)\right),\end{aligned}$$

where for the last inequality we used the bound on $|\tilde{\varphi}|$. But,

$$\frac{d}{dt}\left(\frac{1}{V}\int_M \varphi(t)\Omega\right) = \frac{1}{V}\int_M\dot{\varphi}\,\Omega = \frac{1}{V}\int_M\left(\log\frac{\omega^n}{\Omega}\right)\Omega \leqslant \log\left(\frac{1}{V}\int_M\omega^n\right) = 0,$$

by Jensen's inequality. Hence

$$\int_M \varphi(t)\Omega = \inf_{[0,t]}\int_M\varphi\,\Omega$$

and so we have a uniform upper bound for $\operatorname{tr}_{\omega_0}\omega$. The result then follows by applying Exercise 3.5 and the volume bound of Lemma 4.3. □

EXERCISE 4.6. Prove the claim (4.2.3).

As in the previous section, we have estimates on φ to all orders:

PROPOSITION 4.2. *Let $\varphi = \varphi(t)$ solve (4.2.2) for $t \in [0,\infty)$. Then for each $k = 0, 1, 2, \dots$ there exists a positive constant A_k such that on $[0,\infty)$,*

$$\|\varphi\|_{C^k(M)} \leqslant A_k, \quad \omega_0 + \sqrt{-1}\partial\overline{\partial}\varphi \geqslant \frac{1}{A_0}\omega_0.$$

Now we have estimates for the solution to the Kähler-Ricci flow, we know we have sequential C^∞ convergence of $\varphi(t)$ to some smooth function φ_∞, say (not *a priori* unique). To obtain smooth convergence to a Kähler-Einstein metric, we need a further argument. In the absence of a decay estimate like that of Lemma 4.1, part (ii), we use an argument of Phong-Sturm [**52**] and make use of a functional that decreases along the flow.

Define

$$P(t) = \int_M \dot{\varphi}\, \omega^n.$$

We leave it as an exercise to show that

$$\frac{d}{dt}P(t) = -\frac{1}{n}\int_M |\partial\dot{\varphi}|^2_g \omega^n \leqslant 0. \tag{4.2.4}$$

EXERCISE 4.7. Prove (4.2.4).

From Lemma 4.3, $P(t)$ is bounded. Since it is decreasing, it follows that for a sequence of times $t_i \in [i, i+1]$, we have

$$\left(\int_M |\partial\dot{\varphi}|^2_g \omega^n\right)(t_i) = \left(\int_M \left|\partial \log \frac{\omega^n}{\Omega}\right|^2_g \omega^n\right)(t_i) \to 0.$$

Indeed if not there would exist $\varepsilon > 0$ and infinitely many time intervals $[i_j, i_j+1]$ on which

$$\int_M |\partial\dot{\varphi}|^2_g \omega^n \geqslant \varepsilon,$$

contradicting the fact that

$$P(s) - P(0) = -\frac{1}{n}\int_0^s \int_M |\partial\dot{\varphi}|^2_g \omega^n dt$$

is bounded as $s \to \infty$.

Hence, from the C^∞ estimates, $\varphi(t_i)$ converges (after passing to a subsequence) to a smooth function φ_∞ with

$$\log \frac{\omega^n_\infty}{\Omega} = \text{constant}, \quad \text{for } \omega_\infty := \omega_0 + \sqrt{-1}\partial\overline{\partial}\varphi_\infty,$$

and taking $\sqrt{-1}\partial\overline{\partial}$ gives

$$\mathrm{Ric}(\omega_\infty) = \sqrt{-1}\partial\overline{\partial}\log\Omega = 0.$$

Then $\omega_\infty = \omega_{\mathrm{KE}}$, the unique Kähler-Einstein metric in the class $[\omega_0]$. We have proved smooth convergence of the flow to ω_{KE} for some sequence of times $t_i \to \infty$.

To prove full convergence, we make use of the following exercise.

EXERCISE 4.8. Show that

$$\frac{d^2P}{dt^2} \geqslant C\frac{dP}{dt},$$

for some uniform C (making use of the C^∞ estimates). Hence show that $\dfrac{dP}{dt} \to 0$.

Given this, we obtain smooth convergence using the uniqueness of solutions to the Kähler-Einstein equation (4.2.1). Indeed, suppose for a contradiction that we do not have convergence of $\omega(t)$ to ω_{KE}. Then there exists a sequence of times $t_i \to \infty$ so that, after passing to a subsequence, $\omega(t_i)$ converges in C^∞ to $\omega'_\infty \neq \omega_{\mathrm{KE}}$. But since

$$\frac{dP}{dt} \to 0,$$

it follows from the above argument that $\omega'_\infty \in [\omega_0]$ also satisfies

$$\mathrm{Ric}(\omega'_\infty) = 0,$$

contradicting the uniqueness of Kähler-Einstein metrics in $[\omega_0]$. This completes the proof of Theorem 4.2.

LECTURE 5

The Kähler-Ricci flow on Kähler surfaces, and beyond

In this lecture we will discuss, informally and without proofs, the behavior of the Kähler-Ricci flow on Kähler manifolds of complex dimension two. We will also describe a flow, known as the Chern-Ricci flow, which makes sense on complex manifolds which do not admit Kähler metrics.

5.1. Riemann surfaces

First, let (M, ω_0) be a compact Kähler manifold of complex dimension 1. All such manifolds either have $c_1(M) < 0$, $c_1(M) = 0$ or $c_1(M) > 0$. These correspond topologically to surfaces with genus > 1, genus 1 (a torus) or genus 0 (the 2-sphere).

We know from the previous lecture that if $c_1(M) < 0$ the Kähler-Ricci flow exists for all time with the volume of the manifold tending to infinity. If we rescale the metric so that the volume remains bounded, then the normalized Kähler-Ricci flow converges at infinity to a Kähler-Einstein metric with negative Ricci curvature.

In the case $c_1(M) = 0$ the Kähler-Ricci flow exists for all time and converges at infinity to a Kähler-Einstein metric with zero Ricci curvature.

This only leaves $c_1(M) > 0$ which is precisely the case of $\mathbb{P}^1$. We saw from Example 2.1 that when starting from the standard Fubini-Study metric, the $\mathbb{P}^1$ shrinks to a point along the Kähler-Ricci flow. It is a deep result of Hamilton [**34**] and Chow [**13**] that starting at *any* Kähler metric on $\mathbb{P}^1$, the Kähler-Ricci flow shrinks to a point in finite time, and if rescaled so that the flow exists for all time, converges smoothly to a Kähler-Einstein metric on $\mathbb{P}^1$. The limiting metric is not necessarily the Fubini-Study metric, but is related to it by a biholomorphism.

This is essentially the full picture for the Kähler-Ricci flow on a compact Riemann surface (at least, for smooth initial metrics, cf. [**24, 45**]).

5.2. Kähler surfaces, blowing up and Kodaira dimension

Let M be a compact manifold of complex dimension 2. We call this a *complex surface* (not to be confused with a Riemann surface!) and a *Kähler surface* if it admits a Kähler metric, ω_0, say. There is a classification for complex surfaces, known as the Kodaira-Enriques classification (see [**2**] for example). However, it is much more complicated than the picture for Riemann surfaces, and in fact there are still some gaps to be filled.

One reason that there are "many more" Kähler surfaces than Riemann surfaces comes from the *blow-up procedure*. This is a way of constructing a new complex surface from an old one. We explain this in the simple case of the (non compact) complex manifold $\mathbb{C}^2$. The *blow-up of* $\mathbb{C}^2$ *at* 0 is defined to be

$$\mathrm{Bl}_0\mathbb{C}^2 = \{(z, \ell) \in \mathbb{C}^2 \times \mathbb{P}^1 \mid z \in \ell\}.$$

Recall that points ℓ in $\mathbb{P}^1$ can be regarded as complex lines through the origin in $\mathbb{C}^2$, so this definition makes sense. $\mathrm{Bl}_0\mathbb{C}^2$ is a complex submanifold of $\mathbb{C}^2 \times \mathbb{P}^1$ of codimension 1 (it is given by a single defining equation) and hence is complex manifold of dimension 2. There is a holomorphic map

$$\pi : \mathrm{Bl}_0\mathbb{C}^2 \to \mathbb{C}^2, \qquad \pi(z, \ell) = z,$$

called the *blow-up map.* This map is certainly surjective, and since each non zero element of $\mathbb{C}^2$ lies on a unique line through the origin in $\mathbb{C}^2$, π is injective away from $\pi^{-1}(0)$. Since zero lies in every ℓ in $\mathbb{P}^1$, we have $\pi^{-1}(0) \cong \mathbb{P}^1$. This set is called the *exceptional curve*, which we write as E. The map π is in fact a biholomorphism from $\mathrm{Bl}_0\mathbb{C}^2 - E$ to $\mathbb{C}^2 - \{0\}$, and maps E to 0 (see for example [**28**]).

What we have done here is replace a single point 0 in $\mathbb{C}^2$ with a copy of $\mathbb{P}^1$, which we call the exceptional curve E, which represents all of the directions through 0. This is in fact a local process, and can be performed on any complex surface M with a designated point p to produce a new complex surface Bl_pM of the same dimension, the *blow-up of M at p.* The new surface Bl_pM has "more topology" than M due to this extra $\mathbb{P}^1$, and indeed the second Betti number of Bl_pM is exactly one more than the second Betti number of M.

We can reverse the process of blowing up. If a complex surface M contains a $\mathbb{P}^1$ which looks locally like the $\mathbb{P}^1$ inside $\mathrm{Bl}_0\mathbb{C}^2$ (i.e. same normal bundle) then we say this is an *exceptional curve* E. It's a theorem that there exists a map $\pi : M \to N$ to a new surface N which *blows down* the curve $E \subset M$ to a point $p \in N$.

If M contains no exceptional curves, then we say that M is *minimal.* M can also be called a *minimal model.* Given the results we just stated, it is rather easy to see that given any compact complex surface M we can obtain a minimal model by a finite sequence of blow downs. Indeed, if an exceptional curve exists then blow it down. This process reduces the second Betti number by one and hence must terminate after finitely many steps. This simple algorithm is the baby version of the *minimal model program* and was known to the classical algebraic geometers. Its analogue in higher dimensions is far more complicated and the subject of much recent research (see for example [**3**]).

We now return to the Kähler-Ricci flow. A calculation shows that if we integrate the Ricci curvature of a Kähler metric over an exceptional curve E, we obtain

$$\int_E \mathrm{Ric}(\omega) = 2\pi,$$

and this formula is independent of the choice of Kähler metric and of the surface in which the exceptional curve E is contained (algebraic geometers write this formula as $K \cdot E = -1$). Along the Kähler-Ricci flow we have [**19**]

$$\frac{d}{dt}\int_E \omega = -\int_E \mathrm{Ric}(\omega) = -2\pi.$$

So exceptional curves shrink along the Kähler-Ricci flow. Feldman-Ilmanen-Knopf asked [**19**]: does the Kähler-Ricci flow blow down exceptional curves?

Before answering this question, we make a brief digression to define the important concept of *Kodaira dimension.* Let M be a compact complex manifold of dimension n. Write K for the *canonical bundle of* M, namely the line bundle of holomorphic $(n, 0)$ forms on M. Write $H^0(M, K)$ for the vector space of global holomorphic sections of K. Namely, $H^0(M, K)$ is the vector space of holomorphic $(n, 0)$ forms on M (which could be the set $\{0\}$). Then for $\ell = 1, 2, \ldots$, the space $H^0(M, K^\ell)$ is the space of global holomorphic sections of K^ℓ (the ℓth tensor power). If a line bundle has many holomorphic sections, then its tensor powers will have many more. The Kodaira dimension measures the growth of the dimension of $H^0(M, K^\ell)$ as $\ell \to \infty$.

Define the *Kodaira dimension of* M to be the smallest integer $\mathrm{Kod}(M)$ such that

$$\dim H^0(M, K^\ell) \leqslant C\ell^{\mathrm{Kod}(M)}, \qquad \text{for } \ell \text{ large},$$

with the convention that if $H^0(M, K^\ell) = \{0\}$ then we take $\mathrm{Kod}(M) = -\infty$. It's a fact from algebraic geometry that ℓ must take one of the values $-\infty, 0, 1, 2, \ldots, n$.

We will quote here some basic facts about Kodaira dimension. The first is that on a Riemann surface M we have:

- If $c_1(M) < 0$ then $\mathrm{Kod}(M) = 1$.
- If $c_1(M) = 0$ then $\mathrm{Kod}(M) = 0$.
- If $c_1(M) > 0$ then $\mathrm{Kod}(M) = -\infty$.

Indeed this follows from the fact that $c_1(M) < 0$ corresponds via the Kodaira embedding theorem to the statement that K is *ample*, meaning that K^ℓ has lots of global sections for ℓ large. The condition $c_1(M) > 0$ corresponds to K^{-1} being ample, which implies that K^ℓ has no nonzero sections for $\ell \geqslant 1$. Finally, if a Riemann surface has $c_1(M) = 0$ then K is trivial and $\dim H^0(M, K^\ell) = 1$ for all $\ell \geqslant 1$.

The second fact is that Kodaira dimension has the following additive property:

$$\mathrm{Kod}(M_1 \times M_2) = \mathrm{Kod}(M_1) \times \mathrm{Kod}(M_2).$$

Now we can quickly compute some examples in complex dimension two:

(a) If M is a product of two Riemann surfaces of genus > 1 then $\mathrm{Kod}(M) = 2$.
(b) If M is a product of a torus and a Riemann surface of genus > 1 then $\mathrm{Kod}(M) = 1$.
(c) If M is a product of two tori then $\mathrm{Kod}(M) = 0$.
(d) If M is a product of a $\mathbb{P}^1$ with any other Riemann surface then $\mathrm{Kod}(M) = -\infty$.

Now returning to the Kähler-Ricci flow: if we put a product Kähler-Einstein metric on each of the examples (a)-(d) we notice that the Kähler-Ricci flow exists for all time in cases (a)-(c), whereas in (d) we have collapsing of the $\mathbb{P}^1$ in finite time. Morally speaking: the condition $\mathrm{Kod}(M) = -\infty$ means that there is some "positive curvature" direction which "wants to shrink" along the flow, whereas $\mathrm{Kod}(M) \geqslant 0$ means we have only "zero curvature" or "negative curvature" directions.

Finally, the third fact is that Kodaira dimension is invariant under blow-ups:

$$\mathrm{Kod}(\mathrm{Bl}_p(M)) = \mathrm{Kod}(M).$$

5.3. Behavior of the Kähler-Ricci flow on Kähler surfaces

We now describe the behavior of the Kähler-Ricci flow on a Kähler surface. We break this up into different cases.

5.3.1. Non-minimal Kähler surfaces with $\mathrm{Kod}(M) \neq -\infty$.

We first consider the case when $\mathrm{Kod}(M) \geqslant 0$ (the case $\mathrm{Kod}(M) = -\infty$ is more complicated and will be discussed later). We suppose that M is not minimal - i.e. it has at least one exceptional curve. A result of Song and the author [**64–66**] says, roughly speaking, that the Kähler-Ricci flow blows down exceptional curves finitely many times until obtaining a minimal surface. We state the result somewhat informally:

THEOREM 5.1. *Suppose that M is a compact Kähler surface with* $\mathrm{Kod}(M) \neq -\infty$ *and assume that M contains at least one exceptional curve. Then there exist finitely many disjoint exceptional curves $E_1, \ldots, E_k$ on M and a map $\pi : M \to M_1$ blowing them down. The Kähler-Ricci flow exists on $[0, T)$ for some T with $0 < T < \infty$ and "blows down" $E_1, \ldots, E_k$ and continues on the new manifold M_1. This process repeats finitely many times until we obtain M_ℓ minimal. On M_ℓ the Kähler-Ricci flow exists for all time.*

It should be explained what is meant by the Kähler-Ricci flow "blowing down" exceptional curves $E_1, \ldots, E_k$, since this is the essential content of the result (the fact that the flow exists only for a finite time follows easily from Theorem 2.2, as does the fact that the Kähler-Ricci flow exists for all time on M_ℓ.)

We say that the Kähler-Ricci flow *blows down* $E_1, \ldots, E_k$ if, first, the flow $g(t)$ converges smoothly on compact subsets of $M \setminus \cup E_i$ to a smooth Kähler metric g_T and if $(M, g(t))$ converges globally as a metric space to the metric completion of $(M \setminus \cup E_i, g_T)$. Second, we insist that there exists a smooth solution to the Kähler-Ricci flow on M_1 for $t > T$ which converges as $t \to T^+$ to g_T smoothly on compact subsets away from the points $p_i := \pi(E_i)$. Third, we require that $(M_1, g(t))$ converges globally as a metric space to $(M \setminus \cup E_i, g_T)$ as $t \to T^+$. Here "converges as a metric space" means convergence in the sense of Gromov-Hausdorff (we omit the precise definition here).

The fact that the Kähler-Ricci flow can be restarted on the new manifold M_1 makes use of theorem of Song-Tian [**62**]. The study of the Kähler-Ricci flow in relation to the minimal model program was initiated by Song and Tian [**60**–**62**, **72**] and is known as the *analytic minimal model program* (see also [**41**]).

5.3.2. Minimal surfaces with Kod$(M) \neq -\infty$

Next we discuss the case of what happens on a minimal surface. As stated in the theorem above, the flow exists for all time. Indeed, from some basic algebraic geometry, every minimal M with $\mathrm{Kod}(M) \neq -\infty$ has $-c_1(M)$ nef and we can apply Exercise 2.7 and Theorem 2.2. The behavior of the flow as $t \to \infty$ depends crucially on the Kodaira dimension.

First suppose that $\mathrm{Kod}(M) = 2$. If $c_1(M) < 0$ then, from the results discussed in Lecture 4, the flow converges after normalization to a Kähler-Einstein metric. Otherwise, the canonical bundle is "big and nef" which means that $-c_1(M)$ satisfies a weaker positivity condition. A result of Tsuji [**81**] and Tian-Zhang [**75**] shows that the normalized flow converges to a Kähler-Einstein metric smoothly on compact subsets of $M \setminus V$ where V is a certain subvariety on M.

If $\mathrm{Kod}(M) = 1$ then the manifold is a "properly elliptic surface". Namely, there exists a surjective holomorphic map $f : M \to S$ to a Riemann surface S with the property that $f^{-1}(s)$ is a torus for all but finitely many $s \in S$. It was shown by Song-Tian [**60**] that, in a weak sense, the Kähler-Ricci flow collapses these torus fibers and converges to a "generalized Kähler-Einstein metric" on S. This is a metric whose Ricci curvature is given by the negative of the metric plus some additional terms arising from the non-product structure of M. In the simpler case when M is a product or a smooth fibration (with S necessarily a Riemann surface of genus > 1), it was shown in ([**22, 26, 66**], see also [**29**]) that the flow converges smoothly to the Kähler-Einstein metric on S as $t \to \infty$.

Finally if $\mathrm{Kod}(M) = 0$ then $c_1(M) = 0$ and, by the result discussed in Lecture 4, the Kähler-Ricci flow converges smoothly to a Kähler-Einstein metric with zero Ricci curvature.

Combining these results with that of Theorem 5.1 we see that the Kähler-Ricci flow is largely understood if $\mathrm{Kod}(M) \neq -\infty$. Moreover, the behavior of the flow reflects the geometry of the underlying complex manifold.

5.3.3. The case of $\mathrm{Kod}(M) = -\infty$

We now discuss the more troublesome case when $\mathrm{Kod}(M) = -\infty$. Indeed even in the simple case of $\mathbb{P}^1 \times \mathbb{P}^1$, Exercise 2.5 shows that very different behavior can occur if different initial metrics are chosen.

We will focus on a slightly more complicated example: let M be $\mathbb{P}^2$ blown up at a single point p. Recalling the definition of the blow up of $\mathbb{C}^2$ at the origin, we see that in addition to the map $\pi : \mathrm{Bl}_0\mathbb{C}^2 \to \mathbb{C}^2$ there is another map

$$f : \mathrm{Bl}_0\mathbb{C}^2 = \{(z, \ell) \in \mathbb{C}^2 \times \mathbb{P}^1 \mid z \in \ell\} \to \mathbb{P}^1,$$

given by projection $f(z, \ell) = \ell$ onto the second factor. If we compactify $\mathbb{C}^2$ to $\mathbb{P}^2$ we have maps

$$\begin{array}{l} \mathrm{Bl}_p\mathbb{P}^2 \xrightarrow{\pi} \mathbb{P}^2 \\ \downarrow f \\ \mathbb{P}^1 \end{array}$$

and f is a bundle map whose fibers are isomorphic to $\mathbb{P}^1$. Write $\omega_{\mathbb{P}^2}$ and $\omega_{\mathbb{P}^1}$ for the Fubini-Study metric on $\mathbb{P}^2$ and $\mathbb{P}^1$ respectively.

The Kähler cone of M is given by

$$\mathrm{Ka}(M) = \{x[f^*\omega_{\mathbb{P}^1}] + y[\pi^*\omega_{\mathbb{P}^2}] \mid x, y \in \mathbb{R}^{>0}\},$$

and the first Chern class of M by

$$c_1(M) = 2[\pi^*\omega_{\mathbb{P}^2}] + [f^*\omega_{\mathbb{P}^1}] > 0.$$

There are three different possible behaviors of the Kähler-Ricci flow, depending on where the initial Kähler class $[\omega_0]$ lies, as illustrated by Figure 5.1.

(i) If $[\omega_0]$ lies above the line $y = 2x$ then the Kähler-Ricci flow blows down the exceptional curve in the sense described above, and continues on $\mathbb{P}^2$ [**63–65**].

(ii) If $[\omega_0]$ lies on the line $y = 2x$ containing $c_1(M)$ then the Kähler-Ricci flow shrinks to a point in finite time [**49, 54**]. After rescaling and reparametrizing converges to a Kähler-Ricci soliton (which is a solution of the Kähler-Ricci flow which moves by automorphism) [**9, 15, 39, 51, 74, 76, 87**].

(iii) If $[\omega_0]$ lies below the line $y = 2x$ then the Kähler-Ricci flow contracts the $\mathbb{P}^1$ fibers and converges at least by sequence in the sense of metric spaces to a metric on the base $\mathbb{P}^1$ [**59, 63**].

A general Kähler surface with $\mathrm{Kod}(M) = -\infty$ is either $\mathbb{P}^2$ or a $\mathbb{P}^1$ bundle over a Riemann surface or is obtained by blowing up one of these manifolds. The Kähler-Ricci flow always exhibits one of the three behaviors (i), (ii) or (iii) described above. In (ii), the Kähler-Ricci soliton may be "trivial" - i.e. a Kähler-Einstein metric [**71, 73, 85**].

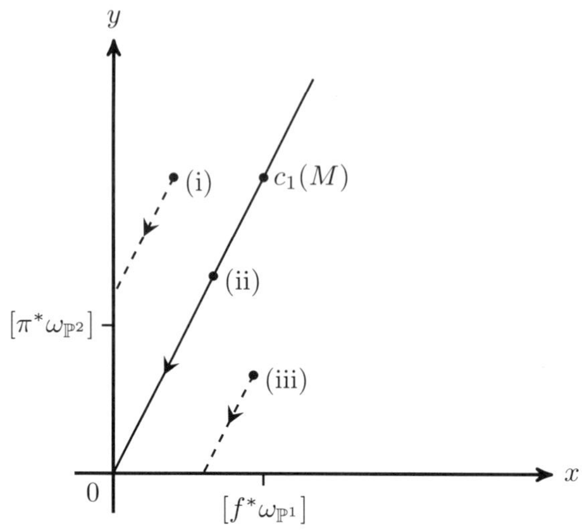

FIGURE 5.1. Three behaviors of the flow on the blow up of $\mathbb{P}^2$

5.3.4. Some open problems

Although much is understood about the Käher-Ricci flow in the case of Kähler surfaces, and the general conjectural picture is now well-laid out, there are a number of difficult problems that remain, and we mention here just a few.

A well-known problem is to understand more precisely the singularity formation when an exceptional curve contracts. Feldman-Ilmanen-Knopf [**19**] conjectured that the blow-up limit, obtaining by rescaling the metric around the singular time, should yield the shrinking non-compact Kähler-Ricci soliton that they constructed. This conjecture was confirmed in the example of Section 5.3.3 assuming symmetric data by Máximo [**44**], using a result of Song [**57**] that in the symmetric case the singularity is of Type I (meaning that the curvature bound $|\text{Rm}| \leqslant C/(T-t)$ holds). Related to this is a folklore conjecture that all finite time singularities of the Kähler-Ricci flow are of Type I. Even the question of whether the bound on the scalar curvature $R \leqslant C/(T-t)$ holds is open (cf. [**86**]).

The result of Theorem 5.1 for the Kähler-Ricci flow makes use of results from algebraic geometry and Kähler surface theory to prove existence of exceptional curves. It would be a long term goal to use the flow to construct algebraic objects such as exceptional curves and give new analytic proofs of results in algebraic geometry. In this direction, it was shown by Collins-Tosatti [**16**] that in general whenever the Kähler-Ricci flow encounters a noncollapsing singularity, the metrics develop singularities precisely along an algebraic variety (proving a conjecture of [**19**]).

Another problem is to understand the global metric behavior of the Kähler-Ricci flow as $t \to \infty$ in the case of a minimal surface with $\text{Kod}(M) = 2$ when $c_1(M)$ is not negative. In the case when the variety V (as discussed above) is a union of disjoint curves of self-intersection -2, it was shown in [**65**] that the Kähler-Ricci flow converges in the sense of metric spaces to an orbifold Kähler-Einstein manifold. The general case of possibly intersecting curves is still open.

It is also an open problem to understand precisely the behavior of the Kähler-Ricci flow as $t \to \infty$ on a minimal surface with $\mathrm{Kod}(M) = 1$ which is not a smooth fibration. The results of Song-Tian **[60]** give information about convergence of the flow in the sense of currents, but it is still unknown exactly what happens to the metrics. Here, difficulties arise from the presence of singular and multiple fibers.

A difficult problem is to understand collapsing along the Kähler-Ricci flow in the case of negative Kodaira dimension. Surprisingly, it is still an open problem to determine the precise behavior of the flow even in the case of $\mathbb{P}^1 \times \mathbb{P}^1$ when one of the fibers collapses. It was shown by Song-Székelyhidi-Weinkove **[59]** that the diameter of the collapsing fiber is bounded above by a multiple of $(T-t)^{1/3}$, but this falls short of the optimal rate of $(T-t)^{1/2}$ (see also **[21]**). It is expected that the blow-up limit is a product with a flat direction and this has been proved under symmetry conditions **[20, 57]**. Underlying this difficulty is the depth of the problem of understanding the Kähler-Ricci flow for a general initial metric on $\mathbb{P}^1$ (the result of Hamilton and Chow), which still has no simple proof. In higher dimensions the problem is far harder, since one could replace the fiber $\mathbb{P}^1$ by a general Fano manifold, where the behavior of the flow is far more difficult to understand (and this goes way beyond the scope of these notes).

Finally, of course, one would like to extend all of these ideas to higher dimensions. Considerable progress is being made **[32, 58, 61, 62, 67]** and this will continue to be a challenging and exciting area of research for many years to come.

5.4. Non-Kähler surfaces and the Chern-Ricci flow

As seen from the last section, the Kähler-Ricci flow is now quite well understood in the case of complex dimension two. Given any Kähler surface, we have a more-or-less complete picture of how the flow will behave (modulo some open problems, as discussed above). This is in contrast to the Ricci flow on general four-manifolds, where despite the success of Ricci flow in three dimensions **[10, 33, 35, 37, 46, 48]** we do not yet have any kind of conjectural picture. Is there a larger class of four-manifolds than Kähler surfaces for which we can say anything?

We consider the class of compact complex surfaces, which include *non-Kähler* surfaces, namely surfaces which do not admit any Kähler metric. For an example, consider the simplest Hopf surface

$$H = (\mathbb{C}^2 - \{0\})/\sim,$$

where $(z^1, z^2) \sim (2z^1, 2z^2)$. This is a compact complex surface, diffeomorphic to $S^3 \times S^1$ via the map

$$z \mapsto \left(\frac{z}{|z|}, |z|\right) \in S^3 \times \mathbb{R}^{>0}/(r \sim 2r) \cong S^3 \times S^1,$$

where we consider S^3 as a subset of $\mathbb{C}^2$ in the usual way. H cannot admit a Kähler metric since its second Betti number vanishes.

All complex surfaces admit Hermitian metrics. Given such a metric g_0, we can consider the associated $(1,1)$ form

$$\omega_0 = \sqrt{-1} g_{i\bar{j}} dz^i \wedge d\bar{z}^j,$$

which is not necessarily closed. We define the *Chern-Ricci form* of ω_0 to be

$$\mathrm{Ric}(\omega_0) = -\sqrt{-1}\partial\bar{\partial} \log \det g_0.$$

This coincides with the usual Ricci curvature if g_0 is Kähler. The form $\mathrm{Ric}(\omega_0)$ is still a closed $(1,1)$-form for Hermitian g_0, but the key point is that in the non-Kähler case $\mathrm{Ric}(\omega_0)$ is not in general equal to the Riemannian Ricci curvature of g_0.

We consider a parabolic flow of Hermitian metrics on a complex surface M:

$$\frac{\partial}{\partial t}\omega = -\mathrm{Ric}(\omega), \quad \omega|_{t=0} = \omega_0. \tag{5.4.1}$$

The formula is the same as for the Kähler-Ricci flow, but we are allowing g_0 to be non-Kähler. The equation (5.4.1) is known as the *Chern-Ricci flow*, and shares many of the properties of the Kähler-Ricci flow [**25**, **27**, **56**, **78**–**80**].

Note that the Chern-Ricci flow is *not* the same as the Ricci flow in general. The Ricci flow starting at a Hermitian metric may immediately become non-Hermitian and little is known about the behavior of the flow. Other examples of Hermitian flows generalizing the Kähler-Ricci flow have been given in [**68**].

The Chern-Ricci flow was first introduced by M. Gill [**25**] in the setting of manifolds with vanishing first Bott-Chern class, which we now explain. We define

$$H^{1,1}_{\mathrm{BC}}(M,\mathbb{R}) = \frac{\{\partial\text{-closed real } (1,1) \text{ forms}\}}{\mathrm{Im}\,\partial\overline{\partial}},$$

which coincides with $H^{1,1}_{\overline{\partial}}(M,\mathbb{R})$ when M is Kähler, as discussed in Lecture 2. We then define the first Bott-Chern class of M by

$$c_1^{\mathrm{BC}}(M) = [\mathrm{Ric}(\omega_0)] \in H^{1,1}_{\mathrm{BC}}(M,\mathbb{R}),$$

and by the same argument as in Lecture 2, this is independent of choice of Hermitian metric ω_0.

Gill [**25**] proved:

THEOREM 5.2. *If M is a compact complex manifold with $c_1^{BC}(M) = 0$ then there exists a unique solution $\omega(t)$ to the Chern-Ricci flow (5.4.1) starting at any Hermitian metric ω_0. As $t \to \infty$,*

$$\omega(t) \to \omega_\infty \ \text{in } C^\infty(M),$$

where $\mathrm{Ric}(\omega_\infty) = 0$.

This is the analogue of Cao's Theorem 4.2 proved in Lecture 4. This result made use of an L^∞ estimate for the complex Monge-Ampère equation in the Hermitian case due to Tosatti and the author [**77**] (see also [**12**, **30**]). Note that in fact this result holds in all dimensions.

An analogue of Theorem 4.1 also exists. As there, it is convenient to consider the normalized flow:

$$\frac{\partial}{\partial t}\omega = -\mathrm{Ric}(\omega) - \omega, \quad \omega|_{t=0} = \omega_0. \tag{5.4.2}$$

It was shown by Tosatti and the author [**78**] that:

THEOREM 5.3. *Let M be a compact complex manifold with $c_1(M) < 0$. Then there exists a unique solution to the Chern-Ricci flow (5.4.1) starting at any Hermitian metric ω_0. As $t \to \infty$, the solution $\omega(t)$ to the normalized flow (5.4.2) satisfies*

$$\omega(t) \to \omega_{\mathrm{KE}} \ \text{in } C^\infty(M),$$

where ω_{KE} is the unique Kähler-Einstein metric on M satisfying $\mathrm{Ric}(\omega_{\mathrm{KE}}) = -\omega_{\mathrm{KE}}$.

Observe that the condition $c_1(M) < 0$ implies that the manifold M is Kähler. The point of this theorem is that the Chern-Ricci flow takes any *non-Kähler* Hermitian metric to the Kähler-Einstein metric.

Furthermore, we have a natural analogue of the maximal existence time theorem [**78**]:

THEOREM 5.4. *Given any Hermitian metric ω_0, there exists a unique maximal solution of the Chern-Ricci flow starting at ω_0 on $[0, T)$, where*

$$T = \sup\left\{ t > 0 \;\middle|\; \begin{array}{c} \text{there exists } \psi \in C^\infty(M) \text{ such that} \\ \omega_0 - t\mathrm{Ric}(\omega_0) + \sqrt{-1}\partial\overline{\partial}\psi > 0 \end{array} \right\}.$$

Although it looks like T depends only on ω_0, it really only depends on the "equivalence class" of ω_0, where we say that two Hermitian metrics are equivalent if their forms differ by the $\partial\overline{\partial}$ of a function. In many cases, it is easy to compute T, just as in the Kähler case.

We return now to the case of complex dimension 2 and impose an additional assumption on ω_0:

$$\sqrt{-1}\partial\overline{\partial}\omega_0 = 0. \tag{5.4.3}$$

This is a natural assumption on complex surfaces, since, by a theorem of Gauduchon [**23**], given any Hermitian metric ω on M there exists a smooth function σ so that $e^\sigma\omega$ satisfies (5.4.3). A metric on a complex surface which satisfies (5.4.3) is called *Gauduchon* (in dimension n, the condition is $\sqrt{-1}\partial\overline{\partial}\omega^{n-1} = 0$.) It is immediate from the definition that the Chern-Ricci flow preserves the condition (5.4.3).

If M is a *minimal* complex surface, then a similar picture as in the Kähler case (Section 5.3.2) is now emerging:

- If $\mathrm{Kod}(M) = 0$ then $c_1^{\mathrm{BC}}(M) = 0$ and Gill's Theorem 5.2 implies that the Chern-Ricci flow exists for all time and converges to a Chern-Ricci flat metric.
- If $\mathrm{Kod}(M) = 1$ and M is non-Kähler, then M is, up to a finite covering, a smooth elliptic bundle over a Riemann surface S. A result of Tosatti-Weinkove-Yang [**80**] says that the normalized Chern-Ricci flow (5.4.2) exists for all time and converges in the sense of metric spaces to an orbifold Kähler-Einstein metric on S.
- If $\mathrm{Kod}(M) = 2$ then M admits a Kähler metric (in fact M is projective algebraic). If $c_1(M) < 0$ and we start the flow from a Hermitian metric, we can apply Theorem 5.3 above to obtain convergence of the normalized Chern-Ricci flow to a Kähler-Einstein metric. Otherwise the canonical bundle is big and nef and a result of Gill [**27**] says that the flow converges smoothly to a Kähler-Einstein metric outside a subvariety, generalizing the results of [**75, 81**].

If M is non-minimal with $\mathrm{Kod}(M) \neq -\infty$, then as in the case of the Kähler-Ricci flow we can ask whether the Chern-Ricci flow "blows down" exceptional curves. This is in general an open problem. However, it was shown in [**78**] that one obtains smooth convergence outside the curves of negative self-intersection. Moreover, the curves contract in the sense of metric spaces if the initial metric ω_0 satisfies the additional assumption that $d\omega_0$ is the exterior derivative of the pull-back of a form from the blow-down manifold [**79**]. It is not difficult to find examples when this condition is satisfied, but we conjecture that one should be able to remove this assumption.

The case of $\mathrm{Kod}(M) = -\infty$ is both more interesting and more difficult. The minimal non-Kähler surfaces with $\mathrm{Kod}(M) = -\infty$ are known as Class VII surfaces. When $b_2 = 0$, M is either a Hopf surface or an Inoue surface. It was shown in [**78**] that on a Hopf surface, the Chern-Ricci flow *collapses* in finite time, meaning that the volume tends to zero. By contrast, on the Inoue surface the flow exists for all time. Explicit examples were given in [**79**] for a family of Hopf surfaces which exhibit collapsing in the sense of Gromov-Hausdorff to S^1. Examples on Inoue surfaces also show collapsing of the normalized flow as $t \to \infty$ [**79**].

When M is a Class VII surface with $b_2 > 0$, the exact behavior of the flow is a mystery, and no explicit examples are known. However, it was shown in [**78**] that the Chern-Ricci flow always collapses in finite time. It is a major open problem to complete the classification of Class VII surfaces when $b_2 > 1$ (for the cases $b_2 = 0, 1$ see [**36, 42, 47, 69, 70**]) and this is a motivating factor for studying non-Kähler surfaces.

APPENDIX A

Solutions to exercises

1.1. $S^2 = \{(x^1, x^2, x^3) \in \mathbb{R}^3 \mid (x^1)^2 + (x^2)^2 + (x^3)^2 = 1\}$ is a complex manifold with charts $(S^2 - \{(0,0,1)\}, w)$ and $(S^2 - \{(0,0,-1)\}, \tilde{w})$ given by

$$w = \frac{x^1 + \sqrt{-1}x^2}{1 - x^3}, \quad \tilde{w} = \frac{x^1 - \sqrt{-1}x^2}{1 + x^3},$$

which are related by $w = 1/\tilde{w}$ on the overlap. On the other hand $\mathbb{P}^1$ has two complex charts $U_0 = \{Z_0 \neq 0\}$ with $z = Z_1/Z_0$ and $U_1 = \{Z_1 \neq 0\}$ with $\tilde{z} = Z_0/Z_1$, which are related by $z = 1/\tilde{z}$. All of the maps $w, \tilde{w}, z, \tilde{z}$ map onto $\mathbb{C}$. Then define a map $S^2 \to \mathbb{P}^1$ by mapping $S^2 - \{0,0,1\} \to U_0$ via $z^{-1} \circ w$ and $S^2 - \{0,0,-1\} \to U_1$ via $\tilde{z}^{-1} \circ \tilde{w}$. This is well-defined and holomorphic with holomorphic inverse, hence a diffeomorphism.

1.2. Let $(U, \tilde{z})$ be another coordinate chart. If $\dfrac{\partial X^i}{\partial \overline{z^j}} = 0$ then on $U \cap \tilde{U}$,

$$\frac{\partial \tilde{X}^k}{\partial \overline{\tilde{z}^\ell}} = \frac{\partial \overline{z^j}}{\partial \overline{\tilde{z}^\ell}} \frac{\partial}{\partial \overline{z^j}} \left(X^i \frac{\partial \tilde{z}^k}{\partial z^i} \right) = \frac{\partial \overline{z^j}}{\partial \overline{\tilde{z}^\ell}} \frac{\partial X^i}{\partial \overline{z^j}} \frac{\partial \tilde{z}^k}{\partial z^i} = 0.$$

1.3. On $U \cap \tilde{U}$, from (1.2.1),

$$g_{i\bar{j}} X^i \overline{Y^j} = \tilde{g}_{k\bar{\ell}} \frac{\partial \tilde{z}^k}{\partial z^i} X^i \overline{\frac{\partial \tilde{z}^\ell}{\partial z^j} Y^j} = \tilde{g}_{k\bar{\ell}} \tilde{X}^k \overline{\tilde{Y}^j}.$$

1.4. Take the inverse of both sides of (1.2.1).

1.5. Assume (1.3.1) holds on U. On $U \cap \tilde{U}$,

$$\begin{aligned}
\frac{\partial}{\partial \tilde{z}^m} \tilde{g}_{p\bar{q}} - \frac{\partial}{\partial \tilde{z}^p} \tilde{g}_{m\bar{q}} &= \frac{\partial}{\partial \tilde{z}^m} \left(g_{i\bar{j}} \frac{\partial z^i}{\partial \tilde{z}^p} \overline{\frac{\partial z^j}{\partial \tilde{z}^q}} \right) - \frac{\partial}{\partial \tilde{z}^p} \left(g_{i\bar{j}} \frac{\partial z^i}{\partial \tilde{z}^m} \overline{\frac{\partial z^j}{\partial \tilde{z}^q}} \right) \\
&= \left(\frac{\partial z^k}{\partial \tilde{z}^m} \frac{\partial}{\partial z^k} g_{i\bar{j}} \right) \frac{\partial z^i}{\partial \tilde{z}^p} \overline{\frac{\partial z^j}{\partial \tilde{z}^q}} + g_{i\bar{j}} \frac{\partial^2 z^i}{\partial \tilde{z}^m \partial \tilde{z}^p} \overline{\frac{\partial z^j}{\partial \tilde{z}^q}} \\
&\quad - \left(\frac{\partial z^k}{\partial \tilde{z}^p} \frac{\partial}{\partial z^k} g_{i\bar{j}} \right) \frac{\partial z^i}{\partial \tilde{z}^m} \overline{\frac{\partial z^j}{\partial \tilde{z}^q}} - g_{i\bar{j}} \frac{\partial^2 z^i}{\partial \tilde{z}^p \partial \tilde{z}^m} \overline{\frac{\partial z^j}{\partial \tilde{z}^q}} \\
&= \left(\frac{\partial}{\partial z^k} g_{i\bar{j}} - \frac{\partial}{\partial z^i} g_{k\bar{j}} \right) \frac{\partial z^k}{\partial \tilde{z}^m} \frac{\partial z^i}{\partial \tilde{z}^p} \overline{\frac{\partial z^j}{\partial \tilde{z}^q}} = 0.
\end{aligned}$$

1.6. To show that it is well-defined tensor, use the general fact that $\dfrac{\partial}{\partial z^i} \dfrac{\partial}{\partial \overline{z^j}}$ acting on functions transforms according to

$$\frac{\partial}{\partial z^i} \frac{\partial}{\partial \overline{z^j}} = \frac{\partial \tilde{z}^k}{\partial z^i} \overline{\frac{\partial \tilde{z}^\ell}{\partial z^j}} \frac{\partial}{\partial \tilde{z}^k} \frac{\partial}{\partial \overline{\tilde{z}^\ell}}, \quad \text{on } U \cap \tilde{U},$$

together with the fact that $\sqrt{-1}\partial\overline{\partial}\log|f|^2 = 0$ for f a nowhere vanishing holomorphic function.

To see that $(g_{i\overline{j}})$ is positive definite, write $\underline{z} = (z^1, \ldots, z^n)$ and compute on e.g. U_0,

$$(g_{i\overline{j}}) = \left(\frac{\delta_{ij} + |\underline{z}|^2\delta_{ij} - z^j\overline{z^i}}{(1+|\underline{z}|^2)^2} \right) \geqslant \left(\frac{\delta_{ij}}{(1+|\underline{z}|^2)^2} \right) > 0,$$

since by the Cauchy-Schwarz inequality $|\underline{z}|^2\delta_{ij} - z^j\overline{z^i}$ is semipositive.

It is immediate from the definition that $\partial_k g_{i\overline{j}} = \partial_i g_{k\overline{j}}$, so g is Kähler.

1.7. $\partial\omega = \left(\partial_k g_{i\overline{j}} - \partial_i g_{k\overline{j}}\right) dz^k \wedge dz^i \wedge d\overline{z}^j$ and hence $\partial\omega = 0$ if and only if ω is Kähler. Taking conjugates, $\overline{\partial}\omega = 0$ if and only if $\partial\omega = 0$ since ω is real. Finally, $d\omega = \partial\omega + \overline{\partial}\omega$ and $\partial\omega$ is of type $(2,1)$ whereas $\overline{\partial}\omega$ is of type $(1,2)$. Hence $d\omega = 0$ if and only if both $\partial\omega$ and $\overline{\partial}\omega$ vanish.

1.8. E.g., for $\nabla_k X^i$, compute

$$\begin{aligned}\frac{\partial}{\partial z^k}X^i + \Gamma^i_{km}X^m &= \frac{\partial \tilde{z}^p}{\partial z^k}\frac{\partial}{\partial \tilde{z}^p}\left(\tilde{X}^\ell \frac{\partial z^i}{\partial \tilde{z}^\ell}\right) + g^{i\overline{q}}\left(\frac{\partial}{\partial z^k} g_{m\overline{q}}\right)\tilde{X}^\ell \frac{\partial z^m}{\partial \tilde{z}^\ell} \\ &= \frac{\partial \tilde{z}^p}{\partial z^k}\frac{\partial z^i}{\partial \tilde{z}^\ell}\frac{\partial \tilde{X}^\ell}{\partial \tilde{z}^p} + \frac{\partial \tilde{z}^p}{\partial z^k}\tilde{X}^\ell \frac{\partial^2 z^i}{\partial \tilde{z}^p \partial \tilde{z}^\ell} \\ &\quad + \tilde{g}^{a\overline{b}}\frac{\partial z^i}{\partial \tilde{z}^a}\overline{\frac{\partial z^q}{\partial \tilde{z}^b}}\frac{\partial \tilde{z}^e}{\partial z^k}\frac{\partial}{\partial \tilde{z}^e}\left(\tilde{g}_{c\overline{d}}\frac{\partial \tilde{z}^c}{\partial z^m}\overline{\frac{\partial \tilde{z}^d}{\partial z^q}}\right)\tilde{X}^\ell \frac{\partial z^m}{\partial \tilde{z}^\ell}\end{aligned}$$

and use the fact that

$$\frac{\partial}{\partial \tilde{z}^e}\left(\frac{\partial \tilde{z}^c}{\partial z^m}\right) = -\frac{\partial \tilde{z}^c}{\partial z^r}\frac{\partial \tilde{z}^s}{\partial z^m}\frac{\partial^2 z^r}{\partial \tilde{z}^e \partial \tilde{z}^s}.$$

1.9. Straightforward calculation.

1.10. Compute at 0,

$$\frac{\partial}{\partial z^k}g_{i\overline{j}} = \frac{\partial}{\partial z^k}\left(\tilde{g}_{p\overline{q}}\frac{\partial \tilde{z}^p}{\partial z^i}\overline{\frac{\partial \tilde{z}^q}{\partial z^j}}\right) = \frac{\partial \tilde{z}^m}{\partial z^k}\left(\frac{\partial}{\partial \tilde{z}^m}\tilde{g}_{p\overline{q}}\right)\frac{\partial \tilde{z}^p}{\partial z^i}\overline{\frac{\partial \tilde{z}^q}{\partial z^j}} + \tilde{g}_{p\overline{q}}\frac{\partial^2 \tilde{z}^p}{\partial z^k \partial z^i}\overline{\frac{\partial \tilde{z}^q}{\partial z^j}}.$$

But at 0 we have $\tilde{g}_{p\overline{q}} = \delta_{pq} = \dfrac{\partial \tilde{z}^p}{\partial z^q}$ and hence

$$\frac{\partial}{\partial z^k}g_{i\overline{j}} = \tilde{\Gamma}^j_{ki}(0) - \tilde{\Gamma}^j_{ki}(0) = 0.$$

1.11. For example, in holomorphic normal coordinates,

$$[\nabla_i, \nabla_{\overline{j}}]b_{\overline{q}} = \partial_i(\partial_{\overline{j}}b_{\overline{q}} - \overline{\Gamma^\ell_{jq}}b_{\overline{\ell}}) - \partial_{\overline{j}}\partial_i b_{\overline{q}} = -(\partial_i\overline{\Gamma^\ell_{jq}})b_{\overline{\ell}} = R_{i\overline{j}}{}^{\overline{\ell}}{}_{\overline{q}}b_{\overline{\ell}}.$$

1.12. (1) Pick coordinates at a point for which $g_{i\overline{j}} = \delta_{ij}$ and $(\beta_{i\overline{j}})$ is a diagonal matrix with eigenvalues $\lambda_1, \ldots, \lambda_n$. Then $n\omega^{n-1}\wedge\beta$ and $g^{i\overline{j}}\beta_{i\overline{j}}\omega^n$ both equal

$$\left(\sum_{i=1}^n \lambda_i\right) n!(\sqrt{-1})^n dz^1 \wedge d\overline{z}^1 \wedge \cdots \wedge dz^n \wedge d\overline{z}^n.$$

(2) follows from (1).

1.13. On $U \cap \tilde{U}$, we have

$$\tilde{a} = \left|\det\left(\frac{\partial z^i}{\partial \tilde{z}^j}\right)\right|^2 a.$$

The exercise follows from the fact that if f is a nowhere vanishing holomorphic function then $\sqrt{-1}\partial\overline{\partial}\log|f|^2 = 0$.

1.14. With the notation of Exercise 1.6,

$$\det\left(\frac{\delta_{ij} + |\underline{z}|^2\delta_{ij} - z^j\overline{z}^i}{(1+|\underline{z}|^2)^2}\right) = \frac{1}{(1+|\underline{z}|^2)^{n+1}},$$

which can be more easily calculated by applying a unitary transformation to $\mathbb{C}^n$ so that $z^2 = \cdots = z^n = 0$. Then $\mathrm{Ric}(\omega_{\mathrm{FS}}) = (n+1)\sqrt{-1}\partial\overline{\partial}\log(1+|\underline{z}|^2) = (n+1)\omega_{\mathrm{FS}}$.

2.1. Adding a constant to f we may assume that f is positive. Applying Stokes' Theorem and Exercise 1.12,

$$0 \geqslant -\int_M f\sqrt{-1}\partial\overline{\partial}f \wedge \omega^{n-1} = \frac{1}{n}\int_M |\partial f|^2\omega^n,$$

so $\partial f = 0$.

2.2. For example, suppose that $\alpha > 0$ and $\alpha < 0$. Then α contains a Kähler metric ω and $-\alpha$ contains a Kähler metric ω'. Then $[\omega+\omega'] = 0$ so $\omega+\omega' = \sqrt{-1}\partial\overline{\partial}f > 0$ for some real-valued function f. This contradicts Exercise 2.1.

2.3. It is immediate from the definition that $\mathrm{Ka}(M)$ is a convex cone. For openness, let $\gamma_1, \ldots, \gamma_m$ be smooth closed $(1,1)$ forms with the property that $[\gamma_1], \ldots, [\gamma_m]$ is a basis for $H^{1,1}_{\overline{\partial}}(M, \mathbb{R})$. If α in $\mathrm{Ka}(M)$ is represented by a Kähler metric ω, then for $\varepsilon_i > 0$ sufficiently small $\omega + \sum_i \varepsilon_i\gamma_i$ is Kähler. Hence $[\alpha] + \sum_i \varepsilon_i[\gamma_i]$ is in $\mathrm{Ka}(M)$ for ε_i sufficiently small.

2.4. For (a) pick product coordinates. (b) follows from (a).

2.5. Follows from Exercise 2.4.

2.6. Let π_E and π_S be the projections onto E and S.

(a) $c_1(M) = -[\pi_S^*\omega_S]$, and $\pi_S^*\omega_S \geqslant 0$, so $T = \infty$.

(b) $\omega(t) = \pi_E^*\omega_E + (1+t)\pi_S^*\omega_S$.

(c) The torus fibers collapse and $\omega(t)/t$ converges to the Kähler-Einstein metric ω_S on S.

2.7. Observation: α is nef if and only if for all $\varepsilon > 0$, we have $\alpha + \varepsilon[\omega_0] > 0$.

(a) If α is nef then by the observation it is immediate that α is in the closure of $\mathrm{Ka}(M)$. Conversely, let α be in the closure of $\mathrm{Ka}(M)$ so that there exist $\alpha_j \in \mathrm{Ka}(M)$ with $\alpha_j \to \alpha$. Let $\beta_1, \ldots, \beta_m$ be smooth closed $(1,1)$ forms so that the $[\beta_i]$ give a basis for $H^{1,1}_{\overline{\partial}}(M, \mathbb{R})$. Then $\alpha - \alpha_j = \sum_i b_{i,j}[\beta_i]$ with $b_{i,j} \to 0$ as $j \to \infty$. Now let $\varepsilon > 0$. For j large enough, we have $\sum_i b_{i,j}\beta_i \geqslant -\varepsilon\omega_0$. Let ω_j in α_j be Kähler. Then $\omega_j + \sum b_{i,j}\beta_j \geqslant -\varepsilon\omega_0$ and $\omega_j + \sum b_{i,j}\beta_j \in \alpha$.

(b) We just have to show that

$$\sup\{t > 0 \mid [\omega_0] - tc_1(M) \text{ is nef}\} \leqslant T.$$

Suppose not. Then $[\omega_0] - (T+\delta)c_1(M)$ is nef for some $\delta > 0$ and so $(1+\varepsilon)[\omega_0] - (T+\delta)c_1(M) > 0$ for all $\varepsilon > 0$. Hence

$$[\omega_0] - \frac{T+\delta}{1+\varepsilon}c_1(M) > 0,$$

a contradiction since we may choose $\varepsilon > 0$ so that $\dfrac{T+\delta}{1+\varepsilon} > T$.

3.1. (cf. [**78**]). Suppose $\omega = \omega(t)$ solves $\dfrac{\partial}{\partial t}\omega = -\mathrm{Ric}(\omega)$. Then if we let φ solve $\frac{\partial}{\partial t}\varphi = \log(\omega^n/\Omega)$ with $\varphi|_{t=0} = 0$ we have

$$\frac{\partial}{\partial t}(\omega - \hat{\omega}_t - \sqrt{-1}\partial\overline{\partial}\varphi) = 0,$$

which implies that $\omega = \hat{\omega}_t + \sqrt{-1}\partial\overline{\partial}\varphi$, with φ solving (3.1.3).
3.2. For (a), suppose f_t does not converge smoothly to f. Then for some k there exists $\varepsilon > 0$ and $t_i \to T_0$ such that $\|f_{t_i} - f\|_{C^k(M)} \geqslant \varepsilon$ for all i. Then applying Arzelà-Ascoli, after passing to a subsequence, f_{t_i} converges smoothly to some function $\tilde{f}$ with $f \neq \tilde{f}$ and this contradicts the fact that f_{t_i} converges to f pointwise. For (b), suppose $|\dot{f}_t| \leqslant A$. Then $f_t + tA$ is nondecreasing and bounded above so converges pointwise to a unique limit. Now apply (a).
3.3. Pick $\psi = \varphi - Bt$ for

$$B = \inf_{M\times[0,S]} \log \frac{\hat{\omega}_t^n}{\Omega} - 1.$$

3.4. Put $Q = \dot{\varphi} - A\varphi$ for A chosen so that $A\hat{\omega}_t \geqslant \chi$. Then compute

$$\left(\frac{\partial}{\partial t} - \Delta\right) Q = \mathrm{tr}_\omega\chi - A\dot{\varphi} + An - A\hat{\omega}_t \leqslant -A\dot{\varphi} + An,$$

so that $\dot{\varphi} \leqslant n$ at a maximum of Q (if achieved at $t_0 > 0$).
3.5. Pick coordinates at a point so that g_0 is the identity and g is diagonal with eigenvalues $\lambda_1, \ldots, \lambda_n$. Then, for example, if $\mathrm{tr}_{\omega_0}\omega \leqslant C_1$, we have $\sum_i \lambda_i \leqslant C_1$ and so $\lambda_i \leqslant C_1$. On the other hand, we have $\lambda_1\lambda_2\cdots\lambda_n \geqslant C^{-1}$, so

$$\frac{1}{\lambda_i} = \frac{\lambda_1\cdots\widehat{\lambda_i}\cdots\lambda_n}{\lambda_1\cdots\lambda_n} \leqslant \frac{C_1^{n-1}}{C^{-1}}$$

where $\widehat{\ }$ means "omit". Hence $\mathrm{tr}_\omega\omega_0 = \sum_i \lambda_i^{-1} \leqslant C_2 := nC_1^{n-1}C$.
3.6. Pick holomorphic coordinates at a point with respect to g_0, so that $\nabla_k^0 = \partial_k$. Then

$$\begin{aligned} g_0^{i\overline{q}}g^{p\overline{j}}g^{k\overline{\ell}}B_{i\overline{j}k}\overline{B_{q\overline{p}\ell}} &= g_0^{i\overline{q}}g^{p\overline{j}}g^{k\overline{\ell}}\left(\partial_i g_{k\overline{j}} - \frac{\partial_k(\mathrm{tr}_{\omega_0}\omega)}{\mathrm{tr}_{\omega_0}\omega}g_{i\overline{j}}\right)\left(\partial_{\overline{q}}g_{p\overline{\ell}} - \frac{\partial_{\overline{\ell}}(\mathrm{tr}_{\omega_0}\omega)}{\mathrm{tr}_{\omega_0}\omega}g_{p\overline{q}}\right) \\ &= (I) + (II) + (III), \end{aligned}$$

where, using the Kähler condition, we have

$$(I) = g_0^{i\overline{q}}g^{p\overline{j}}g^{k\overline{\ell}}\partial_i g_{k\overline{j}}\partial_{\overline{q}}g_{p\overline{\ell}} = g_0^{i\overline{q}}g^{p\overline{j}}g^{k\overline{\ell}}\nabla_k^0 g_{i\overline{j}}\nabla_{\overline{\ell}}^0 g_{p\overline{q}},$$

and

$$\begin{aligned} (II) = &-2\mathrm{Re}\left(g_0^{i\overline{q}}g^{p\overline{j}}g^{k\overline{\ell}}\partial_i g_{k\overline{j}}\frac{\partial_{\overline{\ell}}(\mathrm{tr}_{\omega_0}\omega)}{\mathrm{tr}_{\omega_0}\omega}g_{p\overline{q}}\right) \\ = &-2\mathrm{Re}\left(g^{k\overline{\ell}}g_0^{i\overline{j}}\partial_k g_{i\overline{j}}\frac{\partial_{\overline{\ell}}(\mathrm{tr}_{\omega_0}\omega)}{\mathrm{tr}_{\omega_0}\omega}\right) \\ = &-2\frac{|\partial\mathrm{tr}_{\omega_0}\omega|_g^2}{\mathrm{tr}_{\omega_0}\omega}, \end{aligned}$$

and

$$(III) = g_0^{i\overline{q}}g^{p\overline{j}}g^{k\overline{\ell}}\frac{\partial_k(\mathrm{tr}_{\omega_0}\omega)}{\mathrm{tr}_{\omega_0}\omega}g_{i\overline{j}}\frac{\partial_{\overline{\ell}}(\mathrm{tr}_{\omega_0}\omega)}{\mathrm{tr}_{\omega_0}\omega}g_{p\overline{q}} = \frac{|\partial\mathrm{tr}_{\omega_0}\omega|_g^2}{\mathrm{tr}_{\omega_0}\omega}.$$

4.1. Both $-\mathrm{Ric}(\omega_\infty)$ and $\hat{\omega}_\infty$ lie in $-c_1(M)$ so there exists f with

$$\sqrt{-1}\partial\overline{\partial}\log\hat{\omega}_\infty^n = \hat{\omega}_\infty + \sqrt{-1}\partial\overline{\partial}f.$$

Set $\Omega = \hat{\omega}_\infty^n e^{-f+c}$ with c chosen so that $\int_M \Omega = \int_M \omega_0^n$.

4.2. Similar to the proof in Lecture 3 (see Exercise 3.1) that (2.1.1) is equivalent to (3.1.3).

4.3. At the maximum of φ (if it occurs at $t_0 > 0$) we have $\sqrt{-1}\partial\overline{\partial}\varphi \leqslant 0$ and $\frac{\partial}{\partial t}\varphi \geqslant 0$ and hence from (4.1.5), $\varphi \leqslant \log\frac{\hat{\omega}_t^n}{\Omega} \leqslant C$. The lower bound of φ is similar.

4.4. The only difference compared to the calculation of Lemma 3.3 is that in (3.3.2) there is an extra term coming from the $-\omega$ in $\frac{\partial}{\partial t}\omega = -\mathrm{Ric}(\omega) - \omega$, which yields an additional $\dfrac{-\mathrm{tr}_{\omega_0}\omega}{\mathrm{tr}_{\omega_0}\omega} = -1$.

4.5. For $\varepsilon > 0$, $(\frac{\partial}{\partial t} - \Delta)(f - \varepsilon t) = -\varepsilon < 0$ and hence the maximum of $f - \varepsilon t$ must occur at $t = 0$ giving $f - \varepsilon t \leqslant \sup_M |f_0|$. Let $\varepsilon \to 0$. The lower bound is similar.

4.6. Consider $Q = \log \mathrm{tr}_{\omega_0}\omega - A\varphi$ on $M \times [0,t]$ and show that, for A sufficiently large,

$$\left(\frac{\partial}{\partial t} - \Delta\right) Q \leqslant -\mathrm{tr}_\omega \omega_0 + C,$$

using the fact that $\dot{\varphi}$ is uniformly bounded. If Q achieves a maximum at (x_0, t_0) with $t_0 > 0$ then since φ is bounded we have $(\mathrm{tr}_\omega\omega_0)(x_0,t_0) \leqslant C$ and so $(\mathrm{tr}_{\omega_0}\omega)(x_0,t_0) \leqslant C'$. Hence for any (x,t),

$$(\log \mathrm{tr}_{\omega_0}\omega)(x,t) - A\varphi(x,t) \leqslant Q(x_0,t_0) \leqslant \log C' - A\varphi(x_0,t_0)$$

and the claim follows after exponentiating.

4.7. Since $\frac{\partial}{\partial t}\omega^n = \mathrm{tr}_\omega(\frac{\partial}{\partial t}\omega)\,\omega^n = \Delta\dot{\varphi}\,\omega^n$ we have

$$\frac{d}{dt}P(t) = \int_M \Delta\dot{\varphi}\,\omega^n + \int_M \dot{\varphi}\Delta\dot{\varphi}\,\omega^n = -\frac{1}{n}\int_M |\partial\dot{\varphi}|_g^2\omega^n,$$

using Stokes' Theorem and Exercise 1.12.

4.8. Compute

$$\left(\frac{\partial}{\partial t} - \Delta\right)|\partial\dot{\varphi}|_g^2 = -|\nabla\nabla\dot{\varphi}|_g^2 - |\nabla\overline{\nabla}\dot{\varphi}|_g^2 \leqslant 0,$$

where $|\nabla\nabla\dot{\varphi}|_g^2 = g^{i\overline{j}}g^{k\overline{\ell}}\nabla_i\nabla_k\dot{\varphi}\nabla_{\overline{j}}\nabla_{\overline{\ell}}\dot{\varphi}$ etc. Then

$$\frac{d^2P}{dt^2} \geqslant -\frac{1}{n}\int_M \Delta|\partial\dot{\varphi}|_g^2\omega^n - \frac{1}{n}\int_M |\partial\dot{\varphi}|_g^2\Delta\dot{\varphi}\,\omega^n \geqslant -\frac{C}{n}\int_M |\partial\dot{\varphi}|_g^2\omega^n = C\frac{dP}{dt},$$

since $\Delta\dot{\varphi}$ is uniformly bounded. To show that $dP/dt \to 0$, we use the following elementary fact. If $f : [0,\infty) \to \mathbb{R}$ satisfies the differential inequality $\dot{f} \geqslant Cf$ then if $f \geqslant -\varepsilon$ at t we have $f \geqslant -e^{2C}\varepsilon$ on $[t, t+2]$ (consider fe^{-Ct}). Then since $(dP/dt)(t_i) \to 0$ for $t_i \in [i, i+1]$ it follows that $dP/dt \to 0$.

Bibliography

[1] Aubin, T. *Équations du type Monge-Ampère sur les variétés kählériennes compactes*, Bull. Sci. Math. (2) 102 (1978), no. 1, 63–95 MR494932 (81d:53047)

[2] Barth, W. P., Hulek, K., Peters, C. A. M. and Van de Ven, A. *Compact complex surfaces.* Second edition. Ergebnisse der Mathematik und ihrer Grenzgebiete. Springer-Verlag, Berlin MR2030225 (2004m:14070)

[3] Birkar, C., Cascini, P., Hacon, C. and McKernan, J. *Existence of minimal models for varieties of log general type*, J. Amer. Math. Soc. 23 (2010), no. 2, 405–468 MR2601039 (2011f:14023)

[4] Boucksom, S. and Guedj, V. *Regularizing properties of the Kähler-Ricci flow*, Chapter 4, Lecture Notes in Math. 2086, Springer 2013 MR3185334

[5] Calabi, E. *On Kähler manifolds with vanishing canonical class*, Algebraic geometry and topology. A symposium in honor of S. Lefschetz, pp. 78–89. Princeton University Press, Princeton, N. J., 1957 MR0085583 (19:62b)

[6] Calabi, E. *Improper affine hyperspheres of convex type and a generalization of a theorem by K. Jörgens*, Michigan Math. J. 5 (1958) 105–126 MR0106487 (21:5219)

[7] Cao, H.-D. *Deformation of Kähler metrics to Kähler-Einstein metrics on compact Kähler manifolds*, Invent. Math. 81 (1985), no. 2, 359–372 MR799272 (87d:58051)

[8] Cao, H.-D. *The Kähler-Ricci flow on Fano manifolds*, Chapter 5, Lecture Notes in Math. 2086, Springer 2013 MR3185335

[9] Cao, H.-D. *Existence of gradient Kähler-Ricci solitons*, Elliptic and parabolic methods in geometry (Minneapolis, MN, 1994), 1–16, A.K. Peters, Wellesley, MA, 1996 MR1417944 (98a:53058)

[10] Cao, H.-D. and Zhu, X.-P. *A Complete Proof of the Poincaré and Geometrization Conjectures—Application of the Hamilton-Perelman Theory of the Ricci Flow*, Asian J. Math. 10 (2006), no. 2, 165–492 MR2233789 (2008d:53090)

[11] Chau, A. *Convergence of the Kähler-Ricci flow on noncompact Kähler manifolds*, J. Differential Geom. 66 (2004), no. 2, 211–232 MR2106124 (2005g:53118)

[12] Cherrier, P. *Équations de Monge-Ampère sur les variétés hermitiennes compactes*, Bull. Sci. Math. (2) 111 (1987), no. 4, 343–385 MR921559 (89d:58131)

[13] Chow, B. *The Ricci flow on the 2-sphere*, J. Differential Geom. 33 (1991), no. 2, 325–334 MR1094458 (92d:53036)

[14] Chow, B., Lu, P. and Ni, L. *Hamilton's Ricci flow.* Graduate Studies in Mathematics, 77. American Mathematical Society, Providence, RI; Science Press, New York, 2006. xxxvi+608 pp MR2274812 (2008a:53068)

[15] Collins, T. and Székelyhidi, G. *The twisted Kähler-Ricci flow*, preprint, arXiv:1207.5441

[16] T. C. Collins and V. Tosatti, *Kähler currents and null loci*, Invent. Math. **202** (2015), no. 3, 1167–1198, DOI 10.1007/s00222-015-0585-9. MR3425388

[17] DeTurck, D. M. *Deforming metrics in the direction of their Ricci tensors*, J. Differential Geom. 18 (1983), no. 1, 157–162 MR697987 (85j:53050)

[18] Evans, L.C. *Classical solutions of fully nonlinear, convex, second order elliptic equations*, Comm. Pure Appl. Math 25 (1982), 333–363 MR649348 (83g:35038)

[19] Feldman, M., Ilmanen, T. and Knopf, D. *Rotationally symmetric shrinking and expanding gradient Kähler-Ricci solitons*, J. Differential Geometry 65 (2003), no. 2, 169–209 MR2058261 (2005e:53102)

[20] F. T.-H. Fong, *Kähler-Ricci flow on projective bundles over Kähler-Einstein manifolds*, Trans. Amer. Math. Soc. **366** (2014), no. 2, 563–589, DOI 10.1090/S0002-9947-2013-05726-1. MR3130308

[21] F. T.-H. Fong, *On the collapsing rate of the Kähler-Ricci flow with finite-time singularity*, J. Geom. Anal. **25** (2015), no. 2, 1098–1107, DOI 10.1007/s12220-013-9458-x. MR3319963

[22] F. T.-H. Fong and Z. Zhang, *The collapsing rate of the Kähler-Ricci flow with regular infinite time singularity*, J. Reine Angew. Math. **703** (2015), 95–113, DOI 10.1515/crelle-2013-0043. MR3353543

[23] Gauduchon, P. *Le théorème de l'excentricité nulle*, C. R. Acad. Sci. Paris 285 (1977), 387–390 MR0470920 (57:10664)

[24] Giesen, G. and Topping, P. M. *Existence of Ricci flows of incomplete surfaces*, Comm. Partial Differential Equations 36 (2011), no. 10, 1860–1880 MR2832165 (2012i:53063)

[25] Gill, M. *Convergence of the parabolic complex Monge-Ampère equation on compact Hermitian manifolds*, Comm. Anal. Geom. 19 (2011), no. 2, 277–303 MR2835881 (2012h:32047)

[26] M. Gill, *Collapsing of products along the Kähler-Ricci flow*, Trans. Amer. Math. Soc. **366** (2014), no. 7, 3907–3924, DOI 10.1090/S0002-9947-2013-06073-4. MR3192623

[27] Gill, M. *The Chern-Ricci flow on smooth minimal models of general type*, preprint, arXiv:1307.0066

[28] Griffiths, P. and Harris, J. *Principles of algebraic geometry*, Pure and Applied Mathematics. Wiley-Interscience, New York, 1978 MR507725 (80b:14001)

[29] Gross, M., Tosatti, V. and Zhang, Y. *Collapsing of abelian fibred Calabi-Yau manifolds*, Duke Math J. 162 (2013), no.3, 517–551 MR3024092

[30] Guan, B. and Li, Q. *Complex Monge-Ampère equations and totally real submanifolds*, Adv. Math. 225 (2010), no. 3, 1185–1223 MR2673728 (2011g:32053)

[31] Guedj, V. *Convergence of the Kähler-Ricci flow on a Kähler-Einstein Fano manifold*, Chapter 6, Lecture Notes in Math. 2086, Springer 2013 MR3185336

[32] Guedj, V. and Zeriahi, A. *Regularizing properties of the twisted Kähler-Ricci flow*, preprint, arXiv:1306.4089

[33] Hamilton, R. S. *Three-manifolds with positive Ricci curvature*, J. Differential Geom. 17 (1982), no. 2, 255–306 MR664497 (84a:53050)

[34] Hamilton, R. S. *The Ricci flow on surfaces*, Mathematics and general relativity (Santa Cruz, CA, 1986), 237–262, Contemp. Math., 71, Amer. Math. Soc., Providence, RI, 1988 MR954419 (89i:53029)

[35] Hamilton, R. S. *The formation of singularities in the Ricci flow*, Surveys in differential geometry, Vol. II (Cambridge, MA, 1993), 7–136, Int. Press, Cambridge, MA, 1995 MR1375255 (97e:53075)

[36] Inoue, M. *On surfaces of Class* VII_0, Invent. Math. **24** (1974), 269–310 MR0342734 (49:7479)

[37] Kleiner, B. and Lott, J. *Notes on Perelman's papers*, Geom. Topol. 12 (2008), no. 5, 2587–2855 MR2460872 (2010h:53098)

[38] Kodaira, K. and Morrow, J. *Complex manifolds*, Holt, Rinehart and Winston, Inc., New York-Montreal, Que.-London, 1971. vii+192 pp MR0302937 (46:2080)

[39] Koiso, N. *On rotationally symmetric Hamilton's equation for Kähler-Einstein metrics*, Recent topics in differential and analytic geometry, 327–337, Adv. Stud. Pure Math., 18-I, Academic Press, Boston, MA, 1990 MR1145263 (93d:53057)

[40] Krylov, N. V. *Boundedly nonhomogeneous elliptic and parabolic equations*, Izvestia Akad. Nauk. SSSR 46 (1982), 487–523. English translation in Math. USSR Izv. 20 (1983), no. 3, 459–492 MR661144 (84a:35091)

[41] La Nave, G. and Tian, G *Soliton-type metrics and Kähler-Ricci flow on symplectic quotients*, preprint, arXiv:0903.2413

[42] Li, J., Yau, S.-T. and Zheng, F. *On projectively flat Hermitian manifolds*, Comm. Anal. Geom. 2 (1994), 103–109 MR1312680 (96g:32049)

[43] Lieberman, G. M. *Second order parabolic differential equations*, World Scientific Publishing Co., Inc., River Edge, NJ, 1996 MR1465184 (98k:35003)

[44] D. Máximo, *On the blow-up of four-dimensional Ricci flow singularities*, J. Reine Angew. Math. **692** (2014), 153–171. MR3274550

[45] R. Mazzeo, Y. A. Rubinstein, and N. Sesum, *Ricci flow on surfaces with conic singularities*, Anal. PDE **8** (2015), no. 4, 839–882, DOI 10.2140/apde.2015.8.839. MR3366005

[46] Morgan, J. and Tian, G. *Ricci flow and the Poincaré conjecture*, Clay Mathematics Monographs, 3. American Mathematical Society, Providence, RI; Clay Mathematics Institute, Cambridge, MA, 2007 MR2334563 (2008d:57020)

[47] Nakamura, I. *On surfaces of class* VII_0 *with curves*, Invent. Math. 78 (1984), no. 3, 393–443 MR768987 (86e:32034)

[48] Perelman, G. *The entropy formula for the Ricci flow and its geometric applications*, preprint, arXiv: math.DG/0211159

[49] Perelman, G. unpublished work on the Kähler-Ricci flow

[50] Phong, D. H., Sesum, N. and Sturm, J. *Multiplier ideal sheaves and the Kähler-Ricci flow*, Comm. Anal. Geom. 15 (2007), no. 3, 613–632 MR2379807 (2009a:32037)

[51] Phong, D. H., Song, J., Sturm, J. and Weinkove, B. *On the convergence of the modified Kähler-Ricci flow and solitons*, Comment. Math. Helv. 86 (2011), no. 1, 91–112 MR2745277 (2012d:53218)

[52] Phong, D. H. and Sturm, J. *On stability and the convergence of the Kähler-Ricci flow*, J. Differential Geometry 72 (2006), no. 1, 149–168 MR2215459 (2007a:53128)

[53] Phong, D. H. and Sturm, J. *Lectures on stability and constant scalar curvature*, Current developments in mathematics, 2007, 101–176, Int. Press, Somerville, MA, 2009 MR2532997 (2010j:32042)

[54] Sesum, N. and Tian, G. *Bounding scalar curvature and diameter along the Kähler Ricci flow (after Perelman)*, J. Inst. Math. Jussieu 7 (2008), no. 3, 575–587 MR2427424 (2009c:53092)

[55] Sherman, M. and Weinkove, B. *Interior derivative estimates for the Kähler-Ricci flow*, Pacific J. Math. 257 (2012), no. 2, 491–501 MR2972475

[56] M. Sherman and B. Weinkove, *Local Calabi and curvature estimates for the Chern-Ricci flow*, New York J. Math. **19** (2013), 565–582. MR3119098

[57] Song, J. *Some Type I solutions of Ricci flow with rotational symmetry*, preprint, arXiv:1203.2688

[58] Song, J. *Ricci flow and birational surgery*, preprint, arXiv:1304.2607

[59] Song, J., Székelyhidi, G. and Weinkove, B. *The Kähler-Ricci flow on projective bundles*, Int. Math. Res. Not. 2013 (2013), no. 2, 243–257 MR3010688

[60] Song, J, and Tian, G. *The Kähler-Ricci flow on surfaces of positive Kodaira dimension*, Invent. Math. 170 (2007), no. 3, 609–653 MR2357504 (2008m:32044)

[61] Song, J, and Tian, G. *Canonical measures and Kähler-Ricci flow*, J. Amer. Math. Soc. 25 (2012), no. 2, 303–353 MR2869020

[62] Song, J, and Tian, G. *The Kähler-Ricci flow through singularities*, preprint, arXiv:0909.4898

[63] Song, J. and Weinkove, B. *The Kähler-Ricci flow on Hirzebruch surfaces*, J. Reine Angew. Math. 659 (2011), 141–168 MR2837013 (2012g:53142)

[64] Song, J. and Weinkove, B. *Contracting exceptional divisors by the Kähler-Ricci flow*, Duke Math. J. 162 (2013), no. 2, 367–415 MR3018957

[65] J. Song and B. Weinkove, *Contracting exceptional divisors by the Kähler-Ricci flow II*, Proc. Lond. Math. Soc. (3) **108** (2014), no. 6, 1529–1561, DOI 10.1112/plms/pdt059. MR3218318

[66] Song, J. and Weinkove, B. *An Introduction to the Kähler-Ricci flow*, Chapter 3, Lecture Notes in Math. 2086, Springer 2013 MR3185333

[67] J. Song and Y. Yuan, *Metric flips with Calabi ansatz*, Geom. Funct. Anal. **22** (2012), no. 1, 240–265, DOI 10.1007/s00039-012-0151-1. MR2899688

[68] Streets, J. and Tian, G. *A parabolic flow of pluriclosed metrics*, Int. Math. Res. Not. IMRN 2010, no. 16, 3101–3133 MR2673720 (2011h:53091)

[69] Teleman, A. *Projectively flat surfaces and Bogomolov's theorem on class* VII_0*-surfaces*, Int. J. Math. 5 (1994), 253–264 MR1266285 (95e:32031)

[70] Teleman, A. *Donaldson theory on non-Kählerian surfaces and class VII surfaces with* $b_2 = 1$, Invent. Math. 162 (2005), no. 3, 493–521 MR2198220 (2006i:32020)

[71] Tian, G. *On Calabi's conjecture for complex surfaces with positive first Chern class*, Invent. Math. 101 (1990), no. 1, 101–172 MR1055713 (91d:32042)

[72] Tian, G. *New results and problems on Kähler-Ricci flow*, Géométrie différentielle, physique mathématique, mathématiques et société. II. Astérisque No. 322 (2008), 71–92 MR2521654 (2011a:53131)

[73] Tian, G. and Yau, S.-T. *Kähler-Einstein metrics on complex surfaces with* $C_1 > 0$, Comm. Math. Phys. 112 (1987), no. 1, 175–203 MR904143 (88k:32070)

[74] G. Tian, S. Zhang, Z. Zhang, and X. Zhu, *Perelman's entropy and Kähler-Ricci flow on a Fano manifold*, Trans. Amer. Math. Soc. **365** (2013), no. 12, 6669–6695, DOI 10.1090/S0002-9947-2013-06027-8. MR3105766

[75] Tian, G. and Zhang, Z. *On the Kähler-Ricci flow on projective manifolds of general type*, Chinese Ann. Math. Ser. B 27 (2006), no. 2, 179–192 MR2243679 (2007c:32029)

[76] Tian, G. and Zhu, X. *Convergence of Kähler-Ricci flow*, J. Amer. Math. Soc. 20 (2007), no. 3, 675–699 MR2291916 (2007k:53107)

[77] Tosatti, V. and Weinkove, B. *The complex Monge-Ampère equation on compact Hermitian manifolds*, J. Amer. Math. Soc. 23 (2010), no.4, 1187–1195 MR2669712 (2012c:32055)

[78] V. Tosatti and B. Weinkove, *On the evolution of a Hermitian metric by its Chern-Ricci form*, J. Differential Geom. **99** (2015), no. 1, 125–163. MR3299824

[79] V. Tosatti and B. Weinkove, *The Chern-Ricci flow on complex surfaces*, Compos. Math. **149** (2013), no. 12, 2101–2138, DOI 10.1112/S0010437X13007471. MR3143707

[80] V. Tosatti, B. Weinkove, and X. Yang, *Collapsing of the Chern-Ricci flow on elliptic surfaces*, Math. Ann. **362** (2015), no. 3-4, 1223–1271, DOI 10.1007/s00208-014-1160-1. MR3368098

[81] Tsuji, H. *Existence and degeneration of Kähler-Einstein metrics on minimal algebraic varieties of general type*, Math. Ann. 281 (1988), 123–133 MR944606 (89e:53075)

[82] Tsuji, H. *Generalized Bergmann metrics and invariance of plurigenera*, preprint, arXiv:math.CV/9604228

[83] Yau, S.-T. *A general Schwarz lemma for Kähler manifolds*, Amer. J. Math. 100 (1978), no. 1, 197–203 MR0486659 (58:6370)

[84] Yau, S.-T. *On the Ricci curvature of a compact Kähler manifold and the complex Monge-Ampère equation, I*, Comm. Pure Appl. Math. 31 (1978), 339–411 MR480350 (81d:53045)

[85] Wang, X. and Zhu, X. *Kähler-Ricci solitons on toric manifolds with positive first Chern class*, Adv. Math. 188 (2004), no. 1, 87–103 MR2084775 (2005d:53074)

[86] Zhang, Z. *Scalar curvature behavior for finite-time singularity of Kähler-Ricci flow*, Michigan Math. J. 59 (2010), no. 2, 419–433 MR2677630 (2011j:53128)

[87] X. Zhu, *Kähler-Ricci flow on a toric manifold with positive first Chern class*, Differential geometry, Adv. Lect. Math. (ALM), vol. 22, Int. Press, Somerville, MA, 2012, pp. 323–336. MR3076057

Park City lectures on Eigenfuntions

Steve Zelditch

IAS/Park City Mathematics Series
Volume 22, 2013

Park City lectures on Eigenfuntions

Steve Zelditch

1. Introduction

These lectures are devoted to recent results on the nodal geometry of eigenfunctions

$$(1) \qquad \Delta_g \varphi_\lambda = \lambda^2 \varphi_\lambda$$

of the Laplacian Δ_g of a Riemannian manifold (M^m, g) of dimension m and to associated problems on L^p norms of eigenfunctions (§2.9 and §12). The manifolds are generally assumed to be compact, although the problems can also be posed on non-compact complete Riemannian manifolds. The emphasis of these lectures is on *real analytic* Riemannian manifolds, but we also mention some new results for general C^∞ metrics. Although we mainly discuss the Laplacian, analogous problems and results exist for Schrödinger operators $-\frac{\hbar^2}{2}\Delta_g + V$ for certain potentials V. Moreover, many of the results on eigenfunctions also hold for *quasi-modes* or approximate eigenfunctions defined by oscillatory integrals.

The study of eigenfunctions of Δ_g and $-\frac{\hbar^2}{2}\Delta_g + V$ on Riemannian manifolds is a branch of harmonic analysis. In these lectures, we emphasize high frequency (or semi-classical) asymptotics of eigenfunctions and their relations to the global dynamics of the geodesic flow $G^t : S^*M \to S^*M$ on the unit cosphere bundle of M. Here and henceforth we identity vectors and covectors using the metric. As in [**Ze3**] we give the name "Global Harmonic Analysis" to the use of global wave equation methods to obtain relations between eigenfunction behavior and geodesics. The relations between geodesics and eigenfunctions belongs to the general correspondence principle between classical and quantum mechanics. The correspondence principle has evolved since the origins of quantum mechanics [**Sch**] into a systematic theory of Semi-Classical Analysis and Fourier integral operators, of which [**HoI, HoII, HoIII, HoIV**] and [**Zw**] give systematic presentations; see also §2.11 for further references. Quantum mechanics provides not only the intuition and techniques for the study of eigenfunctions, but in large part also provides the motivation. Readers who are unfamiliar with quantum mechanics are encouraged to read standard texts such as Landau-Lifschitz [**LL**] or Weinberg [**Wei**]. Atoms and molecules are multi-dimensional and difficult to visualize, and there are many efforts to do so in the physics and chemistry literature. Some examples may be found in [**KP, He, Th, SHM**]. Nodal sets of the hydrogen atom have recently been observed using quantum microscopes [**St**]. Here we concentrate on eigenfunctions of the Laplacian; some results on nodal sets of eigenfunctions of Schrödinger operators can be found in [**Jin, HZZ**].

Department of Mathematics, Northwestern University, Evanston, IL 60208, USA
E-mail address: zelditch@math.northwestern.edu
Research partially supported by NSF grant DMS-1206527

Among the fundamental tools in the study of eigenfunctions are parametrix constructions of wave and Poisson kernels, and the method of stationary phase in the real and complex domain. The plan of these lectures is to concentrate at once on applications to nodal sets and L^p norms and refer to Appendices or other texts for the techniques. Parametrix constructions and stationary phase are techniques whose role in spectral asymptotics are as basic as the maximum principle is in elliptic PDE. We do include Appendices on the geodesic flow (§13), on parametrix constructions for the wave group (§14), on general facts and definitions concerning oscillatory integrals (§15) and on spherical harmonics (§16).

1.1. The eigenvalue problem on a compact Riemannian manifold

The (negative) Laplacian Δ_g of (M^m, g) is the unbounded essentially self-adjoint operator on $C_0^\infty(M) \subset L^2(M, dV_g)$ defined by the Dirichlet form

$$D(f) = \int_M |\nabla f|^2 dV_g,$$

where ∇f is the metric gradient vector field and $|\nabla f|$ is its length in the metric g. Also, dV_g is the volume form of the metric. In terms of the metric Hessian Dd,

$$\Delta f = \text{trace } Ddf.$$

In local coordinates,

$$\Delta_g = \frac{1}{\sqrt{g}} \sum_{i,j=1}^n \frac{\partial}{\partial x_i} \left(g^{ij} \sqrt{g} \frac{\partial}{\partial x_j} \right), \tag{2}$$

in a standard notation that we assume the reader is familiar with (see e.g. [**BGM, Ch**] if not).

Remark: We are not always consistent on the sign given to Δ_g. When we work with $\sqrt{-\Delta_g}$ we often define Δ_g to be the opposite of (2) and write $\sqrt{\Delta}$ for notational simplicity. We often omit the subscript g when the metric is fixed. We hope the notational conventions do not cause confusion.

A more geometric definition uses at each point p an orthomormal basis $\{e_j\}_{j=1}^m$ of T_pM and geodesics γ_j with $\gamma_j(0) = p$, $\gamma_j'(0) = e_j$. Then

$$\Delta f(p) = \sum_j \frac{d^2}{dt^2} f(\gamma_j(t)).$$

We refer to [**BGM**] (G.III.12).

EXERCISE 1. *Let $m = 2$ and let γ be a geodesic arc on M. Calculate $(\Delta f)(s, 0)$ in Fermi normal coordinates along γ.*

Background: Define Fermi normal coordinates (s, y) along γ by identifying a small ball bundle of the normal bundle $N\gamma$ along $\gamma(s)$ with its image (a tubular neighborhood of γ) under the normal exponential map, $\exp_{\gamma(s)} y\nu_{\gamma(s)}$. Here, $\nu_{\gamma(s)}$ is the unit normal at $\gamma(s)$ (fix one of the two choices) and $\exp_{\gamma(s)} y\nu_{\gamma(s)}$ is the unit speed geodesic in the direction $\nu_{\gamma(s)}$ of length y.

The focus of these lectures is on the eigenvalue problem (1). As mentioned above, we sometimes multiply Δ and the eigenvalue by -1 for notational simplicity. We assume throughout that φ_λ is L^2-normalized,

$$||\varphi_\lambda||^2_{L^2} = \int_M |\varphi_\lambda|^2 dV = 1.$$

When M is compact, the spectrum of eigenvalues of the Laplacian is discrete there exists an orthonormal basis of eigenfunctions. We fix such a basis $\{\varphi_j\}$ so that

$$\Delta_g \varphi_j = \lambda_j^2 \varphi_j, \qquad \langle \varphi_j, \varphi_k \rangle_{L^2(M)} := \int_M \varphi_j \varphi_k dV_g = \delta_{jk} \tag{3}$$

If $\partial M \neq \emptyset$ we impose Dirichlet or Neumann boundary conditions. Here dV_g is the volume form. When M is compact, the spectrum of Δ_g is a discrete set

$$\lambda_0 = 0 < \lambda_1^2 \leq \lambda_2^2 \leq \cdots \tag{4}$$

repeated according to multiplicity. Note that $\{\lambda_j\}$ denote the *frequencies*, i.e. square roots of Δ-eigenvalues. We mainly consider the behavior of eigenfuntions in the 'high frequency' (or high energy) limit $\lambda_j \to \infty$.

The Weyl law asymptotically counts the number of eigenvalues less than λ,

$$N(\lambda) = \#\{j : \lambda_j \leq \lambda\} = \frac{|B_n|}{(2\pi)^n} Vol(M, g)\lambda^n + O(\lambda^{n-1}). \tag{5}$$

Here, $|B_n|$ is the Euclidean volume of the unit ball and $Vol(M, g)$ is the volume of M with respect to the metric g. The size of the remainder reflects the measure of closed geodesics [**DG, HoIV**]. It is a basic example of global the effect of the global dynamics on the spectrum. See §2.9 and §12 for related results on eigenfunctions.

(1) In the *aperiodic* case where the set of closed geodesics has measure zero, the Duistermaat-Guillemin-Ivrii two term Weyl law states
$$N(\lambda) = \#\{j : \lambda_j \leq \lambda\} = c_m \, Vol(M, g) \, \lambda^m + o(\lambda^{m-1})$$
where $m = \dim M$ and where c_m is a universal constant.

(2) In the *periodic* case where the geodesic flow is periodic (Zoll manifolds such as the round sphere), the spectrum of $\sqrt{\Delta}$ is a union of eigenvalue clusters C_N of the form
$$C_N = \{(\frac{2\pi}{T})(N + \frac{\beta}{4}) + \mu_{Ni}, \; i = 1 \dots d_N\}$$
with $\mu_{Ni} = 0(N^{-1})$. The number d_N of eigenvalues in C_N is a polynomial of degree $m - 1$.

Remark: The proof that the spectrum is discrete is based on the study of spectral kernels such as the heat kernel or Green's function or wave kernel. The standard proof is to show that Δ_g^{-1} (whose kernel is the Green's function, defined on the orthogonal complement of the constant functions) is a compact self-adjoint operator. By the spectral theory for such operators, the eigenvalues of Δ_g^{-1} are discrete, of finite multiplicity, and only accumulate at 0. Although we concentrate on parametrix constructions for the wave kernel, one can construct the Hadamard parametrix for the Green's function in a similar way. Proofs of the above statements can be found in [**GSj, DSj, Zw, HoIII**].

The proof of the integrated and pointwise Weyl law are based on wave equation techniques and Fourier Tauberian theorems. The wave equation techniques mainly involve the construction of parametrices for the fundamental solution of the wave equation and the method of stationary phase. In §15 we review We refer to [**DG, HoIV**] for detailed background.

1.2. Nodal and critical point sets

The main focus of these lectures is on nodal hypersurfaces

$$\mathcal{N}_{\varphi_\lambda} = \{x \in M : \varphi_\lambda(x) = 0\}. \tag{6}$$

The nodal domains are the components of the complement of the nodal set,

$$M \backslash \mathcal{N}_{\varphi_\lambda} = \bigcup_{j=1}^{\mu(\varphi_\lambda)} \Omega_j.$$

For generic metrics, 0 is a regular value of φ_λ of all eigenfunctions, and the nodal sets are smooth non-self-intersecting hypersurfaces [**U**]. Among the main problems on nodal sets are to determine the hypersurface volume $\mathcal{H}^{m-1}(\mathcal{N}_{\varphi_\lambda})$ and ideally how the nodal sets are distributed. Another well-known question is to determine the number $\mu(\varphi_\lambda)$ of nodal domains in terms of the eigenvalue in generic cases. One may also consider the other level sets

$$\mathcal{N}_{\varphi_j}^a = \{x \in M : \varphi_j(x) = a\} \tag{7}$$

and sublevel sets

$$\{x \in M : |\varphi_j(x)| \leq a\}. \tag{8}$$

The zero level is distinguished since the symmetry $\varphi_j \to -\varphi_j$ in the equation preserves the nodal set.

Remark: Nodals sets belong to individual eigenfunctions. To the author's knowledge there do not exist any results on averages of nodal sets over the spectrum in the sense of (5)-(33). That is, we do not know of any asymptotic results concerning the functions

$$Z_f(\lambda) := \sum_{j:\lambda_j \leq \lambda} \int_{\mathcal{N}_{\varphi_j}} f dS,$$

where $\int_{\mathcal{N}_{\varphi_j}} f dS$ denotes the integral of a continuous function f over the nodal set of φ_j. When the eigenvalues are multiple, the sum Z_f depends on the choice of orthonormal basis.

Randomizing by taking Gaussian random combinations of eigenfunctions simplifies nodal problems profoundly, and are studied in many articles (see e.g. [**NS**]).

One would also like to know the "number" and distribution of critical points,

$$\mathcal{C}_{\varphi_j} = \{x \in M : \nabla \varphi_j(x) = 0\}. \tag{9}$$

In fact, the critical point set can be a hypersurface in M, so for counting problems it makes more sense to count the number of critical values,

$$\mathcal{V}_{\varphi_j} = \{\varphi_j(x) : \nabla \varphi_j(x) = 0\}. \tag{10}$$

At this time of writing, there exist few rigorous upper bounds on the number of critical values, so we do not spend much space on them here. If we 'randomize' the problem and consider the average number of critical points (or equivalently values) or random spherical harmonics on the standard $\mathbb{S}^m$, one finds that the random spherical harmonics of degree N (eigenvalue $\simeq N^2$) has $C_m N^m$ critical points in dimension m. This is not surprising since spherical harmonics are harmonic polynomials. In the non-generic case that the critical manifolds are of co-dimension one, the hypersurface volume is calcualated in the real analytic case in [**Ba**]. The upper bound is also given in [**Ze0**].

The frequency λ of an eigenfunction (i.e. the square root of the eigenvalue) is a measure of its "complexity", similar to specifying the degree of a polynomial, and the high frequency limit is the large complexity limit. A sequence of eigenfunctions of increasing frequency oscillates more and more rapidly and the problem is to find its "limit shape". Sequences of eigenfunctions often behave like "Gaussian random waves" but special ones exhibit highly localized oscillation and concentration properties.

1.3. Motivation

Before stating the problems and results, let us motivate the study of eigenfunctions and their high frequency behavior. The eigenvalue problem (1) arises in many areas of physics, for example the theory of vibrating membranes. But renewed motivation to study eigenfunctions comes from quantum mechanics. As is discussed in any textbook on quantum physics or chemistry (see e.g. [**LL, Wei**]), the Schrödinger equation resolves the problem of how an electron can orbit the nucleus without losing its energy in radiation. The classical Hamiltonian equations of motion of a particle in phase space are orbits of Hamilton's equations

$$\begin{cases} \frac{dx_j}{dt} = \frac{\partial H}{\partial \xi_j}, \\ \\ \frac{d\xi_j}{dt} = -\frac{\partial H}{\partial x_j}, \end{cases}$$

where the Hamiltonian

$$H(x,\xi) = \frac{1}{2}|\xi|^2 + V(x) : T^*M \to \mathbb{R}$$

is the total Newtonian kinetic + potential energy. The idea of Schrödinger is to model the electron by a wave function φ_j which solves the eigenvalue problem

$$\hat{H}\varphi_j := (-\frac{\hbar^2}{2}\Delta + V)\varphi_j = E_j(\hbar)\varphi_j, \tag{11}$$

for the Schrödinger operator $\hat{H}$, where V is the potential, a multiplication operator on $L^2(\mathbb{R}^3)$. Here $\hbar$ is Planck's constant, a very small constant. The semi-classical limit $\hbar \to 0$ is mathematically equivalent to the high frequency limit when $V = 0$. The time evolution of an 'energy state' is given by

$$U_\hbar(t)\varphi_j := e^{-i\frac{t}{\hbar}(-\frac{\hbar^2}{2}\Delta + V)}\varphi_j = e^{-i\frac{tE_j(\hbar)}{\hbar}}\varphi_j. \tag{12}$$

The unitary oprator $U_\hbar(t)$ is often called the *propagator*. In the Riemannian case with $V = 0$, the factors of $\hbar$ can be absorbed in the t variable and it suffices to study

$$U(t) = e^{it\sqrt{\Delta}}. \tag{13}$$

An L^2-normalized energy state φ_j defines a probability amplitude, i.e. its modulus square is a probability measure with

$$|\varphi_j(x)|^2 dx = \text{ the probability density of finding the particle at } x\ . \tag{14}$$

According to the physicists, the *observable quantities* associated to the energy state are the probability density (14) and 'more generally' the matrix elements

$$\langle A\varphi_j, \varphi_j\rangle = \int \varphi_j(x) A\varphi_j(x) dV \tag{15}$$

of observables (A is a self adjoint operator, and in these lectures it is assumed to be a pseudo-differential operator). Under the time evolution (12), the factors of $e^{-i\frac{tE_j(\hbar)}{\hbar}}$ cancel and so the particle evolves as if "stationary", i.e. observations of the particle are independent of the time t.

Modeling energy states by eigenfunctions (11) resolves the paradox [1] of particles which are simultaneously in motion and are stationary, but at the cost of replacing the classical model of particles following the trajectories of Hamilton's equations by 'linear algebra', i.e. evolution by (12). The quantum picture is difficult to visualize or understand intuitively. Moreover, it is difficult to relate the classical picture of orbits with the quantum picture of eigenfunctions.

The study of nodal sets was historically motivated in part by the desire to visualize energy states by finding the points where the quantum particle is least likely to be. In fact, just recently (at this time or writing) the nodal sets of the hydrogem atom energy states have become visible to microscopes [**St**].

2. Results

We now introduce the results whose proofs we sketch in the later sections of this article.

2.1. Nodal hypersurface volumes for C^∞ metrics

In the late $70's$, S. T. Yau conjectured that for general C^∞ (M, g) of any dimension m there exist c_1, C_2 depending only on g so that

$$\lambda \lesssim \mathcal{H}^{m-1}(\mathcal{N}_{\varphi_\lambda}) \lesssim \lambda. \tag{16}$$

Here and below $\lesssim$ means that there exists a constant C independent of λ for which the inequality holds. The upper bound of (16) is the analogue for eigenfunctions of the fact that the hypersurface volume of a real algebraic variety is bounded above by its degree. The lower bound is specific to eigenfunctions. It is a strong version of the statement that 0 is not an "exceptional value" of φ_λ. Indeed, a basic result is the following classical result, apparently due to R. Courant (see [**Br**])). It is used to obtain lower bounds on volumes of nodal sets:

PROPOSITION 1. *For any (M, g) there exists a constant $A > 0$ so that every ball of (M, g) od radius greater than $\frac{A}{\lambda}$ contains a nodal point of any eigenfunction φ_λ.*

We sketch the proof in §3.2 for completeness, but leave some of the proof as an exercise to the reader.

The lower bound of (16) was proved for all C^∞ metrics for surfaces, i.e. for $m = 2$ by Brüning [**Br**]. For general C^∞ metrics in dimensions ≥ 3, the known upper and lower bounds are far from the conjecture (16). At present the best lower bound available for general C^∞ metrics of all dimensions is the following estimate of Colding-Minicozzi [**CM**]; a somewhat weaker bound was proved by Sogge-Zelditch

[1] This is an over-simplified account of the stability problem; see [**LS**] for an in-depth account

[**SoZ**] and the later simplification of the proof [**SoZa**] turned out to give the same bound as [**CM**]. We sketch the proof from [**SoZa**].

THEOREM 2.

$$\lambda^{1-\frac{n-1}{2}} \lesssim \mathcal{H}^{m-1}\mathcal{N}_\lambda),$$

The original result of [**SoZ**] is based on lower bounds on the L^1 norm of eigenfunctions. Further work of Hezari-Sogge [**HS**] shows that the Yau lower bound is correct when one has $||\varphi_\lambda||_{L^1} \geq C_0$ for some $C_0 > 0$. It is not known for which (M, g) such an estimate is valid. At the present time, such lower bounds are obtained from upper bounds on the L^4 norm of φ_λ. The study of L^p norms of eigenfunctions is of independent interest and we discuss some recent results which are not directly related to nodal sets In §12 and in §2.9. The study of L^p norms splits into two very different cases: there exists a critical index p_n depending on the dimension of M, and for $p \geq p_n$ the L^p norms of eigenfunctions are closely related to the structure of geodesic loops (see §2.9). For $2 \leq p \leq p_n$ the L^p norms are governed by different geodesic properties of (M, g) which we discuss in §12.

We also recall in §4.4 an interesting upper bound due to R. T. Dong (and Donnelly-Fefferman) in dimension 2, since the techniques of proof of [**Dong**] seem capable of further development.

THEOREM 3. *For C^∞ (M, g) of dimension 2,*

$$\mathcal{H}^1(\mathcal{N}_\lambda) \lesssim \lambda^{3/2}.$$

2.2. Nodal hypersurface volumes for real analytic (M, g)

In 1988, Donnelly-Fefferman [**DF**] proved the conjectured bounds for real analytic Riemannian manifolds (possibly with boundary). We re-state the result as the following

THEOREM 4. *Let (M, g) be a compact real analytic Riemannian manifold, with or without boundary. Then*

$$c_1\lambda \lesssim \mathcal{H}^{m-1}(Z_{\varphi_\lambda}) \lesssim \lambda.$$

See also [**Ba**] for a similar proof that the $\mathcal{H}^{m-1}$ measure of the critical set is $\simeq \lambda$ in the real analytic case.

2.3. Number of intersections of nodal sets with geodesics and number of nodal domains

The Courant nodal domain theorem (see e.g. [**Ch, Ch3**]) asserts that the nth eigenfunction φ_{λ_n} has $\leq n$ nodal domains. This estimate is not sharp (see [**Po**] and §11.2 for recent results) and it is possible to find sequences of eigenfunctions with $\lambda_n \to \infty$ and with a bounded number of nodal domains [**L**]. In fact, it has been pointed out [**Hof**] that we do not even know if a given (M, g) possesses any sequence of eignefunctions φ_n with $\lambda_n \to \infty$ for which the number of nodal domains tends to infinity. A nodal domain always contains a local minimum or maximum, so a necessary condition that the number of nodal domains increases to infinity along sequence of eigenfunctions is that the number of their critical points (or values) also increases to infinity; see [**JN**] for a sequence in which the number of critical points is uniformly bounded.

Some new results give lower bounds on the number of nodal domains on surfaces with an orientation reversing isometry with non-empty fixed point set. The first result is due to Ghosh-Reznikov-Sarnak [**GRS**].

THEOREM 2.1. *Let φ be an even Maass-Hecke L^2 eigenfunction on $\mathbb{X} = SL(2,\mathbb{Z})\backslash\mathbb{H}$. Denote the nodal domains which intersect a compact geodesic segment $\beta \subset \delta = \{iy \mid y > 0\}$ by $N^\beta(\varphi)$. Assume β is sufficiently long and assume the Lindelof Hypothesis for the Maass-Hecke L-functions. Then*

$$N^\beta(\varphi) \gg_\epsilon \lambda_\varphi^{\frac{1}{24}-\epsilon}.$$

We refer to [**GRS**] for background definitions. The strategy of the proof is to first prove that there are many intersections of the nodal set with the vertical geodesic of the modular domain and that the eigenfunction changes sign at many intersections. It follows that the nodal lines intersect the geodesic orthogonally. Using a topological argument, the authors show that the nodal lines must often close up to bound nodal domains.

As is seen from this outline, the main analytic ingredient is to prove that there are many intersections of the nodal line with the geodesic. The study of such intersections in a more general context is closely related to a new series of quantum ergodicity results known as QER (quantum ergodic restriction) theorems [**TZ, TZ2, CTZ**]. We discuss these ingredients below.

In recent work, the author and J. Jung have proved a general kind result in the same direction. The setting is that of a Riemann surface (M, J, σ) with an orienting-reversing involution σ whose fixed point set $\text{Fix}(\sigma)$ is separating, i.e. $M\backslash\text{Fix}(\sigma)$ consists of two connected components. The result is that for any σ-invariant negatively curved metric, and for almost the entire sequence of even or odd eigenfunctions, the number of nodal domains tends to infinity. In fact, the argument only uses ergodicity of the geodesic flow.

. We first explain the hypothesis. When a Riemann surface possesses an orientation-reversing involution $\sigma : M \to M$, Harnack's theorem says that the fixed point set $\text{Fix}(\sigma)$ is a disjoint union

$$H = \gamma_1 \cup \cdots \cup \gamma_k \tag{17}$$

of $0 \leq k \leq g+1$ of simple closed curves. It is possible that $\text{Fix}(\sigma) = \emptyset$ but we assume $k \neq 0$. We also assume that H (17) is a separating set, i.e. $M\backslash H = M_+ \cup M_-$ where $M_+^0 \cap M_-^0 = \emptyset$ (the interiors are disjoint), where $\sigma(M_+) = M_-$ and where $\partial M_+ = \partial M_- = H$.

If $\sigma^* g = g$ where g is a negatively curved metric, then $\text{Fix}(\sigma)$ is a finite union of simple closed geodesics. We denote by $L^2_{even}(M)$ the set of $f \in L^2(M)$ such that $\sigma f = f$ and by $L^2_{odd}(Y)$ the f such that $\sigma f = -f$. We denote by $\{\varphi_j\}$ an orthonormal eigenbasis of Laplace eigenfunctions of $L^2_{even}(M)$, resp. $\{\psi_j\}$ for $L^2_{odd}(M)$.

We further denote by

$$\Sigma_{\varphi_\lambda} = \{x \in \mathcal{N}_{\varphi_\lambda} : d\varphi_\lambda(x) = 0\}$$

the singular set of φ_λ. These are special critical points $d\varphi_j(x) = 0$ which lie on the nodal set Z_{φ_j}.

THEOREM 2.1. *Let (M, g) be a compact negatively curved C^∞ surface with an orientation-reversing isometric involution $\sigma : M \to M$ with $Fix(\sigma)$ separating. Then for any orthonormal eigenbasis $\{\varphi_j\}$ of $L^2_{even}(Y)$, resp. $\{\psi_j\}$ of $L^2_{odd}(M)$,* *one can find a density* 1 *subset* A *of* $\mathbb{N}$ *such that*

$$\lim_{\substack{j \to \infty \\ j \in A}} N(\varphi_j) = \infty,$$

resp.

$$\lim_{\substack{j \to \infty \\ j \in A}} N(\psi_j) = \infty,$$

For odd eigenfunctions, the conclusion holds as long as $Fix(\sigma) \neq \emptyset$. A density one subset $A \subset \mathbf{N}$ is one for which

$$\frac{1}{N} \#\{j \in A, j \leq N\} \to 1, \quad N \to \infty.$$

As the image indicates, the surfaces in question are complexifications of real algebraic curves, with $Fix(\sigma)$ the underlying real curve.

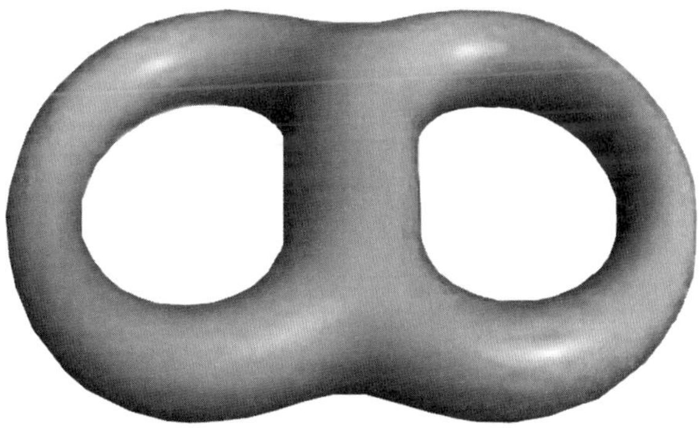

The strategy of the proof is similar to that of Theorem 2.1: we prove that for even or odd eigenfunctions, the nodal sets intersect $\mathrm{Fix}(\sigma)$ in many points with sign changes, and then use a topological argument to conclude that there are many nodal domains. We note that the study of sign changes has been used in [**NPS**] to study nodal sets on surfaces in a different way.

In §2.8 we also discuss upper bounds on nodal intersections in the real analytic case.

2.4. Dynamics of the geodesic or billiard flow

Theorem 2.1 used the hypothesis of ergodicity of the geodesic flow. It is not obvious that nodal sets of eigenfunctions should bear any relation to geodesics, but one of our central themes is that in some ways they do.

In general, there are two broad classes of results on nodal sets and other properties of eigenfunctions:

- Local results which are valid for any local solution of (1), and which often use local arguments. For instance the proof of Proposition 1 is local.
- Global results which use that eigenfunctions are global solutions of (1), or that they satisfy boundary conditions when $\partial M \neq \emptyset$. Thus, they are also satisfy the unitary evolution equation (12). For instance the relation between closed geodesics and the remainder term of Weyl's law is global (5)-(33).

Global results often exploit the relation between classical and quantum mechanics, i.e. the relation between the eigenvalue problem (1)-(12) and the geodesic flow. Thus the results often depend on the dynamical properties of the geodesic flow. The relations between eigenfunctions and the Hamiltonian flow are best established in two extreme cases: (i) where the Hamiltonian flow is completely integrable on an energy surface, or (ii) where it is ergodic. The extremes are illustrated below in the case of (i) billiards on rotationally invariant annulus, (ii) chaotic billiards on a cardioid.

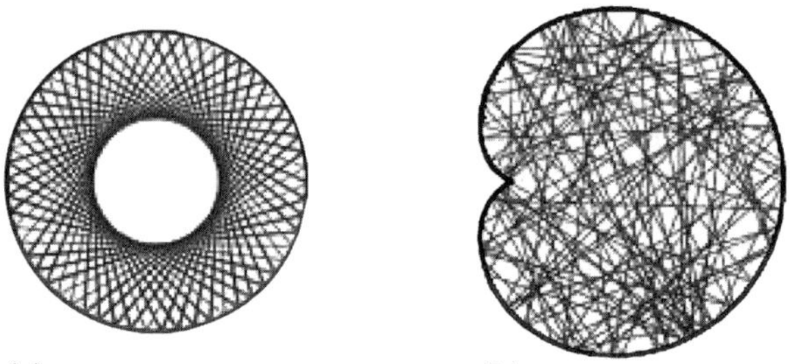

(a) (b)

A random trajectory in the case of ergodic billiards is uniformly distributed, while all trajectories are quasi-periodic in the integrable case.

We do not have the space to review the dynamics of geodesic flows or other Hamiltonian flows. We refer to [**HK**] for background in dynamics and to [**Ze, Ze3, Zw**] for relations between dynamics of geodesic flows and eigenfunctions.

We use the following basic construction: given a measure preserving map (or flow) $\Phi : (X, \mu) \to (T, \mu)$ one can consider the translation operator

$$U_\Phi f(x) = \Phi^* f(x) = f(\Phi(x)), \tag{18}$$

sometimes called the Koopman operator or Perron-Frobenius operator (cf. [**RS, HK**]). It is a unitary operator on $L^2(X, \mu)$ and hence its spectrum lies on the unit circle. Φ is ergodic if and only if U_Φ has the eigenvalue 1 with multiplicity 1, corresponding to the constant functions.

The geodesic (or billiard) flow is the Hamiltonian flow on T^*M generated by the metric norm Hamiltonian or its square,

$$H(x, \xi) = |\xi|_g^2 = \sum_{i,j} g^{ij} \xi_i \xi_j. \tag{19}$$

In PDE one most often uses the $\sqrt{H}$ which is homogeneous of degree 1. The geodesic flow is ergodic when the Hamiltonian flow Φ^t is ergodic on the level set $S^*M = \{H = 1\}$.

2.5. Quantum ergodic restriction theorems and nodal intersections

One of the main themes of these lectures is that ergodicity of the geodesic flow causes eigenfunctions to oscillate rapidly everywhere and in all directions, and hence to have a 'maximal' zero set. We will see this occur both in the real and complex domain. In the real domain (i.e. on M), ergodicity ensures that restrictions of eigenfunctions in the two-dimensional case to geodesics have many zeros along the geodesic. In §5 we will show that such oscillations and zeros are due to the fact that under generic assumptions, restrictions of eigenfunctions to geodesics are 'quantum ergodic' along the geodesic. Roughly this means that they have uniform oscillations at all frequencies below the frequency of the eigenfunction.

We prove that this QER (quantum ergodic restriction) property has the following implications. First, any arc $\beta \subset H$,

$$|\int_\beta \varphi_{\lambda_j} ds| \leq C\lambda_j^{-1/2}(\log \lambda)^{\frac{1}{2}} \tag{20}$$

and

$$\int_\beta |\varphi_{\lambda_j}| ds \geq ||\varphi_{\lambda_j}||_\infty^{-1} ||\varphi_{\lambda_j}||^2_{L^2(\beta)} \geq C\lambda_j^{-1/2} \log \lambda_j. \tag{21}$$

The first inequality is generic while the second uses the QER property. The inequalities are inconsistent if $\varphi_{\lambda_j} \geq 0$ on β, and that shows that eigenfunctions have many sign changes along the geodesic. The estimate also the well-known sup norm bound

$$||\varphi_\lambda||_\infty \leq \lambda^{1/4}/\log \lambda \tag{22}$$

for eigenfunctions on negatively curved surfaces [**Be**].

The same argument shows that the number of singular points of odd eigenfunctions tends to infinity and one can adapt it to prove that the number of critical points of even eigefunctions (on the geodesic) tend to infinity.

2.6. Complexification of M and Grauert tubes

The next series of results concerns 'complex nodal sets', i.e. complex zeros of analytic continuations of eigenfunctions to the complexification of M. It is difficult to draw conclusions about real nodal sets from knowledge of their complexifications. But complex nodal sets are simpler to study than real nodal sets and the results are stronger, just as complex algebraic varieties behave in simpler ways than real algebraic varieties.

A real analytic manifold M always possesses a unique complexification $M_\mathbb{C}$ generalizing the complexification of $\mathbb{R}^m$ as $\mathbb{C}^m$. The complexification is an open complex manifold in which M embeds $\iota : M \to M_\mathbb{C}$ as a totally real submanifold (Bruhat-Whitney)

The Riemannian metric determines a special kind of distance function on $M_\mathbb{C}$ known as a Grauert tube function. In fact, it is observed in [**GS1**] that the Grauert tube function is obtained from the distance function by setting $\sqrt{\rho}(\zeta) = i\sqrt{r^2(\zeta, \bar{\zeta})}$ where $r^2(x, y)$ is the squared distance function in a neighborhood of the diagonal in $M \times M$.

One defines the Grauert tubes $M_\tau = \{\zeta \in M_\mathbb{C} : \sqrt{\rho}(\zeta) \leq \tau\}$. There exists a maximal τ_0 for which $\sqrt{\rho}$ is well defined, known as the Grauert tube radius. For $\tau \leq \tau_0$, M_τ is a strictly pseudo-convex domain in $M_\mathbb{C}$. Since (M, g) is real analytic, the exponential map $\exp_x t\xi$ admits an analytic continuation in t and the imaginary time exponential map

$$E : B^*_\epsilon M \to M_\mathbb{C}, \quad E(x, \xi) = \exp_x i\xi \tag{23}$$

is, for small enough ϵ, a diffeomorphism from the ball bundle $B^*_\epsilon M$ of radius ϵ in T^*M to the Grauert tube M_ϵ in $M_\mathbb{C}$. We have $E^*\omega = \omega_{T^*M}$ where $\omega = i\partial\bar{\partial}\rho$ and where ω_{T^*M} is the canonical symplectic form; and also $E^*\sqrt{\rho} = |\xi|$ [**GS1, LS1**].

It follows that E^* conjugates the geodesic flow on B^*M to the Hamiltonian flow $\exp tH_{\sqrt{\rho}}$ of $\sqrt{\rho}$ with respect to ω, i.e.

$$E(g^t(x,\xi)) = \exp t\Xi_{\sqrt{\rho}}(\exp_x i\xi). \tag{24}$$

In general E only extends as a diffemorphism to a certain maximal radius $\epsilon_{\max}$. We assume throughout that $\epsilon < \epsilon_{\max}$.

2.7. Equidistribution of nodal sets in the complex domain

One may also consider the complex nodal sets

$$\mathcal{N}_{\varphi_j^{\mathbb{C}}} = \{\zeta \in M_\epsilon : \varphi_j^{\mathbb{C}}(\zeta) = 0\}, \tag{25}$$

and the complex critical point sets

$$\mathcal{C}_{\varphi_j^{\mathbb{C}}} = \{\zeta \in M_\epsilon : \partial\varphi_j^{\mathbb{C}}(\zeta) = 0\}. \tag{26}$$

The following is proved in [**Ze5**]:

THEOREM 5. *Assume (M,g) is real analytic and that the geodesic flow of (M,g) is ergodic. Then for all but a sparse subsequence of λ_j,*

$$\frac{1}{\lambda_j}\int_{\mathcal{N}_{\varphi_{\lambda_j}^{\mathbb{C}}}} f\omega_g^{n-1} \to \frac{i}{\pi}\int_{M_\epsilon} f\bar{\partial}\partial\sqrt{\rho}\wedge\omega_g^{n-1}.$$

The proof is based on quantum ergodicity of analytic continuation of eigenfunctions to Grauert tubes and the growth estimates ergodic eigenfunctions satisfy.

We will say that a sequence $\{\varphi_{j_k}\}$ of L^2-normalized eigenfunctions is *quantum ergodic* if

$$\langle A\varphi_{j_k}, \varphi_{j_k}\rangle \to \frac{1}{\mu(S^*M)}\int_{S^*M}\sigma_A d\mu, \quad \forall A \in \Psi^0(M). \tag{27}$$

Here, $\Psi^s(M)$ denotes the space of pseudodifferential operators of order s, and $d\mu$ denotes Liouville measure on the unit cosphere bundle S^*M of (M,g). More generally, we denote by $d\mu_r$ the (surface) Liouville measure on ∂B_r^*M, defined by

$$d\mu_r = \frac{\omega^m}{d|\xi|_g} \quad \text{on } \partial B_r^*M. \tag{28}$$

We also denote by α the canonical action 1-form of T^*M.

2.8. Intersection of nodal sets and real analytic curves on surfaces

In recent work, intersections of nodal sets and curves on surfaces M^2 have been used in a variety of articles to obtain upper and lower bounds on nodal points and domains. The work often is based on restriction theorems for eigenfunctions. Some of the recent articles on restriction theorems and/or nodal intersections are [**TZ, TZ2, GRS, JJ, JJ2, Mar, Yo, Po**].

First we consider a basic upper bound on the number of intersection points:

THEOREM 6. *Let $\Omega \subset \mathbb{R}^2$ be a piecewise analytic domain and let $n_{\partial\Omega}(\lambda_j)$ be the number of components of the nodal set of the jth Neumann or Dirichlet eigenfunction which intersect $\partial\Omega$. Then there exists C_Ω such that $n_{\partial\Omega}(\lambda_j) \le C_\Omega\lambda_j$.*

In the Dirichlet case, we delete the boundary when considering components of the nodal set.

The method of proof of Theorem 6 generalizes from $\partial\Omega$ to a rather large class of real analytic curves $C \subset \Omega$, even when $\partial\Omega$ is not real analytic. Let us call a real analytic curve C a *good* curve if there exists a constant $a > 0$ so that for all λ_j sufficiently large,

$$\frac{\|\varphi_{\lambda_j}\|_{L^2(\partial\Omega)}}{\|\varphi_{\lambda_j}\|_{L^2(C)}} \leq e^{a\lambda_j}. \tag{29}$$

Here, the L^2 norms refer to the restrictions of the eigenfunction to C and to $\partial\Omega$. The following result deals with the case where $C \subset \partial\Omega$ is an *interior* real-analytic curve. The real curve C may then be holomorphically continued to a complex curve $C_{\mathbb{C}} \subset \mathbb{C}^2$ obtained by analytically continuing a real analytic parametrization of C.

THEOREM 7. *Suppose that $\Omega \subset \mathbb{R}^2$ is a C^∞ plane domain, and let $C \subset \Omega$ be a good interior real analytic curve in the sense of (29). Let $n(\lambda_j, C) = \#Z_{\varphi_{\lambda_j}} \cap C$ be the number of intersection points of the nodal set of the j-th Neumann (or Dirichlet) eigenfunction with C. Then there exists $A_{C,\Omega} > 0$ depending only on C, Ω such that $n(\lambda_j, C) \leq A_{C,\Omega}\lambda_j$.*

The proof of Theorem 7 is somewhat simpler than that of Theorem 6, i.e. good interior analytic curves are somewhat simpler than the boundary itself. On the other hand, it is clear that the boundary is good and it is hard to prove that other curves are good. A recent paper of J. Jung shows that many natural curves in the hyperbolic plane are 'good' [**JJ**]. See also [**ElHajT**] for general results on good curves.

The upper bounds of Theorem 6 - 7 are proved by analytically continuing the restricted eigenfunction to the analytic continuation of the curve. We then give a similar upper bound on complex zeros. Since real zeros are also complex zeros, we then get an upper bound on complex zeros. An obvious question is whether the order of magnitude estimate is sharp. Simple examples in the unit disc show that there are no non-trivial lower bounds on numbers of intersection points. But when the dynamics is ergodic we expect to prove an equi-distribution theorem for nodal intersection points (in progress). Ergodicity once again implies that eigenfunctions oscillate as much as possible and therefore saturate bounds on zeros.

Let $\gamma \subset M^2$ be a generic geodesic arc on a real analytic Riemannian surface. For small ϵ, the parametrization of γ may be analytically continued to a strip,

$$\gamma_{\mathbb{C}} : S_\tau := \{t + i\tau \in \mathbb{C} : |\tau| \leq \epsilon\} \to M_\tau.$$

Then the eigenfunction restricted to γ is

$$\gamma_{\mathbb{C}}^* \varphi_j^{\mathbb{C}}(t + i\tau) = \varphi_j(\gamma_{\mathbb{C}}(t + i\tau) \text{ on } S_\tau.$$

Let

$$\mathcal{N}_{\lambda_j}^\gamma := \{(t + i\tau : \gamma_H^* \varphi_{\lambda_j}^{\mathbb{C}}(t + i\tau) = 0\} \tag{30}$$

be the complex zero set of this holomorphic function of one complex variable. Its zeros are the intersection points.

Then as a current of integration,

$$[\mathcal{N}_{\lambda_j}^\gamma] = i\partial\bar{\partial}_{t+i\tau} \log \left|\gamma^* \varphi_{\lambda_j}^{\mathbb{C}}(t + i\tau)\right|^2. \tag{31}$$

The following result is proved in [**Ze6**]:

THEOREM 8. *Let (M, g) be real analytic with ergodic geodesic flow. Then for generic γ there exists a subsequence of eigenvalues λ_{j_k} of density one such that*

$$\frac{i}{\pi \lambda_{j_k}} \partial\bar{\partial}_{t+i\tau} \log \left|\gamma^* \varphi^{\mathbb{C}}_{\lambda_{j_k}}(t+i\tau)\right|^2 \to \delta_{\tau=0} ds.$$

Thus, intersections of (complexified) nodal sets and geodesics concentrate in the real domain– and are distributed by arc-length measure on the real geodesic.

The key point is that

$$\frac{1}{\lambda_{j_k}} \log |\varphi^{\mathbb{C}}_{\lambda_{j_k}}(\gamma(t+i\tau)|^2 \to |\tau|.$$

Thus, the maximal growth occurs along individual (generic) geodesics.

2.9. L^p norms of eigenfunctions

In §2.1 we mentioned that lower bounds on $\mathcal{H}^{n-1}(\mathcal{N}_{\varphi_\lambda}$ are related to lower bounds on $||\varphi_\lambda||_{L^1}$ and to upper bounds on $||\varphi_\lambda||_{L^p}$ for certain p. Such L^p bounds are interesting for all p and depend on the shapes of the eigenfunctions.

In (5) we stated the Weyl law on the number of eigenvalues. There also exists a pointwise local Weyl law which is relevant to the pointwise behavior of eigenfunctions. The pointwise spectral function along the diagonal is defined by

$$E(\lambda, x, x) = N(\lambda, x) := \sum_{\lambda_j \leq \lambda} |\varphi_j(x)|^2. \tag{32}$$

The pointwise Weyl law asserts tht

$$N(\lambda, x) = \frac{1}{(2\pi)^n} |B^n| \lambda^n + R(\lambda, x), \tag{33}$$

where $R(\lambda, x) = O(\lambda^{n-1})$ uniformly in x. These results are proved by studying the cosine transform

$$E(t, x, x) = \sum_{\lambda_j \leq \lambda} \cos t\lambda_j \ |\varphi_j(x)|^2, \tag{34}$$

which is the fundamental (even) solution of the wave equation restricted to the diagonal. Background on the wave equation is given in §14.

We note that the Weyl asymptote $\frac{1}{(2\pi)^n}|B^n|\lambda^n$ is continuous, while the spectral function (32) is piecewise constant with jumps at the eigenvalues λ_j. Hence the remainder must jump at an eigenvalue λ, i.e.

$$R(\lambda, x) - R(\lambda - 0, x) = \sum_{j : \lambda_j = \lambda} |\varphi_j(x)|^2 = O(\lambda^{n-1}). \tag{35}$$

on any compact Riemannian manifold. It follows immediately that

$$\sup_M |\varphi_j| \lesssim \lambda_j^{\frac{n-1}{2}}. \tag{36}$$

There exist (M, g) for which this estimate is sharp, such as the standard spheres. However, as (22) the sup norms are smaller on manifolds of negative curvature. In fact, (36) is very rarely sharp and the actual size of the sup-norms and other L^p norms of eigenfunctions is another interesting problem in global harmonic analysis.

In [**SoZ**] it is proved that if the bound (36) is achieved by some sequence of eigenfunctions, then there must exist a "partial blow-down point" or self-focal point p where a positive measure of directions $\omega \in S_p^*M$ so that the geodesic with initial value (p, ω) returns to p at some time $T(p, \omega)$. Recently the authors have improved the result in the real analytic case, and we sketch the new result in §12.

To state it, we need some further notation and terminology. We only consider real analytic metrics for the sake of simplicity. We call a point p a *self-focal point* or a*blow-down point* if there exists a time $T(p)$ so that $\exp_p T(p)\omega = p$ for all $\omega \in S_p^*M$. Such a point is self-conjugate in a very strong sense. In terms of symplectic geometry, the flowout manifold

$$\Lambda_p = \bigcup_{0 \leq t \leq T(p)} G^t S_p^* M \tag{37}$$

is an embedded Lagrangian submanifold of S^*M whose projection

$$\pi : \Lambda_p \to M$$

has a "blow-down singuarity" at $t = 0, t = T(p)$ (see[**STZ**]). Focal points come in two basic kinds, depending on the first return map

$$\Phi_p : S_p^*M \to S_p^*M, \quad \Phi_p(\xi) := \gamma'_{p,\xi}(T(p)), \tag{38}$$

where $\gamma_{p,\xi}$ is the geodesic defined by the initial data $(p, \xi) \in S_x^*M$. We say that p is a *pole* if

$$\Phi_p = Id : S_p^*M \to S_p^*M.$$

On the other hand, it is possible that $\Phi_p = Id$ only on a codimension one set in S_p^*M. We call such a Φ_p *twisted.*

Examples of poles are the poles of a surface of revolution (in which case all geodesic loops at x_0 are smoothly closed). Examples of self-focal points with fully twisted return map are the four umbilic points of two-dimensional tri-axial ellipsoids, from which all geodesics loop back at time 2π but are almost never smoothly closed. The only smoothly closed directions are the geodesic (and its time reversal) defined by the middle length 'equator'.

At a self-focal point we have a kind of analogue of (18) but not on S^*M but just on S_p^*M. We define the Perron-Frobenius operator at a self-focal point by

$$U_x : L^2(S_x^*M, d\mu_x) \to L^2(S_x^*M, d\mu_x), \quad U_x f(\xi) := \; f(\Phi_x(\xi))\sqrt{J_x(\xi)}. \tag{39}$$

Here, J_x is the Jacobian of the map Φ_x, i.e. $\Phi_x^*|d\xi| = J_x(\xi)|d\xi|$.

The new result of C.D. Sogge and the author is the following:

THEOREM 9. *If (M, g) is real analytic and has maximal eigenfunction growth, then it possesses a self-focal point whose first return map Φ_p has an invariant L^2 function in $L^2(S_p^*M)$. Equivalently, it has an L^1 invariant measure in the class of the Euclidean volume density μ_p on S_p^*M.*

For instance, the twisted first return map at an umbilic point of an ellipsoid has no such finite invariant measure. Rather it has two fixed points, one of which is a source and one a sink, and the only finite invariant measures are delta-functions at the fixed points. It also has an infinite invariant measure on the complement of the fixed points, similar to $\frac{dx}{x}$ on $\mathbb{R}_+$.

The results of [**SoZ, STZ, SoZ2**] are stated for the L^∞ norm but the same results are true for L^p norms above a critical index p_m depending on the dimension (§12). The analogous problem for lower L^p norms is of equal interest, but the geometry of the extremals changes from analogues of zonal harmonics to analogoues of Gaussian beams or highest weight harmonics. For the lower L^p norms there are also several new developments which are discussed in §12.

2.10. Quasi-modes

A significant generalization of eigenfunctions are *quasi-modes*, which are special kinds of approximate solutions of the eigenvalue problem. The two basic types are:

- (i) Lagrangian distributions given by oscillatory integrals which are approximate solutions of the eigenvalue problem, i.e. which satisfy $||(\Delta + \mu_k^2)u_k|| = O(\mu_k^{-p})$ for some $p \geq 0$.
- (ii) Any sequence $\{u_k\}_{k=1}^\infty$ of L^2 normalized solutions of an approximate eigenvalue problem. In [**STZ**] we worked with a more general class of "admissible" quasi-modes. A sequence $\{\psi_\lambda\}$, $\lambda = \lambda_j$, $j = 1, 2, \ldots$ is a sequence of admissible quasimodes if $\|\psi_\lambda\|_2 = 1$ and

$$\|(\Delta + \lambda^2)\psi_\lambda\|_2 + \|S_{2\lambda}^\perp \psi_\lambda\|_\infty = o(\lambda). \tag{40}$$

Here, $S_\mu^\perp$ denotes the projection onto the $[\mu, \infty)$ part of the spectrum of $\sqrt{-\Delta}$, and in what follows $S_\mu = I - S_\mu^\perp$, i.e., $S_\mu f = \sum_{\lambda_j < \mu} e_j(f)$, where $e_j(f)$ is the projection of f onto the eigenspace with eigenvalue λ_j.

The more special type (i) are studied in [**Arn, BB, CV2, K, R1, R2**] and are analogous to Hadamard parametrices for the wave kernel. However, none may exist on a given (M, g). For instance, Gaussian beams [**BB, R1, R2**] are special quasi-modes which concentrate along a closed geodesic. On the standard S^n they exist along any closed geodesic and are eigenfunctions. But one needs a stable elliptic closed geodesic to construct a Gaussian beam and such closed geodesics do not exist when g has negative curvature. They do however exist in many cases, but the associated Gaussian beams are rarely exact eigenfunctions, and are typically just approximate eigenfunctions. More general quasi-modes are constructed from geodesic flow-invariant closed immersed Lagrangian submanifolds $\Lambda \subset S_g^*M$. But such invariant Lagrangian submanifolds are also rare. They do arise in many interesting cases and in particular the unit co-sphere bundle S_g^*M is foliated by flow-invariant Lagrangian tori when the geodesic flow is integrable. this is essentially the definition of complete integrability.

For $p = 1$, the more general type (ii) always exist: it suffices to define u_k as superpositions of eigenfunctions with frequencies λ_j drawn from sufficiently narrow windows $[\lambda, \lambda + O(\lambda^{-p})]$. A model case admissible quasi-modes would be a sequence of L^2-normalized functions $\{\psi_{\lambda_j}\}$ whose $\sqrt{-\Delta}$ spectrum lies in intervals of the form $[\lambda_j - o(1), \lambda_j + o(1)]$ as $\lambda_j \to \infty$. If $p = 1$ then for generic metrics g, it follows from the Weyl law with remainder that there exist $\simeq C_g \lambda^{m-1}$ eigenvalues in the window and one can construct many quasi-modes. For higher values of p existence of refined quasi-modes depends on the number of eigenvalues in very short intervals, and in general one only knows they exist by constructions of type (i).

An important example of such a quasi-mode is a sequence of "shrinking spectral projections", i.e. the L^2-normalized projection kernels

$$\Phi_j^z(x) = \frac{\chi_{[\lambda_j, \lambda_j + \epsilon_j]}(x, z)}{\sqrt{\chi_{[\lambda_j, \lambda_j + \epsilon_j]}(z, z)}}$$

with second point frozen at a point $z \in M$ and with width $\epsilon_j \to 0$. Here, $\chi_{[\lambda_j, \lambda_j + \epsilon_j]}(x, z)$ is the orthogonal projection onto the sum of the eigenspaces V_λ with $\lambda \in [\lambda_j, \lambda_j + \epsilon_j]$ The zonal eigenfunctions of a surface of revolution are examples of such shrinking spectral projections for a sufficiently small ϵ_j, and when z is a partial focus such $\Phi_j^z(x)$ are generalizations of zonal eigenfunctions. On a general Zoll manifold, shrinking spectral projections of widths $\epsilon_j = O(\lambda_j^{-1})$ are the direct analogues of zonal spherical harmonics, and they would satisfy the analog of (40) where $o(\lambda)$ is replaced by the much stronger $O(\lambda^{-1})$.

There are several motivations to consider quasi-modes as well as actual modes (eigenfunctions). First, many results about eigenfunctions automatically hold for quasi-modes as well. Indeed, it is difficult to distinguish modes and quasi-modes when using a wave kernel construction, and many of the methods apply equally to modes and quasi-modes. Second, quasi-modes are often geometrically beautiful. They are not stationary states but retain their shape under propagation by the wave group $U(t) = \exp it\sqrt{\Delta}$ for a "very long time" (essentially the Eherenfest time) before breaking up (see [**Ze2**] for background). They are natural extremals for L^p norms when they exist. Third, they are often the objects needed to close the gap between necessary and sufficient conditions on eigenfunction growth. It is usually interesting to know whether a theorem about eigenfunctions extends to quasi-modes as well.

2.11. Format of these lectures and references to the literature

In keeping with the format of the Park City summer school, we concentrate on the topics of the five lectures rather than give a systematic exposition of the subject. The more detailed account will appear in [**Ze0**]. Various details of the proof are given as Exercises for the reader. The "details" are intended to be stimulating and fundamental, rather than the tedious and routine aspects of proofs often left to readers in textbooks. As a result, the exercises vary widely in difficulty and amount of background assumed. Problems labelled **Problems** are not exercises; they are problems whose solutions are not currently known.

The technical backbone of the semi-classical analysis of eigenfunctions consists of wave equation methods combined with the machinery of Fourier integral operators and Pseudo-differential operators. We do not have time to review this theory. The main results we need are the construction of parametrices for the 'propagator' $E(t) = \cos t\sqrt{\Delta}$ and the Poisson kernel $\exp -\tau\sqrt{\Delta}$. We also need Fourier analysis to construct approximate spectral projections $\rho(\sqrt{\Delta} - \lambda)$ and to prove Tauberian theorems relating smooth expansions and cutoffs.

The books [**GSj, DSj, D2, GS2, GSt2, Sogb, Sogb2, Zw**] give textbook treatments of the semi-classical methods with applications to spectral asymptotics. Somewhat more classical background on the wave equation with many explicit formulae in model cases can be found in [**TI, TII**]. General spectral theory and the relevant functional analysis can also be found in [**RS**]. The series [**HoI, HoII, HoIII, HoIV**] gives a systematic presentation of Fourier integral operator theory:

stationary phase and Tauberian theorems can be found in [**HoI**], Weyl's law and spectral asymptotics can be found in [**HoIII, HoIV**].

In [**Ze0**] the author gives a more systematic presentation of results on nodal sets, L^p norms and other aspects of eigenfunctions. Earlier surveys [**Ze, Ze2, Ze3**] survey related material. Other monographs on Δ-eigenfunctions can be found in [**HL**] and [**Sogb2**]. The methods of [**HL**] mainly involve the local harmonic analysis of eigenfunctions and rely more on classical elliptic estimates, on frequency functions and of one-variable complex analysis. The exposition in [**Sogb2**] is close to the one given here but does not extend to the recent results that we highlight in these lectures and in [**Ze0**].

Acknowledgements Many of the results discussed in these lectures is joint work with C. D. Sogge and/or John. A. Toth. Some of the work in progress is also with B. Hanin and P. Zhou. We also thank E. Potash for his comments on earlier versions.

3. Foundational results on nodal sets

As mentioned above, the nodal domains of an eigenfunction are the connected components of $M\backslash\mathcal{N}_{\varphi_\lambda}$. In the case of a domain with boundary and Dirichlet boundary conditions, the nodal set is defined by taking the closure of the zero set in $M\backslash\partial M$.

The eigenfunction is either positive or negative in each nodal domain and changes sign as the nodal set is crossed from one domain to an adjacent domain. Thus the set of nodal domains can be given the structure of a bi-partite graph [**H**]. Since the eigenfunction has one sign in each nodal domain, it is the ground state eigenfunction with Dirichlet boundary conditions in each nodal domain.

In the case of domains $\Omega \subset \mathbb{R}^n$ (with the Euclidean metric), the Faber-Krahn inequality states that the lowest eigenvalue (ground state eigenvalue, bass note) $\lambda_1(\Omega)$ for the Dirichlet problem has the lower bound,

$$\lambda_1(\Omega) \geq |\Omega|^{-\frac{2}{n}}\, C_n^{\frac{2}{n}}\, j_{\frac{n-2}{2}}, \tag{41}$$

where $|\Omega$ is the Euclidean volume of Ω, $C_n = \frac{\pi^{\frac{n}{2}}}{\Gamma(\frac{n}{2}+1)}$ is the volume of the unit ball in $\mathbb{R}^n$ and where $j_{m,1}$ is the first positive zero of the Bessel function J_m. That is, among all domains of a fixed volume the unit ball has the lowest bass note.

3.1. Vanishing order and scaling near zeros

By the vanishing order $\nu(u, a)$ of u at a is meant the largest positive integer such that $D^\alpha u(a) = 0$ for all $|\alpha| \leq \nu$. A unique continuation theorem shows that the vanishing order of an eigenfunction at each zero is finite. The following estimate is a quantitative version of this fact.

THEOREM 3.1. *(see* [**DF**]*;* [**Lin**] *Proposition 1.2 and Corollary 1.4; and* [**H**] *Theorem 2.1.8.) Suppose that M is compact and of dimension n. Then there exist constants $C(n), C_2(n)$ depending only on the dimension such that the the vanishing order $\nu(u, a)$ of u at $a \in M$ satisfies $\nu(u, a) \leq C(n)\; N(0, 1) + C_2(n)$ for all $a \in B_{1/4}(0)$. In the case of a global eigenfunction, $\nu(\varphi_\lambda, a) \leq C(M, g)\lambda$.*

Highest weight spherical harmonics $C_n(x_1 + ix_2)^N$ on S^2 are examples which vanish at the maximal order of vanishing at the poles $x_1 = x_2 = 0, x_3 = \pm 1$.

The following Bers scaling rule extracts the leading term in the Taylor expansion of the eigenfunction around a zero:

PROPOSITION 3.2. [**Bers, HW2**] *Assume that φ_λ vanishes to order k at x_0. Let $\varphi_\lambda(x) = \varphi_k^{x_0}(x) + \varphi_{k+1}^{x_0} + \cdots$ denote the C^∞ Taylor expansion of φ_λ into homogeneous terms in normal coordinates x centered at x_0. Then $\varphi_k^{x_0}(x)$ is a Euclidean harmonic homogeneous polynomial of degree k.*

To prove this, one substitutes the homogeneous expansion into the equation $\Delta\varphi_\lambda = \lambda^2\varphi_\lambda$ and rescales $x \to \lambda x$, i.e. one applies the dilation operator

$$D_\lambda^{x_0}\varphi_\lambda(u) = \varphi(x_0 + \frac{u}{\lambda}). \tag{42}$$

The rescaled eigenfunction is an eigenfunction of the locally rescaled Laplacian

$$\Delta_\lambda^{x_0} := \lambda^{-2} D_\lambda^{x_0}\Delta_g (D_\lambda^{x_0})^{-1} = \sum_{j=1}^n \frac{\partial^2}{\partial u_j^2} + \cdots \tag{43}$$

in Riemannian normal coordinates u at x_0 but now with eigenvalue 1,

$$\begin{aligned} &D_\lambda^{x_0}\Delta_g(D_\lambda^{x_0})^{-1}\varphi(x_0 + \tfrac{u}{\lambda}) = \lambda^2\varphi(x_0 + \tfrac{u}{\lambda}) \\ &\implies \Delta_\lambda^{x_0}\varphi(x_0 + \tfrac{u}{\lambda}) = \varphi(x_0 + \tfrac{u}{\lambda}). \end{aligned} \tag{44}$$

Since $\varphi(x_0 + \frac{u}{\lambda})$ is, modulo lower order terms, an eigenfunction of a standard flat Laplacian on $\mathbb{R}^n$, it behaves near a zero as a sum of homogeneous Euclidean harmonic polynomials.

In dimension 2, a homogeneous harmonic polynomial of degree N is the real or imaginary part of the unique holomorphic homogeneous polynomial z^N of this degree, i.e. $p_N(r,\theta) = r^N \sin N\theta$. As observed in [**Ch**], there exists a C^1 local diffeormorphism χ in a disc around a zero x_0 so that $\chi(x_0) = 0$ and so that $\varphi_N^{x_0} \circ \chi = p_N$. It follows that the restriction of φ_λ to a curve H is C^1 equivalent around a zero to p_N restricted to $\chi(H)$. The nodal set of p_N around 0 consists of N rays, $\{r(\cos\theta, \sin\theta) : r > 0, p_N|_{S^1}(v) = 0\}$. It follows that the local structure of the nodal set in a small disc around a singular point p is C^1 equivalent to N equi-angular rays emanating from p. We refer to [**HW, HW2, Ch, Ch1, Ch2, Bes**] for futher background and results.

Question Is there any useful scaling behavior of φ_λ around its critical points?

3.2. Proof of Proposition 1

The proofs are based on rescaling the eigenvalue problem in small balls.

PROOF. Fix x_0, r and consider $B(x_0, r)$. If φ_λ has no zeros in $B(x_0, r)$, then $B(x_0, r) \subset D_{j;\lambda}$ must be contained in the interior of a nodal domain $D_{j;\lambda}$ of φ_λ. Now $\lambda^2 = \lambda_1^2(D_{j;\lambda})$ where $\lambda_1^2(D_{j;\lambda})$ is the smallest Dirichlet eigenvalue for the nodal domain. By domain monotonicity of the lowest Dirichlet eigenvalue (i.e. $\lambda_1(\Omega)$ decreases as Ω increases), $\lambda^2 \leq \lambda_1^2(D_{j;\lambda}) \leq \lambda_1^2(B(x_0, r))$. To complete the proof we show that $\lambda_1^2(B(x_0, r)) \leq \frac{C}{r^2}$ where C depends only on the metric. This is proved by comparing $\lambda_1^2(B(x_0, r))$ for the metric g with the lowest Dirichlet Eigenvalue $\lambda_1^2(B(x_0, cr); g_0)$ for the Euclidean ball $B(x_0, cr; g_0)$ centered at x_0 of radius cr with Euclidean metric g_0 equal to g with coefficients frozen at x_0; c is chosen so that $B(x_0, cr; g_0) \subset B(x_0, r, g)$. Again by domain monotonicity, $\lambda_1^2(B(x_0, r, g)) \leq$

$\lambda_1^2(B(x_0, cr; g))$ for $c < 1$. By comparing Rayleigh quotients $\frac{\int_\Omega |df|^2 dV_g}{\int_\Omega f^2 dV_g}$ one easily sees that $\lambda_1^2(B(x_0, cr; g)) \leq C\lambda_1^2(B(x_0, cr; g_0))$ for some C depending only on the metric. But by explicit calculation with Bessel functions, $\lambda_1^2(B(x_0, cr; g_0)) \leq \frac{C}{r^2}$. Thus, $\lambda^2 \leq \frac{C}{r^2}$.

□

For background we refer to [**Ch**].

3.3. A second proof

Another proof is given in [**HL**]: Let u_r denote the ground state Dirichlet eigenfunction for $B(x_0, r)$. Then $u_r > 0$ on the interior of $B(x_0, r)$. If $B(x_0, r) \subset D_{j;\lambda}$ then also $\varphi_\lambda > 0$ in $B(x_0, r)$. Hence the ratio $\frac{u_r}{\varphi_\lambda}$ is smooth and non-negative, vanishes only on $\partial B(x_0, r)$, and must have its maximum at a point y in the interior of $B(x_0, r)$. At this point (recalling that our Δ is minus the sum of squares),

$$\nabla\left(\frac{u_r}{\varphi_\lambda}\right)(y) = 0, \quad -\Delta\left(\frac{u_r}{\varphi_\lambda}\right)(y) \leq 0,$$

so at y,

$$0 \geq -\Delta\left(\frac{u_r}{\varphi_\lambda}\right) = -\frac{\varphi_\lambda \Delta u_r - u_r \Delta \varphi_\lambda}{\varphi_\lambda^2} = -\frac{(\lambda_1^2(B(x_0, r)) - \lambda^2)\varphi_\lambda u_r}{\varphi_\lambda^2}.$$

Since $\frac{\varphi_\lambda u_r}{\varphi_\lambda^2} > 0$, this is possible only if $\lambda_1(B(x_0, r)) \geq \lambda$.

To complete the proof we note that if $r = \frac{A}{\lambda}$ then the metric is essentially Euclidean. We rescale the ball by $x \to \lambda x$ (with coordinates centered at x_0) and then obtain an essentially Euclidean ball of radius r. Then $\lambda_1(B(x_0, \frac{r}{\lambda}) = \lambda\lambda_1 B_{g_0}(x_0, r)$. Therefore we only need to choose r so that $\lambda_1 B_{g_0}(x_0, r) = 1$.

Problem Are the above results true as well for quasi-modes of order zero (§2.10, §15.3)?

3.4. Rectifiability of the nodal set

We recall that the nodal set of an eigenfunction φ_λ is its zero set. When zero is a regular value of φ_λ the nodal set is a smooth hypersurface. This is a generic property of eigenfunctions [**U**]. It is pointed out in [**Bae**] that eigenfunctions can always be locally represented in the form

$$\varphi_\lambda(x) = v(x)\left(x_1^k + \sum_{j=0}^{k-1} x_1^j u_j(x')\right),$$

in suitable coordinates (x_1, x') near p, where φ_λ vanishes to order k at p, where $u_j(x')$ vanishes to order $k - j$ at $x' = 0$, and where $v(x) \neq 0$ in a ball around p. It follows that the nodal set is always countably $n-1$ rectifiable when $\dim M = n$.

4. Lower bounds for $\mathcal{H}^{m-1}(\mathcal{N}_\lambda)$ for C^∞ metrics

In this section we review the lower bounds on $\mathcal{H}^{n-1}(Z_{\varphi_\lambda})$ from [**CM, SoZ, SoZa, HS, HW**]. Here

$$\mathcal{H}^{n-1}(\mathcal{N}_{\varphi_\lambda}) = \int_{Z_{\varphi_\lambda}} dS$$

is the Riemannian surface measure, where dS denotes the Riemannian volume element on the nodal set, i.e. the insert $iota_n dV_g$ of the unit normal into the volume form of (M, g). The main result is:

THEOREM 4.1. *Let (M, g) be a C^∞ Riemannian manifold. Then there exists a constant C independent of λ such that*

$$C\lambda^{1-\frac{n-1}{2}} \leq \mathcal{H}^{n-1}(\mathcal{N}_{\varphi_\lambda}).$$

Remark: In a recent article [**BlSo**], M. Blair and C. Sogge improve this result on manifolds of non-positive curvature by showing that the right side divided by the left side tends to infinity. There exists a related proof using a comparison of diffusion processes in [**Stei**]. The result generalizes in a not completely straighforward way to Schrödinger operators $-\frac{\hbar^2}{2}\Delta_g + V$ for certain potentials V [**ZZh**] (see also [**Jin**] for generalizations of [**DF**] to Schrödinger operators). The new issue is the separation of the domain into classically allowed and forbidden regions. In [**HZZ**] the density of zeros in both regions is studied for random Hermite functions.

We sketch the proof of Theorem 4.1 from [**SoZ, SoZa**]. The starting point is an identity from [**SoZ**] (inspired by an identity in [**Dong**]):

PROPOSITION 4.2. *For any $f \in C^2(M)$,*

$$\int_M |\varphi_\lambda| \, (\Delta_g + \lambda^2) f \, dV_g = 2 \int_{\mathcal{N}_{\varphi_\lambda}} |\nabla_g \varphi_\lambda| \, f \, dS, \tag{45}$$

When $f \equiv 1$ we obtain

COROLLARY 4.3.

$$\lambda^2 \int_M |\varphi_\lambda| \, dV_g = 2 \int_{\mathcal{N}_{\varphi_\lambda}} |\nabla_g \varphi_\lambda| \, f \, dS, \tag{46}$$

EXERCISE 2. *Prove this identity by decomposing M into a union of nodal domains.*

Hint: The nodal domains form a partition of M, i.e.

$$M = \bigcup_{j=1}^{N_+(\lambda)} D_j^+ \cup \bigcup_{k=1}^{N_-(\lambda)} D_k^- \cup \mathcal{N}_\lambda,$$

where the D_j^+ and D_k^- are the positive and negative nodal domains of φ_λ, i.e, the connected components of the sets $\{\varphi_\lambda > 0\}$ and $\{\varphi_\lambda < 0\}$.

Let us assume for the moment that 0 is a regular value for φ_λ, i.e., $\Sigma = \emptyset$. Then each D_j^+ has smooth boundary ∂D_j^+, and so if ∂_ν is the Riemann outward normal derivative on this set, by the Gauss-Green formula we have

$$\begin{aligned} \textstyle\int_{D_j^+} ((\Delta + \lambda^2) f) \, |\varphi_\lambda| \, dV &= \textstyle\int_{D_j^+} ((\Delta + \lambda^2) f) \, \varphi_\lambda \, dV \\ &= \textstyle\int_{D_j^+} f \, (\Delta + \lambda^2) \varphi_\lambda dV - \int_{\partial D_j^+} f \, \partial_\nu \varphi_\lambda \, dS \\ &= \textstyle\int_{\partial D_j^+} f \, |\nabla \varphi_\lambda| \, dS. \end{aligned} \tag{47}$$

We use that $-\partial_\nu \varphi_\lambda = |\nabla \varphi_\lambda|$ since $\varphi_\lambda = 0$ on ∂D_j^+ and φ_λ decreases as it crosses ∂D_j^+ from D_j^+. A similar argument shows that

$$\begin{aligned}(48)\qquad &\textstyle\int_{D_k^-} ((\Delta + \lambda^2) f)\, |\varphi_\lambda|\, dV = - \int_{D_k^-} ((\Delta + \lambda^2) f)\, \varphi_\lambda\, dV \\ &= \textstyle\int f\, \partial_\nu \varphi_\lambda\, dS = \int_{\partial D_k^-} f\, |\nabla \varphi_\lambda|\, dS,\end{aligned}$$

using in the last step that φ_λ increases as it crosses ∂D_k^- from D_k^-.

If we sum these two identities over j and k, we get

$$\begin{aligned}\int_M ((\Delta + \lambda^2) f)\, |\varphi_\lambda|\, dV &= \sum_j \int_{D_j^+} ((\Delta + \lambda^2) f)\, |\varphi_\lambda|\, dV \\ &\quad + \sum_k \int_{D_k^-} ((\Delta + \lambda^2) f)\, |\varphi_\lambda|\, dV \\ &= \sum_j \int_{\partial D_j^+} f\, |\nabla \varphi_\lambda|\, dS \\ &\quad + \sum_k \int_{\partial D_k^-} f\, |\nabla \varphi_\lambda|\, dS = 2 \int_{\mathcal{N}_\lambda} f\, |\nabla \varphi_\lambda|\, dS,\end{aligned}$$

using the fact that $\mathcal{N}_\lambda$ is the disjoint union of the ∂D_j^+ and the disjoint union of the ∂D_k^-.

The lower bound of Theorem 4.1 follows from the identity in Corollary 4.3 and the following lemma:

LEMMA 4.4. *If* $\lambda > 0$ *then*

$$(49)\qquad \|\nabla_g \varphi_\lambda\|_{L^\infty(M)} \lesssim \lambda^{1 + \frac{n-1}{2}} \|\varphi_\lambda\|_{L^1(M)}$$

Here, $A(\lambda) \lesssim B(\lambda)$ means that there exists a constant independent of λ so that $A(\lambda) \leq C B(\lambda)$.

By Lemma 4.4 and Corollary 4.3, we have

$$\begin{aligned}(50)\qquad \lambda^2 \textstyle\int_M |\varphi_\lambda|\, dV = 2 \int_{\mathcal{N}_\lambda} |\nabla_g \varphi_\lambda|_g\, dS &\leqslant 2 |\mathcal{N}_\lambda|\, \|\nabla_g \varphi_\lambda\|_{L^\infty(M)} \\ &\lesssim 2 |\mathcal{N}_\lambda|\, \lambda^{1 + \frac{n-1}{2}} \|\varphi_\lambda\|_{L^1(M)}.\end{aligned}$$

Thus Theorem 4.1 follows from the somewhat curious cancellation of $||\varphi_\lambda||_{L^1}$ from the two sides of the inequality.

Problem Show that Corollary 4.3 and Lemma 4.4 are true modulo $O(1)$ for quasi-modes of order zero (§2.10, §15.3).

4.1. Proof of Lemma 4.4

PROOF. The main point is to construct a designer reproducing kernel K_λ for φ_λ: Let $\hat{\rho} \in C_0^\infty(\mathbb{R})$ satisfy $\rho(0) = \int \hat{\rho}\, dt = 1$. Define the operator

$$(51)\qquad \rho(\lambda - \sqrt{\Delta}) : L^2(M) \to L^2(M)$$

by

$$(52)\qquad \rho(\lambda - \sqrt{\Delta}) f = \int_{\mathbb{R}} \hat{\rho}(t) e^{it\lambda} e^{-it\sqrt{-\Delta}} f\, dt.$$

Then (51) is a function of Δ and has φ_λ as an eigenfunction with eigenvalue $\rho(\lambda - \lambda) = \rho(0) = 1$. Hence,

$$\rho(\lambda - \sqrt{\Delta})\varphi_\lambda = \varphi_\lambda.$$

EXERCISE 3. *Check that* (52) *has the spectral expansion,*

$$\rho(\lambda - \sqrt{\Delta})f = \sum_{j=0}^{\infty} \rho(\lambda - \lambda_j) E_j f, \tag{53}$$

where $E_j f$ is the projection of f onto the λ_j- eigenspace of $\sqrt{-\Delta_g}$. Conclude that (52) *reproduces φ_λ if $\rho(0) = 1$.*

We may choose ρ further so that $\hat{\rho}(t) = 0$ for $t \notin [\epsilon/2, \epsilon]$.

CLAIM If supp $\hat{\rho} \subset [\epsilon/2, \epsilon]$. then the kernel $K_\lambda(x, y)$ of $\rho(\lambda - \sqrt{\Delta})$ for ϵ sufficiently small satisfies

$$|\nabla_g K_\lambda(x, y)| \leqslant C\lambda^{1+\frac{n-1}{2}}. \tag{54}$$

The Claim proves the Lemma, because

$$\begin{array}{rcl} \nabla_x \varphi_\lambda(x) & = & \nabla_x \rho(\lambda - \sqrt{\Delta})\varphi_\lambda(x) \\ & = & \int_M \nabla_x K_\lambda(x, y)\varphi_\lambda(y) dV(y) \\ & \leq & C \sup_{x,y} |\nabla_x K_\lambda(x, y)| \int_M |\varphi_\lambda| dV \\ & \leq & \lambda^{1+\frac{n-1}{2}} ||\varphi_\lambda||_{L^1} \end{array}$$

which implies the lemma.

The gradient estimate on $K_\lambda(x, y)$ is based on the following "parametrix" for the designer reproducing kernel:

PROPOSITION 4.5.

$$K_\lambda(x, y) = \lambda^{\frac{n-1}{2}} a_\lambda(x, y) e^{i\lambda r(x,y)}, \tag{55}$$

where $a_\lambda(x, y)$ is bounded with bounded derivatives in (x, y) and where $r(x, y)$ is the Riemannian distance between points.

PROOF. Let $U(t) = e^{-it\sqrt{\Delta}}$. We may write

$$\rho(\lambda - \sqrt{\Delta}) = \int_{\mathbb{R}} \hat{\rho}(t) e^{it\lambda} U(t, x, y) dt. \tag{56}$$

As reviewed in §14.2, for small t and x, y near the diagonal one may construct the Hadamard parametrix,

$$U(t, x, y) = \int_0^\infty e^{i\theta(r^2(x,y) - t^2)} At, (x, y, \theta) d\theta$$

modulo a smooth remainder (which may be neglected).

EXERCISE 4. *Explain why the remainder may be neglected. How many of the terms in the parametrix construction does one need in the proof of Proposition 4.5? (Hint: if one truncates the amplitude after a finite number of terms in the Hadamard parametrix, the remainder lies in C^k and then the contribution to* (56) *decays as $\lambda \to \infty$.)*

Thus,

$$K_\lambda(x, y) = \int_{\mathbb{R}} \int_0^\infty e^{i\theta(r^2(x,y)-t^2)} e^{it\lambda} \hat{\rho}(t) At, (x, y, \theta) d\theta dt.$$

We change variables $\theta \to \lambda\theta$ to obtain

$$K_\lambda(x, y) = \lambda \int_{\mathbb{R}} \int_0^\infty e^{i\lambda[\theta(r^2(x,y)-t^2)+t]} \hat{\rho}(t) At, (x, y, \lambda\theta) d\theta dt.$$

We then apply stationary phase. The phase is

$$\theta(r^2(x, y) - t^2) + t$$

and the critical point equations are

$$r^2(x, y) = t^2, \quad 2t\theta = 1, \quad (t \in (\epsilon, 2\epsilon)).$$

The power of θ in the amplitude is $\theta^{\frac{n-1}{2}}$. The change of variables thus puts in $\lambda^{\frac{n+1}{2}}$.But we get λ^{-1} from stationary phase with two variables (t, θ).

The value of the phase at the critical point is $e^{it\lambda} = e^{i\lambda r(x,y)}$. The Hessian in (t, θ) is $2t$ and it is invertible. Hence,

$$K_\lambda(x, y) \simeq \lambda^{\frac{n-1}{2}} e^{i\lambda r(x,y)} a(\lambda, x, y),$$

where

$$a \sim a_0 + \lambda^{-1} a_{-1} + \cdots$$

and

$$a_0 = A_0(r(x, y), (x, y, \frac{2}{r(x, y)}).$$

□

Proposition 4.5 implies that $|\nabla_g K_\lambda(x, y)| \leqslant C\lambda^{1+\frac{n-1}{2}}$ by directly differentiating the expression. The extra power of λ comes from the "phase factor" $e^{i\lambda r(x,y)}$.

This concludes the proof of Lemma 4.4. □

Remark: There are many 'reproducing kernels' if one only requires them to reproduce one eigenfunction. A very common choice is the spectral projections operator

$$\Pi_{[\lambda,\lambda+1]}(x, y) = \sum_{j:\lambda_j \in [\lambda,\lambda+1]} \varphi_j(x)\varphi_j(y)$$

for the interval $[\lambda, \lambda+1]$. It reproduces *all* eigenfunction φ_k with $\lambda_k \in [\lambda, \lambda+1]$. This reproducing kernel cannot be used in our application because $\Pi_\lambda(x, x) \simeq \lambda^{n-1}$, as follows from the local Weyl law. Similarly, $\sup_{x,y} |\nabla_x \Pi_\lambda(x, y)| \simeq \lambda^n$. The reader may check these statements on the spectral projections kernel for the standard sphere (§16).

4.2. Modifications

Hezari-Sogge modified the proof Proposition 4.2 in [**HS**] to prove

THEOREM 4.6. *For any C^∞ compact Riemannian manifold, the L^2-normalized eigenfunctions satisfy*

$$\mathcal{H}^{n-1}(\mathcal{N}_{\varphi_\lambda}) \geq C\,\lambda\,||\varphi_\lambda||_{L^1}^2.$$

They first apply the Schwarz inequality to get

$$\lambda^2 \int_M |\varphi_\lambda|\, dV_g \leqslant 2(\mathcal{H}^{n-1}(\mathcal{N}_{\varphi_\lambda}))^{1/2} \left(\int_{Z_{\varphi_\lambda}} |\nabla_g \varphi_\lambda|^2\, dS \right)^{1/2}. \tag{57}$$

They then use the test function

$$f = \left(1 + \lambda^2 \varphi_\lambda^2 + |\nabla_g \varphi_\lambda|_g^2\right)^{\frac{1}{2}} \tag{58}$$

in Proposition 4.2 to show that

$$\int_{\mathcal{N}_{\varphi_\lambda}} |\nabla_g \varphi_\lambda|^2\, dS \leq \lambda^3. \tag{59}$$

See also [**Ar**] for the generalization to the nodal bounds to Dirichlet and Neumann eigenfunctions of bounded domains.

Theorem 4.6 shows that Yau's conjectured lower bound would follow for a sequence of eigenfunctions satisfying $||\varphi_\lambda||_{L^1} \geq C > 0$ for some positive constant C.

4.3. Lower bounds on L^1 norms of eigenfunctions

The following universal lower bound is optimal as (M, g) ranges over all compact Riemannian manifolds.

PROPOSITION 10. *For any (M, g) and any L^2-normalized eigenfunction, $||\varphi_\lambda||_{L^1} \geq C_g \lambda^{-\frac{n-1}{4}}$.*

Remark: There are few results on L^1 norms of eigenfunctions. The reason is probably that $|\varphi_\lambda|^2 dV$ is the natural probability measure associated to eigenfunctions. It is straightforward to show that the expected L^1 norm of random L^2-normalized spherical harmonics of degree N and their generalizations to any (M, g) is a positive constant C_N with a uniform positive lower bound. One expects eigenfunctions in the ergodic case to have the same behavior.

PROBLEM 1. *A difficult but interesting problem would be to show that $||\varphi_\lambda||_{L^1} \geq C_0 > 0$ on a compact hyperbolic manifold. A partial result in this direction would be useful.*

4.4. Dong's upper bound

Let (M, g) be a compact C^∞ Riemannian manifold of dimesion n, let φ_λ be an L^2-normalized eigenfunction of the Laplacian,

$$\Delta \varphi_\lambda = -\lambda^2 \varphi_\lambda,$$

Let

$$q = |\nabla \varphi|^2 + \lambda^2 \varphi^2. \tag{60}$$

In Theorem 2.2 of [**D**], R. T. Dong proves the bound (for M of any dimension n),

$$\mathcal{H}^{n-1}(\mathcal{N} \cap \Omega) \leq \frac{1}{2} \int_{\Omega} |\nabla \log q| + \sqrt{n} vol(\Omega)\lambda + vol(\partial\Omega). \tag{61}$$

He also proves (Theorem 3.3) that on a surface,

$$\Delta \log q \geq -\lambda + 2\min(K, 0) + 4\pi \sum_i (k_i - 1)\delta_{p_i}, \tag{62}$$

where $\{p_i\}$ are the singular points and k_i is the order of p_i. In Dong's notation, $\lambda > 0$. Using a weak Harnack inequality together with (62), Dong proves ([**D**], (25)) that in dimension two,

$$\int_{B_R} |\nabla \log q| \leq C_g R\lambda + C'_g \lambda^2 R^3. \tag{63}$$

Combining with (61) produces the upper bound $\mathcal{H}^1(\mathcal{N} \cap \Omega) \leq \lambda^{3/2}$ in dimension 2.

PROBLEM 2. *To what extent can one generalize these estimates to higher dimensions?*

4.5. Other level sets

Although nodal sets are special, it is of interest to bound the Hausdorff surface measure of any level set $\mathcal{N}^c_{\varphi_\lambda} := \{\varphi_\lambda = c\}$. Let sgn $(x) = \frac{x}{|x|}$.

PROPOSITION 4.7. *For any C^∞ Riemannian manifold, and any $f \in C(M)$ we have,*

$$\int_M f(\Delta + \lambda^2)\, |\varphi_\lambda - c|\, dV + \lambda^2 c \int f\text{sgn}\,(\varphi_\lambda - c) dV = 2 \int_{\mathcal{N}^c_{\varphi_\lambda}} f |\nabla \varphi_\lambda| dS. \tag{64}$$

This identity has similar implications for $\mathcal{H}^{n-1}(\mathcal{N}^c_{\varphi_\lambda})$ and for the equidistribution of level sets.

COROLLARY 4.8. *For $c \in \mathbb{R}$*

$$\lambda^2 \int_{\varphi_\lambda \geqslant c} \varphi_\lambda dV = \int_{\mathcal{N}^c_{\varphi_\lambda}} |\nabla \varphi_\lambda| dS.$$

One can obtain lower bounds on $\mathcal{H}^{n-1}(\mathcal{N}^c_{\varphi_\lambda})$ as in the case of nodal sets. However the integrals of $|\varphi_\lambda|$ no longer cancel out. The numerator is smaller since one only integrates over $\{\varphi_\lambda \geq c\}$. Indeed, $\mathcal{H}^{n-1}(\mathcal{N}^c_{\varphi_\lambda})$ must tend to zero as c tends to the maximum possible threshold $\lambda^{\frac{n-1}{2}}$ for $\sup_M |\varphi_\lambda|$.

The Corollary follows by integrating Δ by parts, and by using the identity,

$$\begin{aligned} \textstyle\int_M |\varphi_\lambda - c| + c \text{ sgn } (\varphi_\lambda - c)\, dV &= \textstyle\int_{\varphi_\lambda > c} \varphi_\lambda dV - \int_{\varphi_\lambda < c} \varphi_\lambda dV \\ &= 2 \textstyle\int_{\varphi_\lambda > c} \varphi_\lambda dV, \end{aligned} \tag{65}$$

since $0 = \int_M \varphi_\lambda dV = \int_{\varphi_\lambda > c} \varphi_\lambda dV + \int_{\varphi_\lambda < c} \varphi_\lambda dV$.

PROBLEM 3. *A difficult problem would be to study $\mathcal{H}^{n-1}(\mathcal{N}^c_{\varphi_\lambda})$ as a function of (c, λ) and try to find thresholds where the behavior changes. For random spherical harmonics, $\sup_M |\varphi_\lambda| \simeq \sqrt{\log \lambda}$ and one would expect the level set volumes to be very small above this height except in special cases.*

4.6. Examples

The lower bound of Theorem 4.1 is far from the lower bound conjectured by Yau, which by Theorem 4 is correct at least in the real analytic case. In this section we go over the model examples to understand why the methds are not always getting sharp results.

4.6.1. *Flat tori.* We have, $|\nabla \sin\langle k, x\rangle|^2 = \cos^2\langle k, x\rangle |k|^2$. Since $\cos\langle k, x\rangle = 1$ when $\sin\langle k, x\rangle = 0$ the integral is simply $|k|$ times the surface volume of the nodal set, which is known to be of size $|k|$. Also, we have $\int_{\mathbf{T}} |\sin\langle k, x\rangle| dx \geq C$. Thus, our method gives the sharp lower bound $\mathcal{H}^{n-1}(Z_{\varphi_\lambda}) \geq C\lambda^1$ in this example.

So the upper bound is achieved in this example. Also, we have $\int_{\mathbf{T}} |\sin\langle k, x\rangle| dx \geq C$. Thus, our method gives the sharp lower bound $\mathcal{H}^{n-1}(Z_{\varphi_\lambda}) \geq C\lambda^1$ in this example. Since $\cos\langle k, x\rangle = 1$ when $\sin\langle k, x\rangle = 0$ the integral is simply $|k|$ times the surface volume of the nodal set, which is known to be of size $|k|$.

4.6.2. *Spherical harmonics on S^2.* For background on spherical harmonics we refer to §16.

The L^1 of Y_0^N norm can be derived from the asymptotics of Legendre polynomials

$$P_N(\cos\theta) = \sqrt{2}(\pi N \sin\theta)^{-\frac{1}{2}} \cos\left((N + \frac{1}{2})\theta - \frac{\pi}{4} \right) + O(N^{-3/2})$$

where the remainder is uniform on any interval $\epsilon < \theta < \pi - \epsilon$. We have

$$||Y_0^N||_{L^1} = 4\pi \sqrt{\frac{(2N+1)}{2\pi}} \int_0^{\pi/2} |P_N(\cos r)| dv(r) \sim C_0 > 0,$$

i.e. the L^1 norm is asymptotically a positive constant. Hence $\int_{Z_{Y_0^N}} |\nabla Y_0^N| ds \simeq C_0 N^2$. In this example $|\nabla Y_0^N|_{L^\infty} = N^{\frac{3}{2}}$ saturates the sup norm bound. The length of the nodal line of Y_0^N is of order λ, as one sees from the rotational invariance and by the fact that P_N has N zeros. The defect in the argument is that the bound $|\nabla Y_0^N|_{L^\infty} = N^{\frac{3}{2}}$ is only obtained on the nodal components near the poles, where each component has length $\simeq \frac{1}{N}$.

EXERCISE 5. *Calculate the L^1 norms of (L^2-normalized) zonal spherical harmonics and Gaussian beams.*

The left image is a zonal spherical harmonic of degree N on S^2: it has high peaksof height $\sqrt{N}$ at the north and south poles. The right image is a Gaussian beam: its height along the equator is $N^{1/4}$ and then it has Gaussian decay transverse to the equator.

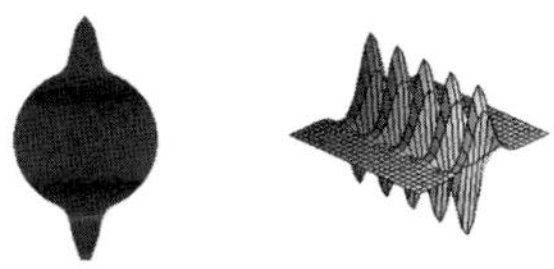

Gaussian beams

Gaussian beams are Gaussian shaped lumps which are concentrated on $\lambda^{-\frac{1}{2}}$ tubes $\mathcal{T}_{\lambda^{-\frac{1}{2}}}(\gamma)$ around closed geodesics and have height $\lambda^{\frac{n-1}{4}}$. We note that their L^1 norms decrease like $\lambda^{-\frac{(n-1)}{4}}$, i.e. they saturate the L^p bounds of [**Sog**] for small

p. In such cases we have $\int_{Z_{\varphi_\lambda}} |\nabla \varphi_\lambda| dS \simeq \lambda^2 ||\varphi_\lambda||_{L^1} \simeq \lambda^{2-\frac{n-1}{4}}$. It is likely that Gaussian beams are minimizers of the L^1 norm among L^2-normalized eigenfunctions of Riemannian manifolds. Also, the gradient bound $||\nabla \varphi_\lambda||_{L^\infty} = O(\lambda^{\frac{n+1}{2}})$ is far off for Gaussian beams, the correct upper bound being $\lambda^{1+\frac{n-1}{4}}$. If we use these estimates on $||\varphi_\lambda||_{L^1}$ and $||\nabla \varphi_\lambda||_{L^\infty}$, our method gives $\mathcal{H}^{n-1}(Z_{\varphi_\lambda}) \geq C\lambda^{1-\frac{n-1}{2}}$, while λ is the correct lower bound for Gaussian beams in the case of surfaces of revolution (or any real analytic case). The defect is again that the gradient estimate is achieved only very close to the closed geodesic of the Gaussian beam. Outside of the tube $\mathcal{T}_{\lambda^{-\frac{1}{2}}}(\gamma)$ of radius $\lambda^{-\frac{1}{2}}$ around the geodesic, the Gaussian beam and all of its derivatives decay like $e^{-\lambda d^2}$ where d is the distance to the geodesic. Hence $\int_{Z_{\varphi_\lambda}} |\nabla \varphi_\lambda| dS \simeq \int_{Z_{\varphi_\lambda} \cap \mathcal{T}_{\lambda^{-\frac{1}{2}}}(\gamma)} |\nabla \varphi_\lambda| dS$. Applying the gradient bound for Gaussian beams to the latter integral gives $\mathcal{H}^{n-1}(Z_{\varphi_\lambda} \cap \mathcal{T}_{\lambda^{-\frac{1}{2}}}(\gamma)) \geq C\lambda^{1-\frac{n-1}{2}}$, which is sharp since the intersection $Z_{\varphi_\lambda} \cap \mathcal{T}_{\lambda^{-\frac{1}{2}}}(\gamma)$ cuts across γ in $\simeq \lambda$ equally spaced points (as one sees from the Gaussian beam approximation).

5. Quantum ergodic restriction theorem for Dirichlet or Neumann data

QER (quantum ergodic restriction) theorems for Dirichlet data assert the quantum ergodicity of restrictions $\varphi_j|_H$ of eigenfunctions or their normal derivatives to hypersurfaces $H \subset M$. In this section we briefly review the QER theorem for hypersurfaces of [**TZ2, CTZ**]. For lack of space, we must assume the reader's familiarity with quantum ergodicity on the global manifold M. We refer to [**Ze2, Ze3, Ze6, Zw**] for recent expositions.

5.1. Quantum ergodic restriction theorems for Dirichlet data

Roughly speaking, the QER theorem for Dirichlet data says that restrictions of eigenfunctions to hypersurfaces $H \subset M$ for (M, g) with ergodic geodesic flow are quantum ergodic along H as long as H is *asymmetric* for the geodesic flow. Here we note that a tangent vector ξ to H of length ≤ 1 is the projection to TH of two unit tangent vectors $\xi_\pm$ to M. The $\xi_\pm = \xi + r\nu$ where ν is the unit normal to H and $|\xi|^2 + r^2 = 1$. There are two possible signs of r corresponding to the two choices of "inward" resp. "outward" normal. Asymmetry of H with respect to the geodesic flow G^t means that the two orbits $G^t(\xi_\pm)$ almost never return at the same time to the same place on H. A generic hypersurface is asymmetric [**TZ2**]. We refer to [**TZ2**] (Definition 1) for the precise definition of "positive measure of microlocal reflection symmetry" of H. By asymmetry we mean that this measure is zero.

We write $h_j = \lambda_j^{-\frac{1}{2}}$ and employ the calculus of semi-classical pseudo-differential operators [**Zw**] where the pseudo-differential operators on H are denoted by $a^w(y, hD_y)$ or $Op_{h_j}(a)$. The unit co-ball bundle of H is denoted by B^*H.

THEOREM 5.1. *Let (M, g) be a compact surface with ergodic geodesic flow, and let $H \subset M$ be a closed curve which is asymmetric with respect to the geodesic flow. Then there exists a density-one subset S of $\mathbb{N}$ such that for $a \in S^{0,0}(T^*H \times [0, h_0))$,*

$$\lim_{j \to \infty; j \in S} \langle Op_{h_j}(a) \varphi_{h_j}|_H, \varphi_{h_j}|_H \rangle_{L^2(H)} = \omega(a),$$

where

$$\omega(a) = \frac{4}{vol(S^*M)} \int_{B^*H} a_0(s, \sigma) \, (1 - |\sigma|^2)^{-\frac{1}{2}} \, ds d\sigma.$$

In particular this holds for multiplication operators f.

There is a similar result for normalized Neumann data. The normalized Neumann data of an eigenfunction along H is denoted by

$$\lambda_j^{-\frac{1}{2}} D_\nu \varphi_j|_H. \tag{66}$$

Here, $D_\nu = \frac{1}{i}\partial_\nu$ is a fixed choice of unit normal derivative.

We define the microlocal lifts of the Neumann data as the linear functionals on semi-classical symbols $a \in S^0_{sc}(H)$ given by

$$\mu_h^N(a) := \int_{B^*H} a\, d\Phi_h^N := \langle Op_H(a) h D_\nu \varphi_h|_H, h D_\nu \varphi_h|_H\rangle_{L^2(H)}.$$

THEOREM 5.2. *Let (M,g) be a compact surface with ergodic geodesic flow, and let $H \subset M$ be a closed curve which is asymmetric with respect to the geodesic flow. Then there exists a density-one subset S of $\mathbb{N}$ such that for $a \in S^{0,0}(T^*H \times [0,h_0))$,*

$$\lim_{h_j \to 0^+; j \in S} \mu_h^N(a) \to \omega(a),$$

where

$$\omega(a) = \frac{4}{vol(S^*M)} \int_{B^*H} a_0(s,\sigma)\,(1-|\sigma|^2)^{\frac{1}{2}}\, ds d\sigma.$$

In particular this holds for multiplication operators f.

5.2. Quantum ergodic restriction theorems for Cauchy data

Our application is to the hypersurface $H = \text{Fix}(\sigma)$ (17) given by the fixed point set of the isometric involution σ. Such a hypersurface (i.e. curve) fails to be asymmetric. However there is a quantum ergodic restriction theorem for Cauchy data in [**CTZ**] which does apply and shows that the even eigenfunctions are quantum ergodic along H, hence along each component γ.

The normalized Cauchy data of an eigenfunction along γ is denoted by

$$CD(\varphi_h) := \{(\varphi_h|_\gamma,\ h D_\nu \varphi_h|_\gamma)\}. \tag{67}$$

Here, D_ν is a fixed choice of unit normal derivative. The first component of the Cauchy data is called the Dirichlet data and the second is called the Neumann data.

THEOREM 5.3. *Assume that (M,g) has an orientation reversing isometric involution with separating fixed point set H. Let γ be a component of H. Let φ_h be the sequence of even ergodic eigenfunctions. Then,*

$$\langle Op_\gamma(a)\varphi_h|_\gamma, \varphi_h|_\gamma\rangle_{L^2(\gamma)}$$

$$\to_{h\to 0^+} \tfrac{4}{2\pi Area(M)} \textstyle\int_{B^*\gamma} a_0(s,\sigma)(1-|\sigma|^2)^{-1/2} ds d\sigma.$$

In particular, this holds when $Op_\gamma(a)$ is multiplication by a smooth function f.

Here we use the semi-classical notation $h_j = \lambda_\varphi^{-\frac{1}{4}}$. ergodic along γ, but we do not use this result here. We refer to [**TZ2, CTZ, Zw**] for background and for notation concerning pseudo-differential operators.

We further define the microlocal lifts of the Neumann data as the linear functionals on semi-classical symbols $a \in S^0_{sc}(\gamma)$ given by

$$\mu_h^N(a) := \int_{B^*\gamma} a\, d\Phi_h^N := \langle Op_\gamma(a) h D_\nu \varphi_h|_\gamma, h D_\nu \varphi_h|_\gamma\rangle_{L^2(\gamma)}.$$

We also define the *renormalized microlocal lifts* of the Dirichlet data by

$$\mu_h^D(a) := \int_{B^*\gamma} a\, d\Phi_h^{RD} := \langle Op_\gamma(a)(1 + h^2\Delta_\gamma)\varphi_h|_\gamma, \varphi_h|_\gamma\rangle_{L^2(\gamma)}.$$

Here, $h^2\Delta_\gamma$ denotes the negative tangential Laplacian $-h^2\frac{d^2}{ds^2}$ for the induced metric on γ, so that the symbol $1 - |\sigma|^2$ of the operator $(1 + h^2\Delta_\gamma)$ vanishes on the tangent directions $S^*\gamma$ of γ. Finally, we define the microlocal lift $d\Phi_h^{CD}$ of the Cauchy data to be the sum

$$d\Phi_h^{CD} := d\Phi_h^N + d\Phi_h^{RD}. \tag{68}$$

Let $B^*\gamma$ denote the unit "ball-bundle" of γ (which is the interval $\sigma \in (-1, 1)$ at each point s), s denotes arc-length along γ and σ is the dual symplectic coordinate. The first result of [**CTZ**] relates QE (quantum ergodicity) on M to quantum ergodicity on a hypersurface γ. A sequence of eigenfunctions is QE globally on M if

$$\langle A\varphi_{j_k}, \varphi_{j_k}\rangle \to \frac{1}{\mu(S^*M)} \int_{S^*M} \sigma_A d\mu,$$

where $d\mu_L$ is Liouville measure, i.e. the measure induced on the co-sphere bundle by the symplectic volume measure and the Hamiltonian $H(x, \xi) = |\xi|_g$.

THEOREM 5.4. *Assume that $\{\varphi_h\}$ is a quantum ergodic sequence of eigenfunctions on M. Then the sequence $\{d\Phi_h^{CD}\}$ (68) of microlocal lifts of the Cauchy data of φ_h is quantum ergodic on γ in the sense that for any $a \in S^0_{sc}(\gamma)$,*

$$\langle Op_H(a)hD_\nu\varphi_h|_\gamma, hD_\nu\varphi_h|_\gamma\rangle_{L^2(\gamma)} + \langle Op_\gamma(a)(1 + h^2\Delta_\gamma)\varphi_h|_\gamma, \varphi_h|_\gamma\rangle_{L^2(\gamma)}$$

$$\to_{h\to 0^+} \tfrac{4}{\mu(S^*M)} \textstyle\int_{B^*\gamma} a_0(s, \sigma)(1 - |\sigma|^2)^{1/2} ds d\sigma$$

where a_0 is the principal symbol of $Op_\gamma(a)$.

When applied to even eigenfunctions under an orientation-reversing isometric involution with separating fixed point set, the Neumann data vanishes, and we obtain

COROLLARY 5.1. *Let (M, g) have an orientation-reversing isometric involution with separating fixed point set H and let γ be one of its components. Then for any sequence of even quantum ergodic eigenfunctions of (M, g),*

$$\langle Op_\gamma(a)(1 + h^2\Delta_\gamma)\varphi_h|_\gamma, \varphi_h|_\gamma\rangle_{L^2(\gamma)}$$

$$\to_{h\to 0^+} \tfrac{4}{\mu(S^*M)} \textstyle\int_{B^*\gamma} a_0(s, \sigma)(1 - |\sigma|^2)^{1/2} ds d\sigma$$

For applications to zeros along γ, we need a limit formula for the integrals $\int_\gamma f\varphi_h^2 ds$, i.e. a quantum ergodicity result for for Dirichlet data. We invert the operator $(1 + h^2\Delta_\gamma)$ and obtain

THEOREM 5.5. *Assume that $\{\varphi_h\}$ is a quantum ergodic sequence on M. Then, there exists a sub-sequence of density one as $h \to 0^+$ such that for all $a \in S^0_{sc}(\gamma)$,*

$$\left\langle (1 + h^2\Delta_\gamma + i0)^{-1} Op_\gamma(a) hD_\nu\varphi_h|_H, hD_\nu\varphi_h|_\gamma \right\rangle_{L^2(\gamma)} + \langle Op_\gamma(a)\varphi_h|_\gamma, \varphi_h|_\gamma\rangle_{L^2(\gamma)}$$
$$\to_{h\to 0^+} \tfrac{4}{2\pi Area(M)} \textstyle\int_{B^*\gamma} a_0(s, \sigma)(1 - |\sigma|^2)^{-1/2} ds d\sigma.$$

Theorem 5.3 follows from Theorem 5.5 since the Neumann term drops out (as before) under the hypothesis of Corollary 5.1.

6. Counting intersections of nodal sets and geodesics

As discussed in the introduction §2.5, the QER results can be used to obtain results on intersections of nodal sets with geodesics in dimension two. In general, we do not know how to use interesection results to obtain lower bounds on numbers of nodal domains unless we assume a symmetry condition on the surface. But begin with general results on intersection that do not assume any symmetries.

THEOREM 6.1. *Let (M, g) be a C^∞ compact negatively curved surface, and let H be a closed curve which is asymmetric with respect to the geodesic flow. Then for any orthonormal eigenbasis $\{\varphi_j\}$ of Δ-eigenfunctions of (M, g), there exists a density 1 subset A of $\mathbb{N}$ such that*

$$\begin{cases} \lim_{\substack{j\to\infty \\ j\in A}} \# \mathcal{N}_{\varphi_j} \cap H = \infty \\ \\ \lim_{\substack{j\to\infty \\ j\in A}} \# \{x \in H : \partial_\nu \varphi_j(x) = 0\} = \infty. \end{cases}$$

Furthermore, there are an infinite number of zeros where $\varphi_j|_H$ (resp. $\partial_\nu \varphi_j|_H$) changes sign.

We now add the assumption of a symmetry as discussed in the introduction in §2.3.

THEOREM 6.2. *Let (M, g) be a compact negatively curved C^∞ surface with an orientation-reversing isometric involution $\sigma : M \to M$ with $Fix(\sigma)$ separating. Let $\gamma \subset Fix(\sigma)$. Then for any orthonormal eigenbasis $\{\varphi_j\}$ of $L^2_{even}(M)$, resp. $\{\psi_j\}$ of $L^2_{odd}(M)$, one can find a density 1 subset A of $\mathbb{N}$ such that*

$$\begin{cases} \lim_{\substack{j\to\infty \\ j\in A}} \# \mathcal{N}_{\varphi_j} \cap \gamma = \infty \\ \\ \lim_{\substack{j\to\infty \\ j\in A}} \# \Sigma_{\psi_j} \cap \gamma = \infty. \end{cases}$$

Furthermore, there are an infinite number of zeros where $\varphi_j|_H$ (resp. $\partial_\nu \psi_j|_H$) changes sign.

We now sketch the proof.

6.1. Kuznecov sum formula on surfaces

The first step is to use an old result [**Ze9**] on the asymptotics of the 'periods' $\int_\gamma f \varphi_j ds$ of eigenfunctions over closed geodesics when f is a smooth function.

THEOREM 6.3. [**Ze9**] *(Corollary 3.3) Let $f \in C^\infty(\gamma)$. Then there exists a constant $c > 0$ such that,*

$$\sum_{\lambda_j < \lambda} \left| \int_\gamma f \varphi_j ds \right|^2 = c \left| \int_\gamma f ds \right|^2 \sqrt{\lambda} + O_f(1).$$

There is a similar result for the normal derivative ∂_ν of eigenfunctions along γ.

THEOREM 6.4. *Let $f \in C^\infty(\gamma)$. Then there exists a constant $c > 0$ such that,*

$$\sum_{\lambda_j < \lambda} \left| \lambda_j^{-1/2} \int_\gamma f \partial_\nu \varphi_j ds \right|^2 = c \left| \int_\gamma f ds \right|^2 \sqrt{\lambda} + O_f(1).$$

These 'Kuznecov sum formulae' do not imply individual results about asymptotic periods of the full sequence of eigenfunction. However, because the terms are positive, there must exists a subsequence of eigenfunctions φ_j of natural density one so that, for all $f \in C^\infty(\gamma)$,

$$\text{(69)} \qquad \left\{ \begin{array}{l} \left| \int_\gamma f \varphi_j ds \right| \\ \\ \lambda_j^{-\frac{1}{2}} \left| \int_\gamma f \partial_\nu \varphi_j ds \right| \end{array} \right. = O_f(\lambda_j^{-1/4} (\log \lambda_j)^{1/2})$$

Indeed, this follows by Chebychev's inequality. Denote by $N(\lambda)$ the number of eigenfunctions in $\{j \mid \lambda < \lambda_j < 2\lambda\}$. Then for each f,

$$\frac{1}{N(\lambda)} |\{j \mid \lambda < \lambda_j < 2\lambda, \left| \int_{\gamma_i} f \varphi_j ds \right|^2 \geq \lambda_j^{-1/2} \log \lambda_j\}| = O_f(\frac{1}{\log \lambda}).$$

It follows that the upper density of exceptions to (69) tends to zero. We then choose a countable dense set $\{f_n\}$ and apply the diagonalization argument of [**Ze4**] (Lemma 3) or [**Zw**] Theorem 15.5 step (2)) to conclude that there exists a density one subsequence for which (69) holds for all $f \in C^\infty(\gamma)$. The same holds for the normal derivative.

We then use the argument sketched in §2.5 of the introduction. If we combine (69) with the QER result (21), we conclude that a full density subsequence of eigenfunctions of a negatively curved surface must have an unbounded number of sign changing zeros along γ. In the asymmetric case we use Theorem 5.1 while in the symmetric case we use Theorem 5.5 in the QER step.

Problem The QER step and the Kuznecov extend to manifolds with boundary (although the Kuznecov sum formula of [**Ze9**] has so far only been proved in the boundaryless case). But the main obstacle to generalizing the results on intersections and nodal domains is that the logarithmic improvement in (22) has not been generalized to the boundary problems with chaotic billiards (to the author's knowledge).

7. Counting nodal domains

We now sketch the proof of Theorem 2.1. Granted that there are many zeros of even eigenfunctions along each component γ of Fix(σ) and many zeros of normal derivatives of odd eigenfunctions, the remaining point is to relate such zeros to nodal domains. The nodal sets cross the geodesic transversally but may link up in complicated ways far from the separating set. In dimension two, use use the Euler inequality for graphs to relate numbers of intersection points and numbers of nodal domains.

Problem: The QER and Kuznecov sum formula results are valid in all dimensions. It is only the topological step which is simpler in dimension two. Can one find any generalizations of this argument to higher dimensions?

In dimension two we can construct an embedded graph from the nodal set $\mathcal{N}_{\varphi_\lambda}$ as follows.

(1) For each embeded circle which does not intersect γ, we add a vertex.

(2) Each singular point is a vertex.
(3) If $\gamma \not\subset \mathcal{N}_{\varphi_\lambda}$, then each intersection point in $\gamma \cap \mathcal{N}_{\varphi_\lambda}$ is a vertex.
(4) Edges are the arcs of $\mathcal{N}_{\varphi_\lambda}$ ($\mathcal{N}_{\varphi_\lambda} \cup \gamma$, when φ_λ is even) which join the vertices listed above.

We thus obtain a graph embeded into the surface M. The *faces* f of G are the connected components of $M \backslash V(G) \cup \bigcup_{e \in E(G)} e$. The set of faces is denoted $F(G)$. An edge $e \in E(G)$ is *incident to* f if the boundary of f contains an interior point of e. Every edge is incident to at least one and to at most two faces; if e is incident to f then $e \subset \partial f$. The faces are not assumed to be cells and the sets $V(G), E(G), F(G)$ are not assumed to form a CW complex.

Now let $v(\varphi_\lambda)$ be the number of vertices, $e(\varphi_\lambda)$ be the number of edges, $f(\varphi_\lambda)$ be the number of faces, and $m(\varphi_\lambda)$ be the number of connected components of the graph. Then by Euler's formula (Appendix F, [**G**]),

$$v(\varphi_\lambda) - e(\varphi_\lambda) + f(\varphi_\lambda) - m(\varphi_\lambda) \geq 1 - 2g_M \tag{70}$$

where g_M is the genus of the surface. We use this inequality to give a lower bound for the number of nodal domains for even and odd eigenfunctions.

In the odd case, we get a lower bound using the large number of singular points of the odd eigenfunctions. Note that since an odd eigenfunction vanishes on $\mathrm{Fix}(\sigma)$, the points where $\partial_\nu \psi_j = 0$ belong to the singular set.

We claim that for an odd eigenfunction ψ_j,

$$N(\psi_j) \geq \# \left(\Sigma_{\psi_j} \cap \gamma\right) + 2 - 2g_M,$$

Proof: for an odd eigenfunction ψ_j, $\gamma \subset \mathcal{N}_{\psi_j}$. Therefore $f(\psi_j) = N(\psi_j)$. Let $n(\psi_j) = \# \Sigma_{\psi_j} \cap \gamma$ be the number of singular points on γ. These points correspond to vertices having degree at least 4 on the graph, hence

$$\begin{aligned} 0 &= \sum_{x:vertices} \deg(x) - 2e(\psi_j) \\ &\geq 2\left(v(\psi_j) - n(\psi_j)\right) + 4n(\psi_j) - 2e(\psi_j). \end{aligned} \tag{71}$$

Therefore

$$e(\psi_j) - v(\psi_j) \geq n(\psi_j),$$

and plugging into (70) with $m(\psi_j) \geq 1$, we obtain

$$N(\psi_j) \geq n(\psi_j) + 2 - 2g_M.$$

Now consider an even eigenfunction φ_j. In this case (following the idea of [**GRS**]), we relate the number of intersection points of the nodal set with γ to the number of 'inert nodal domains', i.e. nodal domains which are invariant under the isometric involution, $\sigma U = U$. We claim that,

$$N(\varphi_j) \geq \frac{1}{2} \# \left(\mathcal{N}_{\varphi_j} \cap \gamma\right) + 1 - g_M.$$

Proof: For an even eigenfunction φ_j, let $N_{in}(\varphi_j)$ be the number of 'inert' nodal domain U. Let $N_{sp}(\varphi_j)$ be the number of the rest (split nodal domains). From the assumption that $Fix(\sigma)$ is separating, inert domains intersect γ in a finite non-empty union of relative open intervals, and $\mathrm{Fix}(\sigma)$ divides each inert nodal domain into two connected components. Hence each inert nodal domain may correspond to two faces on the graph, depending on whether the nodal domain intersects γ or not. Therefore $f(\varphi_j) \leq 2N_{in}(\varphi_j) + N_{sp}(\varphi_j)$.

Each point in $\mathcal{N}_{\varphi_j} \cap \gamma$ corresponds to a vertex having degree at least 4 on the graph. The nodal intersections are non-singuar points. Hence by the same reasoning as the odd case, we have

$$N(\varphi_j) \geq N_{in} + \frac{1}{2} N_{sp}(\varphi_j) \geq \frac{f(\varphi_j)}{2} \geq \frac{n(\varphi_j)}{2} + 1 - g_M$$

where $n(\varphi_j) = \#\mathcal{N}_{\varphi_j} \cap \gamma$.

This concludes the proof of Theorem 2.1.

8. Analytic continuation of eigenfuntions to the complex domain

We next discuss three results that use analytic continuation of eigenfunctions to the complex domain. First is the Donnelly-Fefferman volume bound Theorem 4. We sketch a somewhat simplified proof which will appear in more detail in [**Ze0**]. Second we discuss the equidistribution theory of nodal sets in the complex domain in the ergodic case [**Ze5**] and in the completely integrable case [**Ze8**]. Third, we discuss nodal intersection bounds. This includes bounds on the number of nodal lines intersecting the boundary in [**TZ**] for the Dirichlet or Neuman problem in a plane domain, the number (and equi-distribution) of nodal intersections with geodesics in the complex domain [**Ze6**] and results on nodal intersections and nodal domains for the modular surface

8.1. Grauert tubes

. As examples, we have:

- $M = \mathbb{R}^m/\mathbb{Z}^m$ is $M_{\mathbb{C}} = \mathbb{C}^m/\mathbb{Z}^m$.
- The unit sphere S^n defined by $x_1^2 + \cdots + x_{n+1}^2 = 1$ in $\mathbb{R}^{n+1}$ is complexified as the complex quadric $S^2_{\mathbb{C}} = \{(z_1, \ldots, z_n) \in \mathbb{C}^{n+1} : z_1^2 + \cdots + z_{n+1}^2 = 1\}$.
- The hyperboloid model of hyperbolic space is the hypersurface in $\mathbb{R}^{n+1}$ defined by

$$\mathbb{H}^n = \{x_1^2 + \cdots x_n^2 - x_{n+1}^2 = -1, \;\; x_n > 0\}.$$

Then,

$$H^n_{\mathbb{C}} = \{(z_1, \ldots, z_{n+1}) \in \mathbb{C}^{n+1} : z_1^2 + \cdots z_n^2 - z_{n+1}^2 = -1\}.$$

- Any real algebraic subvariety of $\mathbb{R}^m$ has a similar complexification.
- Any Lie group G (or symmetric space) admits a complexification $G_{\mathbb{C}}$.

Let us consider examples of holomorphic continuations of eigenfunctions:

- On the flat torus $\mathbb{R}^m/\mathbb{Z}^m$, the real eigenfunctions are $\cos\langle k, x\rangle, \sin\langle k, x\rangle$ with $k \in 2\pi\mathbb{Z}^m$. The complexified torus is $\mathbb{C}^m/\mathbb{Z}^m$ and the complexified eigenfunctions are $\cos\langle k, \zeta\rangle, \sin\langle k, \zeta\rangle$ with $\zeta = x + i\xi$.
- On the unit sphere S^m, eigenfunctions are restrictions of homogeneous harmonic functions on $\mathbb{R}^{m+1}$. The latter extend holomorphically to holomorphic harmonic polynomials on $\mathbb{C}^{m+1}$ and restrict to holomorphic function on $S^m_{\mathbb{C}}$.
- On $\mathbf{H}^m$, one may use the hyperbolic plane waves $e^{(i\lambda+1)\langle z, b\rangle}$, where $\langle z, b\rangle$ is the (signed) hyperbolic distance of the horocycle passing through z and b to 0. They may be holomorphically extended to the maximal tube of radius $\pi/4$.

- On compact hyperbolic quotients $\mathbf{H}^m/\Gamma$, eigenfunctions can be then represented by Helgason's generalized Poisson integral formula [**H**],

$$\varphi_\lambda(z) = \int_B e^{(i\lambda+1)\langle z,b\rangle} dT_\lambda(b).$$

Here, $z \in D$ (the unit disc), $B = \partial D$, and $dT_\lambda \in \mathcal{D}'(B)$ is the boundary value of φ_λ, taken in a weak sense along circles centered at the origin 0. To analytically continue φ_λ it suffices to analytically continue $\langle z, b\rangle$. Writing the latter as $\langle \zeta, b\rangle$, we have:

(72) $$\varphi_\lambda^{\mathbb{C}}(\zeta) = \int_B e^{(i\lambda+1)\langle \zeta,b\rangle} dT_\lambda(b).$$

The (L^2-normalizations of the) modulus squares

(73) $$|\varphi_j^{\mathbb{C}}(\zeta)|^2 : M_\epsilon \to \mathbb{R}_+$$

are sometimes known as Husimi functions. They are holomorphic extensions of L^2-normalized functions but are not themselves L^2 normalized on M_ϵ. However, as will be discussed below, their L^2 norms may on the Grauert tubes (and their boundaries) can be determined. One can then ask how the mass of the normalized Husimi function is distributed in phase space, or how the L^p norms behave.

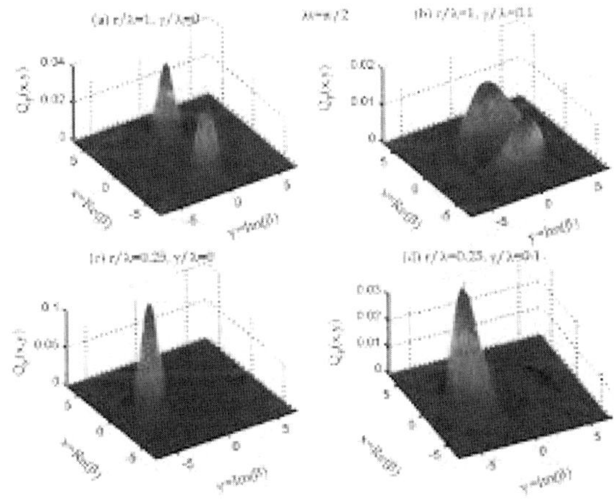

8.2. Weak * limit problem for Husimi measures in the complex domain One of the general problems of quantum dynamics is to determine all of the weak* limits of the sequence,

$$\{\frac{|\varphi_j^{\mathbb{C}}(z)|^2}{||\varphi_j^{\mathbb{C}}||_{L^2(\partial M_\epsilon)}} d\mu_\epsilon\}_{j=1}^\infty.$$

Here, $d\mu_\epsilon$ is the natural measure on ∂M_ϵ corresponding to the contact volume form on $S^*_\epsilon M$. Recall that a sequence μ_n of probability measures on a compact space X is said to converge weak* to a measure μ if $\int_X f d\mu_n \to \int_X f d\mu$ for all $f \in C(X)$. We refer to Theorem 10.3 for the ergodic case. In the integrable case one has localization results, which are not presented here.

8.3. Background on currents and PSH functions

We next consider logarithms of Husimi functions, which are PSH = (pluri-subharmonic) functions on M_ϵ. A function f on a domain in a complex manifold is PSH if $i\partial\bar{\partial}f$ is a *positive (1,1) current*. That is, $i\partial\bar{\partial}f$ is a singular form of type $\sum_{i\bar{j}} a_{i\bar{j}} dz^i \wedge d\bar{z}^k$ with $(a_{j\bar{k}})$ postive definte Hermitian. If f is a local holomorphic function, then $\log|f(z)|$ is PSH and $i\partial\bar{\partial}\log|f(z)| = [Z_f]$. General references are [**GH, HoC**].

In Theorem 5, we regard the zero set $[Z_f]$ as a *current of integration*, i.e. as a linear functional on $(m-1, m-1)$ forms ψ

$$\langle [Z_{\varphi_j}], \psi\rangle = \int_{Z_{\varphi_j}} \psi.$$

Recall that a *current* is a linear functional (distribution) on smooth forms. We refer to [**GH**] for background. On a complex manifold one has (p,q) forms with p dz_j and q $d\bar{z}_k$'s. In (96) we use the Kähler hypersurface volume form ω_g^{m-1} (where $\omega_g = i\partial\bar{\partial}\rho$) to make Z_{φ_j} into a measure:

$$\langle [Z_{\varphi_j}], f\rangle = \int_{Z_{\varphi_j}} f\omega_g^{m-1}, \quad (f \in C(M)).$$

A sequence of $(1,1)$ currents E_k converges weak* to a current E if $\langle E_k, \psi\rangle \to \langle E, \psi\rangle$ for all smooth $(m-1, m-1)$ forms. Thus, for all f

$$[Z_{\varphi_j}] \to i\partial\bar{\partial}\sqrt{\rho} \iff \int_{Z_{\varphi_j}} f\omega^{m-1} \to i\int_{M_\epsilon} f\partial\bar{\partial}\sqrt{\rho}\wedge\omega^{m-1,m-1}.$$

8.4. Poincaré-Lelong formula

One of the two key reasons for the gain in simplicity is that there exists a simple analytical formula for the delta-function on the nodal set. The *Poincaré-Lelong formula* gives an exact formula for the delta-function on the zero set of φ_j

$$i\partial\bar{\partial}\log|\varphi_j^{\mathbb{C}}(z)|^2 = [\mathcal{N}_{\varphi_j^{\mathbb{C}}}]. \tag{74}$$

Thus, if ψ is an $(n-1, n-1)$ form,

$$\int_{\mathcal{N}_{\varphi_j^{\mathbb{C}}}} \psi = \int_{M_\epsilon} \psi \wedge i\partial\bar{\partial}\log|\varphi_j^{\mathbb{C}}(z)|^2.$$

8.5. Pluri-subharmonic functions and compactness

In the real domain, we have emphasized the problem of finding weak* limits of the probability measures (14) and of their microlocal lifts or Wigner measures in phase space. The same problem exists in the complex domain for the sequence of Husimi functions (73). However, there also exists a new problem involving the sequence of normalized logarithms

$$\{u_j := \frac{1}{\lambda_j}\log|\varphi_j^{\mathbb{C}}(z)|^2\}_{j=1}^{\infty}. \tag{75}$$

A key fact is that this sequence is pre-compact in $L^p(M_\epsilon)$ for all $p < \infty$ and even that

$$\{\frac{1}{\lambda_j}\nabla\log|\varphi_j^{\mathbb{C}}(z)|^2\}_{j=1}^{\infty}. \tag{76}$$

is pre-compact in $L^1(M_\epsilon)$.

LEMMA 8.1. *(Hartog's Lemma; (see* [**HoI**, Theorem 4.1.9]*): Let $\{v_j\}$ be a sequence of subharmonic functions in an open set $X \subset \mathbb{R}^m$ which have a uniform upper bound on any compact set. Then either $v_j \to -\infty$ uniformly on every compact set, or else there exists a subsequence v_{j_k} which is convergent to some $u \in L^1_{loc}(X)$. Further, $\limsup_n u_n(x) \leq u(x)$ with equality almost everywhere. For every compact subset $K \subset X$ and every continuous function f,*

$$\limsup_{n\to\infty} \sup_K (u_n - f) \leq \sup_K (u - f).$$

In particular, if $f \geq u$ and $\epsilon > 0$, then $u_n \leq f + \epsilon$ on K for n large enough.

8.6. A general weak* limit problem for logarithms of Husimi functions

The study of exponential growth rates gives rise to a new kind new weak* limit problem for complexified eigenfunctions.

PROBLEM 8.2. *Find the weak* limits G on M_ϵ of sequences*

$$\frac{1}{\lambda_{j_k}} \log |\varphi_{j_k}^{\mathbb{C}}(z)|^2 \to G??$$

(The limits are actually in L^1 and not just weak.)

See Theorems 10.4 , 10.5 and 10.7 for the solution to this problem (modulo sparse subsequences) in the ergodic case.

Here is a general Heuristic principle to pin down the possible G: If

$$\frac{1}{\lambda_{j_k}} \log |\varphi_{j_k}^{\mathbb{C}}(z)|^2 \to G(z)$$

then

$$|\varphi_{j_k}^{\mathbb{C}}(z)|^2 \simeq e^{\lambda_j G(z)}(1 + \text{SOMETHING SMALLER }) \; (\lambda_j \to \infty).$$

But $\Delta_{\mathbb{C}}|\varphi_{j_k}^{\mathbb{C}}(z)|^2 = \lambda_{j_k}^2 |\varphi_{j_k}^{\mathbb{C}}(z)|^2$, so we should have

CONJECTURE 8.3. *Any limit G as above solves the Hamilton-Jacobi equation,*

$$(\nabla_{\mathbb{C}} G)^2 = 1.$$

(Note: The weak* limits of $\frac{|\varphi_j^{\mathbb{C}}(z)|^2}{||\varphi_j^{\mathbb{C}}||_{L^2(\partial M_\epsilon)}} d\mu_\epsilon$ must be supported in $\{G = G_{\max}\}$ (i.e. in the set of maximum values).

9. Poisson operator and Szegö operators on Grauert tubes

9.1. Poisson operator and analytic Continuation of eigenfunctions

The half-wave group of (M, g) is the unitary group $U(t) = e^{it\sqrt{\Delta}}$ generated by the square root of the positive Laplacian. Its Schwartz kernel is a distribution on $\mathbb{R} \times M \times M$ with the eigenfunction expansion

$$U(t, x, y) = \sum_{j=0}^{\infty} e^{it\lambda_j} \varphi_j(x) \varphi_j(y). \tag{77}$$

By the Poisson operator we mean the analytic continuation of$U(t)$ to positive imaginary time,

$$e^{-\tau\sqrt{\Delta}} = U(i\tau). \tag{78}$$

The eigenfunction expansion then converges absolutely to a real analytic function on $\mathbb{R}_+ \times M \times M$.

Let $A(\tau)$ denote the operator of analytic continuation of a function on M to the Grauert tube M_τ. Since

$$U_{\mathbb{C}}(i\tau)\varphi_\lambda = e^{-\tau\lambda}\varphi_\lambda^{\mathbb{C}}, \tag{79}$$

it is simple to see that

$$A(\tau) = U_{\mathbb{C}}(i\tau)e^{\tau\sqrt{\Delta}} \tag{80}$$

where $U_{\mathbb{C}}(i\tau, \zeta, y)$ is the analytic continuation of the Poisson kernel in x to M_τ. In terms of the eigenfunction expansion, one has

$$U(i\tau, \zeta, y) = \sum_{j=0}^{\infty} e^{-\tau\lambda_j}\varphi_j^{\mathbb{C}}(\zeta)\varphi_j(y), \quad (\zeta, y) \in M_\epsilon \times M. \tag{81}$$

This is a very useful observation because $U_{\mathbb{C}}(i\tau)e^{\tau\sqrt{\Delta}}$ is a Fourier integral operator with complex phase and can be related to the geodesic flow. The analytic continuability of the Poisson operator to M_τ implies that every eigenfunction analytically continues to the same Grauert tube.

9.2. Analytic continuation of the Poisson wave group

The analytic continuation of the Possion-wave kernel to M_τ in the x variable is discussed in detail in [**Ze8**] and ultimately derives from the analysis by Hadamard of his parametrix construction. We only briefly discuss it here and refer to [**Ze8**] for further details. In the case of Euclidean $\mathbb{R}^n$ and its wave kernel $U(t, x, y) = \int_{\mathbb{R}^n} e^{it|\xi|}e^{i\langle\xi, x-y\rangle}d\xi$ which analytically continues to $t + i\tau, \zeta = x + ip \in \mathbb{C}_+ \times \mathbb{C}^n$ as the integral

$$U_{\mathbb{C}}(t + i\tau, x + ip, y) = \int_{\mathbb{R}^n} e^{i(t+i\tau)|\xi|}e^{i\langle\xi, x+ip-y\rangle}d\xi.$$

The integral clearly converges absolutely for $|p| < \tau$.

Exact formulae of this kind exist for S^m and $\mathbf{H}^m$. For a general real analytic Riemannian manifold, there exists an oscillatry integral expression for the wave kernel of the form,

$$U(t, x, y) = \int_{T_y^*M} e^{it|\xi|_{g_y}}e^{i\langle\xi, \exp_y^{-1}(x)\rangle}A(t, x, y, \xi)d\xi \tag{82}$$

where $A(t, x, y, \xi)$ is a polyhomogeneous amplitude of order 0. The holomorphic extension of (82) to the Grauert tube $|\zeta| < \tau$ in x at time $t = i\tau$ then has the form

$$U_{\mathbb{C}}(i\tau, \zeta, y) = \int_{T_y^*} e^{-\tau|\xi|_{g_y}}e^{i\langle\xi, \exp_y^{-1}(\zeta)\rangle}A(t, \zeta, y, \xi)d\xi \quad (\zeta = x + ip). \tag{83}$$

9.3. Complexified spectral projections

The next step is to holomorphically extend the spectral projectors $d\Pi_{[0,\lambda]}(x,y) = \sum_j \delta(\lambda - \lambda_j)\varphi_j(x)\varphi_j(y)$ of $\sqrt{\Delta}$. The complexified diagonal spectral projections measure is defined by

$$d_\lambda \Pi^{\mathbb{C}}_{[0,\lambda]}(\zeta,\bar{\zeta}) = \sum_j \delta(\lambda - \lambda_j)|\varphi_j^{\mathbb{C}}(\zeta)|^2. \tag{84}$$

Henceforth, we generally omit the superscript and write the kernel as $\Pi^{\mathbb{C}}_{[0,\lambda]}(\zeta,\bar{\zeta})$. This kernel is not a tempered distribution due to the exponential growth of $|\varphi_j^{\mathbb{C}}(\zeta)|^2$. Since many asymptotic techniques assume spectral functions are of polynomial growth, we simultaneously consider the damped spectral projections measure

$$d_\lambda P^{\tau}_{[0,\lambda]}(\zeta,\bar{\zeta}) = \sum_j \delta(\lambda - \lambda_j)e^{-2\tau\lambda_j}|\varphi_j^{\mathbb{C}}(\zeta)|^2, \tag{85}$$

which is a temperate distribution as long as $\sqrt{\rho}(\zeta) \le \tau$. When we set $\tau = \sqrt{\rho}(\zeta)$ we omit the τ and put

$$d_\lambda P_{[0,\lambda]}(\zeta,\bar{\zeta}) = \sum_j \delta(\lambda - \lambda_j)e^{-2\sqrt{\rho}(\zeta)\lambda_j}|\varphi_j^{\mathbb{C}}(\zeta)|^2. \tag{86}$$

The integral of the spectral measure over an interval I gives

$$\Pi_I(x,y) = \sum_{j:\lambda_j \in I} \varphi_j(x)\varphi_j(y).$$

Its complexification gives the spectral projections kernel along the anti-diagonal,

$$\Pi_I(\zeta,\bar{\zeta}) = \sum_{j:\lambda_j \in I} |\varphi_j^{\mathbb{C}}(\zeta)|^2, \tag{87}$$

and the integral of (85) gives its temperate version

$$P_I^{\tau}(\zeta,\bar{\zeta}) = \sum_{j:\lambda_j \in I} e^{-2\tau\lambda_j}|\varphi_j^{\mathbb{C}}(\zeta)|^2, \tag{88}$$

or in the crucial case of $\tau = \sqrt{\rho}(\zeta)$,

$$P_I(\zeta,\bar{\zeta}) = \sum_{j:\lambda_j \in I} e^{-2\sqrt{\rho}(\zeta)\lambda_j}|\varphi_j^{\mathbb{C}}(\zeta)|^2, \tag{89}$$

9.4. Poisson operator as a complex Fourier integral operator

The damped spectral projection measure $d_\lambda\, P^{\tau}_{[0,\lambda]}(\zeta,\bar{\zeta})$ (85) is dual under the real Fourier transform in the t variable to the restriction

$$U(t + 2i\tau, \zeta, \bar{\zeta}) = \sum_j e^{(-2\tau + it)\lambda_j}|\varphi_j^{\mathbb{C}}(\zeta)|^2 \tag{90}$$

to the anti-diagonal of the mixed Poisson-wave group. The adjoint of the Poisson kernel $U(i\tau, x, y)$ also admits an anti-holomorphic extension in the y variable. The sum (90) are the diagonal values of the complexified wave kernel

$$\begin{aligned} U(t + 2i\tau, \zeta, \bar{\zeta}') &= \int_M U(t + i\tau, \zeta, y)E(i\tau, y, \bar{\zeta}')dV_g(x) \\ &= \textstyle\sum_j e^{(-2\tau + it)\lambda_j}\varphi_j^{\mathbb{C}}(\zeta)\overline{\varphi_j^{\mathbb{C}}(\zeta')}. \end{aligned} \tag{91}$$

We obtain (91) by orthogonality of the real eigenfunctions on M.

Since $U(t+2i\tau,\zeta,y)$ takes its values in the CR holomorphic functions on ∂M_τ, we consider the Sobolev spaces $\mathcal{O}^{s+\frac{n-1}{4}}(\partial M_\tau)$ of CR holomorphic functions on the boundaries of the strictly pseudo-convex domains M_ϵ, i.e.

$$\mathcal{O}^{s+\frac{m-1}{4}}(\partial M_\tau) = W^{s+\frac{m-1}{4}}(\partial M_\tau) \cap \mathcal{O}(\partial M_\tau),$$

where W_s is the sth Sobolev space and where $\mathcal{O}(\partial M_\epsilon)$ is the space of boundary values of holomorphic functions. The inner product on $\mathcal{O}^0(\partial M_\tau)$ is with respect to the Liouville measure

$$d\mu_\tau = (i\partial\bar{\partial}\sqrt{\rho})^{m-1} \wedge d^c\sqrt{\rho}. \tag{92}$$

We then regard $U(t+i\tau,\zeta,y)$ as the kernel of an operator from $L^2(M) \to \mathcal{O}^0(\partial M_\tau)$. It equals its composition $\Pi_\tau \circ U(t+i\tau)$ with the Szegö projector

$$\Pi_\tau : L^2(\partial M_\tau) \to \mathcal{O}^0(\partial M_\tau)$$

for the tube M_τ, i.e. the orthogonal projection onto boundary values of holomorphic functions in the tube.

This is a useful expression for the complexified wave kernel, because $\tilde{\Pi}_\tau$ is a complex Fourier integral operator with a small wave front relation. More precisely, the real points of its canonical relation form the graph Δ_Σ of the identity map on the symplectic one $\Sigma_\tau \subset T^*\partial M_\tau$ spanned by the real one-form $d^c\rho$, i.e.

$$\Sigma_\tau = \{(\zeta; rd^c\rho(\zeta)),\quad \zeta \in \partial M_\tau,\ r>0\} \subset T^*(\partial M_\tau). \tag{93}$$

We note that for each τ, there exists a symplectic equivalence $\Sigma_\tau \simeq T^*M$ by the map $(\zeta, rd^c\rho(\zeta)) \to (E_{\mathbb{C}}^{-1}(\zeta), r\alpha)$, where $\alpha = \xi \cdot dx$ is the action form (cf. [**GS2**]).

The following result was first stated by Boutet de Monvel[**Bou**] and has been proved in detail in [**Ze8, L, Ste**].

THEOREM 9.1. *$\Pi_\epsilon \circ U(i\epsilon) : L^2(M) \to \mathcal{O}(\partial M_\epsilon)$ is a complex Fourier integral operator of order $-\frac{m-1}{4}$ associated to the canonical relation*

$$\Gamma = \{(y,\eta,\iota_\epsilon(y,\eta)\} \subset T^*M \times \Sigma_\epsilon.$$

Moreover, for any s,

$$\Pi_\epsilon \circ U(i\epsilon) : W^s(M) \to \mathcal{O}^{s+\frac{m-1}{4}}(\partial M_\epsilon)$$

is a continuous isomorphism.

In [**Ze8**] we give the following sharpening of the sup norm estimates of [**Bou**]:

PROPOSITION 9.2. *Suppose (M,g) is real analytic. Then*

$$\sup_{\zeta\in M_\tau} |\varphi_\lambda^{\mathbb{C}}(\zeta)| \leq C\lambda^{\frac{m+1}{2}} e^{\tau\lambda}, \qquad \sup_{\zeta\in M_\tau} |\frac{\partial\varphi_\lambda^{\mathbb{C}}(\zeta)}{\partial\zeta_j}| \leq C\lambda^{\frac{m+3}{2}} e^{\tau\lambda}$$

The proof follows easily from the fact that the complexified Poisson kernel is a complex Fourier integral operator of finite order. The estimates can be improved further.

9.5. Toeplitz dynamical construction of the wave group

There exists an alternative to the parametrix constructions of Hadamard-Riesz, Lax, Hörmander and others which are reviewed in §14. It is useful for constructing the wave group $U(t)$ for large t, when it is awkward to use the group property $U(t/N)^N = U(t)$. As in Theorem 9.1 we denote by $U(i\epsilon)$ the operator with kernel $U(i\epsilon, \zeta, y)$ with $\zeta \in \partial M_\epsilon, y \in M$. We also denote by $U^*(i\epsilon) : \mathcal{O}(\partial M_\epsilon) \to L^2(M)$ the adjoint operator. Further, let

$$T_{g^t} : L^2(\partial M_\epsilon, d\mu_\epsilon) \to L^2(\partial M_\epsilon, d\mu_\epsilon)$$

be the unitary translation operator

$$T_{g^t} f(\zeta) = f(g^t(\zeta))$$

where $d\mu_\epsilon$ is the contact volume form on ∂M_ϵ and g^t is the Hamiltonian flow of $\sqrt{\rho}$ on M_ϵ.

PROPOSITION 9.3. *There exists a symbol $\sigma_{\epsilon,t}$ such that*

$$U(t) = U^*(i\epsilon)\sigma_{\epsilon,t} T_{g^t} U(i\epsilon).$$

The proof of this Proposition is to verify that the right side is a Fourier integral operator with canonical relation the graph of the geodesic flow. One then constructs $\sigma_{\epsilon,t}$ so that the symbols match. The proof is given in [**Ze6**]. Related constructions are given in [**G1, BoGu**].

10. Equidistribution of complex nodal sets of real ergodic eigenfunctions

We now consider global results when hypotheses are made on the dynamics of the geodesic flow. The main purpose of this section is to sketch the proof of Theorem 5. Use of the global wave operator brings into play the relation between the geodesic flow and the complexified eigenfunctions, and this allows one to prove gobal results on nodal hypersurfaces that reflect the dynamics of the geodesic flow. In some cases, one can determine not just the volume, but the limit distribution of complex nodal hypersurfaces. Since we have discussed this result elsewhere [**Ze6**] we only briefly review it here.

The complex nodal hypersurface of an eigenfunction is defined by

$$Z_{\varphi_\lambda^{\mathbb{C}}} = \{\zeta \in M_{\epsilon_0} : \varphi_\lambda^{\mathbb{C}}(\zeta) = 0\}. \tag{94}$$

As discussed in §8.3, there exists a natural current of integration over the nodal hypersurface in any Grauert tube M_ϵ with $\epsilon < \epsilon_0$, given by

$$\langle [Z_{\varphi_\lambda^{\mathbb{C}}}], \varphi \rangle = \frac{i}{2\pi} \int_{M_\epsilon} \partial\bar{\partial} \log |\varphi_\lambda^{\mathbb{C}}|^2 \wedge \varphi = \int_{Z_{\varphi_\lambda^{\mathbb{C}}}} \varphi, \quad \varphi \in \mathcal{D}^{(m-1,m-1)}(M_\epsilon). \tag{95}$$

In the second equality we used the Poincaré-Lelong formula (§8.4). We recall that $\mathcal{D}^{(m-1,m-1)}(M_\epsilon)$ stands for smooth test $(m-1, m-1)$-forms with support in $M*_\epsilon$.

The nodal hypersurface $Z_{\varphi_\lambda^{\mathbb{C}}}$ also carries a natural volume form $|Z_{\varphi_\lambda^{\mathbb{C}}}|$ as a complex hypersurface in a Kähler manifold. By Wirtinger's formula, it equals the restriction of $\frac{\omega_g^{m-1}}{(m-1)!}$ to $Z_{\varphi_\lambda^{\mathbb{C}}}$. Hence, one can regard $Z_{\varphi_\lambda^{\mathbb{C}}}$ as defining the measure

$$\langle |Z_{\varphi_\lambda^{\mathbb{C}}}|, \varphi \rangle = \int_{Z_{\varphi_\lambda^{\mathbb{C}}}} \varphi \frac{\omega_g^{m-1}}{(m-1)!}, \quad \varphi \in C(B_\epsilon^* M). \tag{96}$$

We prefer to state results in terms of the current $[Z_{\varphi_\lambda^{\mathbb{C}}}]$ since it carries more information.

We re-state Theorem 5 as follows:

THEOREM 10.1. *Let (M, g) be real analytic, and let $\{\varphi_{j_k}\}$ denote a quantum ergodic sequence of eigenfunctions of its Laplacian Δ. Let M_{ϵ_0} be the maximal Grauert tube around M. Let $\epsilon < \epsilon_0$. Then:*

$$\frac{1}{\lambda_{j_k}}[Z_{\varphi_{j_k}^{\mathbb{C}}}] \to \frac{i}{\pi}\partial\bar{\partial}\sqrt{\rho} \text{ weakly in } \mathcal{D}'^{(1,1)}(M_\epsilon),$$

in the sense that, for any continuous test form $\psi \in \mathcal{D}^{(m-1,m-1)}(M_\epsilon)$, we have

$$\frac{1}{\lambda_{j_k}}\int_{Z_{\varphi_{j_k}^{\mathbb{C}}}} \psi \to \frac{i}{\pi}\int_{M_\epsilon} \psi \wedge \partial\bar{\partial}\sqrt{\rho}.$$

Equivalently, for any $\varphi \in C(M_\epsilon)$,

$$\frac{1}{\lambda_{j_k}}\int_{Z_{\varphi_{j_k}^{\mathbb{C}}}} \varphi\frac{\omega_g^{m-1}}{(m-1)!} \to \frac{i}{\pi}\int_{M_\epsilon} \varphi\partial\bar{\partial}\sqrt{\rho} \wedge \frac{\omega_g^{m-1}}{(m-1)!}.$$

10.1. Sketch of the proof

The first step is to find a nice way to express $\varphi_j^{\mathbb{C}}$ on $M_{\mathbb{C}}$. Very often, when we analytically continue a function, we lose control over its behavior. The trick is to observe that the complexified wave group analytically continues the eigenfunctions. Recall that $U(t) = \exp it\sqrt{\Delta}$. Define the Poisson operator as

$$U(i\tau) = e^{-\tau\sqrt{\Delta}}.$$

Note that

$$U(i\tau)\varphi_j = e^{-\tau\lambda_j}\varphi_j.$$

The next step is to analytically continue $U(i\tau, x, y)$ in $x \to z \in M_\tau$.

The complexified Poisson kernel is define by

$$U(i\tau, \zeta, y) = \sum_{j=0}^{\infty} e^{-\tau\lambda_j}\varphi_j^{\mathbb{C}}(\zeta)\varphi_j(y).$$

It is holomorphic in $\zeta \in M_\tau$, i.e. when $\sqrt{\rho}(\zeta) < \tau$. But the main point is that it remains a Fourier integral operator after analytic continuation:

THEOREM 10.2. *(Hadamard, Boutet de Monvel, Z, M. Stenzel, G. Lebeau) $U(i\epsilon, z, y) : L^2(M) \to H^2(\partial M_\epsilon)$ is a complex Fourier integral operator of order $-\frac{m-1}{4}$ quantizing the complexified exponential map $\exp_y i\epsilon\frac{\eta}{|\eta|} : S_\epsilon^* \to \partial M_\epsilon$.*

We first observe that

$$U(i\tau)\varphi_{\lambda_j} = e^{-\tau\lambda_j}\varphi_{\lambda_j}^{\mathbb{C}}.$$

This follows immediately by integrating

$$U(i\tau, \zeta, y) = \sum_{k=0}^{\infty} e^{-\tau\lambda_k}\varphi_k^{\mathbb{C}}(\zeta)\varphi_k(y)$$

against φ_j and using orthogonality.

But we know that $U(i\tau)\varphi_{\lambda_j}$ is a Fourier integral operator. It is a fact that such an operator can only change L^2 norms by powers of λ_j. So

$$||U(i\tau)\varphi_{\lambda_j}||^2_{L^2(\partial M_\epsilon)}$$

has polynomial growth in λ_j and therefore we have,

$$||\varphi_{\lambda_j}||^2_{L^2(\partial M_\epsilon)} = \lambda^{somepower} e^{\tau \lambda_j}.$$

The power is relevant because we are taking the normalized logarithm.

The first step is to prove quantum ergodicity of the complexified eigenfunctions:

THEOREM 10.3. *Assume the geodesic flow of (M, g) is ergodic. Then*

$$\frac{|\varphi^\epsilon_{j_k}(z)|^2}{||\varphi^\epsilon_{j_k}||^2_{L^2(\partial M_\epsilon)}} \to 1, \quad weak * on \quad C(\partial M_\epsilon),$$

along a density one subsquence of λ_j. I.e. for any continuous V,

$$\int_{\partial M_\epsilon} V \frac{|\varphi^\epsilon_{j_k}(z)|^2}{||\varphi^\epsilon_{j_k}||^2_{L^2(\partial M_\epsilon)}} dvol \to \int_{\partial M_\epsilon} V dvol.$$

Thus, Husimi measures tend to 1 weakly as measures. We then apply Hartogs' Lemma 8.1 to obtain,

LEMMA 10.4. *We have: For all but a sparse subsequence of eigenvalues,*

$$\frac{1}{\lambda_{j_k}} \log \frac{|\varphi^\epsilon_{j_k}(z)|^2}{||\varphi^\epsilon_{j_k}||^2_{L^2(\partial M_\epsilon)}} \to 0, \quad in\ L^1(M_\epsilon).$$

This is almost obvious from the QE theorem. The limit is ≤ 0 and it were < 0 on a set of positive measure it would contradict

$$\frac{|\varphi^\epsilon_{j_k}(z)|^2}{||\varphi^\epsilon_{j_k}||^2_{L^2(\partial M_\epsilon)}} \to 1.$$

Combine Lemma 10.4 with Poincare- Lelong:

$$\frac{1}{\lambda_{j_k}}[Z_{j_k}] = i\partial\bar{\partial} \log |\varphi^{\mathbb{C}}_{j_k}|^2.$$

We get

$$\frac{1}{\lambda_{j_k}} \partial\bar{\partial} \log |\varphi^{\mathbb{C}}_{j_k}|^2 \sim \frac{1}{\lambda_{j_k}} \partial\bar{\partial} \log ||\varphi^{\mathbb{C}}_{j_k}||^2_{L^2(\partial M_\epsilon)} \text{ weak * on } \quad M_\epsilon.$$

To complete proof we need to prove:

$$\frac{1}{\lambda_j} \log ||\varphi^{\mathbb{C}}_j||^2_{\partial M_\epsilon} \to 2\epsilon. \tag{97}$$

But $U(i\epsilon) = e^{-\epsilon\lambda_j}\varphi^{\mathbb{C}}_j$, hence $||\varphi^{\mathbb{C}}_\lambda||^2_{L^2(\partial M_\epsilon)}$ equals $e^{2\epsilon\lambda_j}$ times

$$\langle U(i\epsilon)\varphi_\lambda, U(i\epsilon)\varphi_\lambda \rangle = \langle U(i\epsilon)^* U(i\epsilon)\varphi_\lambda, \varphi_\lambda \rangle.$$

But $U(i\epsilon)^* U(i\epsilon)$ is a pseudodifferential operator of order $\frac{n-1}{2}$. Its symbol $|\xi|^{-\frac{n-1}{2}}$ doesn't contribute to the logarithm.

We now provide more details on each step.

10.2. Growth properties of complexified eigenfunctions

In this section we prove Lemma 10.4 in more detail. We state it in combination with (97).

THEOREM 10.5. *If the geodesic flow is ergodic, then for all but a sparse subsequence of* λ_j,

$$\frac{1}{\lambda_{j_k}} \log |\varphi_{j_k}^{\mathbb{C}}(z)|^2 \to \sqrt{\rho} \text{ in } L^1(M_\epsilon).$$

The Grauert tube function is a maximal PSH function with bound $\leq \epsilon$ on M_ϵ. Hence Theorem 10.5 says that ergodic eigenfunctions have the maximum exponential growth rate possible for any eigenfunctions.

A key object in the proof is the sequence of functions $U_\lambda(x,\xi) \in C^\infty(M_\epsilon)$ defined by

$$(98)\qquad \begin{cases} U_\lambda(x,\xi) := \frac{\varphi_\lambda^{\mathbb{C}}(x,\xi)}{\rho_\lambda(x,\xi)}, \quad (x,\xi) \in M_\epsilon, \quad \text{where} \\ \\ \rho_\lambda(x,\xi) := ||\varphi_\lambda^{\mathbb{C}}|_{\partial M_{|\xi|_g}}||_{L^2(\partial M_{|\xi|_g})} \end{cases}$$

Thus, $\rho_\lambda(x,\xi)$ is the L^2-norm of the restriction of $\varphi_\lambda^{\mathbb{C}}$ to the sphere bundle $\{\partial M_\epsilon\}$ where $\epsilon = |\xi|_g$. U_λ is of course not holomorphic, but its restriction to each sphere bundle is CR holomorphic there, i.e.

$$(99)\qquad u_\lambda^\epsilon = U_\lambda|_{\partial M_\epsilon} \in \mathcal{O}^0(\partial M_\epsilon).$$

Our first result gives an ergodicity property of holomorphic continuations of ergodic eigenfunctions.

LEMMA 10.6. *Assume that* $\{\varphi_{j_k}\}$ *is a quantum ergodic sequence of* Δ*-eigenfunctions on* M *in the sense of (27). Then for each* $0 < \epsilon < \epsilon_0$,

$$|U_{j_k}|^2 \to \frac{1}{\mu_1(S^*M)} \sqrt{\rho}^{-m+1}, \quad \text{weakly in } L^1(M_\epsilon, \omega^m).$$

We note that $\omega^m = r^{m-1} dr d\omega dvol(x)$ in polar coordinates, so the right side indeed lies in L^1. The actual limit function is otherwise irrelevant. The next step is to use a compactness argument to obtain strong convergence of the normalized logarithms of the sequence $\{|U_\lambda|^2\}$. The first statement of the following lemma immediately implies the second.

LEMMA 10.7. *Assume that* $|U_{j_k}|^2 \to \frac{1}{\mu_\epsilon(\partial M_\epsilon)} \sqrt{\rho}^{-m+1}$, *weakly in* $L^1(M_\epsilon, \omega^m)$. *Then:*

(1) $\frac{1}{\lambda_{j_k}} \log |U_{j_k}|^2 \to 0$ *strongly in* $L^1(M_\epsilon)$.
(2) $\frac{1}{\lambda_j} \partial\bar{\partial} \log |U_{j_k}|^2 \to 0$, *weakly in* $\mathcal{D}'^{(1,1)}(M_\epsilon)$.

Separating out the numerator and denominator of $|U_j|^2$, we obtain that

$$(100)\qquad \frac{1}{\lambda_{j_k}} \partial\bar{\partial} \log |\varphi_{j_k}^{\mathbb{C}}|^2 - \frac{2}{\lambda_{j_k}} \partial\bar{\partial} \log \rho_{\lambda_{j_k}} \to 0, \quad (\lambda_{j_k} \to \infty).$$

The next lemma shows that the second term has a weak limit:

LEMMA 10.8. *For* $0 < \epsilon < \epsilon_0$,

$$\frac{1}{\lambda j_k} \log \rho_{\lambda j_k}(x,\xi) \to \sqrt{\rho}, \quad in \quad L^1(M_\epsilon) \ as \ \lambda j_k \to \infty.$$

Hence,

$$\frac{1}{\lambda_{j_k}} \partial\bar{\partial} \log \rho_{\lambda_{j_k}} \to \partial\bar{\partial}\sqrt{\rho} \quad (\lambda_j \to \infty) \ weakly \ in \ \mathcal{D}'(M_\epsilon).$$

It follows that the left side of (100) has the same limit, and that will complete the proof of Theorem 10.1.

10.3. Proof of Lemma 10.6 and Theorem 10.3

We begin by proving a weak limit formula for the CR holomorphic functions u_λ^ϵ defined in (99) for fixed ϵ.

LEMMA 10.9. *Assume that* $\{\varphi_{j_k}\}$ *is a quantum ergodic sequence. Then for each* $0 < \epsilon < \epsilon_0$,

$$|u_{j_k}^\epsilon|^2 \to \frac{1}{\mu_\epsilon(\partial M_\epsilon)}, \quad weakly \ in \ L^1(\partial M_\epsilon, d\mu_\epsilon).$$

That is, for any $a \in C(\partial M_\epsilon)$,

$$\int_{\partial M_\epsilon} a(x,\xi)|u_{j_k}^\epsilon((x,\xi)|^2 d\mu_\epsilon \to \frac{1}{\mu_\epsilon(\partial M_\epsilon)} \int_{\partial M_\epsilon} a(x,\xi) d\mu_\epsilon.$$

PROOF. It suffices to consider $a \in C^\infty(\partial M_\epsilon)$. We then consider the Toeplitz operator $\Pi_\epsilon a \Pi_\epsilon$ on $\mathcal{O}^0(\partial M_\epsilon)$. We have,

$$\begin{aligned} \langle \Pi_\epsilon a \Pi_\epsilon u_j^\epsilon, u_j^\epsilon \rangle &= e^{2\epsilon\lambda_j} \, ||\varphi_\lambda^{\mathbb{C}}||^{-2}_{L^2(\partial M_\epsilon)} \langle \Pi_\epsilon a \Pi_\epsilon U(i\epsilon)\varphi_j, U(i\epsilon)\varphi_j \rangle_{L^2(\partial M_\epsilon)} \\ &= e^{2\epsilon\lambda_j} \, ||\varphi_\lambda^{\mathbb{C}}||^{-2}_{L^2(\partial M_\epsilon)} \langle U(i\epsilon)^* \Pi_\epsilon a \Pi_\epsilon U(i\epsilon)\varphi_j, \varphi_j \rangle_{L^2(M)}. \end{aligned} \tag{101}$$

It is not hard to see that $U(i\epsilon)^*\Pi_\epsilon a \Pi_\epsilon U(i\epsilon)$ is a pseudodifferential operator on M of order $-\frac{m-1}{2}$ with principal symbol $\tilde{a}|\xi|_g^{-\frac{m-1}{2}}$, where $\tilde{a}$ is the (degree 0) homogeneous extension of a to $T^*M - 0$. The normalizing factor $e^{2\epsilon\lambda_j} \, ||\varphi_\lambda^{\mathbb{C}}||^{-2}_{L^2(\partial B_\epsilon^* M)}$ has the same form with $a = 1$. Hence, the expression on the right side of (101) may be written as

$$\frac{\langle U(i\epsilon)^* \Pi_\epsilon a \Pi_\epsilon U(i\epsilon)\varphi_j, \varphi_j \rangle_{L^2(M)}}{\langle U(i\epsilon)^* \Pi_\epsilon U(i\epsilon)\varphi_j, \varphi_j \rangle_{L^2(M)}}. \tag{102}$$

By the standard quantum ergodicity result on compact Riemannian manifolds with ergodic geodesic flow (see [**Shn, Ze4, CV**] for proofs and references) we have

$$\frac{\langle U(i\epsilon)^* \Pi_\epsilon a \Pi_\epsilon U(i\epsilon)\varphi_{j_k}, \varphi_{j_k} \rangle_{L^2(M)}}{\langle U(i\epsilon)^* \Pi_\epsilon U(i\epsilon)\varphi_{j_k}, \varphi_{j_k} \rangle_{L^2(M)}} \to \frac{1}{\mu_\epsilon(\partial M_\epsilon)} \int_{\partial M_\epsilon} a d\mu_\epsilon. \tag{103}$$

More precisely, the numerator is asymptotic to the right side times $\lambda^{-\frac{m-1}{2}}$, while the denominator has the same asymptotics when a is replaced by 1. We also use that $\frac{1}{\mu_\epsilon(\partial M_\epsilon)} \int_{\partial M_\epsilon} a d\mu_\epsilon$ equals the analogous average of $\tilde{a}$ over ∂M_ϵ. Taking the ratio produces (103).

Combining (101), (103) and the fact that

$$\langle \Pi_\epsilon a \Pi_\epsilon u_j^\epsilon, u_j^\epsilon \rangle = \int_{\partial B_\epsilon^* M} a|u_j^\epsilon|^2 d\mu_\epsilon$$

completes the proof of the lemma.

□

We now complete the proof of Lemma 10.6, i.e. we prove that

$$\int_{M_\epsilon} a|U_{j_k}|^2\omega^m \to \frac{1}{\mu_\epsilon(\partial M_\epsilon)}\int_{M_\epsilon} a\sqrt{\rho}^{-m+1}\omega^m \tag{104}$$

for any $a \in C(M_\epsilon)$. It is only necessary to relate the surface Liouville measures $d\mu_r$ (28) to the Kähler volume measure. One may write $d\mu_r = \frac{d}{dt}|_{t=r}\chi_t\omega^m$, where χ_t is the characteristic function of $M_t = \{\sqrt{\rho} \leq t\}$. By homogeneity of $|\xi|_g$, $\mu_r(\partial M_r) = r^{m-1}\mu_\epsilon(\partial M_\epsilon)$. If $a \in C(M_\epsilon)$, then $\int_{M_\epsilon} a\omega^m = \int_0^\epsilon\{\int_{\partial M_r} a d\mu_r\}dr$. By Lemma 10.9, we have

(105)

$$\begin{array}{lll} \int_{M_\epsilon} a|U_{j_k}|^2\omega^m = \int_0^\epsilon\{\int_{\partial M_r} a|u_{j_k}^r|^2 d\mu_r\}dr & \to & \int_0^\epsilon\{\frac{1}{\mu_r(\partial B_r^*)}\int_{\partial M_r} a d\mu_r\}dr \\ & = & \frac{1}{\mu_\epsilon(\partial M_\epsilon)}\int_{M_\epsilon} ar^{-m+1}\omega^m, \\ & \Longrightarrow & w^* - \lim_{\lambda\to\infty}|U_{j_k}|^2 \\ & = & \frac{1}{\mu_1(\partial M_\epsilon)}\sqrt{\rho}^{-m+1}. \end{array}$$

10.4. Proof of Lemma 10.8

In fact, one has

$$\frac{1}{\lambda}\log\rho_\lambda(x,\xi) \to \sqrt{\rho}, \quad \text{uniformly in } \; M_\epsilon \; \text{ as } \lambda \to \infty.$$

PROOF. Again using $U(i\epsilon)\varphi_\lambda = e^{-\lambda\epsilon}\varphi_\lambda^{\mathbb{C}}$, we have:

$$\begin{aligned} \rho_\lambda^2(x,\xi) &= \langle \Pi_\epsilon\varphi_\lambda^{\mathbb{C}}, \Pi_\epsilon\varphi_\lambda^{\mathbb{C}}\rangle_{L^2(\partial B_\epsilon^* M)} \quad (\epsilon = |\xi|_{g_x}) \\ &= e^{2\lambda\epsilon}\langle \Pi_\epsilon U(i\epsilon)\varphi_\lambda, \Pi_\epsilon U(i\epsilon)\varphi_\lambda\rangle_{L^2(\partial B_\epsilon^* M)} \\ &= e^{2\lambda\epsilon}\langle U(i\epsilon)^*\Pi_\epsilon U(i\epsilon)\varphi_\lambda, \varphi_\lambda\rangle_{L^2(M)}. \end{aligned} \tag{106}$$

Hence,

$$\frac{2}{\lambda}\log\rho_\lambda(x,\xi) = 2|\xi|_{g_x} + \frac{1}{\lambda}\log\langle U(i\epsilon)^*\Pi_\epsilon U(i\epsilon)\varphi_\lambda, \varphi_\lambda\rangle. \tag{107}$$

The second term on the right side is the matrix element of a pseudo-differential operator, hence is bounded by some power of λ. Taking the logarithm gives a remainder of order $\frac{\log\lambda}{\lambda}$.

□

10.5. Proof of Lemma 10.7

PROOF. We wish to prove that

$$\psi_j := \frac{1}{\lambda_j}\log|U_j|^2 \to 0 \;\text{ in } L^1(M_\epsilon).$$

As we have said, this is almost obvious from Lemmas 10.6 and 10.9. If the conclusion is not true, then there exists a subsequence ψ_{j_k} satisfying $||\psi_{j_k}||_{L^1(B_\epsilon^* M)} \geq \delta > 0$. To obtain a contradiction, we use Lemma 8.1.

To see that the hypotheses are satisfied in our example, it suffices to prove these statements on each surface ∂M_ϵ with uniform constants independent of ϵ. On the surface ∂M_ϵ, $U_j = u_j^\epsilon$. By the Sobolev inequality in $\mathcal{O}^{\frac{m-1}{4}}(\partial M_\epsilon)$, we have

$$\begin{aligned} \sup_{(x,\xi)\in\partial M_\epsilon} |u_j^\epsilon(x,\xi)| &\leq \lambda_j^m ||u_j^\epsilon(x,\xi)||_{L^2(\partial M_\epsilon)} \\ &\leq \lambda_j^m. \end{aligned}$$

Taking the logarithm, dividing by λ_j, and combining with the limit formula of Lemma 10.8 proves (i) - (ii).

We now settle the dichotomy above by proving that the sequence $\{\psi_j\}$ does not tend uniformly to $-\infty$ on compact sets. That would imply that $\psi_j \to -\infty$ uniformly on the spheres ∂M_ϵ for each $\epsilon < \epsilon_0$. Hence, for each ϵ, there would exist $K > 0$ such that for $k \geq K$,

$$\frac{1}{\lambda_{j_k}} \log |u_{j_k}^\epsilon(z)| \leq -1. \tag{108}$$

However, (108) implies that

$$|u_{j_k}(z)| \leq e^{-2\lambda_{j_k}} \quad \forall z \in \partial M_\epsilon ,$$

which is inconsistent with the hypothesis that $|u_{j_k}^\epsilon(z)| \to 1$ in $\mathcal{D}'(\partial M_\epsilon)$.

Therefore, there must exist a subsequence, which we continue to denote by $\{\psi_{j_k}\}$, which converges in $L^1(M_{\epsilon_0})$ to some $\psi \in L^1(M_{\epsilon_0})$. Then,

$$\psi(z) = \limsup_{k\to\infty} \psi_{j_k} \leq 2|\xi|_g \qquad \text{(a.e)} .$$

Now let

$$\psi^*(z) := \limsup_{w\to z} \psi(w) \leq 0$$

be the upper-semicontinuous regularization of ψ. Then ψ^* is plurisubharmonic on M_ϵ and $\psi^* = \psi$ almost everywhere.

If $\psi^* \leq 2|\xi|_g - \delta$ on a set U_δ of positive measure, then $\psi_{j_k}(\zeta) \leq -\delta/2$ for $\zeta \in U_\delta$, $k \geq K$; i.e.,

$$|\psi_{j_k}(\zeta)| \leq e^{-\delta\lambda_{j_k}}, \qquad \zeta \in U_\delta, \ \ k \geq K. \tag{109}$$

This contradicts the weak convergence to 1 and concludes the proof. □

11. Intersections of nodal sets and analytic curves on real analytic surfaces

It is often possible to obtain more refined results on nodal sets by studying their intersections with some fixed (and often special) hypersurface. This has been most successful in dimension two. In §11.1 we discuss upper bounds on the number of intersection points of the nodal set with the bounary of a real analytic plane domain and more general 'good' analytic curves. To obtain lower bounds or asymptotics, we need to add some dynamical hypotheses. In case of ergodic geodesic flow, we can obtain equidistribution theorems for intersections of nodal sets and geodesics on surfaces. The dimensional restriction is due to the fact that the results are partly based on the quantum ergodic restriction theorems of [**TZ, TZ2**], which concern restrictions of eigenfunctions to hypersurfaces. Nodal sets and geodesics have complementary dimensions and intersect in points, and therefore it makes sense to count the number of intersections. But we do not yet have a mechanism for studying restrictions to geodesics when $\dim M \geq 3$.

11.1. Counting nodal lines which touch the boundary in analytic plane domains

In this section, we review the results of [**TZ**] giving upper bounds on the number of intersections of the nodal set with the boundary of an analytic (or more generally piecewise analytic) plane domain. One may expect that the results of this section can also be generalized to higher dimensions by measuring codimension two nodal hypersurface volumes within the boundary.

Thus we would like to count the number of nodal lines (i.e. components of the nodal set) which touch the boundary. Here we assume that 0 is a regular value so that components of the nodal set are either loops in the interior (closed nodal loops) or curves which touch the boundary in two points (open nodal lines). It is known that for generic piecewise analytic plane domains, zero is a regular value of all the eigenfunctions φ_{λ_j}, i.e. $\nabla \varphi_{\lambda_j} \neq 0$ on $\mathcal{N}_{\varphi_{\lambda_j}}$ [**U**]; we then call the nodal set regular. Since the boundary lies in the nodal set for Dirichlet boundary conditions, we remove it from the nodal set before counting components. Henceforth, the number of components of the nodal set in the Dirichlet case means the number of components of $\mathcal{N}_{\varphi_{\lambda_j}} \backslash \partial\Omega$.

We now sketch the proof of Theorems 6 in the case of Neumann boundary conditions. By a piecewise analytic domain $\Omega^2 \subset \mathbb{R}^2$, we mean a compact domain with piecewise analytic boundary, i.e. $\partial\Omega$ is a union of a finite number of piecewise analytic curves which intersect only at their common endpoints. Such domains are often studied as archtypes of domains with ergodic billiards and quantum chaotic eigenfunctions, in particular the Bunimovich stadium or Sinai billiard.

For the Neumann problem, the boundary nodal points are the same as the zeros of the boundary values $\varphi_{\lambda_j}|_{\partial\Omega}$ of the eigenfunctions. The number of boundary nodal points is thus twice the number of open nodal lines. Hence in the Neumann case, the Theorem follows from:

THEOREM 11.1. *Suppose that $\Omega \subset \mathbb{R}^2$ is a piecewise real analytic plane domain. Then the number $n(\lambda_j) = \#\mathcal{N}_{\varphi_{\lambda_j}} \cap \partial\Omega$ of zeros of the boundary values $\varphi_{\lambda_j}|_{\partial\Omega}$ of the jth Neumann eigenfunction satisfies $n(\lambda_j) \leq C_\Omega \lambda_j$, for some $C_\Omega > 0$.*

This is a more precise version of Theorem 6 since it does not assume that 0 is a regular value. We prove Theorem 11.1 by analytically continuing the boundary values of the eigenfunctions and counting *complex zeros and critical points* of analytic continuations of Cauchy data of eigenfunctions. When $\partial\Omega \in C^\omega$, the eigenfunctions can be holomorphically continued to an open tube domain in $\mathbb{C}^2$ projecting over an open neighborhood W in $\mathbb{R}^2$ of Ω which is independent of the eigenvalue. We denote by $\Omega_{\mathbb{C}} \subset \mathbb{C}^2$ the points $\zeta = x + i\xi \in \mathbb{C}^2$ with $x \in \Omega$. Then $\varphi_{\lambda_j}(x)$ extends to a holomorphic function $\varphi_{\lambda_j}^{\mathbb{C}}(\zeta)$ where $x \in W$ and where $|\xi| \leq \epsilon_0$ for some $\epsilon_0 > 0$.

Assuming $\partial\Omega$ real analytic, we define the (interior) complex nodal set by

$$\mathcal{N}_{\varphi_{\lambda_j}}^{\mathbb{C}} = \{\zeta \in \Omega_{\mathbb{C}} : \varphi_{\lambda_j}^{\mathbb{C}}(\zeta) = 0\}.$$

THEOREM 11.2. *Suppose that $\Omega \subset \mathbb{R}^2$ is a piecewise real analytic plane domain, and denote by $(\partial\Omega)_{\mathbb{C}}$ the union of the complexifications of its real analytic boundary components.*

(1) *Let* $n(\lambda_j, \partial\Omega_{\mathbb{C}}) = \#Z^{\partial\Omega_{\mathbb{C}}}_{\varphi_{\lambda_j}}$ *be the number of complex zeros on the complex boundary. Then there exists a constant* $C_\Omega > 0$ *independent of the radius of* $(\partial\Omega)_{\mathbb{C}}$ *such that* $n(\lambda_j, \partial\Omega_{\mathbb{C}}) \leq C_\Omega \lambda_j$.

The theorems on real nodal lines and critical points follow from the fact that real zeros and critical points are also complex zeros and critical points, hence

$$n(\lambda_j) \leq n(\lambda_j, \partial\Omega_{\mathbb{C}}). \tag{110}$$

All of the results are sharp, and are already obtained for certain sequences of eigenfunctions on a disc (see §4.6).

To prove 11.2, we represent the analytic continuations of the boundary values of the eigenfunctions in terms of layer potentials. Let $G(\lambda_j, x_1, x_2)$ be any 'Green's function' for the Helmholtz equation on Ω, i.e. a solution of $(-\Delta - \lambda_j^2)G(\lambda_j, x_1, x_2) = \delta_{x_1}(x_2)$ with $x_1, x_2 \in \bar{\Omega}$. By Green's formula,

$$\varphi_{\lambda_j}(x, y) = \int_{\partial\Omega} \big(\partial_\nu G(\lambda_j, q, (x, y))\varphi_{\lambda_j}(q) - G(\lambda_j, q, (x, y))\partial_\nu \varphi_{\lambda_j}(q)\big)\, d\sigma(q), \tag{111}$$

where $(x, y) \in \mathbb{R}^2$, where $d\sigma$ is arc-length measure on $\partial\Omega$ and where ∂_ν is the normal derivative by the interior unit normal. Our aim is to analytically continue this formula.

In the case of Neumann eigenfunctions φ_λ in Ω,

$$\varphi_{\lambda_j}(x, y) = \int_{\partial\Omega} \frac{\partial}{\partial \nu_q} G(\lambda_j, q, (x, y)) u_{\lambda_j}(q) d\sigma(q), \quad (x, y) \in \Omega^o \text{ (Neumann)}. \tag{112}$$

To obtain concrete representations we need to choose G. We choose the real ambient Euclidean Green's function S

$$S(\lambda_j, \xi, \eta; x, y) = -Y_0(\lambda_j r((x, y); (\xi, \eta))), \tag{113}$$

where $r = \sqrt{zz^*}$ is the distance function (the square root of r^2 above) and where Y_0 is the Bessel function of order zero of the second kind. The Euclidean Green's function has the form

$$S(\lambda_j, \xi, \eta; x, y) = A(\lambda_j, \xi, \eta; x, y) \log \frac{1}{r} + B(\lambda_j, \xi, \eta; x, y), \tag{114}$$

where A and B are entire functions of r^2. The coefficient $A = J_0(\lambda_j r)$ is known as the Riemann function.

By the 'jumps' formulae, the double layer potential $\frac{\partial}{\partial \nu_{\tilde{q}}} S(\lambda_j, \tilde{q}, (x, y))$ on $\partial\Omega \times \bar{\Omega}$ restricts to $\partial\Omega \times \partial\Omega$ as $\frac{1}{2}\delta_q(\tilde{q}) + \frac{\partial}{\partial \nu_{\tilde{q}}} S(\lambda_j, \tilde{q}, q)$ (see e.g. [**TI, TII**]). Hence in the Neumann case the boundary values u_{λ_j} satisfy,

$$u_{\lambda_j}(q) = 2 \int_{\partial\Omega} \frac{\partial}{\partial \nu_{\tilde{q}}} S(\lambda_j, \tilde{q}, q) u_{\lambda_j}(\tilde{q}) d\sigma(\tilde{q}) \text{ (Neumann)}. \tag{115}$$

We have,

$$\frac{\partial}{\partial \nu_{\tilde{q}}} S(\lambda_j, \tilde{q}, q) = -\lambda_j Y_1(\lambda_j r) \cos \angle(q - \tilde{q}, \nu_{\tilde{q}}). \tag{116}$$

It is equivalent, and sometimes more convenient, to use the (complex valued) Euclidean outgoing Green's function $\mathrm{Ha}_0^{(1)}(kz)$, where $\mathrm{Ha}_0^{(1)} = J_0 + iY_0$ is the Hankel function of order zero. It has the same form as (114) and only differs by

the addition of the even entire function J_0 to the B term. If we use the Hankel free outgoing Green's function, then in place of (116) we have the kernel
(117)

$$\begin{aligned} N(\lambda_j, q(s), q(s')) &= \tfrac{i}{2}\partial_{\nu_y} \operatorname{Ha}_0^{(1)}(\lambda_j|q(s)-y|)|_{y=q(s')} \\ &= -\tfrac{i}{2}\lambda_j \operatorname{Ha}_1^{(1)}(\lambda_j|q(s)-q(s')|)\cos\angle(q(s')-q(s),\nu_{q(s')}), \end{aligned}$$

and in place of (115) we have the formula

$$u_{\lambda_j}(q(t)) = \int_0^{2\pi} N(\lambda_j, q(s), q(t))\, u_{\lambda_j}(q(s))ds. \tag{118}$$

The next step is to analytically continue the layer potential representations (115) and (118). The main point is to express the analytic continuations of Cauchy data of Neumann and Dirichlet eigenfunctions in terms of the real Cauchy data. For brevity, we only consider (115) but essentially the same arguments apply to the free outgoing representation (118).

As mentioned above, both $A(\lambda_j, \xi, \eta, x, y)$ and $B(\lambda_j, \xi, \eta, x, y)$ admit analytic continuations. In the case of A, we use a traditional notation $R(\zeta, \zeta^*, z, z^*)$ for the analytic continuation and for simplicity of notation we omit the dependence on λ_j.

The details of the analytic continuation are complicated when the curve is the boundary, and they simplify when the curve is interior. So we only continue the sketch of the proof in the interior case.

As above, the arc-length parametrization of C is denoed by by $q_C : [0, 2\pi] \to C$ and the corresponding arc-length parametrization of the boundary, $\partial\Omega$, by $q : [0, 2\pi] \to \partial\Omega$. Since the boundary and C do not intersect, the logarithm

$$\log r^2(q(s); q_C^C(t))$$

is well defined and the holomorphic continuation of equation (118) is given by:

$$\varphi_{\lambda_j}^{\mathbb{C}}(q_C^{\mathbb{C}}(t)) = \int_0^{2\pi} N(\lambda_j, q(s), q_C^{\mathbb{C}}(t))\, u_{\lambda_j}(q(s))d\sigma(s), \tag{119}$$

From the basic formula (117) for $N(\lambda_j, q, q_C)$ and the standard integral formula for the Hankel function $\operatorname{Ha}_1^{(1)}(z)$, one easily gets an asymptotic expansion in λ_j of the form:

$$\begin{aligned} N(\lambda_j, q(s), q_C^{\mathbb{C}}(t)) &= e^{i\lambda_j r(q(s); q_C^{\mathbb{C}}(t))} \sum_{m=0}^{k} a_m(q(s), q_C^{\mathbb{C}}(t))\, \lambda_j^{1/2-m} \\ &\quad + O(e^{i\lambda_j r(q(s); q_C^{\mathbb{C}}(t))}\, \lambda_j^{1/2-k-1}). \end{aligned} \tag{120}$$

Note that the expansion in (120) is valid since for interior curves,

$$C_0 := \min_{(q_C(t), q(s)) \in C \times \partial\Omega} |q_C(t) - q(s)|^2 > 0.$$

Then, $\operatorname{Re} r^2(q(s); q_C^C(t)) > 0$ as long as

$$|\operatorname{Im} q_C^{\mathbb{C}}(t)|^2 < C_0. \tag{121}$$

So, the principal square root of r^2 has a well-defined holomorphic extension to the tube (121) containing C. We have denoted this square root by r in (120).

Substituting (120) in the analytically continued single layer potential integral formula (119) proves that for $t \in A(\epsilon)$ and $\lambda_j > 0$ sufficiently large,
(122)

$$\varphi_{\lambda_j}^{\mathbb{C}}(q_C^{\mathbb{C}}(t)) = 2\pi\lambda_j^{1/2} \int_0^{2\pi} e^{i\lambda_j r(q(s):q_C^{\mathbb{C}}(t))} a_0(q(s), q_C^{\mathbb{C}}(t))(1+O(\lambda_j^{-1}))\, u_{\lambda_j}(q(s))d\sigma(s).$$

Taking absolute values of the integral on the RHS in (122) and applying the Cauchy-Schwartz inequality proves

LEMMA 11.3. *For $t \in [0, 2\pi] + i[-\epsilon, \epsilon]$ and $\lambda_j > 0$ sufficiently large*

$$|\varphi_{\lambda_j}^{\mathbb{C}}(q_C^{\mathbb{C}}(t))| \leq C_1 \lambda_j^{1/2} \exp \lambda_j \left(\max_{q(s) \in \partial\Omega} \mathrm{Re}\, ir(q(s); q_C^{\mathbb{C}}(t))\right) \cdot \|u_{\lambda_j}\|_{L^2(\partial\Omega)}.$$

From the pointwise upper bounds in Lemma 11.3, it is immediate that

$$\text{(123)}\quad \log \max_{q_C^{\mathbb{C}}(t) \in Q_C^{\mathbb{C}}(A(\epsilon))} |\varphi_{\lambda_j}^{\mathbb{C}}(q_C^{\mathbb{C}}(t))| \leq C_{\max}\lambda_j + C_2 \log \lambda_j + \log \|u_{\lambda_j}\|_{L^2(\partial\Omega)},$$

where,

$$C_{\max} = \max_{(q(s), q_C^{\mathbb{C}}(t)) \in \partial\Omega \times Q_C^{\mathbb{C}}(A(\epsilon))} \mathrm{Re}\, ir(q(s); q_C^{\mathbb{C}}(t)).$$

Finally, we use that $\log \|u_{\lambda_j}\|_{L^2(\partial\Omega)} = O(\lambda_j)$ by the assumption that C is a good curve and apply Proposition 11.4 to get that $n(\lambda_j, C) = O(\lambda_j)$.

The following estimate, suggested by Lemma 6.1 of Donnelly-Fefferman [**DF**], gives an upper bound on the number of zeros in terms of the growth of the family:

PROPOSITION 11.4. *Suppose that C is a good real analytic curve in the sense of (29). Normalize u_{λ_j} so that $||u_{\lambda_j}||_{L^2(C)} = 1$. Then, there exists a constant $C(\epsilon) > 0$ such that for any $\epsilon > 0$,*

$$n(\lambda_j, Q_C^{\mathbb{C}}(A(\epsilon/2))) \leq C(\epsilon) \max_{q_C^{\mathbb{C}}(t) \in Q_C^{\mathbb{C}}(A(\epsilon))} \log |u_{\lambda_j}^{\mathbb{C}}(q_C^{\mathbb{C}}(t))|.$$

PROOF. Let G_ϵ denote the Dirichlet Green's function of the 'annulus' $Q_C^{\mathbb{C}}(A(\epsilon))$. Also, let $\{a_k\}_{k=1}^{n(\lambda_j, Q_C^{\mathbb{C}}(A(\epsilon/2)))}$ denote the zeros of $u_{\lambda_j}^{\mathbb{C}}$ in the sub-annulus $Q_C^{\mathbb{C}}(A(\epsilon/2))$. Let $U_{\lambda_j} = \frac{u_{\lambda_j}^{\mathbb{C}}}{||u_{\lambda_j}^{\mathbb{C}}||_{Q_C^{\mathbb{C}}(A(\epsilon))}}$ where $||u||_{Q_C^{\mathbb{C}}(A(\epsilon))} = \max_{\zeta \in Q_C^{\mathbb{C}}(A(\epsilon))} |u(\zeta)|$. Then,

$$\begin{aligned}\log |U_{\lambda_j}(q_C^{\mathbb{C}}(t))| &= \int_{Q_C^{\mathbb{C}}((A(\epsilon/2)))} G_\epsilon(q_C^{\mathbb{C}}(t), w)\partial\bar{\partial} \log |u_{\lambda_j}^{\mathbb{C}}(w)| + H_{\lambda_j}(q_C^{\mathbb{C}}(t)) \\ &= \sum_{a_k \in Q_C^{\mathbb{C}}(A(\epsilon/2)): u_{\lambda_j}^{\mathbb{C}}(a_k)=0} G_\epsilon(q_C^{\mathbb{C}}(t), a_k) + H_{\lambda_j}(q_C^{\mathbb{C}}(t)),\end{aligned}$$

since $\partial\bar{\partial} \log |u_{\lambda_j}^{\mathbb{C}}(w)| = \sum_{a_k \in C_{\mathbb{C}}: u_{\lambda_j}^{\mathbb{C}}(a_k)=0} \delta_{a_k}$. Moreover, the function H_{λ_j} is subharmonic on $Q_C^{\mathbb{C}}(A(\epsilon))$ since

$$\begin{aligned}\partial\bar{\partial} H_{\lambda_j} &= \partial\bar{\partial} \log |U_{\lambda_j}(q_C^{\mathbb{C}}(t))| - \sum_{a_k \in Q_C^{\mathbb{C}}(A(\epsilon/2)): u_{\lambda_j}^{\mathbb{C}}(a_k)=0} \partial\bar{\partial} G_\epsilon(q_C^{\mathbb{C}}(t), a_k) \\ &= \sum_{a_k \in Q_C^{\mathbb{C}}(A(\epsilon)) \backslash Q_C^{\mathbb{C}}(A(\epsilon/2))} \delta_{a_k} > 0.\end{aligned}$$

So, by the maximum principle for subharmonic functions,

$$\max_{Q_C^{\mathbb{C}}(A(\epsilon))} H_{\lambda_j}(q_C^{\mathbb{C}}(t)) \leq \max_{\partial Q_C^{\mathbb{C}}(A(\epsilon))} H_{\lambda_j}(q_C^{\mathbb{C}}(t)) = \max_{\partial Q_C^{\mathbb{C}}(A(\epsilon))} \log |U_{\lambda_j}(q_C^{\mathbb{C}}(t))| = 0.$$

It follows that

$$\log |U_{\lambda_j}(q_C^{\mathbb{C}}(t))| \leq \sum_{a_k \in Q_C^{\mathbb{C}}(A(\epsilon/2)): u_{\lambda_j}^{\mathbb{C}}(a_k)=0} G_\epsilon(q_C^{\mathbb{C}}(t), a_k), \tag{124}$$

hence that

$$\begin{aligned} &\max_{q_C^{\mathbb{C}}(t) \in Q_C^{\mathbb{C}}(A(\epsilon/2))} \log |U_{\lambda_j}(q_C^{\mathbb{C}}(t))| \\ &\leq \left(\max_{z,w \in Q_C^{\mathbb{C}}(A(\epsilon/2))} G_\epsilon(z,w)\right) \; n(\lambda_j, Q_C^{\mathbb{C}}(A(\epsilon/2))). \end{aligned} \tag{125}$$

Now $G_\epsilon(z,w) \leq \max_{w \in Q_C^{\mathbb{C}}(\partial A(\epsilon))} G_\epsilon(z,w) = 0$ and $G_\epsilon(z,w) < 0$ for $z, w \in Q_C^{\mathbb{C}}(A(\epsilon/2))$. It follows that there exists a constant $\nu(\epsilon) < 0$ so that

$$\max_{z,w \in Q_C^{\mathbb{C}}(A(\epsilon/2))} G_\epsilon(z,w) \leq \nu(\epsilon).$$

Hence,

$$\max_{q_C^{\mathbb{C}}(t) \in Q_C^{\mathbb{C}}(A(\epsilon/2))} \log |U_{\lambda_j}(Q_C^{\mathbb{C}}(t))| \leq \nu(\epsilon) \; n(\lambda_j, Q_C^{\mathbb{C}}(A(\epsilon/2))). \tag{126}$$

Since both sides are negative, we obtain

$$\begin{aligned} &n(\lambda_j, Q_C^{\mathbb{C}}(A(\epsilon/2))) \leq \frac{1}{|\nu(\epsilon)|} \left| \max_{q_C^{\mathbb{C}}(t) \in Q_C^{\mathbb{C}}(A(\epsilon/2))} \log |U_{\lambda_j}(q_C^{\mathbb{C}}(t))| \right| \\ &\leq \frac{1}{|\nu(\epsilon)|} \left(\max_{q_C^{\mathbb{C}}(t) \in Q_C^{\mathbb{C}}(A(\epsilon))} \log |u_{\lambda_j}^{\mathbb{C}}(q_C^{\mathbb{C}}(t))| - \max_{q_C^{\mathbb{C}}(t) \in Q_C^{\mathbb{C}}(A(\epsilon/2))} \log |u_{\lambda_j}^{\mathbb{C}}(q_C^{\mathbb{C}}(t))| \right) \\ &\leq \frac{1}{|\nu(\epsilon)|} \; \max_{q_C^{\mathbb{C}}(t) \in Q_C^{\mathbb{C}}(A(\epsilon))} \log |u_{\lambda_j}^{\mathbb{C}}(q_C^{\mathbb{C}}(t))|, \end{aligned} \tag{127}$$

where in the last step we use that $\max_{q_C^{\mathbb{C}}(t) \in Q_C^{\mathbb{C}}(A(\epsilon/2))} \log |u_{\lambda_j}^{\mathbb{C}}(q_C^{\mathbb{C}}(t))| \geq 0$, which holds since $|u_{\lambda_j}^{\mathbb{C}}| \geq 1$ at some point in $Q_C^{\mathbb{C}}(A(\epsilon/2))$. Indeed, by our normalization, $\|u_{\lambda_j}\|_{L^2(C)} = 1$, and so there must already exist points on the real curve C with $|u_{\lambda_j}| \geq 1$. Putting $C(\epsilon) = \frac{1}{|\nu(\epsilon)|}$ finishes the proof. □

This completes the proof of Theorem 7.

11.2. Application to Pleijel's conjecture

I. Polterovich [**Po**] observed that Theorem 6 can be used to prove an old conjecture of A. Pleijel regarding Courant's nodal domain theorem, which says that the number n_k of nodal domains (components of $\Omega \backslash Z_{\varphi_{\lambda_k}}$) of the kth eigenfunction satisfies $n_k \leq k$. Pleijel improved this result for Dirichlet eigefunctions of plane domains: For any plane domain with Dirichlet boundary conditions, $\limsup_{k \to \infty} \frac{n_k}{k} \leq \frac{4}{j_1^2} \simeq 0.691...$, where j_1 is the first zero of the J_0 Bessel function. He conjectured that the same result should be true for a free membrane, i.e. for Neumann boundary conditions. This was recently proved in the real analytic case by I. Polterovich [**Po**]. His argument is roughly the following: Pleijel's original argument applies to all nodal domains which do not touch the boundary, since the eigenfunction is a Dirichlet eigenfunction in such a nodal domain. The argument does not apply to nodal domains which touch the boundary, but by the Theorem above the number of such domains is negligible for the Pleijel bound.

11.3. Equidistribution of intersections of nodal lines and geodesics on surfaces

We fix $(x, \xi) \in S^*M$ and let

$$\gamma_{x,\xi} : \mathbb{R} \to M, \quad \gamma_{x,\xi}(0) = x, \quad \gamma'_{x,\xi}(0) = \xi \in T_xM \tag{128}$$

denote the corresponding parametrized geodesic. Our goal is to determine the asymptotic distribution of intersection points of $\gamma_{x,\xi}$ with the nodal set of a highly eigenfunction. As usual, we cannot cope with this problem in the real domain and therefore analytically continue it to the complex domain. Thus, we consider the intersections

$$\mathcal{N}_{\lambda_j}^{\gamma_{x,\xi}^{\mathbb{C}}} = Z_{\varphi_j^{\mathbb{C}}} \cap \gamma_{x,\xi}^{\mathbb{C}}$$

of the complex nodal set with the (image of the) complexification of a generic geodesic If

$$S_\epsilon = \{(t + i\tau \in \mathbb{C} : |\tau| \leq \epsilon\} \tag{129}$$

then $\gamma_{x,\xi}$ admits an analytic continuation

$$\gamma_{x,\xi}^{\mathbb{C}} : S_\epsilon \to M_\epsilon. \tag{130}$$

In other words, we consider the zeros of the pullback,

$$\{\gamma_{x,\xi}^* \varphi_\lambda^{\mathbb{C}} = 0\} \subset S_\epsilon.$$

We encode the discrete set by the measure

$$[\mathcal{N}_{\lambda_j}^{\gamma_{x,\xi}^{\mathbb{C}}}] = \sum_{(t+i\tau):\ \varphi_j^{\mathbb{C}}(\gamma_{x,\xi}^{\mathbb{C}}(t+i\tau))=0} \delta_{t+i\tau}. \tag{131}$$

We would like to show that for generic geodesics, the complex zeros on the complexified geodesic condense on the real points and become uniformly distributed with respect to arc-length. This does not always occur: as in our discussion of QER theorems, if $\gamma_{x,\xi}$ is the fixed point set of an isometric involution, then "odd" eigenfunctions under the involution will vanish on the geodesic. The additional hypothesis is that QER holds for $\gamma_{x,\xi}$. The following is proved ([**Ze3**]):

THEOREM 11.5. *Let (M^2, g) be a real analytic Riemannian surface with ergodic geodesic flow. Let $\gamma_{x,\xi}$ satisfy the QER hypothesis. Then there exists a subsequence of eigenvalues λ_{j_k} of density one such that for any $f \in C_c(S_\epsilon)$,*

$$\lim_{k\to\infty} \sum_{(t+i\tau):\ \varphi_j^{\mathbb{C}}(\gamma_{x,\xi}^{\mathbb{C}}(t+i\tau))=0} f(t + i\tau) = \int_{\mathbb{R}} f(t)dt.$$

In other words,

$$\text{weak}^* \lim_{k\to\infty} \frac{i}{\pi\lambda_{j_k}} [\mathcal{N}_{\lambda_j}^{\gamma_{x,\xi}^{\mathbb{C}}}] = \delta_{\tau=0},$$

in the sense of weak* convergence on $C_c(S_\epsilon)$. Thus, the complex nodal set intersects the (parametrized) complexified geodesic in a discrete set which is asymptotically (as $\lambda \to \infty$) concentrated along the real geodesic with respect to its arclength.

This concentration- equidistribution result is a 'restricted' version of the result of §10. As noted there, the limit distribution of complex nodal sets in the ergodic case is a singular current $dd^c\sqrt{\rho}$. The motivation for restricting to geodesics is that restriction magnifies the singularity of this current. In the case of a geodesic, the singularity is magnified to a delta-function; for other curves there is additionally a smooth background measure.

The assumption of ergodicity is crucial. For instance, in the case of a flat torus, say $\mathbb{R}^2/L$ where $L \subset \mathbb{R}^2$ is a generic lattice, the real eigenfunctions are $\cos\langle\lambda, x\rangle, \sin\langle\lambda, x\rangle$ where $\lambda \in L^*$, the dual lattice, with eigenvalue $-|\lambda|^2$. Consider a geodesic $\gamma_{x,\xi}(t) = x + t\xi$. Due to the flatness, the restriction $\sin\langle\lambda, x_0 + t\xi_0\rangle$ of the eigenfunction to a geodesic is an eigenfunction of the Laplacian $-\frac{d^2}{dt^2}$ of submanifold metric along the geodesic with eigenvalue $-\langle\lambda, \xi_0\rangle^2$. The complexification of the restricted eigenfunction is $\sin\langle\lambda, x_0 + (t + i\tau)\xi_0\rangle|$ and its exponent of its growth is $\tau|\langle\frac{\lambda}{|\lambda|}, \xi_0\rangle|$, which can have a wide range of values as the eigenvalue moves along different rays in L^*. The limit current is $i\partial\bar{\partial}$ applied to the limit and thus also has many limits

The proof involves several new principles which played no role in the global result of §10 and which are specific to geodesics. However, the first steps in the proof are the same as in the global case. By the Poincaré-Lelong formula, we may express the current of summation over the intersection points in (131) in the form,

$$[\mathcal{N}_{\lambda_j}^{\gamma_{x,\xi}^{\mathbb{C}}}] = i\partial\bar{\partial}_{t+i\tau} \log \left|\gamma_{x,\xi}^* \varphi_{\lambda_j}^{\mathbb{C}}(t + i\tau)\right|^2. \tag{132}$$

Thus, the main point of the proof is to determine the asymptotics of

$$\frac{1}{\lambda_j} \log \left|\gamma_{x,\xi}^* \varphi_{\lambda_j}^{\mathbb{C}}(t + i\tau)\right|^2.$$

When we freeze τ we put

$$\gamma_{x,\xi}^{\tau}(t) = \gamma_{x,\xi}^{\mathbb{C}}(t + i\tau). \tag{133}$$

PROPOSITION 11.6. *(Growth saturation) If $\{\varphi_{j_k}\}$ satisfies QER along any arcs of $\gamma_{x,\xi}$, then in $L^1_{loc}(S_\tau)$, we have*

$$\lim_{k\to\infty} \frac{1}{\lambda_{j_k}} \log \left|\gamma_{x,\xi}^{\tau *} \varphi_{\lambda_{j_k}}^{\mathbb{C}}(t + i\tau)\right|^2 = |\tau|.$$

Proposition 11.6 immediately implies Theorem 11.5 since we can apply $\partial\bar{\partial}$ to the L^1 convergent sequence $\frac{1}{\lambda_{j_k}} \log \left|\gamma_{x,\xi}^* \varphi_{\lambda_{j_k}}^{\mathbb{C}}(t + i\tau)\right|^2$ to obtain $\partial\bar{\partial}|\tau|$.

The upper bound in Proposition 11.6 follows immediately from the known global estimate

$$\lim_{k\to\infty} \frac{1}{\lambda_j} \log |\varphi_{j_k}(\gamma_{x,\xi}^{\mathbb{C}}(\zeta)| \leq |\tau|$$

on all of ∂M_τ. Hence the difficult point is to prove that this growth rate is actually obtained upon restriction to $\gamma_{x,\xi}^{\mathbb{C}}$. This requires new kinds of arguments related to the QER theorem.

- Complexifications of restrictions of eigenfunctions to geodesics have incommensurate Fourier modes, i.e. higher modes are exponentially larger than lower modes.
- The quantum ergodic restriction theorem in the real domain shows that the Fourier coefficients of the top allowed modes are 'large' (i.e. as large as the lower modes). Consequently, the L^2 norms of the complexified eigenfunctions along arcs of $\gamma_{x,\xi}^{\mathbb{C}}$ achieve the lower bound of Proposition 11.6.

- Invariance of Wigner measures along the geodesic flow implies that the Wigner measures of restrictions of complexified eigenfunctions to complexified geodesics should tend to constant multiples of Lebesgue measures dt for each $\tau > 0$. Hence the eigenfunctions everywhere on $\gamma_{x,\xi}^{\mathbb{C}}$ achieve the growth rate of the L^2 norms.

These principles are most easily understood in the case of periodic geodesics. We let $\gamma_{x,\xi} : S^1 \to M$ parametrize the geodesic with arc-length (where $S^1 = \mathbb{R}/L\mathbb{Z}$ where L is the length of $\gamma_{x,\xi}$).

LEMMA 11.7. *Assume that $\{\varphi_j\}$ satsifies QER along the periodic geodesic $\gamma_{x,\xi}$. Let $||\gamma_{x,\xi}^{\tau *}\varphi_j^{\mathbb{C}}||^2_{L^2(S^1)}$ be the L^2-norm of the complexified restriction of φ_j along $\gamma_{x,\xi}^{\tau}$. Then,*

$$\lim_{\lambda_j \to \infty} \frac{1}{\lambda_j} \log ||\gamma_{x,\xi}^{\tau *}\varphi_j^{\mathbb{C}}||^2_{L^2(S^1)} = |\tau|.$$

To prove Lemma 11.7, we study the orbital Fourier series of $\gamma_{x,\xi}^{\tau *}\varphi_j$ and of its complexification. The orbital Fourier coefficients are

$$\nu_{\lambda_j}^{x,\xi}(n) = \frac{1}{L_\gamma} \int_0^{L_\gamma} \varphi_{\lambda_j}(\gamma_{x,\xi}(t)) e^{-\frac{2\pi i n t}{L_\gamma}} \, dt,$$

and the orbital Fourier series is

$$\varphi_{\lambda_j}(\gamma_{x,\xi}(t)) = \sum_{n \in \mathbb{Z}} \nu_{\lambda_j}^{x,\xi}(n) e^{\frac{2\pi i n t}{L_\gamma}}. \tag{134}$$

Hence the analytic continuation of $\gamma_{x,\xi}^{\tau *}\varphi_j$ is given by

$$\varphi_{\lambda_j}^{\mathbb{C}}(\gamma_{x,\xi}(t + i\tau)) = \sum_{n \in \mathbb{Z}} \nu_{\lambda_j}^{x,\xi}(n) e^{\frac{2\pi i n (t + i\tau)}{L_\gamma}}. \tag{135}$$

By the Paley-Wiener theorem for Fourier series, the series converges absolutely and uniformly for $|\tau| \leq \epsilon_0$. By "energy localization" only the modes with $|n| \leq \lambda_j$ contribute substantially to the L^2 norm. We then observe that the Fourier modes decouple, since they have different exponential growth rates. We use the QER hypothesis in the following way:

LEMMA 11.8. *Suppose that $\{\varphi_{\lambda_j}\}$ is QER along the periodic geodesic $\gamma_{x,\xi}$. Then for all $\epsilon > 0$, there exists $C_\epsilon > 0$ so that*

$$\sum_{n : |n| \geq (1-\epsilon)\lambda_j} |\nu_{\lambda_j}^{x,\xi}(n)|^2 \geq C_\epsilon.$$

Lemma 11.8 implies Lemma 11.7 since it implies that for any $\epsilon > 0$,

$$\sum_{n : |n| \geq (1-\epsilon)\lambda_j} |\nu_{\lambda_j}^{x,\xi}(n)|^2 e^{-2n\tau} \geq C_\epsilon e^{2\tau(1-\epsilon)\lambda_j}.$$

To go from asymptotics of L^2 norms of restrictions to Proposition 11.6 we then use the third principle:

PROPOSITION 11.9. *(Lebesgue limits) If $\gamma_{x,\xi}^{*}\varphi_j \neq 0$ (identically), then for all $\tau > 0$ the sequence*

$$U_j^{x,\xi,\tau} = \frac{\gamma_{x,\xi}^{\tau *}\varphi_j^{\mathbb{C}}}{||\gamma_{x,\xi}^{\tau *}\varphi_j^{\mathbb{C}}||_{L^2(S^1)}}$$

is QUE with limit measure given by normalized Lebesgue measure on S^1.

The proof of Proposition 11.6 is completed by combining Lemma 11.7 and Proposition 11.9. Theorem 11.5 follows easily from Proposition 11.6.

The proof for non-periodic geodesics is considerably more involved, since one cannot use Fourier analysis in quite the same way.

11.4. Real zeros and complex analysis

PROBLEM 4. *An important but apparently rather intractable problem is, how to obtain information on the real zeros from knowledge of the complex nodal distribution? There are several possible approaches:*

- *Try to intersect the nodal current with the current of integration over the real points $M \subset M_\epsilon$. I.e. try to slice the complex nodal set with the real domain.*

- *Thicken the real slice slightly by studying the behavior of the nodal set in M_ϵ as $\epsilon \to 0$. The sharpest version is to try to re-scale the nodal set by a factor of λ^{-1} to zoom in on the zeros which are within λ^{-1} of the real domain. They may not be real but at least one can control such "almost real" zeros. Try to understand (at least in real dimension 2) how the complex nodal set 'sprouts' from the real nodal set. How do the connected components of the real nodal set fit together in the complex nodal set?*

- *Intersect the nodal set with geodesics. This magnifies the singularity along the real domain and converts nodal sets to isolated points.*

12. L^p norms of eigenfuncions

In §4.3 we pointed out that lower bounds on $||\varphi_\lambda||_{L^1}$ lead to improved lower bounds on Hausdorff measures of nodal sets. In this section we consider general L^p-norm problems for eigenfunctions.

12.1. Generic upper bounds on L^p norms

We have already explained that the pointwise Weyl law (33) and remainder jump estimate (35) leads to the general sup norm bound for L^2-normalized eigenfunctions,

$$||\varphi_\lambda||_{L^\infty} \leq C_g \lambda^{\frac{m-1}{2}}, \quad (m = \dim M). \tag{136}$$

The upper bound is achieved by zonal spherical harmonics. In [**Sog**] (see also [**Sogb, Sogb2**]) C.D. Sogge proved general L^p bounds:

THEOREM 12.1. *(Sogge, 1985)*

$$\sup_{\varphi \in V_\lambda} \frac{||\varphi||_p}{||\varphi||_2} = O(\lambda^{\delta(p)}), \quad 2 \leqslant p \leqslant \infty \tag{137}$$

where

$$\delta(p) = \begin{cases} n(\frac{1}{2} - \frac{1}{p}) - \frac{1}{2}, & \frac{2(n+1)}{n-1} \leqslant p \leqslant \infty \\ \frac{n-1}{2}(\frac{1}{2} - \frac{1}{p}), & 2 \leqslant p \leqslant \frac{2(n+1)}{n-1}. \end{cases} \tag{138}$$

The upper bounds are sharp in the class of all (M, g) and are saturated on the round sphere:

- For $p > \frac{2(n+1)}{n-1}$, zonal (rotationally invariant) spherical harmonics saturate the L^p bounds. Such eigenfunctions also occur on surfaces of revolution.

- For L^p for $2 \leq p \leq \frac{2(n+1)}{n-1}$ the bounds are saturated by highest weight spherical harmonics, i.e. Gaussian beam along a stable elliptic geodesic. Such eigenfunctions also occur on surfaces of revolution.

The zonal has high L^p norm due to its high peaks on balls of radius $\frac{1}{N}$. The balls are so small that they do not have high L^p norms for small p. The Gaussian beams are not as high but they are relatively high over an entire geodesic.

12.2. Lower bounds on L^1 norms

The L^p upper bounds are the only known tool for obtaining lower bounds on L^1 norms. We now prove Proposition 10:

PROOF. Fix a function $\rho \in \mathcal{S}(\mathbb{R})$ having the properties that $\rho(0) = 1$ and $\hat{\rho}(t) = 0$ if $t \notin [\delta/2, \delta]$, where $\delta > 0$ is smaller than the injectivity radius of (M, g). If we then set

$$T_\lambda f = \rho(\sqrt{-\Delta} - \lambda) f,$$

we have that $T_\lambda \varphi_\lambda = \varphi_\lambda$. Also, by Lemma 5.1.3 in [**Sogb**], T_λ is an oscillatory integral operator of the form

$$T_\lambda f(x) = \lambda^{\frac{n-1}{2}} \int_M e^{i\lambda r(x,y)} a_\lambda(x,y) f(y) dy,$$

with $|\partial^\alpha_{x,y} a_\lambda(x,y)| \leqslant C_\alpha$. Consequently, $||T_\lambda \varphi_\lambda||_{L^\infty} \leqslant C\lambda^{\frac{n-1}{2}} ||\varphi_\lambda||_{L^1}$, with C independent of λ, and so

$$1 = ||\varphi_\lambda||^2_{L^2} = \langle T\varphi_\lambda, \varphi_\lambda \rangle \leq ||T\varphi_\lambda||_{L^\infty} ||\varphi_\lambda||_{L^1} \leq C\lambda^{\frac{n-1}{2}} ||\varphi_\lambda||^2_{L^1}.$$

We can give another proof based on the eigenfunction estimates (Theorem 12.1), which say that

$$||\varphi_\lambda||_{L^p} \leqslant C\lambda^{\frac{(n-1)(p-2)}{4p}}, \quad 2 < p \leqslant \frac{2(n+1)}{n-1}.$$

If we pick such a $2 < p < \frac{2(n+1)}{n-1}$, then by Hölder's inequality, we have

$$1 = ||\varphi_\lambda||^{1/\theta}_{L^2} \leqslant ||\varphi_\lambda||_{L^1} \, ||\varphi_\lambda||^{\frac{1}{\theta}-1}_{L^p} \leqslant ||\varphi_\lambda||_{L^1} \big(C\lambda^{\frac{(n-1)(p-2)}{4p}} \big)^{\frac{1}{\theta}-1},$$

$$\theta = \frac{p}{p-1}\left(\frac{1}{2} - \frac{1}{p}\right) = \frac{(p-2)}{2(p-1)},$$

which implies $||\varphi_\lambda||_{L^1} \geqslant c\lambda^{-\frac{n-1}{4}}$, since $(1 - \frac{1}{\theta})\frac{(n-1)(p-2)}{4p} = \frac{n-1}{4}$. □

We remark that this lowerbound for $||\varphi_\lambda||_{L^1}$ is sharp on the standard sphere, since L^2-normalized highest weight spherical harmonics of degree k with eigenvalue $\lambda^2 = k(k + n - 1)$ have L^1-norms which are bounded above and below by $k^{(n-1)/4}$ as $k \to \infty$. Similarly, the L^p-upperbounds that we used in the second proof of this L^1-lowerbound is also sharp because of these functions.

12.3. Riemannian manifolds with maximal eigenfunction growth

Although the general sup norm bound (136) is achieved by some sequences of eigenfunctions on some Riemannian manifolds (the standard sphere or a surface of revolution), it is very rare that (M, g) has such sequences of eigenfunctions. We say that such (M, g) have maximal eigenfunction growth. In a series of articles [**SoZ, STZ, SoZ2**], ever more stringent conditions are given on such (M, g). We now go over the results.

Denote the eigenspaces by

$$V_\lambda = \{\varphi : \Delta\varphi = -\lambda^2\varphi\}.$$

We measure the growth rate of L^p norms by

$$L^p(\lambda, g) = \sup_{\varphi \in V_\lambda : ||\varphi||_{L^2}=1} ||\varphi||_{L^p}. \tag{139}$$

Definition: Say that (M, g) has maximal L^p eigenfunction growth if it possesses a sequence of eigenfunctions $\varphi_{\lambda_{j_k}}$ which saturates the L^p bounds. When $p = \infty$ we say that it has maximal sup norm growth.

PROBLEM 12.2.

- *Characterize (M, g) with maximal L^∞ eigenfunction growth. The same sequence of eigenfunctions should saturate all L^p norms with $p \geq p_n := \frac{2(n+1)}{n-1}$.*
- *Characterize (M, g) with maximal L^p eigenfunction growth for $2 \leq p \leq \frac{2(n+1)}{n-1}$.*
- *Characterize (M, g) for which $||\varphi_\lambda||_{L^1} \geq C > 0$.*

In [**SoZ**], it was shown that (M, g) of maximal L^p eigenfunction growth for $p \geq p_n$ have self-focal points. The terminology is non-standard and several different terms are used.

Definition:

We call a point p a *self-focal point* or *blow-down point* if all geodesics leaving p loop back to p at a common time T. That is, $\exp_p T\xi = p$ (They do not have to be closed geodesics.)

We call a point p a partial self-focal point if there exists a positive measure in S_x^*M of directions ξ which loop back to p.

The poles of a surface of revolution are self-focal and all geodesics close up smoothly (i.e. are closed geodesics). The umbilic points of an ellipsoid are self-focal but only two directions give smoothly closed geodesics (one up to time reversal).

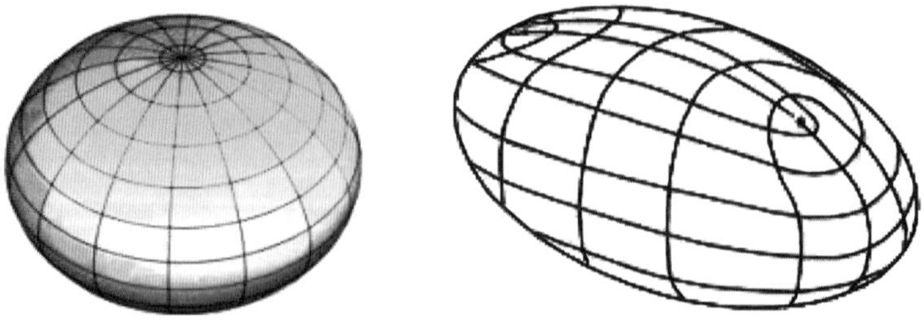

In [**SoZ**] is proved:

THEOREM 12.3. *Suppose (M, g) is a C^∞ Riemannian manifold with maximal eigenfunction growth, i.e. having a sequence $\{\varphi_{\lambda_{j_k}}\}$ of eigenfunctions which achieves (saturates) the bound $||\varphi_{\lambda_{j_k}}||_{L^\infty} \geq C_0 \lambda_{j_k}^{(n-1)/2}$ for some $C_0 > 0$ depending only on (M, g).*

Then there must exist a point $x \in M$ for which the set

$$\mathcal{L}_x = \{\xi \in S_x^* M : \exists T : \exp_x T\xi = x\} \tag{140}$$

of directions of geodesic loops at x has positive measure in S_x^*M. Here, $\exp$ is the exponential map, and the measure $|\Omega|$ of a set Ω is the one induced by the metric g_x on T_x^*M. For instance, the poles x_N, x_S of a surface of revolution (S^2, g) satisfy $|\mathcal{L}_x| = 2\pi$.

Theorem 12.3, Theorem 9, as well as the results of [**SoZ, STZ**], are proved by studying the remainder term $R(\lambda, x)$ in the pointwise Weyl law,

$$N(\lambda, x) = \sum_{j:\lambda_j \leq \lambda} |\varphi_j(x)|^2 = C_m \lambda^m + R(\lambda, x). \tag{141}$$

The first term $N_W(\lambda) = C_m \lambda^m$ is called the Weyl term. It is classical that the remainder is of one lower order, $R(\lambda, x) = O(\lambda^{m-1})$. The relevance of the remainder to maximal eigenfunction growth is through the following well-known Lemma (see e.g. [**SoZ**]):

LEMMA 12.4. *Fix $x \in M$. Then if $\lambda \in spec\sqrt{-\Delta}$*

$$\sup_{\varphi \in V_\lambda} \frac{|\varphi(x)|}{||\varphi||_2} = \sqrt{R(\lambda, x) - R(\lambda - 0, x)}. \tag{142}$$

Here, for a right continuous function $f(x)$ we denote by $f(x + 0) - f(x - 0)$ the jump of f at x. Thus, Theorem 12.3 follows from

THEOREM 12.5. *Let $R(\lambda, x)$ denote the remainder for the local Weyl law at x. Then*

$$R(\lambda, x) = o(\lambda^{n-1}) \text{ } if \text{ } |\mathcal{L}_x| = 0. \tag{143}$$

Additionally, if $|\mathcal{L}_x| = 0$ then, given $\varepsilon > 0$, there is a neighborhood $\mathcal{N}$ of x and a $\Lambda =< \infty$, both depending on ε so that

$$|R(\lambda, y)| \leqslant \varepsilon \lambda^{n-1}, \; y \in \mathcal{N}, \; \lambda \geqslant \Lambda. \tag{144}$$

12.4. Theorem 9

However, Theorem 12.3 is not sharp: on a tri-axial ellipsoid (three distinct axes), the umbilic points are self-focal points. But the eigenfunctions which maximize the sup-norm only have L^∞ norms of order $\lambda^{\frac{n-1}{2}} / \log \lambda$. An improvement is given in [**STZ**].

Recently, Sogge and the author have further improved the result in the case of real analytic (M, g) In this case $|\mathcal{L}_x| > 0$ implies that $\mathcal{L}_x = S_x^* M$ and the geometry simplifies. In [**SoZ2**], we prove Theorem 9, which we restate in terms of the jumps of the remainder:

THEOREM 12.6. *Assume that U_x has no invariant L^2 function for any x. Then*

$$N(\lambda + o(1), x) - N(\lambda, x) = o(\lambda^{n-1}), \quad \text{uniformly in } x.$$

Equivalently,

$$R(\lambda + o(1), x) - R(\lambda, x) = o(\lambda^{n-1}) \tag{145}$$

uniformly in x.

Before discussing the proof we note that the conclusion gives very stringent conditions on (M, g). First, there are topological restrictions on manifolds possessing a self-focal point. If (M, g) has a focal point x_0 then the rational cohomology $H^*(M, \mathbb{Q})$ has a single generator (Berard-Bergery). But even in this case there are many open problems:

PROBLEM 12.7. *All known examples of (M, g) with maximal eigenfunction growth have completely integrable geodesic flow, and indeed quantum integrable Laplacians. Can one prove that maximum eigenfunction growth only occurs in the integrable case? Does it only hold if there exists a point p for which $\Phi_p = Id$?*

A related purely geometric problem: Do there exist (M, g) with $\dim M \geq 3$ possessing self-focal points wth $\Phi_x \neq Id$. I.e. do there exist generalizations of umbilic points of ellipsoids in dimension two. There do not seem to exist any known examples; higher dimensional ellipsoids do not seem to have such points.

Despite these open questions, Theorem 12.6 is in a sense sharp. If there exists a self-focal point p with a smooth invariant function, then one can construct a *quasi-mode* of order zero which lives on the flow-out Lagrangian

$$\Lambda_p := \bigcup_{t \in [0,T]} G^t S^*_p M$$

where G^t is the geodesic flow and T is the minimal common return time. The 'symbol' is the flowout of the smooth invariant density.

PROPOSITION 12.8. *Suppose that (M, g) has a point p which is a self-focal point whose first return map Φ_x at the return time T is the identity map of $S^*_p M$. Then there exists a quasi-model of order zero associated to the sequence $\{\frac{2}{\pi}Tk + \frac{\beta}{2} : k = 1, 2, 3, \dots\}$ which concentrates microlocally on the flow-out of $S^*_p M$. (See §12.8 for background and more precise information).*

12.5. Sketch of proof of Theorem 9

We first outline the proof. A key issue is the uniformity of remainder estimates of $R(\lambda, x)$ as x varies. Intuitively it is obvious that the main points of interest are the self-focal points. But at this time of writing, we cannot exclude the possibility, even in the real analytic setting, that there are an infinite number of such points with twisted return maps. Points which isolated from the set of self-focal points are easy to deal with, but there may be non-self-focal points which lie in the closure of the self-focal points. We introduce some notation.

Definition: We say that $x \in M$

- is an $\mathcal{L}$ point ($x \in \mathcal{L}$) if $\mathcal{L}_x = \pi^{-1}(x) \simeq S^*_x M$. Thus, x is a self-focal point.
- is a $\mathcal{CL}$ point ($x \in \mathcal{CL}$) if $x \in \mathcal{L}$ and $\Phi_x = Id$. Thus, all of the loops at x are smoothly closed.
- is a $\mathcal{TL}$ point ($x \in \mathcal{TL}$) if $x \in \mathcal{L}$ but $\Phi_x \neq Id$, i.e. x is a twisted self-focal point. Equivalently, $\mu_x\{\xi \in \mathcal{L}_x : \Phi_x(\xi) = \xi\} = 0$; All directions are loop directions, but almost none are directions of smoothly closed loops.

To prove Theorem 9 we may (and henceforth will) assume that $\mathcal{CL} = \emptyset$. Thus, $\mathcal{L} = \mathcal{TL}$. We also let $\overline{\mathcal{L}}$ denote the closure of the set of self-focal points. At this time of writing, we do not know how to exclude that $\overline{\mathcal{L}} = M$, i.e. that the set of self-focal points is dense.

PROBLEM 12.9. *Prove (or disprove) that if $\mathcal{CL} = \emptyset$ and if (M, g) is real analytic, then $\mathcal{L}$ is a finite set.*

We also need to further specify times of returns. It is well-known and easy to prove that if all $\xi \in \pi^{-1}(x)$ are loop directions, then the time $T(x, \xi)$ of first return is constant on $\pi^{-1}(x)$. This is because an analytic function is constant on its critical point set.

Definition: We say that $x \in M$

- is a $\mathcal{TL}_T$ point ($x \in \mathcal{TL}_T$) if $x \in \mathcal{TL}$ and if $T(x, \xi) \leq T$ for all $\xi \in \pi^{-1}(x)$. We denote the set of such points by $\mathcal{TL}_T$.

LEMMA 12.10. *If (M, g) is real analytic, then $\mathcal{TL}_T$ is a finite set.*

There are several ways to prove this. One is to consider the set of all loop points,

$$\mathcal{E} = \{(x, \xi) \in TM : \exp_x \xi = x\},$$

where as usual we identify vectors and co-vectors with the metric. Then at a self-focal point p, $\mathcal{E} \cap T_pM$ contains a union of spheres of radii $kT(p)$, $k = 1, 2, 3, \ldots$. The condition that $\Phi_p \neq I$ can be used to show that each sphere is a component of $\mathcal{E}$, i.e. is isolated from the rest of $\mathcal{E}$. Hence in the compact set $B_T^*M = \{(x, \xi) : |\xi| \leq T\}$, there can only exist a finite number of such components. Another way to prove it is to show that any limit point x, with $p_j \to x$ and $p_j \in \mathcal{TL}_T$ must be a $\mathcal{TL}_T$ point whose first return map is the identity. Both proofs involve the study of Jacobi fields along the looping geodesics.

To outline the proof, let $\hat{\rho} \in C_0^\infty$ be an even function with $\hat{\rho}(0) = 1$, $\rho(\lambda) > 0$ for all $\lambda \in \mathbb{R}$, and $\hat{\rho}_T(t) = \hat{\rho}(\frac{t}{T})$. The classical cosine Tauberian method to determine Weyl asymptotics is to study

One starts from the smoothed spectral expansion [**DG, SV**]

(146)
$$\begin{aligned} \rho_T * dN(\lambda, x) &= \int_{\mathbb{R}} \hat{\rho}(\tfrac{t}{T}) e^{i\lambda t} U(t, x, x) dt \\ &= a_0 \lambda^{n-1} + a_1 \lambda^{n-2} + \lambda^{n-1} \textstyle\sum_{j=1}^\infty R_j(\lambda, x, T) + o_T(\lambda^{n-1}), \end{aligned}$$

with uniform remainder in x. The sum over j is a sum over charts needed to parametrize the canonical relation of $U(t, x, y)$, i.e. the graph of the geodesic flow. By the usual parametrix construction for $U(t) = e^{it\sqrt{\Delta}}$, one proves that there exist phases $\tilde{t}_j$ and amplitudes a_{j0} such that

$$R_j(\lambda, x, T) \simeq \lambda^{n-1} \int_{S_x^*M} e^{i\lambda \tilde{t}_j(x,\xi)} \left((\hat{\rho}_T a_{j0}) \right) |d\xi| + O(\lambda^{n-2}). \tag{147}$$

As in [**DG, Saf, SV**] we use polar coordinates in T^*M, and stationary phase in $dtdr$ to reduce to integrals over S_x^*M. The phase $\tilde{t}_j$ is the value of the phase $\varphi_j(t, x, x, \xi)$ of $U(t, x, x)$ at the critical point. The loop directions are those ξ such that $\nabla_\xi \tilde{t}_j(x, \xi) = 0$.

EXERCISE 6. *Show that $\rho_T * dN(\lambda, x)$ is a semi-classical Lagrangian distribution in the sense of* §15. *What is its principal symbol?*

To illustrate the notation, we consider a flat torus $\mathbb{R}^n/\Gamma$ with $\Gamma \subset \mathbb{R}^n$ a full rank lattice. As is well-known, the wave kernel then has the form

$$U(t,x,y) = \sum_{\gamma\in\Gamma} \int_{\mathbb{R}^n} e^{i\langle x-y-\gamma,\xi\rangle} e^{it|\xi|} d\xi.$$

Thus, the indices j may be taken to be the lattice points $\gamma \in \Gamma$, and

$$\rho_T * dN(\lambda,x) \quad = \quad \textstyle\sum_{\gamma\in\Gamma} \int_{\mathbb{R}} \int_0^\infty \int_{S^{n-1}} \hat{\rho}(\frac{t}{T}) e^{ir\langle\gamma,\omega\rangle} e^{itr} e^{-it\lambda} r^{n-1} dr dt d\omega$$

We change variables $r \to \lambda r$ to get a full phase $\lambda(r\langle\gamma,\omega\rangle + tr - t)$. The stationary phase points in (r,t) are $\langle\gamma,\omega\rangle = t$ and $r = 1$. Thus,

$$\tilde{t}_\gamma(x,\omega) = \langle\gamma,\omega\rangle.$$

The geometric interpretation of $t^*_\gamma(x,\omega)$ is that it is the value of t for which the geodesic $\exp_x t\omega = x + t\omega$ comes closest to the representative $x+\gamma$ of x in the γth chart. Indeed, the line $x+t\omega$ is 'closest' to $x+\gamma$ when $t\omega$ closest to γ, since $|\gamma - t\omega|^2 = |\gamma|^2 - 2t\langle\gamma,\omega\rangle + t^2$. On a general (M,g) without conjugate points,

$$\tilde{t}_\gamma(x,\omega) = \langle \exp_x^{-1} \gamma x, \omega\rangle.$$

12.6. Size of the remainder at a self-focal point

The first key observation is that (147) takes a special form at a self-focal point. At a self-focal point x define U_x as in (39). Also define

$$U_x^\pm(\lambda) = e^{i\lambda T_x^\pm} U_x^\pm. \tag{148}$$

The following observation is due to Safarov [**Saf**] (see also [**SV**]).

LEMMA 12.11. *Suppose that x is a self-focal point. If $\hat{\rho} = 0$ in a neighborhood of $t = 0$ then*

$$\rho_T' * N(\lambda,x) = \lambda^{n-1} \sum_{k\in\mathbb{Z}\backslash 0} \int_{S_x^* M} \hat{\rho}(\frac{kT(\xi)}{T}) \overline{U_x(\lambda)^k \cdot 1} d\xi + O(\lambda^{n-2}). \tag{149}$$

Here is the main result showing that $R(\lambda,x)$ is small at the self-focal points if there do not exist invariant L^2 functions. $T_x^{(k)}(\xi)$ is the kth return time of ξ for Φ_x.

PROPOSITION 12.12. *Assume that x is a self-focal point and that U_x has no invariant L^2 function. Then, for all $\eta > 0$, there exists T so that*

$$\frac{1}{T}\left| \int_{S_x^* M} \sum_{k=0}^\infty \hat{\rho}(\frac{T_x^{(k)}(\xi)}{T}) U_x^k \cdot 1 |d\xi| \right| \leq \eta. \tag{150}$$

This is a simple application of the von Neumann mean ergodic theorem to the unitary operator U_x. Indeed, $\frac{1}{N}\sum_{k=0}^N U_x^k \to P_x$, where $P_x : L^2(S_x^* M) \to L_0^2(S_x^* M)$ is the orthogonal projections onto the invariant L^2 functions for U_x. By our assumption, $P_x = 0$.

Proposition 12.12 is not apriori uniform as x varies over self-focal points, since there is no obvious relation between Φ_x at one self-focal point and another. It would of course be uniform if we knew that there only exist a finite number of self-focal points. As mentioned above, this is currently unknown. However, there is a second mechanism behind Proposition 12.12. Namely, if the first common return time $T_x^{(1)}(\xi)$ is larger than T, then there is only one term $k = 0$ in the sum with the cutoff $\hat{\rho}_T$ and the sum is $O(\frac{1}{T})$.

12.7. Decomposition of the remainder into almost loop directions and far from loop directions

We now consider non-self-focal points. Then the function $\tilde{t}_j(x,\xi)$ has almost no critical points in S^*_xM.

Pick $f \geq 0 \in C_0^\infty(\mathbb{R})$ which equals 1 on $|s| \leq 1$ and zero for $|s| \geq 2$ and split up the jth term into two terms using $f(\epsilon^{-2}|\nabla_\xi \tilde{t}_j|^2)$ and $1 - f(\epsilon^{-2}|\nabla_\xi \tilde{t}_j|^2)$:

$$R_j(\lambda, x, T) = R_{j1}(\lambda, x, T, \epsilon) + R_{j2}(\lambda, x, T, \epsilon), \tag{151}$$

where

$$R_{j1}(\lambda, x, T, \epsilon) := \int_{S^*_xM} e^{i\lambda \tilde{t}_j} f(\epsilon^{-1}|\nabla_\xi \tilde{t}_j(x,\xi)|^2)(\hat{\rho}(T_x(\xi)))a_0(T_x(\xi), x, \xi)d\xi \tag{152}$$

The second term R_{j2} comes from the $1-f(\epsilon^{-2}|\nabla_\xi T_x(\xi)|^2)$ term. By one integration by parts, one easily has

LEMMA 12.13. *For all $T > 0$ and $\epsilon \geq \lambda^{-\frac{1}{2}} \log \lambda$ we have*

$$\sup_{x \in M} |R_2(\lambda, x, T, \epsilon)| \leq C(\epsilon^2\lambda)^{-1}.$$

The f term involves the contribution of the almost-critical points of $\tilde{t}_j$. They are estimated by the measure of the almost-critical set.

LEMMA 12.14. *There exists a uniform positive constant C so that for all (x, ϵ),*

$$|R_{j1}(x;\epsilon)| \leq C\mu_x\left(\{\xi : 0 < |\nabla_\xi \tilde{t}_j(x,\xi)|^2 < \epsilon^2\}\right), \tag{153}$$

12.8. Points in $M \backslash \overline{\mathcal{TL}}$

If x is isolated from $\overline{\mathcal{TL}}$ then there is a uniform bound on the size of the remainder near x.

LEMMA 12.15. *Suppose that $x \notin \overline{\mathcal{TL}}$. Then given $\eta > 0$ there exists a ball $B(x, r(x,\eta))$ with radius $r(x,\eta) > 0$ and $\epsilon > 0$ so that*

$$\sup_{y \in B(x, r(x,\eta))} |R(\lambda, y, \epsilon)| \leq \eta.$$

Indeed, we pick $r(x,\eta)$ so that the closure of $B(x, r(x,\eta))$ is disjoint from $\overline{\mathcal{TL}}$. Then the one-parameter family of functions

$$F_\epsilon(y) =\to \mu_y\left(\{\xi \in S^*_yM : 0 < |\nabla_\xi \tilde{t}_j(y,\xi)|^2 < \epsilon^2\}\right)$$

is decreasing to zero as $\epsilon \to 0$ for each y. By Dini's theorem, the family tends to zero uniformly on $\overline{B(x, r(x,\eta))}$.

12.9. Perturbation theory of the remainder

So far, we have good remainder estimates at each self-focal point and in balls around points isolated from the self-focal points. We still need to deal with the uniformity issues as p varies among self-focal points and points in $\overline{\mathcal{TL}}$.

We now compare remainders at nearby points. Although $R_j(\lambda, x, T)$ is oscillatory, the estimates on R_{j1} and the ergodic estimates do not use the oscillatory factor $e^{i\lambda\tilde{t}}$, which in fact is only used in Lemma 20. Hence we compare absolute remainders $|R|(x,T)|$, i.e. where we take the absolute under the integral sign. They are independent of λ. The integrands of the remainders vary smoothly with the base point and only involve integrations over different fibers S^*_xM of $S^*M \to M$.

LEMMA 12.16. *We have,*

$$||R|(x,T) - |R|(y,T)| \leq Ce^{aT} dist(x,y).$$

Indeed, we write the difference as the integral of its derivative. The derivative involves the change in Φ_x^n as x varies over iterates up to time T and therefore is estimated by the sup norm e^{aT} of the first derivative of the geodesic flow up to time T. If we choose a ball of radius δe^{-aT} around a focal point, we obtain'

COROLLARY 12.17. *For any $\eta > 0$, $T > 0$ and any focal point $p \in \mathcal{TL}$ there exists $r(p,\eta)$ so that*

$$\sup_{y \in \overline{B(p,r(p,\eta))}} |R(\lambda,y,T)| \leq \eta.$$

To complete the proof of Theorem 9 we prove

LEMMA 12.18. *Let $x \in \overline{\mathcal{TL}} \backslash \mathcal{TL}$. Then for any $\eta > 0$ there exists $r(x,\eta) > 0$ so that*

$$\sup_{y \in \overline{B(x,r(x,\eta))}} |R(\lambda,y,T)| \leq \eta.$$

Indeed, let $p_j \to x$ with $T(p_j) \to \infty$. Then the remainder is given at each p_j by the left side of (149). But for any fixed T, the first term of (149) has at most one term for j sufficiently large. Since the remainder is continuous, the remainder at x is the limit of the remainders at p_j and is therefore $O(T^{-1}) + O(\lambda^{-1})$.

By the perturbation estimate, one has the same remainder estimate in a sufficiently small ball around x.

12.10. Conclusions

- No eigenfunction $\varphi_j(x)$ can be maximally large at a point x which is $\geq \lambda_j^{-\frac{1}{2}} \log \lambda_j$ away from the self-focal points.
- When there are no invariant measures, φ_j also cannot be large at a self-focal point.
- If φ_j is not large at any self-focal point, it is also not large nera a self-focal point.

13. Appendix on the phase space and the geodesic flow

Classical mechanics takes place in phase space T^*M (the cotangent bundle). Let (x,ξ) be Darboux coordinates on T^*M, i.e. x_j are local coordinates on M and ξ_j are the functions on T^*M which pick out components with respect to dx_j. Hamilton's equations for the Hamiltonian $H(x,\xi) = \frac{1}{2}|\xi|^2 + V(x) : T^*M \to \mathbb{R}$ are

$$\frac{dx}{dt} = \frac{\partial H}{\partial \xi}, \quad \frac{d\xi}{dt} = -\frac{\partial H}{\partial \xi}.$$

The Hamilton flow is $\Phi^t(x,\xi) = (x_t,\xi_t)$ with $(x_0,\xi_0) = (x,\xi)$. Let us recall the symplectic interpreation of Hamilton's equations.

The cotangent bundle T^*M of any manifold M carries a canonical action form α and symplectic form $\omega = d\alpha$. Given any local coordinates x_j and associated frame dx_j for T^*M, we put: $\alpha = \sum_j \xi_j dx_j$. The form is independent of the choice of coordinates and is called the action form.

It is invariantly defined as follows: At a covector $\xi \in T_x^*M$ define $\alpha_{x,\xi}(X) = \xi(\pi_* X)$ where X is a tangent vector to T^*M at (x, ξ) and $\pi : T^*M \to M$ is the projection $(x, \xi) \to x$. Another way to define α is that for any 1-form η, viewed as a section of $T * M$, $\eta^* \alpha = \eta$.

By definition $\omega = d\alpha = \sum_{j=1}^n dx_j \wedge d\xi_j$.

A symplectic form is a non-degenerate closed 2-form. Thus it is an anti-symmetric bilinear form on tangent vectors at each point $(x, \xi) \in T^*M$. We note that

- $\omega(\frac{\partial}{\partial x_j}, \frac{\partial}{\partial x_k}) = 0$; one says that the vector space Span$\{\frac{\partial}{\partial x_j}\}$ at each $(x, \xi) \in T^*M$ is Lagrangian;
- $\omega(\frac{\partial}{\partial x_j}, \frac{\partial}{\partial \xi_k}) = \delta_{jk}$; one says that $\frac{\partial}{\partial x_j}, \frac{\partial}{\partial \xi_j}$ are symplectically paired;
- $\omega(\frac{\partial}{\partial \xi_j}, \frac{\partial}{\partial \xi_k}) = 0$. Thus, the vector space Span$\{\frac{\partial}{\partial \xi_j}\}$ is Lagrangian.

A Hamiltonian is a smooth function $H(x, \xi)$ on T^*M. We say it is homogeneous of degree p if $H(x, r\xi) = r^p H(x, \xi)$ for $r > 0$.

The Hamiltonian vector field Ξ_H of H is the symplectic gradient of H. That is, one takes dH, a 1-form on T^*M and uses the symplectic form ω to convert it to a vector field. That is,

$$\omega(\Xi_H, \cdot) = dH.$$

We claim:

$$\Xi_H = \sum_{j=1}^n (\frac{\partial H}{\partial \xi_j} \frac{\partial}{\partial x_j} - \frac{\partial H}{\partial x_j} \frac{\partial}{\partial \xi_j}).$$

We note that $dH = \frac{\partial H}{\partial \xi_j} d\xi_j + \frac{\partial H}{\partial x_j} dx_j$, so the equation $\omega(\Xi_H, \cdot) = dH$ is equivalent to:

- $\omega(\frac{\partial}{\partial \xi_j}, \cdot) = -dx_j$;

The flow of the Hamiltonian vector field is the one-parameter group

$$\Phi^t : T^*M \to T^*M$$

defined by

$$\Phi^t(x_0, \xi_0) = (x_t, \xi_t)$$

where (x_t, ξ_t) solve the ordinary differential equation:

$$\begin{cases} \frac{dx_j}{dt} = \frac{\partial H}{\partial \xi_j}; \\ \\ \frac{d\xi_j}{dt} = -\frac{\partial H}{\partial x_j}. \end{cases},$$

with initial conditions $x(0) = x_0, \xi(0) = \xi_0$.

The symplectic form induces the following Lie bracket on functions:

$$\{f, g\}(x) = \Xi_f(g) = \sum_{j=1}^n (\frac{\partial f}{\partial \xi_j} \frac{\partial g}{\partial x_j} - \frac{\partial f}{\partial x_j} \frac{\partial g}{\partial \xi_j}).$$

We see that $\{f, g\} = -\{g, f\}$.

Consider the coordinate functions x_j, ξ_k on T^*M. Exercise: Show:

- $\{x_j, x_k\} = 0$;
- $\{x_j, \xi_k\} = \delta_{jk}$;
- $\{\xi_j, \xi_k\} = 0$.

These are the "canonical commutation relations".

Classical phase space = cotangent bundle T^*M of M, equipped with its canonical symplectic form $\sum_i dx_i \wedge d\xi_i$. The metric defines the Hamiltonian $H(x,\xi) = |\xi|_g = \sqrt{\sum_{ij=1}^n g^{ij}(x)\xi_i\xi_j}$ on T^*M, where $g_{ij} = g(\frac{\partial}{\partial x_i}, \frac{\partial}{\partial x_j})$, $[g^{ij}]$ is the inverse matrix to $[g_{ij}]$. Hamilton's equations:

$$\begin{cases} \frac{dx_j}{dt} = \frac{\partial H}{\partial \xi_j} \\ \\ \frac{d\xi_j}{dt} = -\frac{\partial H}{\partial x_j}. \end{cases}$$

Its flow is the 'geodesic flow'

$$G^t : S_g^* M \to S_g^* M$$

restricted to the energy surface $\{H = 1\} := S_g^* M$.

14. Appendix: Wave equation and Hadamard parametrix

The Cauchy problem for the wave equation on $\mathbb{R} \times M$ ($\dim M = n$) is the initial value problem (with Cauchy data f, g)

$$\begin{cases} \Box u(t,x) = 0, \\ \\ u(0,x) = f, \;\; \frac{\partial}{\partial t} u(0,x) = g(x), \end{cases}.$$

The solution operator of the Cauchy problem (the "propagator") is the wave group,

$$\mathcal{U}(t) = \begin{pmatrix} \cos t\sqrt{\Delta} & \frac{\sin t\sqrt{\Delta}}{\sqrt{\Delta}} \\ \\ \sqrt{\Delta}\sin t\sqrt{\Delta} & \cos t\sqrt{\Delta} \end{pmatrix}.$$

The solution of the Cauchy problem with data (f, g) is $\mathcal{U}(t)\begin{pmatrix} f \\ g \end{pmatrix}$.

- Even part $\cos t\sqrt{\Delta}$ which solves the initial value problem

$$(154) \qquad \begin{cases} (\frac{\partial}{\partial t}^2 - \Delta)u = 0 & \\ u|_{t=0} = f & \frac{\partial}{\partial t} u|_{t=0} = 0 \end{cases}$$

- Odd part $\frac{\sin t\sqrt{\Delta}}{\sqrt{\Delta}}$ is the operator solving

$$(155) \qquad \begin{cases} (\frac{\partial}{\partial t}^2 - \Delta)u = 0 & \\ u|_{t=0} = 0 & \frac{\partial}{\partial t} u|_{t=0} = g \end{cases}$$

The forward half-wave group is the solution operator of the Cauchy problem

$$(\frac{1}{i}\frac{\partial}{\partial t} - \sqrt{-\Delta})u = 0, \quad u(0,x) = u_0.$$

The solution is given by

$$u(t,x) = U(t)u_0(x),$$

with

$$U(t) = e^{it\sqrt{-\Delta}}$$

the unitary group on $L^2(M)$ generated by the self-adjoint elliptic operator $\sqrt{-\Delta}$.

A fundamental solution of the wave equation is a solution of

$$\Box E(t, x, y) = \delta_0(t)\delta_x(y).$$

The right side is the Schwartz kernel of the identity operator on $\mathbb{R} \times M$.

There exists a unique fundamental solution with support in the forward light cone, called the advanced (or forward) propagator. It is given by

$$E_+(t) = H(t)\frac{\sin t\sqrt{\Delta}}{\sqrt{\Delta}},$$

where $H(t) = \mathbf{1}_{t\geq 0}$ is the Heaviside step function.

14.1. Hormander parametrix

We would like to construct a parametrix of the form

$$\int_{T_x^*M} e^{i\langle \exp_y^{-1} x, \eta\rangle} e^{it|\eta|_y} A(t, x, y, \eta) d\eta.$$

This is a homogeneous Fourier integral operator kernel (see §15).

Hörmander actually constructs one of the form

$$\int_{T_x^*M} e^{i\psi(x,y,\eta)} e^{it|\eta|} A(t, x, y, \eta) d\eta,$$

where ψ solves the Hamilton Jacobi Cauchy problem,

$$\begin{cases} q(x, d_x\psi(x, y, \eta)) = q(y, \eta), \\ \psi(x, y, \eta) = 0 \iff \langle x - y, \eta\rangle = 0, \\ d_x\psi(x, y, \eta) = \eta, \quad (\text{for } x = y \end{cases}$$

The question is whether $\langle \exp_y^{-1} x, \eta\rangle$ solves the equations for ψ. Only the first one is unclear. We need to understand $\nabla_x\langle \exp_y^{-1} x, \eta\rangle$. We are only interested in the norm of the gradient at x but it is useful to consider the entire expression. If we write $\eta = \rho\omega$ with $|\omega|_y = 1$, then ρ can be eliminated from the equation by homogeneity. We fix $(y, \eta) \in S_y^*M$ and consider $\exp_y : T_yM \to M$. We wish to vary $\exp_y^{-1} x(t)$ along a curve. Now the level sets of $\langle \exp_y^{-1} x, \eta\rangle$ define a notion of local 'plane waves' of (M, g) near y. They are actual hyperplanes normal to ω in flat $\mathbb{R}^n$ and in any case are far different from distance spheres. Having fixed (y, η), $\nabla_x\langle \exp_y^{-1} x, \omega\rangle$ are normal to the plane waves defined by (y, η). To determine the length we need to see how $\nabla_x\langle \exp_y^{-1} x, \omega\rangle$ changes in directions normal to plane waves.

The level sets of $\langle \exp_y^{-1} x, \eta\rangle$ are images under $\exp_y$ of level sets of $\langle \xi, \eta\rangle = C$ in T_yM. These are parallel hyperplanes normal to η. The radial geodesic in the direction η is of course normal to the exponential image of the hyperplanes. Hence, this radial geodesic is parallel to $\langle \exp_y^{-1} x, \eta\rangle$ when $\exp_y t\eta = x$. It follows that $|\nabla_x\langle \exp_y^{-1} x, \eta\rangle$ at this point equals $\frac{\partial}{\partial t}\langle \exp_y^{-1} \exp_y t\frac{\eta}{|\eta|}, \eta\rangle = t|\eta|_y$. Hence $|\nabla_x\langle \exp_y^{-1} x, \eta\rangle|_x = 1$ at such points.

14.2. Wave group: $r^2 - t^2$

We now review the construction of a Hadamard parametrix,

$$U(t)(x,y) = \int_0^\infty e^{i\theta(r^2(x,y)-t^2)} \sum_{k=0}^\infty W_k(x,y)\theta^{\frac{n-3}{2}-k} d\theta \qquad (t < \mathrm{inj}(M,g)) \tag{156}$$

where $U_o(x,y) = \Theta^{-\frac{1}{2}}(x,y)$ is the volume 1/2-density, where the higher coefficients are determined by transport equations, and where θ^r is regularized at 0 (see below). This formula is only valid for times $t < inj(M,g)$ but using the group property of $U(t)$ it determines the wave kernel for all times. It shows that for fixed (x,t) the kernel $U(t)(x,y)$ is singular along the distance sphere $S_t(x)$ of radius t centered at x, with singularities propagating along geodesics. It only represents the singularity and in the analytic case only converges in a neighborhood of the characteristic conoid.

Closely related but somewhat simpler is the even part of the wave kernel, $\cos t\sqrt{\Delta}$ which solves the initial value problem

$$\begin{cases} (\frac{\partial}{\partial t}^2 - \Delta)u = 0 \\ u|_{t=0} = f \qquad \frac{\partial}{\partial t}u|_{t=0} = 0 \end{cases} \tag{157}$$

Similar, the odd part of the wave kernel, $\frac{\sin t\sqrt{\Delta}}{\sqrt{\Delta}}$ is the operator solving

$$\begin{cases} (\frac{\partial}{\partial t}^2 - \Delta)u = 0 \\ u|_{t=0} = 0 \qquad \frac{\partial}{\partial t}u|_{t=0} = g \end{cases} \tag{158}$$

These kernels only really involve Δ and may be constructed by the Hadamard-Riesz parametrix method. As above they have the form

$$\int_0^\infty e^{i\theta(r^2-t^2)} \sum_{j=0}^\infty W_j(x,y)\theta_{reg}^{\frac{n-1}{2}-j} d\theta \mod C^\infty \tag{159}$$

where W_j are the Hadamard-Riesz coefficients determined inductively by the transport equations

$$\begin{aligned} &\tfrac{\Theta'}{2\Theta}W_0 + \tfrac{\partial W_0}{\partial r} = 0 \\ &4ir(x,y)\{(\tfrac{k+1}{r(x,y)} + \tfrac{\Theta'}{2\Theta})W_{k+1} + \tfrac{\partial W_{k+1}}{\partial r}\} = \Delta_y W_k. \end{aligned} \tag{160}$$

The solutions are given by:

$$\begin{aligned} &W_0(x,y) = \Theta^{-\frac{1}{2}}(x,y) \\ &W_{j+1}(x,y) = \Theta^{-\frac{1}{2}}(x,y) \textstyle\int_0^1 s^k \Theta(x,x_s)^{\frac{1}{2}} \Delta_2 W_j(x,x_s) ds \end{aligned} \tag{161}$$

where x_s is the geodesic from x to y parametrized proportionately to arc-length and where Δ_2 operates in the second variable.

According to [**GS**], page 171,

$$\int_0^\infty e^{i\theta\sigma}\theta_+^\lambda d\lambda = ie^{i\lambda\pi/2}\Gamma(\lambda+1)(\sigma+i0)^{-\lambda-1}.$$

One has,

$$\int_0^\infty e^{i\theta(r^2-t^2)}\theta_+^{\frac{d-3}{2}-j} d\theta = ie^{i(\frac{n-1}{2}-j)\pi/2}\Gamma(\frac{n-3}{2}-j+1)(r^2-t^2+i0)^{j-\frac{n-3}{2}-2} \tag{162}$$

Here there is a problem when n is odd since $\Gamma(\frac{n-3}{2}-j+1)$ has poles at the negative integers.

One then uses

$$\Gamma(\alpha+1-k)=(-1)^{k+1}(-1)^{[\alpha]}\frac{\Gamma(\alpha+1-[\alpha])\Gamma([\alpha]+1-\alpha)}{\alpha+1}\frac{1}{\alpha-[\alpha]}\frac{1}{\Gamma(k-\alpha)}.$$

We note that

$$\Gamma(z)\Gamma(1-z)=\frac{\pi}{\sin\pi z}.$$

Here and above t^{-n} is the distribution defined by $t^{-n}=Re(t+i0)^{-n}$ (see [**Be**], [G.Sh., p.52,60].) We recall that $(t+i0)^{-n}=e^{-i\pi\frac{n}{2}}\frac{1}{\Gamma(n)}\int_0^\infty e^{itx}x^{n-1}dx$.

We also need that $(x+i0)^\lambda$ is entire and

$$(x+i0)^\lambda=\begin{cases} e^{i\pi\lambda}|x|^\lambda, & x<0 \\ x_+^\lambda, & x>0. \end{cases}$$

The imaginary part cancels the singularity of $\frac{1}{\alpha-[\alpha]}$ as $\alpha\to\frac{d-3}{2}$ when $n=2m+1$. There is no singularity in even dimensions. In odd dimensions the real part is $\cos\pi\lambda x_-^\lambda+x_+^\lambda$ and we always seem to have a pole in each term!

But in any dimension, the imaginary part is well-defined and we have

$$\frac{\sin t\sqrt{\Delta}}{\sqrt{\Delta}}(x,y)=C_o sgn(t)\sum_{j=0}^\infty(-1)^j w_j(x,y)\frac{(r^2-t^2)_-^{j-\frac{n-3}{2}-1}}{4^j\Gamma(j-\frac{n-3}{2})} \mod C^\infty \tag{163}$$

By taking the time derivative we also have,

$$\cos t\sqrt{\Delta}(x,y)=C_o|t|\sum_{j=0}^\infty(-1)^j w_j(x,y)\frac{(r^2-t^2)_-^{j-\frac{n-3}{2}-2}}{4^j\Gamma(j-\frac{n-3}{2}-1)} \mod C^\infty \tag{164}$$

where C_o is a universal constant and where $W_j=\tilde{C}_o e^{-ij\frac{\pi}{2}}4^{-j}w_j(x,y),$.

14.3. Exact formula in spaces of constant curvature

The Poisson kernel of $\mathbb{R}^{n+1}$ is the kernel of $e^{-t\sqrt{\Delta}}$, given by

$$\begin{aligned} K(t,x,y) &= t^{-n}(1+|\tfrac{x-y}{t}|^2)^{-\frac{n+1}{2}} \\ &= t\,(t^2+|x-y|^2)^{-\frac{n+1}{2}}. \end{aligned}$$

It is defined only for $t>0$, although formally it appears to be odd.

Thus, the kernel of $e^{it\sqrt{\Delta}}$ is

$$U(t,x,y) = (it)\,(|x-y|^2-t^2)^{-\frac{n+1}{2}}.$$

One would conjecture that the Poisson kernel of any Riemannian manifold would have the form

$$K(t,x,y)=t\sum_{j=0}^\infty(t^2+r(x,y)^2)^{-\frac{n+1}{2}+j}U_j(x,y) \tag{165}$$

for suitable U_j.

14.4. $\mathbb{S}^n$

One can determine the kernel of $e^{it\sqrt{\Delta}}$ on $\mathbb{S}^n$ from the Poisson kernel of the unit ball $B \subset \mathbb{R}^{n+1}$. We recall that the Poisson integral formula for the unit ball is:

$$u(x) = C_n \int_{\mathbb{S}^n} \frac{1-|x|^2}{|x-\omega'|^2} f(\omega') dA(\omega').$$

Write $x = r\omega$ with $|\omega| = 1$ to get:

$$P(r,\omega,\omega') = \frac{1-r^2}{(1-2r\langle\omega,\omega'\rangle + r^2)^{\frac{n+1}{2}}}.$$

A second formula for $u(r\omega)$ is

$$u(r,\omega) = r^{A-\frac{n-1}{2}} f(\omega),$$

where $A = \sqrt{\Delta + (\frac{n-1}{4})^2}$. This follows from by writing the equation $\Delta_{\mathbb{R}^{n+1}} u = 0$ as an Euler equation:

$$\{r^2 \frac{\partial^2}{\partial r^2} + nr\frac{\partial}{\partial r} - \Delta_{\mathbb{S}^n}\}u = 0.$$

Therefore, the Poisson operator e^{-tA} with $r = e^{-t}$ is given by

$$\begin{aligned} P(t,\omega,\omega') &= C_n \frac{\sinh t}{(\cosh t - \cos r(\omega,\omega'))^{\frac{n+1}{2}}} \\ &= C_n \frac{\partial}{\partial t} \frac{1}{(\cosh t - \cos r(\omega,\omega'))^{\frac{n-1}{2}}}. \end{aligned}$$

Here, $r(\omega,\omega')$ is the distance between points of $\mathbb{S}^n$.

We analytically continue the expressions to $t > 0$ and obtain the wave kernel as a boundary value:

$$e^{itA} = \lim_{\epsilon \to 0^+} C_n i \sin t(\cosh\epsilon\cos t - i\sinh\epsilon\sin t - \cos r(\omega,\omega')^{-\frac{n+1}{2}}$$

$$= \lim_{\epsilon \to 0^+} C_n i \sinh(it-\epsilon)(\cosh(it-\epsilon) - \cos r(\omega,\omega')^{-\frac{n+1}{2}}.$$

If we formally put $\epsilon = 0$ we obtain:

$$e^{itA} = C_n i \sin t(\cos t - \cos r(\omega,\omega'))^{-\frac{n+1}{2}}.$$

This expression is singular when $\cos t = \cos r$. We note that $r \in [0,\pi]$ and that it is singular on the cut locus $r = \pi$. Also, $\cos : [0,\pi] \to [-1,1]$ is decreasing, so the wave kernel is singular when $t = \pm r$ if $t \in [-\pi,\pi]$.

When n is even, the expression appears to be pure imaginary but that is because we need to regularize it on the set $t = \pm r$. When n is odd, the square root is real if $\cos t \geq \cos r$ and pure imaginary if $\cos t < \cos r$.

We see that the kernels of $\cos tA$, $\frac{\sin tA}{A}$ are supported inside the light cone $|r| \leq |t|$. On the other hand, e^{itA} has no such support property (it has infinite propagation speed). On odd dimensional spheres, the kernels are supported on the distance sphere (sharp Huyghens phenomenon).

The Poisson kernel of the unit sphere is then

$$e^{-tA} = C_n \sinh t(\cosh t - \cos r(\omega,\omega'))^{-\frac{n+1}{2}}.$$

It is singular on the complex characteristic conoid when $\cosh t - \cos r(\zeta,\bar{\zeta}') = 0$.

14.5. Analytic continuation into the complex

If we write out the eigenfunction expansions of $\cos t\sqrt{\Delta}(x,y)$ and $\frac{\sin t\sqrt{\Delta}}{\sqrt{\Delta}}(x,y)$ for $t = i\tau$, we would not expect convergence since the eigenvalues are now exponentially growing. Yet the majorants argument seems to indicate that these wave kernels admit an analytic continuation into a complex neighborhood of the complex characteristic conoid. Define the characteristic conoid in $\mathbb{R} \times M \times M$ by $r(x,y)^2 - t^2 = 0$. For simplicity of visualization, assume x is fixed. Then analytically the conoid to $\mathbb{C} \times M_{\mathbb{C}} \times M_{\mathbb{C}}$. By definition $(\zeta, \bar{\zeta}, 2\tau$ lies on the complexified conoid. That is, the series also converge after analytic continuation, again if $r^2 - t^2$ is small. If $t = i\tau$ then we need $r(\zeta, y)^2 + \tau^2$ to be small, which either forces $r(\zeta, y)^2$ to be negative and close to τ or else forces both τ and $r(\zeta, y)$ to be small.

If we wish to use orthogonality relations on M to sift out complexifications of eigenfunctions, then we need $U(i\tau, \zeta, y)$ to be holomorphic in ζ no matter how far it is from y!. So far, we do not have a proof that $U(i\tau, \zeta, y)$ is globally holomorphic for $\zeta \in M_\tau$ for every y.

Regimes of analytic continuation. Let $E(t,x,y)$ be any of the above kernels. Then analytically continue to $E(i\tau, \zeta, \bar{\zeta}')$ where $r(\zeta, \zeta')^2 + \tau^2$ is small. For instance if $\zeta' = \zeta$ and $\sqrt{\rho}(\zeta) = \frac{\tau}{2}$, then $r(\zeta, \zeta')^2 + \tau^2 = 0$.

Is there a neighborhood of the characteristic conoid into which the analytic continuation is possible? We need to have

$$r^2(\zeta, \zeta') \pm \tau^2 << \epsilon.$$

If we analytically continue in ζ and anti-analytically continue in ζ', we seem to get a neighborhood of the conoid.

We would like to analytically continue the Hadamard parametrix to a small neighborhood of the characteristic conoid. It is singular on the conoid.

15. Appendix: Lagrangian distributions, quasi-modes and Fourier integral operators

In this section, we go over the definitions of Lagrangian distributions, both semi-classical and homogeneous. A very detailed treatment of the homogeneous Lagrangian distributions (and Fourier integral operators) can be found in [**HoIV**]. The semi-classical case is almost the same and the detailed treatments can be found in [**D, DSj, GSj, CV2, Zw**]. We also continue the discussion in §2.10 of quasi-modes.

15.1. Semi-classical Lagrangian distributions and Fourier integral operators

Semi-classical Fourier integral operators with large parameter $\lambda = \frac{1}{\hbar}$ are operators whose Schwartz kernels are defined by semi-classical Lagrangian distributions,

$$I_\lambda(x,y) = \int_{\mathbb{R}^N} e^{i\lambda\varphi(x,y,\theta)} a(\lambda, x, y, \theta) d\theta.$$

More generally, semi-classical Lagrangian distributions are defined by oscillatory integrals (see [**D**]),

$$u(x,\hbar) = \hbar^{-N/2} \int_{\mathbb{R}^N} e^{\frac{i}{\hbar}\varphi(x,\theta)} a(x,\theta,\hbar) d\theta. \tag{166}$$

We assume that $a(x, \theta, \hbar)$ is a semi-classical symbol,

$$a(x, \theta, \hbar) \sim \sum_{k=0}^{\infty} \hbar^{\mu+k} a_k(x, \theta).$$

The critical set of the phase is given by

$$C_\varphi = \{(x, \theta) : d_\theta \varphi = 0\}.$$

The phase is called non-degenerate if

$$d(\frac{\partial \varphi}{\partial \theta_1}), \dots, d(\frac{\partial \varphi}{\partial \theta_N})$$

are independent on C_φ. Thus, the map

$$\varphi'_\theta := \left((\frac{\partial \varphi}{\partial \theta_1}), \dots, (\frac{\partial \varphi}{\partial \theta_N}) \right) : X \times \mathbb{R}^N \to \mathbb{R}^N$$

is locally a submersion near 0 and $(\varphi'_\theta)^{-1}(0)$ is a manifold of codimension N whose tangent space is $\ker D\varphi'_\theta$. Then

$$T_{(x_0,\theta_0)} C_\varphi = \ker d_{x,\theta} d_\theta \varphi.$$

We write a tangent vector to $M \times \mathbb{R}^N$ as $(\delta_x, \delta_\theta)$. The kernel of

$$D\varphi'_\theta = \begin{pmatrix} \varphi''_{\theta x} & \varphi''_{\theta\theta} \end{pmatrix}$$

is $T_{(x,\theta)} C_\varphi$. I.e. $(\delta_x, \delta_\theta) \in TC_\varphi$ if and only if $\varphi''_{\theta x} \delta_x + \varphi''_{\theta\theta} \delta_\theta = 0$. Indeed, φ'_θ is the defining function of C_φ and $d\varphi_\theta$ is the defining function of TC_φ. From [**HoIV**] Definition 21.2.5: The number of linearly independent differentials $d\frac{\partial \varphi}{\partial \theta}$ at a point of C_φ is $N - e$ where e is the excess. Then $C \to \Lambda$ is locally a fibration with fibers of dimension e. So to find the excess we need to compute the rank of $\begin{pmatrix} \varphi''_{x\theta} & \varphi''_{\theta\theta} \end{pmatrix}$ on $T_{x,\theta}(\mathbb{R}^N \times M)$.

Non-degeneracy is thus the condition that

$$\begin{pmatrix} \varphi''_{\theta x} & \varphi''_{\theta\theta} \end{pmatrix} \text{ is surjective on } C_\varphi \iff \begin{pmatrix} \varphi''_{\theta x} \\ \\ \varphi''_{\theta\theta} \end{pmatrix} \text{ is injective on } C_\varphi.$$

If φ is non-degenerate, then $\iota_\varphi(x, \theta) = (x, \varphi'_x(x, \theta))$ is an immersion from $C_\varphi \to T^*X$. Note that

$$d\iota_\varphi(\delta_x, \delta_\theta) = (\delta_x, \varphi''_{xx} \delta_x + \varphi''_{x\theta} \delta_\theta).$$

So if $\begin{pmatrix} \varphi''_{\theta x} \\ \\ \varphi''_{\theta\theta} \end{pmatrix}$ is injective, then $\delta_\theta = 0$.

If $(\lambda_1, \dots, \lambda_n)$ are any local coordinates on C_φ, extended as smooth functions in neighborhood, the delta-function on C_φ is defined by

$$d_{C_\varphi} := \frac{|d\lambda|}{|D(\lambda, \varphi'_\theta)/D(x, \theta)|} = \frac{dvol_{T_{x_0} M} \otimes dvol_{\mathbb{R}^N}}{d\frac{\partial \varphi}{\partial \theta_1} \wedge \cdots \wedge d\frac{\partial \varphi}{\partial \theta_N}}$$

where the denominator can be regarded as the pullback of $dVol_{\mathbb{R}^N}$ under the map

$$d_{\theta,x} d_\theta \varphi(x_0, \theta_0).$$

The symbol $\sigma(\nu)$ of a Lagrangian (Fourier integral) distributions is a section of the bundle $\Omega_{\frac{1}{2}} \otimes \mathcal{M}_{\frac{1}{2}}$ of the bundle of half-densities (tensor the Maslov line bundle). In terms of a Fourier integral representation it is the square root $\sqrt{d_{C_\varphi}}$

of the delta-function on C_φ defined by $\delta(d_\theta \varphi)$, transported to its image in T^*M under ι_φ

Definition: The principal symbol $\sigma_u(x_0, \xi_0)$ is

$$\sigma_u(x_0, \xi_0) = a_0(x_0, \xi_0)\sqrt{d_{C_\varphi}}.$$

It is a $\frac{1}{2}$ density on $T_{(x_0,\xi_0)}\Lambda_\varphi$ which depends on the choice of a density on $T_{x_0}M$).

15.2. Homogeneous Fourier integral operators

A homogeneous Fourier integral operator $A : C^\infty(X) \to C^\infty(Y)$ is an operator whose Schwartz kernel may be represented by an oscillatory integral

$$K_A(x, y) = \int_{\mathbb{R}^N} e^{i\varphi(x,y,\theta)} a(x, y, \theta) d\theta$$

where the phase φ is homogeneous of degree one in θ. We assume $a(x, y, \theta)$ is a zeroth order classical polyhomogeneous symbol with $a \sim \sum_{j=0}^\infty a_j$, a_j homogeneous of degree $-j$. We refer to [**DSj, GS2**] and especially to [**HoIV**] for background on Fourier integral operators. We use the notation $I^m(X \times Y, C)$ for the class of Fourier integral operators of order m with wave front set along the canonical relation C, and $WF'(F)$ to denote the canonical relation of a Fourier integral operator F.

When

$$\iota_\varphi : C_\varphi \to \Lambda_\varphi \subset T^*(X, Y), \quad \iota_\varphi(x, y, \theta) = (x, d_x\varphi, y, -d_y\varphi)$$

is an embedding, or at least an immersion, the phase is called non-degenerate. Less restrictive, although still an ideal situation, is where the phase is clean. This means that the map $\iota_\varphi : C_\varphi \to \Lambda_\varphi$, where Λ_φ is the image of ι_φ, is locally a fibration with fibers of dimension e. From [**HoIV**] Definition 21.2.5, the number of linearly independent differentials $d\frac{\partial \varphi}{\partial \theta}$ at a point of C_φ is $N - e$ where e is the excess.

We a recall that the order of $F : L^2(X) \to L^2(Y)$ in the non-degenerate case is given in terms of a local oscillatory integral formula by $m + \frac{N}{2} - \frac{n}{4}$,, where $n = \dim X + \dim Y$, where m is the order of the amplitude, and N is the number of phase variables in the local Fourier integral representation (see [**HoIV**], Proposition 25.1.5); in the general clean case with excess e, the order goes up by $\frac{e}{2}$ ([**HoIV**], Proposition 25.1.5'). Further, under clean composition of operators of orders m_1, m_2, the order of the composition is $m_1 + m_2 - \frac{e}{2}$ where e is the so-called excess (the fiber dimension of the composition); see [**HoIV**], Theorem 25.2.2.

The definition of the principal symbol is essentially the same as in the semi-classical case. As discussed in [**SV**], (see (2.1.2) and ((2.2.5) and Definition 2.7.1)), if an oscillatory integral is represented as

$$\mathcal{I}_{\varphi,a}(t, x, y) = \int e^{i\varphi(t,x,y,\eta)} a(t, x, y, \eta)\zeta(t, x, y, \eta) d_\varphi(t, x, y, \eta) d\eta,$$

where

$$d_\varphi(t, x, y, \eta) = |\det \varphi_{x,\eta}|^{\frac{1}{2}}$$

and where the number of phase variables equals the number of x variables, then the phase is non-degenerate if and only if $(\varphi_{x\eta}(t, x, y, \eta)$ is non-singular. Then $a_0|\det \varphi_{x,\eta}|^{-\frac{1}{2}}$ is the symbol.

The behavior of symbols under pushforwards and pullbacks of Lagrangian submanifolds are described in [**GS2**], Chapter IV. 5 (page 345). The main statement

(Theorem 5.1, loc. cit.) states that the symbol map $\sigma : I^m(X, \Lambda) \to S^m(\Lambda)$ has the following pullback-pushforward properties under maps $f : X \to Y$ satisfying appropriate transversality conditions,

$$\left\{ \begin{array}{l} \sigma(f^*\nu) = f^*\sigma(\nu), \\ \\ \sigma(f_*\mu) = f_*\sigma(\mu), \end{array} \right. \tag{167}$$

Here, $f_*\sigma(\mu)$ is integration over the fibers of f when f is a submersion. In order to define a pushforward, f must be a "morphism" in the language of [**GSt2**], i.e. must be accompanied by a map $r(x) : |\bigwedge|^{\frac{1}{2}} TY_{f(x)} \to |\bigwedge|^{\frac{1}{2}} TX_x$, or equivalently a half-density on $N^*(\text{graph}(f))$, the co-normal bundle to the graph of f which is constant long the fibers of $N^*(\text{graph}(f)) \to \text{graph}(f))$.

15.3. Quasi-modes

We consider here Lagrangian quasi-modes of order zero, i.e. semi-classical oscillatory integrals (166) which solves $\Delta u_k = O(1)$. If we pass Δ under the integral sign we obtain a leading order term whose amplitude contains the factor $k^2 |\nabla_x \varphi(x, \theta)|^2$. The integral of this term must vanish and hence must vanish on the stationary phase set $\nabla_\theta \varphi(x, \theta) = 0$. Hence φ must generate a Lagrangian submanifold $\Lambda_\varphi \subset S^*M$. Such a Lagrangian submanifold must be invariant under the geodesic flow $G^t : S^*M \to S^*M$ because the Hamilton vector field Ξ_H of a function H which is constant on a Lagrangian submanifold is tangent to Λ. The Lagrangian submanifold may only be locally defined and its global extension might be dense in S^*M. The amplitude a must then be assumed to be compactly supported and the distribution is $O(k^{-M})$ for all M outside of its support.

The $O(k)$ term has an invariant interpretation as $\mathcal{L}_{\Xi_H} a_0 \sqrt{d_{C_\varphi}}$, the Lie derivative of the principal symbol, which is a half-density on Λ_φ. To define a quasi-mode of order zero, the principal symbol must define a global half-density invariant under the geodesic flow. Here we suppress the role of the Maslov bundle and refer to [**D, DSj, Zw**] for its definition.

To obtain a quasi-mode of higher order, one must solve recursively a sequence of transport equations, which are homogeneous equations of the form $\mathcal{L}_{\Xi_H} a_{j+1} = \mathcal{D}_j a_j$ for various operators $\mathcal{D}_k$. In general there are obstructions to solving the inhomogeneous equations. The integral of the left side over Λ with respect to the invariant density equals zero, and therefore so must the right side. Here we express all half-densities in terms of the invariant one.

In sum, a zeroth order quasi-mode is an oscillatory integral (166) quantizing a pair (Λ, σ) where $\Lambda \subset S^*M$ is a closed invariant Lagrangian submanifold and σ is a G^t-invariant half-density along it.

There are additional quasi-modes associated to isotropic submanifolds of S^*M, i.e. manifolds on which the symplectic form vanishes but which have dimension $< \dim M$. Gaussian beams are of this kind. We refer to [**BB, R1, R2**] for their definition. They are constructed along stable elliptic closed geodesics, and may be regarded as Lagrangian distributions with complex phase.

In the case of toric completely integrable systems, the joint eigenfunctions are automatically semi-classical Lagrangian distributions. Indeed, they may be expressed as Fourier coefficients of the unitary Fourier integral operator quantizing the torus action. It is also true that joint eigenfunctions in the general case of quantum integrable systems are Lagrangian. We refer to [**TZ3**] for background and references.

We should emphasize that it is very rare that eigenfunctions of the Laplacian are quasi-modes or that quasi-modes are genuine eigenfunctions of the Laplacian. In §16.2 we will see that the standard eigenfunctions on S^2 are Lagrangian quasi-modes.

16. Appendix on Spherical Harmonics

Spherical harmonics furnish the extremals for L^p norms of eigenfunctions φ_λ as (M, g) ranges over Riemannian manifolds and φ_λ ranges over its eigenfunctions. They are not unique in this respect: surfaces of revolution and their higher dimensional analogues also give examples where extremal eigenfunction bounds are achieved. In this appendix we review the definition and properties of spherical harmonics.

Eigenfunctions of the Laplacian Δ_{S^n} on the standard sphere S^n are restrictions of harmonic homogeneous polynomials on $\mathbb{R}^{n+1}$.

Let $\Delta_{\mathbb{R}^{n+1}} = -(\frac{\partial^2}{\partial x_1^2} + \cdots + \frac{\partial^2}{\partial x_{n+1}^2})$ denote the Euclidean Laplacian. In polar coordinates (r, ω) on $\mathbb{R}^{n+1}$, we have $\Delta_{\mathbb{R}^{n+1}} = -\left(\frac{\partial^2}{\partial r^2} + \frac{n}{r}\frac{\partial}{\partial r}\right) + \frac{1}{r^2}\Delta_{S^n}$. A polynomial $P(x) = P(x_1, \ldots, x_{n+1})$ on $\mathbb{R}^{n+1}$ is called:

- homogeneous of degree k if $P(rx) = r^k P(x)$. We denote the space of such polynomials by $\mathcal{P}_k$. A basis is given by the monomials
$$x^\alpha = x_1^{\alpha_1} \cdots x_{n+1}^{\alpha_{n+1}}, \quad |\alpha| = \alpha_1 + \cdots + \alpha_{n+1} = k.$$
- Harmonic if $\Delta_{\mathbb{R}^{n+1}} P(x) = 0$. We denote the space of harmonic homogeneous polynomials of degree k by $\mathcal{H}_k$.

Suppose that $P(x)$ is a homogeneous harmonic polynomial of degree k on $\mathbb{R}^{n+1}$. Then,
$$0 = \Delta_{\mathbb{R}^{n+1}} P = -\{\frac{\partial^2}{\partial r^2} + \frac{n}{r}\frac{\partial}{\partial r}\} r^k P(\omega) + \frac{1}{r^2}\Delta_{S^n} P(\omega)$$
$$\implies \Delta_{S^n} P(\omega) = (k(k-1) + nk) P(\omega).$$
Thus, if we restrict $P(x)$ to the unit sphere S^n we obtain an eigenfunction of eigenvalue $k(n+k-1)$. Let $\mathcal{H}_k \subset L^2(S^n)$ denote the space of spherical harmonics of degree k. Then:

- $L^2(S^n) = \bigoplus_{k=0}^\infty \mathcal{H}_k$. The sum is orthogonal.
- $Sp(\Delta_{S^n}) = \{\lambda_k^2 = k(n+k-1)\}$.
- $\dim \mathcal{H}_k$ is given by
$$d_k = \binom{n+k-1}{k} - \binom{n+k-3}{k-2}$$

The Laplacian Δ_{S^n} is quantum integrable. For simplicity, we restrict to S^2. Then the group $SO(2) \subset SO(3)$ of rotations around the x_3-axis commutes with the Laplacian. We denote its infinitesimal generator by $L_3 = \frac{\partial}{i\partial\theta}$. The standard basis of spherical harmonics is given by the joint eigenfunctions ($|m| \leq k$)
$$\begin{cases} \Delta_{S^2} Y_m^k = k(k+1) Y_m^k; \\ \frac{\partial}{i\partial\theta} Y_m^k = m Y_m^k. \end{cases}$$

Two basic spherical harmonics are:

- The highest weight spherical harmonic Y^k_k. As a homogeneous polynomial it is given up to a normalizing constant by $(x_1 + ix_2)^k$ in $\mathbb{R}^3$ with coordinates (x_1, x_2, x_3). It is a 'Gaussian beam' along the equator $\{x_3 = 0\}$, and is also a quasi-mode associated to this stable elliptic orbit.

- The zonal spherical harmonic Y^k_0. It may be expressed in terms of the orthogonal projection $\Pi_k : L^2(S^2) \to \mathcal{H}_k$.

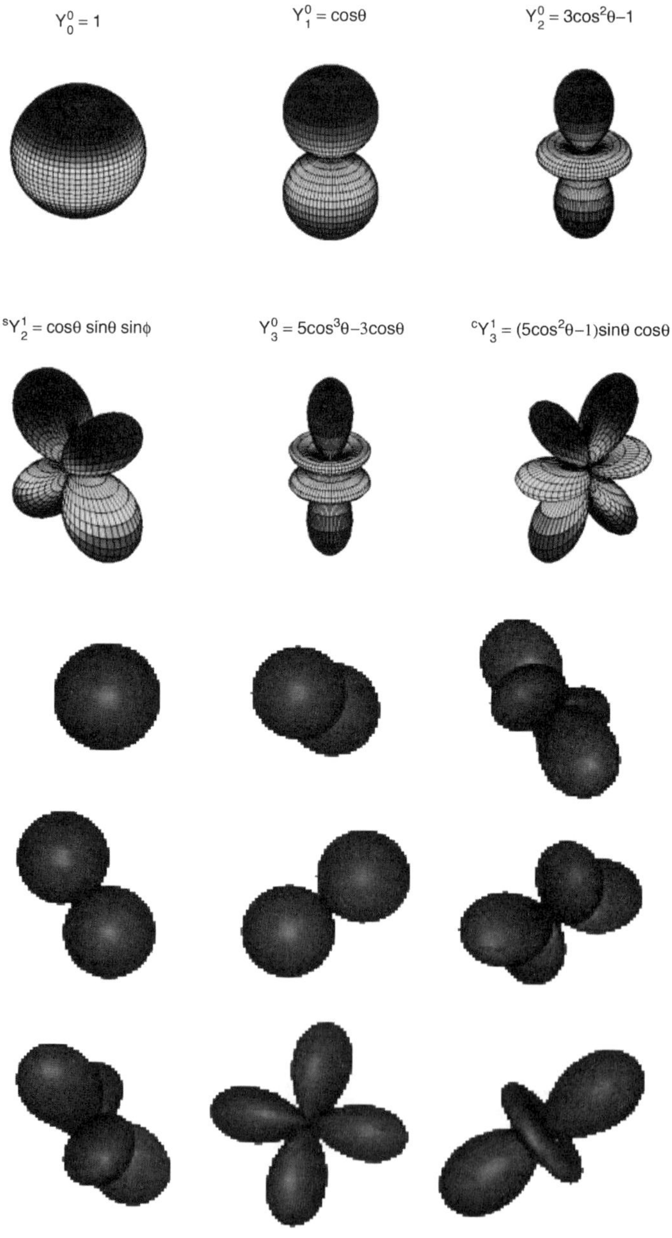

We now explain the last statement: For any n, the kernel $\Pi_k(x,y)$ of Π_k is defined by

$$\Pi_k f(x) = \int_{S^n} \Pi_k(x,y) f(y) dS(y),$$

where dS is the standard surface measure. If $\{Y_m^k\}$ is an orthonormal basis of $\mathcal{H}_k$ then

$$\Pi_k(x,y) = \sum_{m=1}^{d_k} Y_m^k(x) Y_m^k(y).$$

Thus for each y, $\Pi_k(x,y) \in \mathcal{H}_k$. We can L^2 normalize this function by dividing by the square root of

$$||\Pi_k(\cdot,y)||_{L^2}^2 = \int_{S^n} \Pi_k(x,y)\Pi_k(y,x) dS(x) = \Pi_k(y,y).$$

We note that $\Pi_k(y,y) = C_k$ since it is rotationally invariant and $O(n+1)$ acts transitively on S^n. Its integral is $\dim \mathcal{H}_k$, hence, $\Pi_k(y,y) = \frac{1}{Vol(S^n)} \dim \mathcal{H}_k$. Hence the normalized projection kernel with 'peak' at y_0 is

$$Y_0^k(x) = \frac{\Pi_k(x,y_0)\sqrt{Vol(S^n)}}{\sqrt{\dim \mathcal{H}_k}}.$$

Here, we put y_0 equal to the north pole $(0, 0 \cdots, 1)$. The resulting function is called a zonal spherical harmonic since it is invariant under the group $O(n+1)$ of rotations fixing y_0.

One can rotate $Y_0^k(x)$ to $Y_0^k(g \cdot x)$ with $g \in O(n+1)$ to place the 'pole' or 'peak point' at any point in S^2.

16.1. Highest weight spherical harmonics

As mentioned above, the highest weight spherical harmonic is an extremal for the L^6 norm and for all L^p norms with $2 < p < 6$ it is the unique extremal. Let us verify that it achieves the maximum.

We claim that $||(x+iy)^k||_{L^2(S^2)} \sim k^{-1/4}$. Indeed we compute it using Gaussian integrals:

$$\int_{\mathbb{R}^3} (x^2+y^2)^k e^{-(x^2+y^2+z^2)} = ||(x+iy)^k||_{L^2(S^2)}^2 \int_0^\infty r^{2k} e^{-r^2} r^2 dr,$$

$$\int_{\mathbb{R}^3} (x^2+y^2)^k e^{-(x^2+y^2+z^2)} = \int_{\mathbb{R}^2} (x^2+y^2)^k e^{-(x^2+y^2)}$$

$$= \int_0^\infty r^{2k} e^{-r^2} r dr, \implies ||(x+iy)^k||_{L^2(S^2)}^2 = \frac{\Gamma(k+1)}{\Gamma(k+\frac{3}{2})} \sim k^{-1/2}.$$

Thus, $k^{1/4}(x+iy)^k$ is the L^2-normalized highest weight spherical harmonic. It achieves its L^∞ norm at $(1,0,0)$ where it has size $k^{1/4}$.

To see that it is an extremal for L^p for $2 \leq p \leq 6$, we use Gaussian integrals:

$$\int_{\mathbb{R}^3} (x^2+y^2)^{3k} e^{-(x^2+y^2+z^2)} = ||(x+iy)^k||_{L^6(S^2)}^6 \int_0^\infty r^{6k} e^{-r^2} r^2 dr,$$

$$\int_{\mathbb{R}^3} (x^2+y^2)^{3k} e^{-(x^2+y^2+z^2)} = \int_{\mathbb{R}^2} (x^2+y^2)^{3k} e^{-(x^2+y^2)} = \int_0^\infty r^{6k} e^{-r^2} r dr,$$

$$\implies ||(x+iy)^k||_{L^6(S^2)}^6 = \frac{\Gamma(6k+1)}{\Gamma(6k+\frac{3}{2})} \sim k^{-1/2}.$$

Hence, the L^6 norm of $k^{1/4}(x+iy)^k$ equals

$$k^{1/4}k^{-1/12} = k^{1/6}.$$

Since $\lambda_k \sim k$ and $\delta(6) = \frac{1}{6}$ in dimension 2, we see that it is an extremal.

Problem: If a surface (M^2, g) has maximal L^p growth for $2 < p < 6$, must it have a Gaussian beam? Here we may insist that the Gaussian beam be a sequence of eigenfunctions or we may relax the definition and allow it to be a quasi-mode. We might also ask, if there exists a quasi-mode Gaussian beam, does (M^2, g) has maximal L^p growth for $2 < p < 6$?

16.2. Spherical harmonics as quasi-modes

The normalized joint eigenfunctions on the standard sphere are given by

$$Y_m^N(\theta,\varphi) = \sqrt{(2N+1)\frac{(N-m)!}{(N+m)!}}P_m^N(\cos\varphi)e^{im\theta}, \tag{168}$$

where

$$P_m^N(\cos\varphi) = \frac{1}{2\pi}\int_0^{2\pi}(i\sin\varphi\cos\theta+\cos\varphi)^N e^{-im\theta}d\theta$$

are the Legendre polynomials.

To obtain a Lagrangian distribution, we consider a sequence of Y_m^N with $\frac{m}{N} \to C$ for some C. I.e. we consider pairs $k(m_0, N_0)$ lying on a ray in the lattice in $\mathbb{Z}^2$ of (m, N) with $|m| \leq N$.

The Lagrangian submanifold of T^*S^2 associated to Y_m^N with $m/N \to c$ is the torus in S^*S^2 defined by

$$p_\theta(x,\xi) := \langle \xi, \frac{\partial}{\partial\theta}\rangle = c.$$

This is a level set of the Clairaut integral

$$p_\theta : S^*S^2 \to [-1, 1].$$

Examples:

- (i) The Lagrangian submanifold associated to the zonal spherical harmonic is the "meridian torus" consisting of geodesics from the north to south poles.
- (ii) The Gaussian beam is associated to the unit vectors along the equator– a degenerate Lagrangian torus of dimension 1.

Note that we also express the standard spherical harmonics as

$$Y_m^N(x) = \int_0^{2\pi}\Pi_N(x, r_\theta y)e^{-im\theta}d\theta,$$

where r_θ are rotations around the third axis. Since $\Pi_N(x,y)$ is a semi-classical Lagrangian quasi-mode, this also exhibits Y_m^N as one.

Bibliography

[Ar] S. Ariturk, Lower bounds for nodal sets of Dirichlet and Neumann eigenfunctions. Comm. Math. Phys. 317 (2013), no. 3, 817–825 (arXiv:1110.6885).

[Arn] V. I. Arnol' d, Modes and quasimodes. (Russian) Funkcional. Anal. i Prilozen. 6 (1972), no. 2, 12–20.

[BB] V.M.Babic, V.S. Buldyrev: *Short-Wavelength Diffraction Theory*, Springer Series on Wave Phenomena 4, Springer-Verlag, New York (1991)

[Ba] L. Bakri, Critical set of eigenfunctions of the Laplacian, J. Geom. Phys. 62, No. 10, 2024–2037 (2012) (arXiv:1008.1699.)

[Bae] C. Bär, On nodal sets for Dirac and Laplace operators. Comm. Math. Phys. 188 (1997), no. 3, 709–721.

[Be] P. Bérard, On the wave equation without conjugate points, Math. Zeit. **155** (1977), 249–276.

[BGM] M. Berger, P. Gauduchon, and E. Mazet, Le spectre d'une variété riemannienne. Lecture Notes in Mathematics, Vol. 194 Springer-Verlag, Berlin-New York 1971.

[Bers] L. Bers, Local behavior of solutions of general linear elliptic equations. Comm. Pure Appl. Math. 8 (1955), 473–496.

[Bes] G. Besson, Propriétées génériques des fonctions propres et multiplicité. Comment. Math. Helv. 64 (1989), no. 4, 542–588.

[BlSo] M. D. Blair and C. D. Sogge, On Kakeya-Nikodym averages, L^p-norms and lower bounds for nodal sets of eigenfunctions in higher dimensions. J. Eur. Math. Soc. (JEMS) 17 (2015), no. 10, 2513–2543 (arXiv:1301.7468).

[Bou] L. Boutet de Monvel, Convergence dans le domaine complexe des séries de fonctions propres. C. R. Acad. Sci. Paris Sér. A-B 287 (1978), no. 13, A855–A856.

[BoGu] L. Boutet de Monvel and V. Guillemin, *The Spectral Theory of Toeplitz Operators*, Ann. Math. Studies 99, Princeton Univ. Press, Princeton, 1981.

[BoSj] L. Boutet de Monvel and J. Sjöstrand, Sur la singularité des noyaux de Bergman et de Szegö, *Asterisque* 34–35 (1976), 123–164.

[Br] J. Brüning, Über Knoten von Eigenfunktionen des Laplace-Beltrami Operators", Math. Z. 158 (1978), 15–21.

[Ch] I. Chavel, *Eigenvalues in Riemannian geometry.* Including a chapter by Burton Randol. With an appendix by Jozef Dodziuk. Pure and Applied Mathematics, 115. Academic Press, Inc., Orlando, FL, 1984.

[Ch1] S. Y. Cheng, Eigenfunctions and eigenvalues of Laplacian. Differential geometry (Proc. Sympos. Pure Math., Vol. XXVII, Stanford Univ., Stanford, Calif., 1973), Part 2, pp. 185–193. Amer. Math. Soc., Providence, R.I., 1975.

[Ch2] ———, Eigenvalue comparison theorems and its geometric applications. Math. Z. 143 (1975), no. 3, 289–297.

[Ch3] ———, Eigenfunctions and nodal sets. Comment. Math. Helv. 51 (1976), no. 1, 43–55.

[CTZ] H. Christianson, J. A. Toth and S. Zelditch, Quantum ergodic restriction for Cauchy data: Interior QUE and restricted QUE, Math Res. Lett. 20 (2013), no. 3, 1–11 (arXiv:1205.0286).

[CM] T. H. Colding and W. P. Minicozzi II, Lower bounds for nodal sets of eigenfunctions. Comm. Math. Phys. 306 (2011), no. 3, 777–784.

[CM2] T. Colding and W. P. Minicozzi II, Volumes for eigensections. Geom. Dedicata 102 (2003), 19–24.

[CV] Y. Colin de Verdiére, Ergodicité et fonctions propres du laplacien. Comm. Math. Phys. 102 (1985), no. 3, 497–502.

[CV2] Y.Colin de Verdiere, Quasi-modes sur les varietes Riemanniennes compactes, Invent.Math. 43 (1977), 15–52.

[DSj] M. Dimassi and J. Sjöstrand, *Spectral asymptotics in the semi-classical limit.* London Mathematical Society Lecture Note Series, 268. Cambridge University Press, Cambridge, 1999.

[Dong] R-T. Dong, Nodal sets of eigenfunctions on Riemann surfaces. J. Differential Geom. 36 (1992), no. 2, 493–506.

[DF] H. Donnelly and C. Fefferman, Nodal sets of eigenfunctions on Riemannian manifolds. Invent. Math. 93 (1988), no. 1, 161–183.

[DF2] H. Donnelly and C. Fefferman, Growth and geometry of eigenfunctions of the Laplacian. Analysis and partial differential equations, 635–655, Lecture Notes in Pure and Appl. Math., 122, Dekker, New York, 1990.

[DF3] H. Donnelly and C. Fefferman, Nodal sets for eigenfunctions of the Laplacian on surfaces. J. Amer. Math. Soc. 3 (1990), no. 2, 333–353.

[D] J. J. Duistermaat, Oscillatory integrals, Lagrange immersions and unfolding of singularities. Comm. Pure Appl. Math. 27 (1974), 207–281.

[D2] J. J. Duistermaat, Fourier integral operators. Reprint of the 1996 original. Modern Birkhäuser Classics. Birkhäuser/Springer, New York, 2011.

[DG] J.J.Duistermaat and V.Guillemin, The spectrum of positive elliptic operators and periodic bicharacteristics, Inv.Math. 24 (1975), 39–80.

[DH] J. J. Duistermaat and L. Hörmander, Fourier integral operators. II. Acta Math. 128 (1972), no. 3–4, 183–269.

[ElHajT] L. El-Hajj, J. A.Toth Intersection bounds for nodal sets of planar Neumann eigenfunctions with interior analytic curves, J. Diff. Geom. 100 (2015), no. 1, 1–53 (arXiv:1211.3395).

[F] H. Federer, Geometric measure theory. Die Grundlehren der mathematischen Wissenschaften, Band 153 Springer-Verlag New York Inc., New York 1969.

[GeSh] I. M. Gel'fand and G.E. Shilov,*Generalized functions. Vol. I: Properties and operations.* Academic Press, New York-London 1964.

[GRS] A. Ghosh, A. Reznikov, and P. Sarnak, Nodal domains of Maass forms I, Geom. Funct. Anal. 23, No. 5, 1515–1568 (2013) (arXiv:1207.6625).

[GH] P. Griffiths and J. Harris, *Principles of Algebraic Geometry*, Wiley-Interscience, New York, 1978.

[GrSj] A. Grigis and J. Sjöstrand, *Microlocal analysis for differential operators. An introduction.* London Mathematical Society Lecture Note Series, 196. Cambridge University Press, Cambridge, 1994.

[GSj] A. Grigis and J. Sjöstrand, *Microlocal analysis for differential operators. An introduction.* London Mathematical Society Lecture Note Series, 196. Cambridge University Press, Cambridge, 1994.

[G] M. Grohe, Fixed-point definability and polynomial time on graphs with excluded minors. J. ACM 59 (2012), no. 5

[GS] V. Guillemin and S. Sternberg, Geometric asymptotics. Mathematical Surveys, No. 14. American Mathematical Society, Providence, R.I., 1977.

[G1] Guillemin, Victor Toeplitz operators in n dimensions. Integral Equations Operator Theory 7 (1984), no. 2, 145–205.

[GS1] V. Guillemin and M. Stenzel, Grauert tubes and the homogeneous Monge-Ampère equation. J. Differential Geom. 34 (1991), no. 2, 561–570.

[GS2] V. Guillemin and M. Stenzel, Grauert tubes and the homogeneous Monge-Ampère equation. II. J. Differential Geom. 35 (1992), no. 3, 627–641.

[GS2] V. Guillemin and S. Sternberg, Geometric asymptotics. Mathematical Surveys, No. 14. American Mathematical Society, Providence, R.I., 1977.

[GSt2] V. Guillemin and S. Sternberg, Semi-classical Analysis (preprint, 2012).

[HL] Q. Han and F. H. Lin: *Nodal sets of solutions of Elliptic Differential Equations*, book in preparation (online at http://www.nd.edu/qhan/nodal.pdf).

[HHHZ] X. Han, A. Hassell, H. Hezari and S. Zelditch, Completeness of boundary traces of eigenfunctions, Proc. Lond. Math. Soc. (3) 111 (2015), no. 3, 749–773.

[HZZ] B. Hanin, P. Zhou and S. Zelditch, Nodal Sets of Random Eigenfunctions for the Isotropic Harmonic Oscillator, International Mathematics Research Notices, Vol. 2015, No. 13, pp. 4813–4839, (2015).

[HaS] R. Hardt and L. Simon, Nodal sets for solutions of elliptic equations. J. Differential Geom. 30 (1989), no. 2, 505–522.

[HW] P. Hartman and A. Wintner, On the local behavior of solutions of non-parabolic partial differential equations. Amer. J. Math. 75, (1953). 449–476.

[HW2] P. Hartman and W. Wintner, On the local behavior of solutions of non-parabolic partial differential equations. III. Approximations by spherical harmonics. Amer. J. Math. 77 (1955), 453–474.

[HK] B. Hasselblatt and A. Katok, *A first course in dynamics. With a panorama of recent developments.* Cambridge University Press, New York, 2003

[H] B. Helffer, On spectral minimal partitions: a survey. Milan J. Math. 78 (2010), no. 2, 575–590.

[He] E. J. Heller, http://www.ericjhellergallery.com.

[HS] H. Hezari and C. D. Sogge, A natural lower bound for the size of nodal sets, Analysis and PDE vol. 5 (2012), 1133–1137 (arXiv:1107.3440).

[Hof] T. Hoffmann-Ostenhof, Problems for the number of nodal domains in "Geometric Aspects of Spectral Theory xiv" Oberwolfach Reports 33, page 2008–270, 2012.

[HoI] L. Hörmander, *The analysis of linear partial differential operators. I. Distribution theory and Fourier analysis.* Classics in Mathematics. Springer-Verlag, Berlin, 2003.

[HoII] L. Hörmander, *The analysis of linear partial differential operators. II. Differential operators with constant coefficients.* Classics in Mathematics. Springer-Verlag, Berlin, 2005.

[HoIII] L. Hörmander, *The analysis of linear partial differential operators. III. Pseudo-differential operators.* Classics in Mathematics. Springer, Berlin, 2007.

[HoIV] L. Hörmander, *The analysis of linear partial differential operators.* IV. Fourier integral operators. Reprint of the 1994 edition. Classics in Mathematics. Springer-Verlag, Berlin, 2009.

[HoC] L. Hörmander, *Notions of convexity.* . Modern Birkhäuser Classics. Birkhäuser Boston, Inc., Boston, MA, 2007.

[JN] D. Jakobson and N. Nadirashvili, Eigenfunctions with few critical points. J. Differential Geom. 53 (1999), no. 1, 177–182.

[Jin] Long Jin, Semiclassical Cauchy Estimates and Applications, to appear in Trans. AMS (arXiv:1302.5363).

[JJ] J. Jung, Sharp bounds for the intersection of nodal lines with certain curves. J. Eur. Math. Soc. (JEMS) 16 (2014), no. 2, 273-288 (arXiv: 1108.2335).

[JJ2] J. Jung, Quantitative quantum ergodicity and the nodal domains of Maass-Hecke cusp forms, (arXiv:1301.6211).

[JZ] J. Jung and S. Zelditch, Number of nodal domains and singular points of eigenfunctions of negatively curved surfaces with an isometric involution, J. Diff. Geom. 102 (2016) 37–66 (arXiv:1310.2919).

[KP] M. Karplus and R. N. Porter, *Atoms and molecules an introduction for students of physical chemistry* , W. A. Benjamin, New York (1970).

[K] J. B. Keller, Corrected Bohr-Sommerfeld quantum conditions for nonseparable systems. Ann. Physics 4 1958 180–188.

[Kua] I. Kukavica, Nodal volumes for eigenfunctions of analytic regular elliptic problems. J. Anal. Math. 67 (1995), 269–280.

[L] G. Lebeau, The complex Poisson kernel on a compact analytic Riemannian manifold, preprint (2013).

[LS1] L. Lempert and R. Szöke, Global solutions of the homogeneous complex Monge-Ampère equation and complex structures on the tangent bundle of Riemannian manifolds. Math. Ann. 290 (1991), no. 4, 689–712.

[L] H. Lewy, On the minimum number of domains in which the nodal lines of spherical harmonics divide the sphere, Comm. Partial Differential Equations 2 (1977), no. 12, 1233–1244.

[LL] L. D. Landau and E. M. Lifshitz, *Quantum mechanics: non-relativistic theory.* Course of Theoretical Physics, Vol. 3. Addison-Wesley Series in Advanced Physics. Pergamon Press Ltd., London-Paris; for U.S.A. and Canada: Addison-Wesley Publishing Co., Inc., Reading, Mass; 1958.

[LS] E. H. Lieb and R. Seiringer, The stability of matter in quantum mechanics. Cambridge University Press, Cambridge, 2010.

[Lin] F.H. Lin, Nodal sets of solutions of elliptic and parabolic equations. Comm. Pure Appl. Math. 44 (1991), no. 3, 287–308.

[Man2] D. Mangoubi, The Volume of a Local Nodal Domain, J. Topol. Anal. 2 (2010), no. 2, 259–275 (arXiv:0806.3327).

[Man3] D. Mangoubi, On the inner radius of a nodal domain. Canad. Math. Bull. 51 (2008), no. 2, 249–260.

[Mar] S. Marshall, Geodesic restrictions of eigenfunctions on arithmetic surfaces, Duke Math. J. (to appear) (arXiv:1204.0781).

[NS] F. Nazarov and M. Sodin, On the number of nodal domains of random spherical harmonics. Amer. J. Math. 131 (2009), no. 5, 1337–1357.

[NPS] F. Nazarov, L. Polterovich, and M. Sodin, Sign and area in nodal geometry of Laplace eigenfunctions. Amer. J. Math. 127 (2005), no. 4, 879–910..

[Po] I. Polterovich, Pleijel's nodal domain theorem for free membranes, Proc. Amer. Math. Soc. 137 (2009), no. 3, 1021–1024 (arXiv:0805.1553).

[R1] J.Ralston, On the construction of quasi-modes associated with stable periodic orbits, Comm.Math.Phys.51 (1976),219–242.

[R2] J.Ralston, Approximate eigenfunctions of the Laplacian, J.Diff.Geom. 12 (1977),87–100.

[RS] M. Reed and B. Simon,*Methods of modern mathematical physics. I. Functional analysis.* Second edition. Academic Press, Inc. [Harcourt Brace Jovanovich, Publishers], New York, 1980.

[Saf] Y. Safarov, Asymptotics of a spectral function of a positive elliptic operator without a nontrapping condition. (Russian) Funktsional. Anal. i Prilozhen. 22 (1988), no. 3,53–65, 96 translation in Funct. Anal. Appl. 22 (1988), no. 3, 213–223 (1989).

[SV] Y. Safarov and D. Vassiliev, *The asymptotic distribution of eigenvalues of partial differential operators. Translated from the Russian manuscript by the authors.* Translations of Mathematical Monographs, 155. American Mathematical Society, Providence, RI, 1997.

[Shn] A. I. Shnirelman, Ergodic properties of eigenfunctions, *Usp. Mat. Nauk.* 29/6 (1974), 181–182.

[Sch] E. Schrödinger, Quantisierung als Eigenwertproblem, Annalen der Physik (1926).

[Sj] J. Sjöstrand, *Singularités analytiques microlocales.* Astérisque, 95, 1–166, Astérisque, 95, Soc. Math. France, Paris, 1982.

[Sogb] C. D. Sogge, *Fourier integrals in classical analysis.* Cambridge Tracts in Mathematics, 105. Cambridge University Press, Cambridge, 1993.

[Sogb2] C. D. Sogge, *Hangzhou lectures on eigenfunctions of the Laplacian*, Annals of Math. Studies 188, Princeton University Press (2014).

[Sog] C. D. Sogge, Concerning the L^p norm of spectral clusters for second-order elliptic operators on compact manifolds, J. Funct. Anal. 77 (1988), 123–138.

[SoZ] C. D. Sogge and S. Zelditch, Lower bounds on the hypersurface measure of nodal sets, Math. Research Letters 18 (2011), 27–39 (arXiv:1009.3573).

[SoZa] C. D. Sogge and S. Zelditch, Lower bounds on the Hausdorff measure of nodal sets II, Math. Res. Lett. 19 (2012), no. 6, 1361–1364 (arXiv:1208.2045).

[SoZ2] C. D. Sogge and S. Zelditch, Riemannian manifolds with maximal eigenfunction growth. Duke Math. J. 114 (2002), no. 3, 387–437

[SoZ3] C. D. Sogge and S. Zelditch, Focal points and sup norms of eigenfunctions on real analytic Riemannian manifolds, to appear in Rev. Mat. Iberoam. (preprint, 2013).

[SoZ4] C.D. Sogge and S. Zelditch, On eigenfunction restriction estimates and L^4-bounds for compact surfaces with nonpositive curvature, Advances in analysis: the legacy of Elias M. Stein, 447–461, Princeton Math. Ser., 50, Princeton Univ. Press, Princeton, NJ, 2014. (arXiv:1108.2726).

[SoZ5] C.D. Sogge and S. Zelditch, Concerning the L^4 norms of typical eigenfunctions on compact surfaces, Recent developments in geometry and analysis, Vol 23, 407–423, Higher Ed. Press and Int. Press, Beijing-Boston (2013) (arXiv:1011.0215).

[STZ] C.D. Sogge, J. A. Toth and S. Zelditch, About the blowup of quasimodes on Riemannian manifolds. J. Geom. Anal. 21 (2011), no. 1, 150–173.

[Stei] S. Steinerberger, Lower bounds on nodal sets of eigenfunctions via the heat flow, Comm. Partial Differential Equations 39 (2014), no. 12, 2240–2261 (arXiv:1301.3371).

[Ste] Stenzel, Matthew B. On the analytic continuation of the Poisson kernel. Manuscripta Math. 144 (2014), no. 1–2, 253–276.

[Ste2] M. Stenzel, The Poisson transform on a compact real analytic Riemannian manifold. Monatsh. Math. 178 (2015), no. 2, 299–309.

[St] A. S. Stodolna, A. Rouzée, F. L'epine, S. Cohen, F. Robicheaux, A. Gijsbertsen, J. H. Jungmann, C. Bordas, M. J. J. Vrakking. Hydrogen Atoms under Magnification: Direct Observation of the Nodal Structure of Stark States. Physical Review Letters, 2013; 110 (21).

[SHM] RM Stratt, NC Handy, WH Miller, On the quantum mechanical implications of classical ergodicity, - The Journal of Chemical Physics, 1979.

[TI] M. E. Taylor, *Partial differential equations I. Basic theory.* Second edition. Applied Mathematical Sciences, 115. Springer, New York, 2011.

[TII] M. E. Taylor, *Partial differential equations II. Qualitative studies of linear equations.* Second edition. Applied Mathematical Sciences, 116. Springer, New York, 2011.

[Th] B. Thaller, *Visual quantum mechanics. Selected topics with computer-generated animations of quantum-mechanical phenomena.* Springer-Verlag–TELOS, New York, 2000; *Advanced visual quantum mechanics.* Springer-Verlag, New York, 2005.

[TZ] J. A. Toth and S. Zelditch, Counting Nodal Lines Which Touch the Boundary of an Analytic Domain, Jour. Diff. Geom. 81 (2009), 649–686 (arXiv:0710.0101).

[TZ2] J. A. Toth and S. Zelditch, Quantum ergodic restriction theorems, II: manifolds without boundary, Geom. Funct. Anal. 23 (2013), no. 2, 715–775 (arXiv:1104.4531).

[TZ3] J. A. Toth and S. Zelditch, L^p norms of eigenfunctions in the completely integrable case. Ann. Henri Poincaré 4 (2003), no. 2, 343–368.

[U] K. Uhlenbeck, Generic properties of eigenfunctions. Amer. J. Math. 98 (1976), no. 4, 1059–1078.

[Wei] S. Weinberg, *Lectures on quantum mechanics.* Cambridge University Press, Cambridge, 2013.

[Y1] S.T. Yau, Survey on partial differential equations in differential geometry. *Seminar on Differential Geometry*, pp. 3–71, Ann. of Math. Stud., 102, Princeton Univ. Press, Princeton, N.J., 1982.

[Y2] ———, Open problems in geometry. *Differential geometry: partial differential equations on manifolds* (Los Angeles, CA, 1990), 1–28, Proc. Sympos. Pure Math., 54, Part 1, Amer. Math. Soc., Providence, RI, 1993.

[Y] A. Yoshikawa, On Maslov's canonical operator. Hokkaido Math. J. 4 (1975), 8–38.

[Yo] M. P. Young, The quantum unique ergodicity conjecture for thin sets, Adv. Math. 286 (2016), 958–1016 (arXiv:1306.1554).

[W] Wikipedia article on atomic orbitals, http://en.wikipedia.org/wiki/Atomicorbital.

[Ze0] S. Zelditch, CMBS lectures on eigenfunctions, in preparation.

[Ze] S. Zelditch, Eigenfunctions and Nodal sets, Surveys in Differential Geometry volume 18 (2013), pp. 237–308 (arXiv:1205.2812.)

[Ze2] S. Zelditch, Recent developments in mathematical quantum chaos. Current developments in mathematics, 2009, 115–204, Int. Press, Somerville, MA, 2010.

[Ze3] S. Zelditch, Local and global analysis of eigenfunctions on Riemannian manifolds. Handbook of geometric analysis. No. 1, 545–658, Adv. Lect. Math. (ALM), 7, Int. Press, Somerville, MA, 2008.

[Ze4] S. Zelditch, Uniform distribution of eigenfunctions on compact hyperbolic surfaces. Duke Math. J. 55 (1987), no. 4, 919–941

[Ze5] S. Zelditch,, Complex zeros of real ergodic eigenfunctions. Invent. Math. 167 (2007), no. 2, 419–443.

[Ze6] S. Zelditch, Ergodicity and intersections of nodal sets and eigenfunctions on real analytic surfaces, to appear in J. Diff. Geom. (arXiv:1210.0834).

[Ze7] S. Zelditch, Pluri-potential theory on Grauert tubes of real analytic Riemannian manifolds, I. Spectral geometry, 299–339, Proc. Sympos. Pure Math., 84, Amer. Math. Soc., Providence, RI, 2012.

[Ze8] S. Zelditch, Complex zeros of quantum integrable eigenfunctions, (in preparation).

[Ze9] S. Zelditch, Kuznecov sum formulae and Szegö limit formulae on manifolds. Comm. Partial Differential Equations 17 (1992), no. 1–2, 221–260.

[ZZh] P. Zhou and S. Zelditch, Nodal sets for Schrödinger eigenfunctions (in preparation).

[Zw] M. Zworski, *Semiclassical analysis.* Graduate Studies in Mathematics, 138. American Mathematical Society, Providence, RI, 2012.

Critical Metrics for Riemannian Curvature Functionals

Jeff A. Viaclovsky

IAS/Park City Mathematics Series
Volume 22, 2013

Critical Metrics for Riemannian Curvature Functionals

Jeff A. Viaclovsky

Introduction

The goal of these lectures is to gain an understanding of critical points of certain Riemmannian functionals. The starting point will be the (normalized) Einstein-Hilbert functional:

$$\tilde{\mathcal{E}}(g) = Vol(g)^{\frac{2-n}{n}} \int_M R_g dV_g,$$

where R_g is the scalar curvature. The Euler-Lagrange equations of $\tilde{\mathcal{E}}(g)$ are

$$Ric(g) = \lambda \cdot g, \tag{0.1}$$

where Ric denotes the Ricci tensor, and λ is a constant. A Riemannian manifold (M, g) satisfying (0.1) is called an *Einstein manifold.*

In Lecture 1, we will study the first and second variation of the functional $\tilde{\mathcal{E}}(g)$, and give an analysis of the Jacobi operator on transverse-traceless tensors. In Lecture 2, we will study conformal variations, discuss the Lichnerowicz eigenvalue estimate and Obata's Theorem, and give a survey of the Yamabe Problem. In Lecture 3, we will introduce an important splitting of the space of symmetric 2-tensors into pure-trace directions, transverse-traceless directions, and "diffeomorphism" directions. An infinitesimal version of the Ebin-Palais slice theorem will show that the diffeomorphism directions can be ignored. It will follow that critical points of the Einstein-Hilbert functional are in general saddle points. This leads one naturally to define the smooth Yamabe invariant of a manifold, or σ-constant.

Next, in Lecture 4, we will study the space of Einstein metrics modulo diffeomorphism, and use Fredholm theory to construct a map between finite-dimensional spaces called the *Kuranishi map* whose zero set is locally in one-to-one correspondence with the moduli space of Einstein metrics (locally). We will also discuss some basic rigidity results for Einstein metrics.

The topic in Lecture 5 will be quadratic curvature functionals in dimension four, that is linear combinations of the following quadratic curvature functionals:

$$\mathcal{W}(g) = \int_M |W_g|^2 dV_g, \quad \rho(g) = \int_M |Ric_g|^2 dV_g, \quad \mathcal{S}(g) = \int_M R_g^2 dV_g.$$

Department of Mathematics, University of Wisconsin, Madison, WI 53706
E-mail address: jeffv@math.wisc.edu
The author would like to thank the organizers of this PCMI Program for the invitation to give these lectures. Also, thanks are due to Michael T. Lock, who served as teaching assistant, and who helped make many significant improvements to these lectures.

Einstein metrics are critical for these functionals, and we will give a discussion of some known results about Einstein metrics. There are also many non-Einstein critical metrics for various linear combinations of these functionals. There is a special family of critical metrics for $\mathcal{W}$ known as anti-self-dual metrics. In Lecture 6, we will study the deformation theory of such metrics, and discuss local properties of the moduli space and existence of the Kuranishi map. We will also discuss several other interesting properties of anti-self-dual metrics.

In Lecture 7, we will discuss some rigidity and stability results regarding critical metrics for quadratic curvature functionals, which are joint work with Matt Gursky. As mentioned above, critical points of the Einstein-Hilbert functional in general have a saddle-point structure. However, critical points for certain quadratic functionals have a nicer local variational structure, see Theorem 7.1. Several rigidity results will also be discussed (which will be crucial in the final lecture).

In Lecture 8, we will study a special class of metrics called asymptotically locally Euclidean metrics (ALE), and present several hyperkähler examples. We will also discuss a result about non-collapsed limits of Einstein metrics: with certain geometric assumptions, a subsequence will converge to an orbifold Einstein metric. A natural question is whether one can reverse this process; that is, can one start with an orbifold Einstein metric, use Ricci-flat ALE metrics to resolve the singularities, and find an Einstein metric on the resolution? We will discuss a recent result of Biquard in the asymptotically hyperbolic Einstein setting which says that this is possible, provided that a certain obstruction vanishes.

We will next discuss a generalization of the Einstein condition, called B^t-flat metrics, give several examples, and discuss an analogous orbifold convergence result which is joint work with Gang Tian. In Lecture 9, we will present some of the key points of the proof of this result, and also give a discussion an Einstein metric on $\mathbb{CP}^2 \# 2\overline{\mathbb{CP}}^2$ found by Chen-LeBrun-Weber.

Finally, in Lecture 10 we will discuss an existence theorem for critical metrics on certain 4-manifolds which is joint work with Matt Gursky. The general idea is to "glue" together two metrics which are critical for a functional to get an "approximate" critical metric, and then find conditions so that one can perturb to an actual solution. Theorem 10.1 produces critical metrics for specific functionals on the manifolds $\mathbb{CP}^2 \# \overline{\mathbb{CP}}^2, \mathbb{CP}^2 \# 2\overline{\mathbb{CP}}^2$, and $S^2 \times S^2 \# S^2 \times S^2$.

This is an expanded version of lectures the author gave at the PCMI Program in Geometric Analysis in Park City from July 16–19, 2013.

LECTURE 1

The Einstein-Hilbert functional

1. Notation and conventions

The notation $\nabla_X Y$ will denote the covariant derivative on a Riemannian manifold (M, g). In a coordinate system $\{x^i\}$, $i = 1 \dots n$, the Christoffel symbols are defined by

$$\nabla_{\partial_i} \partial_j = \Gamma_{ij}^k \partial_k, \tag{1.1}$$

where ∂_i denotes the ith coordinate tangent vector field. The Christoffel symbols can be expressed in terms of the metric as

$$\Gamma_{ij}^k = \frac{1}{2} g^{kl} \Big(\partial_i g_{jl} + \partial_j g_{il} - \partial_l g_{ij} \Big). \tag{1.2}$$

The curvature tensor as a $(1,3)$-tensor is given by

$$R(X,Y)Z = \nabla_X \nabla_Y Z - \nabla_Y \nabla_X Z - \nabla_{[X,Y]} Z, \tag{1.3}$$

and in coordinates our convention is

$$R(\partial_i, \partial_j)\partial_k = R_{ijk}{}^l \partial_l. \tag{1.4}$$

The curvature tensor as a $(0,4)$-tensor is given by

$$Rm(X, Y, Z, W) \equiv -g\big(R(X,Y)Z, W\big), \tag{1.5}$$

and in coordinates

$$R_{ijkl} = Rm(\partial_i, \partial_j, \partial_k, \partial_l). \tag{1.6}$$

Note that our convention is

$$R_{ijlk} = R_{ijk}{}^m g_{ml}. \tag{1.7}$$

That is, we lower the upper index to the *third* position (warning: some authors to lower this index to a different position). The components of the Ricci tensor are given by

$$R_{ij} = R_{lij}{}^l = g^{lm} R_{limj} = R_{ji}, \tag{1.8}$$

and the scalar curvature is

$$R = g^{pq} R_{pq} = g^{pq} g^{lm} R_{lpmq}. \tag{1.9}$$

2. First variation

We let $\mathcal{M}$ denote the space of Riemannian metrics on a manifold M:

$$\mathcal{M} = \{g \in \Gamma(S^2(T^*M)),\ g \text{ is positive definite}\}, \tag{1.10}$$

where $\Gamma(S^2(T^*M))$ denotes the space of smooth symmetric covariant 2-tensors on M. The (unnormalized) Einstein-Hilbert functional $\mathcal{E} : \mathcal{M} \to \mathbb{R}$ is defined by

$$\mathcal{E}(g) = \int_M R_g dV_g. \tag{1.11}$$

This is a *Riemannian functional* in the sense that it is invariant under diffeomorphisms:

$$\mathcal{E}(\varphi^* g) = \mathcal{E}(g). \tag{1.12}$$

Next, we compute the Euler-Lagrange equations of the unnormalized functional:

PROPOSITION 1.1. *If M is closed and $n \geq 3$, then a metric $g \in \mathcal{M}$ is critical for $\mathcal{E}$ if and only if g is Ricci-flat.*

PROOF. Let $g(t)$ be a variation, with $h = g'(0)$. Then

$$\begin{aligned} \mathcal{E}(g(t))' &= \int_M (R_{g(t)} dV_{g(t)})' \\ &= \int_M R'_{g(t)} dV_{g(t)} + R_{g(t)} (dV_{g(t)})' \end{aligned} \tag{1.13}$$

Recall the formula for the linearization of the scalar curvature

$$(R_{g(t)})'\big|_{t=0} = -\Delta(trh) + \delta^2 h - R_{lp} h^{lp}, \tag{1.14}$$

where δ^2 is the double-divergence operator defined in coordinates by

$$\delta^2 h = \nabla^i \nabla^j h_{ij}, \tag{1.15}$$

and Δ is the Laplacian on functions (note that we use the analysts' Laplacian, which has negative eigenvalues). We also recall the formula for the linearization of the volume element

$$(dV_{g(t)})'\big|_{t=0} = \frac{1}{2} tr_g(h) dV_g. \tag{1.16}$$

Next, we evaluate (1.13) at $t = 0$, and consider $\mathcal{E}'_g$ as a mapping on symmetric tensors. Substituting the formulas (1.14) and (1.16) into (1.13), and integrating by parts, we obtain

$$\begin{aligned} \mathcal{E}'_g(h) &= \int_M \Big(R' + \frac{R}{2} tr_g(h) \Big) dV_g \\ &= \int_M \Big(-\Delta(trh) + \delta^2 h - R^{lp} h_{lp} + \frac{R}{2} tr_g(h) \Big) dV_g \\ &= \int_M \Big((-R^{lp} + \frac{R}{2} g^{lp}) h_{lp} \Big) dV_g. \end{aligned}$$

If this vanishes for all variations h, then

$$Ric = \frac{R}{2} g.$$

If $n > 2$, taking a trace, we find that $R = 0$, so (M, g) is Ricci-flat. □

REMARK 1.2. If $n = 2$ then $\mathcal{E}$ has zero variation, thus is constant. This is not surprising in view of the Gauss-Bonnet Theorem:

$$\int_{M^2} K_g dV_g = 2\pi\chi(M^2), \tag{1.17}$$

where $K_g = R_g/2$ is the Gaussian curvature, and χ denotes the Euler characteristic.

EXERCISE 1.3. (i) Prove (1.14). (Hint: first prove that (1.2) linearizes to

$$(\Gamma_{ij}^k)' = \frac{1}{2}g^{kl}\Big(\nabla_i h_{jl} + \nabla_j h_{il} - \nabla_l h_{ij}\Big). \tag{1.18}$$

Next, write out a formula for the scalar curvature in terms of Christoffel symbols, and use (1.18). Note: these computations are much simpler if one works in a normal coordinate system, since the Christoffel symbols vanish at the base point in normal coordinates.)
(ii) Prove (1.16) using that the volume element is locally $dV_g = \sqrt{\det(g_{ij})} \cdot dx$.

2.1. Diffeomorphism invariance ⇒ Bianchi identity

Define the divergence operator $\delta : \Gamma(S^2(T^*M)) \to \Gamma(T^*M)$ by

$$(\delta h)_j = g^{pq}\nabla_p h_{qj}. \tag{1.19}$$

The tensor that arises in the above calculation

$$G = -Ric + \frac{R}{2}g, \tag{1.20}$$

is known as the Einstein tensor. By the contracted second Bianchi identity, it is divergence-free. This is actually a consequence of diffeomorphism invariance of the functional. To see this, let ϕ_t be a path of diffeomorphisms, and let $g_t = \phi_t^* g$. Then $g' = \mathcal{L}_X g$, where X is the tangent vector field of this 1-parameter group of diffeomorphisms at $t = 0$, and $\mathcal{L}$ is the Lie derivative operator. Integrating by parts:

$$\int_M \langle G, \mathcal{L}_X g\rangle dV_g = -\int_M \langle 2\delta G, X\rangle dV_g,$$

for any vector field X, which implies that $\delta G = 0$. Moreover, this argument shows that if $\mathcal{F}$ is *any* Riemannian functional, then $\delta\nabla\mathcal{F} = 0$.

3. Normalized functional

The functional $\mathcal{E}$ is not scale-invariant for $n \geq 3$. To account for this, we define the normalized Einstein-Hilbert functional by

$$\tilde{\mathcal{E}}(g) = Vol(g)^{\frac{2-n}{n}} \int_M R_g dV_g. \tag{1.21}$$

PROPOSITION 1.4. *A metric g is critical for $\tilde{\mathcal{E}}$ under all conformal variations (those of the form $h = f \cdot g$ for $f : M \to \mathbb{R}$) if and only if g has constant scalar curvature. Furthermore, a metric $g \in \mathcal{M}$ is critical for $\tilde{\mathcal{E}}$ if and only if g is Einstein, that is, $Ric(g) = \lambda \cdot g$ for some constant $\lambda \in \mathbb{R}$.*

PROOF. We compute

$$\tilde{\mathcal{E}}'(h) = Vol(g)^{\frac{2-n}{n}}\Big(\frac{2-n}{n}Vol(g)^{-1}\int_M \frac{1}{2}(tr_g h)dV_g \cdot \int_M R_g dV_g\Big) + Vol(g)^{\frac{2-n}{n}}\int_M \Big(-R^{lp} + \frac{R}{2}g^{lp}\Big)h_{lp}dV_g.$$

If $g(t) = f(t)g$, then

$$\tilde{\mathcal{E}}'(h) = \frac{n-2}{2n}Vol(g)^{\frac{2-n}{n}}\Big(\int_M (tr_g h)(R_g - \overline{R})dV_g\Big),$$

where $\overline{R}$ denotes the average scalar curvature. If this is zero for an arbitrary function $tr_g h$, then R_g must be constant. The full variation then simplifies to

$$\tilde{\mathcal{E}}'(h) = Vol(g)^{\frac{2-n}{n}}\int_M \Big(-R^{lp} + \frac{R}{n}g^{lp}\Big)h_{lp}dV_g. \tag{1.22}$$

If this vanishes for all variations, then the traceless Ricci tensor must vanish, so (M, g) is Einstein. □

4. Second variation

Since the functional is scale invariant, from now on we will always restrict to variations satisfying

$$\int_M tr_g(h)dV_g = 0. \tag{1.23}$$

PROPOSITION 1.5. *Let g be Einstein with $Ric(g) = \lambda \cdot g$. Then the second derivative of $\tilde{\mathcal{E}}$ at $t = 0$ is given by*

$$\tilde{\mathcal{E}}'' = Vol(g)^{\frac{2-n}{n}}\Big\{\frac{2-n}{2}\lambda\int_M |h|^2 dV_g + \int_M \langle G'(h), h\rangle dV_g\Big\}. \tag{1.24}$$

PROOF. The proof is left as an exercise. An important point is that the second derivative of a functional is well-defined at a critical point (it only depends on the tangent to the variation). □

EXERCISE 1.6. Prove the formula for the linearization of the Ricci tensor,

$$\begin{aligned}(Ric')_{ij} = \frac{1}{2}\Big(&-\Delta h_{ij} + \nabla_i(\delta h)_j + \nabla_j(\delta h)_i - \nabla_i\nabla_j(tr_g h)\\ &- 2R_{iljp}h^{lp} + R_i^p h_{jp} + R_j^p h_{ip}\Big),\end{aligned} \tag{1.25}$$

where $\Delta : \Gamma(S^2(T^*M)) \to \Gamma(S^2(T^*M))$ is the rough Laplacian defined by

$$(\Delta h)_{ij} = g^{pq}\nabla_p\nabla_q h_{ij}. \tag{1.26}$$

(Hint: write out a formula for the Ricci tensor in terms of Christoffel symbols, and use (1.18). Use normal coordinates to simplify the computation.)

Next, letting

$$(Rm * h)_{ij} = R_{iljp}h^{lp}, \tag{1.27}$$

equation (1.24) can be rewritten as follows.

PROPOSITION 1.7. *Let g be Einstein with $Ric(g) = \lambda \cdot g$. Then the second derivative of $\tilde{\mathcal{E}}$ at $t = 0$ is given by*

$$\tilde{\mathcal{E}}'' = Vol(g)^{\frac{2-n}{n}} \int_M \langle h, Jh \rangle dV_g, \tag{1.28}$$

where $J : \Gamma(S^2(T^*M)) \to \Gamma(S^2(T^*M))$ is the operator

$$Jh = \frac{1}{2}\Delta h - \frac{1}{2}\mathcal{L}(\delta h) + (\delta^2 h)g - \frac{1}{2}(\Delta tr_g h)g - \frac{\lambda}{2}(tr_g h)g + Rm * h. \tag{1.29}$$

5. Transverse-traceless variations

We next have the following definition.

DEFINITION 1.8. *A symmetric 2-tensor h is called transverse-traceless (TT for short) if $\delta_g h = 0$ and $tr_g(h) = 0$.*

The second variation formula simplifies considerably for TT variations:

PROPOSITION 1.9. *If h is transverse-traceless, then*

$$\tilde{\mathcal{E}}'' = Vol(g)^{\frac{2-n}{n}} \left\{ \int_M \left\langle h, \frac{1}{2}\Delta h + Rm * h \right\rangle dV_g \right\}. \tag{1.30}$$

The first term is manifestly negative, which shows that critical metrics for $\tilde{\mathcal{E}}$ always have a saddle point structure. In other words, modulo a finite dimensional space, $\tilde{\mathcal{E}}$ is locally strictly *maximizing* in TT directions.

5.1. The case of constant curvature

If (M, g) has constant sectional curvature, then

$$R_{ijkl} = k_0(g_{ik}g_{jl} - g_{jk}g_{il}). \tag{1.31}$$

The above second variation formula for TT tensors simplifies to

$$\tilde{\mathcal{E}}'' = Vol(g)^{\frac{2-n}{n}} \left\{ \int_M \left\langle h, \frac{1}{2}\Delta h - k_o h \right\rangle dV_g \right\}. \tag{1.32}$$

This immediately yields:

COROLLARY 1.10. *Let (M^n, g) have constant sectional curvature k_0. If $k_0 > 0$ and $n > 2$, then the second variation is strictly negative when restricted to transverse-traceless variations. If $k_0 = 0$, then the second variation is strictly negative except for parallel h.*

For $n = 2$, as pointed out above, we know that $\tilde{\mathcal{E}}$ is constant; thus our computation shows that if h is TT then h is identically zero on S^2, and h must be parallel on T^2. The parallel sections in the case $k_0 = 0$ correspond to deformations of the flat structure, this will be discussed in more detail in Lecture 4 (also see Exercise 4.16 below).

5.2. Hyperbolic manifolds

In the hyperbolic case, we have

LEMMA 1.11. *If (M^n, g) is compact and hyperbolic, then the smallest eigenvalue of the rough Laplacian on TT tensors is at least n.*

PROOF. Exercise. Hint: start with the inequality

$$\int_M |\nabla_i h_{jk} - \nabla_j h_{ik}|^2 dV_g \geq 0, \tag{1.33}$$

integrate by parts, commute covariant derivatives, etc. □

This immediately yields

COROLLARY 1.12. *If (M^n, g) is hyperbolic and $n > 2$, then $\tilde{\mathcal{E}}$ is locally strictly maximizing in TT directions.*

Define the operator $d^{\nabla} : \Gamma(S^2_0(T^*M)) \to \Gamma(\Lambda^2 \otimes T^*M)$ by

$$(d^{\nabla} h)_{ijk} = \nabla_i h_{jk} - \nabla_j h_{ik}, \tag{1.34}$$

which is called the *Codazzi* operator. From the proof of Lemma 1.11, the eigentensors corresponding to the least eigenvalue of the rough Laplacian are exactly the Codazzi tensor, that is, $d^{\nabla} h = 0$. These yield kernel elements of the second variation when $n = 2$, this will be examined in more detail below in Lecture 4.

5.3. The case of $S^2 \times S^2$

A nice example with saddle-point structure in the TT-directions is given by the product metric on $S^2 \times S^2$. Let $\pi_i : S^2 \times S^2 \to S^2$ denote the projection onto the ith factor for $i = 1, 2$. The product metric is $g = g_1 + g_2$ where $g_i = \pi_i^* g_{S^2}$, and g_{S^2} is the round metric on S^2 of constant Gaussian curvature equal to 1.

PROPOSITION 1.13. *On $S^2 \times S^2$ with the product metric $g_1 + g_2$, the lowest eigenvalue of the operator $\frac{1}{2}\Delta h + Rm * h$ on TT tensors is -1. The corresponding eigenspace is 1-dimensional, and is spanned by $h = g_1 - g_2$. The next largest eigenvalue is 1.*

PROOF. This is left as an exercise, with the following hint: decompose a traceless symmetric 2-tensor as

$$h = \overset{\circ}{h}_1 + \frac{f}{m} g_1 + \hat{h} + \overset{\circ}{h}_2 - \frac{f}{m} g_2, \tag{1.35}$$

where h_i is the pull-back of a tensor from the ith factor, $\overset{\circ}{h}_i$ is its trace-free part, for $i = 1, 2$, and $\hat{h}$ are the mixed directions. Since the curvature tensor is given by

$$R_{ijkl} = (g_1)_{ik}(g_1)_{jl} - (g_1)_{jk}(g_1)_{il} + (g_2)_{ik}(g_2)_{jl} - (g_2)_{jk}(g_2)_{il}, \tag{1.36}$$

the eigenvalue equation reduces to three separate equations on the pieces in the decomposition (1.35), which can be analyzed separately. For more details, see for example [**GV11**, Proposition 7.9]. □

EXERCISE 1.14. (i) Find a constant scalar curvature deformation of the product metric corresponding to $h = g_1 - g_2$, and which increases the functional $\tilde{\mathcal{E}}$.
(ii) Show that $\alpha_1 \odot \alpha_2$ ($\odot$ = symmetric product), where α_i are 1-forms dual to Killing fields are eigentensors with eigenvalue 1.

LECTURE 2

Conformal geometry

1. Conformal variations

We will next look at *conformal variations*, that is, those variations of the form $h = f \cdot g$, for a function $f : M \to \mathbb{R}$.

PROPOSITION 2.1. *Assume that g has constant scalar curvature, and let $h = fg$ with $\int_M f dV_g = 0$. Then*

$$\tilde{\mathcal{E}}'' = Vol(g)^{\frac{2-n}{n}} \frac{2-n}{2} \Big\{ \int_M \Big((n-1)\Delta f + R_g f \Big) f dV_g \Big\}. \tag{2.1}$$

PROOF. Assuming g is Einstein, this follows from the above formulas, but it is easy to verify that this holds more generally for constant scalar curvature metrics, the calculation is left as an exercise. □

Let (S^n, g_S) denote the unit n-sphere with round metric g_S. Note that the Ricci tensor satisfies $Ric(g_S) = (n-1)g_S$. An immediate corollary is the following

COROLLARY 2.2. *Let g have constant scalar curvature. If $R_g \leq 0$, then $\tilde{\mathcal{E}}$ is locally strictly minimizing in the conformal direction. If $R_g > 0$ and $Ric \geq (n-1)g$, then the same is true, unless g is isometric to (S^n, g_S).*

The case $R_g \leq 0$ is obvious and the case $R_g > 0$ will follow from a result due to Lichnerowicz which we will discuss next. Let λ_1 denote the lowest non-trivial eigenvalue of the Laplacian on functions, that is $\Delta u = -\lambda_1 u$. We have the following eigenvalue estimate which was proved by Lichnerowicz in 1958 in [**Lic58**], and for which the equality case was characterized by Obata in 1962 [**Oba62**]:

THEOREM 2.3 (Lichnerowicz-Obata). *If a compact manifold (M^n, g) satisfies*

$$Ric \geq (n-1) \cdot g, \tag{2.2}$$

then $\lambda_1 \geq n$, with equality if and only if (M^n, g) is isometric to (S^n, g_S).

PROOF. We only give an outline of the proof. First, commuting covariant derivatives, write

$$\int_M (\Delta f)^2 dV_g = \int_M |\nabla^2 f|^2 dV_g + \int_M Ric(\nabla f, \nabla f) dV_g, \tag{2.3}$$

and then use the matrix inequality $|A|^2 \geq (1/n)(tr(A))^2$. If $\lambda_1 = n$, then equality in this inequality implies that there is a non-trivial solution of the equation

$$\nabla^2 f = \frac{\Delta f}{n} g = -f \cdot g. \tag{2.4}$$

This implies that along any unit-speed geodesic,

$$f(s) = A\cos(s) + B\sin(s), \tag{2.5}$$

where s is the arc-length from a fixed point P_+. If we choose the point P_+ to be a maximum of f, then $f(s) = A\cos(s)$ along any geodesic through P_+. One then uses this information to construct an isometry with (S^n, g_S). For more details, see [**Oba62**] and also [**Küh88**] for an excellent exposition. □

1.1. Conformal variations on S^n

On (S^n, g_S), eigenfunctions corresponding to the eigenvalue n yield directions with $\tilde{\mathcal{E}}'' = 0$. There is a nice geometric explanation for this fact:

PROPOSITION 2.4. *Let ϕ_t be a 1-parameter group of conformal automorphisms of g_S. Then*

$$\frac{d}{dt}(\phi_t^* g_S)|_{t=0} = fg, \tag{2.6}$$

where f is an eigenfunction satisfying $\Delta f = -nf$.

PROOF. We leave this as an exercise, with the following hint: use the Hodge decomposition to write any 1-form α dual to a conformal vector field as $\alpha = df + \omega$, with ω divergence free. Apply the conformal Killing operator to α and use the resulting equation to show that the trace-free Hessian of f vanishes, and that ω is Killing. □

2. Global conformal minimization

Actually, it turns out that something much stronger is true for Einstein metrics:

THEOREM 2.5. *An Einstein metric (M^n, g) is the unique global minimizer of $\tilde{\mathcal{E}}$ in its conformal class (up to scaling), unless (M, g) is isometric to (S^n, g_S). In this case, any critical point is the pull-back of g_S under a conformal diffeomorphism.*

This will be proved below. The first key point in the proof is the following theorem of Obata:

THEOREM 2.6 ([**Oba72**]). *If (M^n, g) is Einstein, then g is the unique constant scalar curvature metric in its conformal class (up to scaling), unless (M, g) is isometric to (S^n, g_S), in which case all critical points are the pull-back of g_S under a conformal diffeomorphism.*

PROOF. To prove this, assume that $\hat{g}$ is a constant scalar curvature metric which is conformal to g. Letting E denote the traceless Ricci tensor, we recall the transformation formula: if $g = \phi^{-2}\hat{g}$, then

$$E_g = E_{\hat{g}} + (n-2)\phi^{-1}\big(\nabla^2\phi - (\Delta\phi/n)\hat{g}\big), \tag{2.7}$$

where n is the dimension, and the covariant derivatives are taken with respect to $\hat{g}$. Since g is Einstein, we have

$$E_{\hat{g}} = (2-n)\phi^{-1}\big(\nabla^2\phi - \frac{1}{n}(\Delta\phi)\hat{g}\big). \tag{2.8}$$

Integrating,

$$\begin{aligned}\int_M \phi|E_{\hat{g}}|^2 dV_{\hat{g}} &= (2-n)\int_M \phi E_{\hat{g}}^{ij}\left\{\phi^{-1}(\nabla^2\phi - \frac{1}{n}(\Delta\phi)\hat{g})_{ij}\right\} dV_{\hat{g}}\\ &= (2-n)\int_M E_{\hat{g}}^{ij}\nabla^2\phi_{ij} dV_{\hat{g}}\\ &= (n-2)\int_M (\nabla_j E_{\hat{g}}^{ij}\cdot\nabla_i\phi) dV_{\hat{g}} = 0,\end{aligned}$$

by the Bianchi identity. Consequently, $\hat{g}$ is also Einstein. If $\hat{g}$ is not a constant multiple of g, then (M, g) admits a nonconstant solution of the equation

$$\nabla^2\phi = \frac{\Delta\phi}{n} g. \tag{2.9}$$

Taking a divergence of this equation, it follows that $\phi + c$, where c is a constant, is an eigenfunction of the Laplacian with eigenvalue n, so (M, g) is isometric to (S^n, g_S) by the same argument in Theorem 2.3 given above. □

We will next take a slight detour and discuss the Yamabe Problem, before returning to the proof of Theorem 2.5.

3. Green's function metric and mass

A key idea in the final resolution of the Yamabe Problem is the following construction of an asymptotically flat metric, called the Green's function metric. First, we define an asymptotically flat metric:

DEFINITION 2.7. A complete Riemannian manifold (X^n, g) is called *asymptotically flat* or *AF* of order τ if there exists a diffeomorphism $\psi : X\backslash K \to (\mathbb{R}^n\backslash B(0,R))$ where K is a compact subset of X, and such that under this identification,

$$(\psi_* g)_{ij} = \delta_{ij} + O(\rho^{-\tau}), \tag{2.10}$$

$$\partial^{|k|}(\psi_* g)_{ij} = O(\rho^{-\tau-k}), \tag{2.11}$$

for any partial derivative of order k, as $r \to \infty$, where ρ is the distance to some fixed basepoint.

The conformal Laplacian is the operator:

$$\Box u = -4\frac{n-1}{n-2}\Delta u + Ru. \tag{2.12}$$

If (M, g) is compact and $R > 0$, then for any $p \in M$, there is a unique positive solution to the equation

$$\begin{aligned}\Box G &= 0 \quad \text{on } M\setminus\{p\}\\ G &= \rho^{2-n}(1 + o(1))\end{aligned} \tag{2.13}$$

as $\rho \to 0$, where ρ is geodesic distance to the basepoint p. This function G is called the *Green's function for the conformal Laplacian.*

EXERCISE 2.8. Show that if $\tilde{g} = u^{\frac{4}{n-2}} g$, then

$$\Box_g u = R_{\tilde{g}} u^{\frac{n+2}{n-2}}. \tag{2.14}$$

Denote $N = M \setminus \{p\}$ with metric $g_N = G^{\frac{4}{n-2}} g_M$. From Exercise 2.8, g_N is scalar-flat. From a more careful expansion of the Green's function, it is possible to show that g_N is also asymptotically flat, but we omit the proof.

The mass of an AF space is defined by

$$\text{mass}(g_N) = \lim_{R\to\infty} \frac{1}{\omega_{n-1}} \int_{S(R)} \sum_{i,j} (\partial_i g_{ij} - \partial_j g_{ii})(\partial_i \lrcorner \, dV_g), \tag{2.15}$$

where $\omega_{n-1} = Vol(S^{n-1})$, and $S(R)$ denotes the sphere of radius R. It was shown in [**Bar86**] that if $\tau > (n-2)/2$, then this mass is well-defined, that is, it is independent of the coordinate system chosen around infinity. The mass is consequently a geometric invariant of an AF metric, and plays an important rôle in the final resolution of the Yamabe Problem, which we discuss next.

4. The Yamabe Problem

By Hölder's inequality, the functional $\tilde{\mathcal{E}}$ is bounded from below when restricted to any fixed conformal class. It is then natural to minimize in the conformal direction:

$$Y(M,[g]) = \inf_{\tilde{g}\in[g]} \tilde{\mathcal{E}}(\tilde{g}). \tag{2.16}$$

This is called the *conformal Yamabe invariant.*

THEOREM 2.9. *If (M^n, g) is compact, then there exists a conformal metric $\tilde{g} \in [g]$ which has constant scalar curvature, and which minimizes $\tilde{\mathcal{E}}$ in its conformal class.*

OUTLINE OF PROOF. For any conformal class, Aubin showed that

$$Y(M,[g]) \leq \tilde{\mathcal{E}}(g_S). \tag{2.17}$$

The idea of the proof of this estimate is to choose a conformal factor which is spherical in an ϵ-neighborhood of a point, and zero everywhere else (this is called a "bubble"). Expanding the Yamabe energy of this test function in the parameter ϵ then yields a leading term which is exactly the Yamabe energy of the spherical metric.

Next, one shows that if this inequality is strict, then a solution exists. This step is now considered "trivial" by experts, but in fact this took a long time to figure out. Yamabe's original paper [**Yam60**] contains a serious mistake on this point, this was fixed by Trudinger [**Tru68**], and then optimized by Aubin [**Aub76**].

The more difficult step is to show that if $(M,[g])$ is not conformally diffeomorphic to $(S^n, [g_S])$ then the inequality (2.17) is strict. In case $n \geq 6$ and g is not locally conformally flat, this was proved by Aubin [**Aub76, Aub98**] by basing the above test function at a point where the Weyl tensor does not vanish. The locally conformally flat case was proved by Schoen-Yau [**SY88**] using ideas involving the developing map. The case $n \leq 6$ was proved by Schoen [**Sch84**]. The main idea is the following. Instead of making the above test function be zero away from the bubble, Schoen's idea was to instead choose the conformal factor to be the Green's function for the conformal Laplacian away from the bubble. The mass of the associated asympotically flat metric arises as the next term in the expansion of the Yamabe energy of this test function, so the result follows from the positive mass theorem of Schoen-Yau [**SY79, SY81, Sch89**]. □

We next return to the global minimization statement in Theorem 2.5:

PROOF OF THEOREM 2.5. The uniqueness follows from Theorem 2.6. For the minimization statement, of course, we know a minimizer exists from the resolution of the Yamabe problem, but there is an "easy" proof in the Einstein case. In the negative or zero scalar curvature case, one can apply a standard argument from the calculus of variations to show that a minimizing sequence converges. In the positive case, scale so that $Ric = (n-1)g$. Then

$$\tilde{\mathcal{E}}(g) = n(n-1)Vol(g)^{2/n}. \tag{2.18}$$

By Bishops' volume comparison theorem, $Vol(M,g) \leq Vol(S^n, g_S)$ with equality if and only if g is isometric to g_S. So if g is not isometric to g_S, we have

$$Y(M,[g]) = \inf_{\tilde{g}\in[g]} \tilde{\mathcal{E}}(g) < \tilde{\mathcal{E}}(g_S). \tag{2.19}$$

As discussed above, this estimate implies that a minimizing sequence converges (no bubbles are possible).

Finally, the case of (S^n, g_S) takes some extra work. One needs to suitably renormalize a minimizing sequence using the conformal group to obtain a minimizing sequence which converges, see [**LP87**, Proposition 4.6] for an argument due to Karen Uhlenbeck.

□

REMARK 2.10. An important question is if the set of unit volume constant scalar curvature metrics in a conformal class is compact if the manifold is not conformally diffeomorphic to the sphere. This is true in dimensions $n \leq 24$ [**KMS09**]. Surprisingly, it is false in higher dimensions [**Bre08, BM09**].

5. Generalizations of the Yamabe Problem

We mention that there are many other Yamabe-type conformal deformation problems which also have variational characterizations. We describe one such example next. Define the *Schouten tensor* by

$$A_g = \frac{1}{n-2}\left(Ric_g - \frac{R_g}{2(n-1)}g\right). \tag{2.20}$$

Consider the functional

$$\tilde{\mathcal{F}}_{\sigma_2}(g) = Vol(g)^{\frac{4}{n}-1}\int_M \sigma_2(g^{-1}A_g)dV_g, \tag{2.21}$$

where σ_2 denotes the second elementary symmetric function of the eigenvalues. This functional has a nice conformal variational property, analogous to that for the Einstein-Hilbert functional.

THEOREM 2.11 ([**Via00**]). *If $n \neq 4$, a metric g is a critical for $\tilde{\mathcal{F}}_{\sigma_2}$ under all conformal variations if and only if*

$$\sigma_2(g^{-1}A_g) = C, \tag{2.22}$$

for some constant C.

One may also generalize the Yamabe problem by asking if it is possible to conformally deform a metric so that (2.22) is satisfied. Note that, in contrast to the Yamabe equation which is semi-linear, equation (2.22) is a fully nonlinear equation, and some assumption must be made on the conformal structure to ensure that the equation is elliptic. There has been much progress on this σ_2 problem, see for example [**CGY02a, CGY02b, GW06, GV03, STW07**]. More generally, one can consider other symmetric functions of the eigenvalues, and for the kth elementary function, this is known as the σ_k-Yamabe Problem. The locally conformally flat case has been solved for all k, see [**GW03, LL03**]. This has also been solved for the case $k > n/2$, see [**GV07**]. There have been many other related works involving various symmetric functions of the eigenvalues, we refer the reader to [**Via06**] for a more detailed description and other references.

For solving conformal deformation problems, we note that parabolic methods also play an important rôle, see for example [**Bre05, Bre07, GW03, STW07**]. Another generalization of the Yamabe Problem is to the class of higher order equations, and deals with prescribing Q-curvature, which is a higher order generalization of the scalar curvature. We will not discuss this further, and refer the reader to [**BG08, DM08**] for details about the notion of Q-curvature.

The Yamabe Problem can also be generalized to the setting of orbifolds [**AB03, AB04, Aku12**]. This turns out to be more subtle than the Yamabe Problem on manifolds – there in fact exist conformal classes on compact orbifolds which do not contain any constant scalar curvature metrics. For example, the conformal compactifications of hyperkähler ALE metrics and also the conformal classes of certain Bochner-Kähler metrics on weighted projective spaces do not admit any solution of the orbifold Yamabe Problem [**Via10, Via13a**].

LECTURE 3

Diffeomorphisms and gauging

1. Splitting

We begin by discussing a decomposition of the space of symmetric 2-tensors; some references for this material are [**BE69, Bes08**]. We let $\mathcal{K} : T^*M \to S^2_0(T^*M)$ be the conformal Killing operator

$$(\mathcal{K}\alpha)_{ij} = \nabla_i \alpha_j + \nabla_j \alpha_i - \frac{2}{n}(\delta\alpha) g_{ij}. \tag{3.1}$$

Also, consider the operator $\Box : \Gamma(T^*M) \to \Gamma(T^*M)$, defined by $\Box = \delta\mathcal{K}$ where $\delta : \Gamma(S^2(T^*M)) \to \Gamma(T^*M)$ is the divergence defined by

$$(\delta h)_j = g^{pq} \nabla_p h_{iq}. \tag{3.2}$$

EXERCISE 3.1. (i) Show that the operator $\Box$ is elliptic and self-adjoint.
(ii) Prove that the kernel of $\Box$ is exactly the space of conformal Killing forms, i.e., they satisfy $\mathcal{K}\alpha = 0$.

LEMMA 3.2. *The space of symmetric* 2*-tensors admits the following orthogonal decomposition:*

$$S^2(T^*M) = \{f \cdot g\} \oplus \{\mathcal{K}(\alpha)\} \oplus \{\delta h = 0, tr_g(h) = 0\}. \tag{3.3}$$

PROOF. Given $h \in S^2_0(T^*M)$, consider the 1-form δh. By Fredholm theory, the equation $\Box\alpha = \delta h$ has a solution if and only if δh is orthogonal to the kernel of the adjoint operator, which is exactly the space of conformal Killing 1-forms (by the exercise). If κ is any conformal Killing 1-form, then

$$\int_M \langle \delta h, \kappa \rangle = \frac{1}{2} \int_M \langle h, \mathcal{K}\kappa \rangle = 0.$$

So the equation $\Box\alpha = \delta h$ has a solution, which proves that $h - \mathcal{K}\alpha$ is divergence-free. $\Box$

1.1. Another decomposition

The orthogonal decomposition given in (3.3) implies the decomposition

$$S^2(T^*M) = \{f \cdot g\} + \{\mathcal{L}(\alpha)\} \oplus \{\delta h = 0, tr_g(h) = 0\}. \tag{3.4}$$

PROPOSITION 3.3. *If* (M, g) *is Einstein, with* $Ric = \lambda \cdot g$*, then this latter decomposition is a direct sum, unless* (M, g) *is isometric to* (S^n, g_S)*.*

PROOF. We need to show that the spaces $\{f \cdot g\}$ and $\{\mathcal{L}(\alpha)\}$ have intersection $\{0\}$. So if $\mathcal{L}(\alpha) = f \cdot g$, then taking a trace, we have

$$2\delta\alpha = nf,$$

which implies that $\mathcal{K}(\alpha) = 0$. Taking a divergence of this equation, we have

$$\begin{aligned}\nabla_i(\nabla_i\alpha_j + \nabla_j\alpha_i - (2/n)(\delta\alpha)g_{ij}) &= \Delta\alpha_j + \nabla_i\nabla_j\alpha_i - (2/n)\nabla_j(\delta\alpha)\\ &= \Delta\alpha_j + \left(1 - \frac{2}{n}\right)\nabla_j(\delta\alpha) + \lambda\alpha_j.\end{aligned}$$

Next, recall the Bochner formula for 1-forms

$$(\Delta\alpha)_i = -(\Delta_H\alpha)_i + R_{ip}g^{pj}\alpha_j, \tag{3.5}$$

where Δ_H is the Hodge Laplacian. This yields that

$$\Delta\alpha = -(d\delta_H + \delta_H d)\alpha + \lambda\alpha,$$

where δ_H is the Hodge divergence (which is the negative of our divergence). Putting these together, we obtain

$$\Box\alpha = -2\left(\frac{n-1}{n}\right)d\delta_H\alpha - \delta_H d\alpha + 2\lambda\alpha = 0. \tag{3.6}$$

Next, pairing (3.6) with α and integrating,

$$-2\left(\frac{n-1}{n}\right)\int_M |\delta\alpha|^2 dV_g - \int_M |d\alpha|^2 dV_g + 2\lambda\int_M |\alpha|^2 dV_g = 0.$$

This implies that $\alpha = 0$ if $\lambda < 0$ (so any conformal Killing field vanishes for a negative Einstein metric). If $\lambda = 0$, we see that $\delta\alpha = 0$ and $d\alpha = 0$. In particular, α is a Killing 1-form, and we are done.

In the case $\lambda > 0$, applying a divergence to (3.6) yields

$$2\left(\frac{n-1}{n}\right)\Delta(\delta\alpha) + 2\lambda(\delta\alpha) = 0. \tag{3.7}$$

By Lichnerowicz' Theorem, this implies that $\delta\alpha = 0$ unless (M, g) is isometric to (S^n, g_S), so α is Killing. $\Box$

2. Second variation as a bilinear form

From Proposition 1.5, let us recall the second variation is

$$\tilde{\mathcal{E}}''(h,h) = Vol(g)^{\frac{2-n}{n}}\int_M \langle h, Jh\rangle dV_g, \tag{3.8}$$

where J is the operator

$$Jh = \frac{2-n}{2}\lambda h + G'h. \tag{3.9}$$

Using polarization, the Hessian of $\tilde{\mathcal{E}}$ is the bilinear form given by

$$\tilde{\mathcal{E}}''(h_1,h_2) = Vol(g)^{\frac{2-n}{n}}\int_M \langle h_1, Jh_2\rangle dV_g. \tag{3.10}$$

PROPOSITION 3.4. *The decomposition*

$$S^2(T^*M) = \{f\cdot g\} \oplus \{\mathcal{L}(\alpha)\} \oplus \{\delta h = 0, tr_g(h) = 0\} \tag{3.11}$$

is orthogonal with respect to $\tilde{\mathcal{E}}''(\cdot,\cdot)$.

PROOF. First, $\tilde{\mathcal{E}}''(\mathcal{L}(\alpha),\cdot) = 0$ from diffeomorphism invariance. So we just need to check that

$$\tilde{\mathcal{E}}''(f\cdot g, z) = 0 \tag{3.12}$$

if z is TT. To see this,

$$\begin{aligned}\tilde{\mathcal{E}}''(f\cdot g, z) &= Vol(g)^{\frac{2-n}{n}}\int_M \langle f\cdot g, Jz\rangle dV_g\\ &= Vol(g)^{\frac{2-n}{n}}\int_M \langle f\cdot g, \frac{1}{2}\Delta z + Rm * z\rangle dV_g\\ &= Vol(g)^{\frac{2-n}{n}}\int_M f(R_{ijip}z_{jp})dV_g = 0.\end{aligned}$$

□

To summarize: if h is any symmetric 2-tensor, then decompose h as

$$h = f\cdot g + \mathcal{L}\alpha + z, \tag{3.13}$$

where z is TT. Then

$$\tilde{\mathcal{E}}''(h,h) = \tilde{\mathcal{E}}''(f\cdot g, f\cdot g) + \tilde{\mathcal{E}}''(z,z). \tag{3.14}$$

So we have shown that to check the second variation, we really only need to consider conformal variations and TT variations separately.

3. Ebin-Palais slice theorem (infinitesimal version)

The above discussion was at the level of the "tangent space to the space of Riemannian metrics at g". We will next transfer this statement directly to the space of Riemannian metrics near g modulo diffeomorphism.

THEOREM 3.5. *The local behavior of $\tilde{\mathcal{E}}$, when considered as a map on $\mathcal{M}/\mathcal{D}$ (the space of Riemannian metrics modulo diffeomorphism), is determined by the conformal and TT directions (to second order).*

The main tool for this is the following infinitesimal version of a "slice" theorem due to Ebin-Palais. The notation $C^{k,\alpha}$ will denote the space of Hölder continuous mappings (or tensors) with $0 < \alpha < 1$.

THEOREM 3.6. *For each metric g_1 in a sufficiently small $C^{\ell+1,\alpha}$-neighborhood of g ($\ell \geq 1$), there is a $C^{\ell+2,\alpha}$-diffeomorphism $\varphi : M \to M$ such that*

$$\tilde{\theta} \equiv \varphi^* g_1 - g \tag{3.15}$$

satisfies

$$\delta_g\Big(\tilde{\theta} - \frac{1}{n}tr_g(\tilde{\theta})g\Big) = 0. \tag{3.16}$$

PROOF. Let $\{\omega_1,\ldots,\omega_\kappa\}$ denote a basis of the space of conformal Killing forms with respect to g. Consider the map

$$\mathcal{N} : C^{\ell+2,\alpha}(TM)\times\mathbb{R}^\kappa\times C^{\ell+1,\alpha}(S^2(T^*M)) \to C^{\ell,\alpha}(T^*M) \tag{3.17}$$

given by

$$\mathcal{N}(X,v,\theta) = \mathcal{N}_\theta(X,v) = \Big(\delta_g\big[\overbrace{\varphi^*_{X,1}(g+\theta)}^{\circ}\big] + \sum_i v_i\omega_i\Big), \tag{3.18}$$

where $\varphi_{X,1}$ denotes the diffeomorphism obtained by following the flow generated by the vector field X for unit time, and $\circ$ denotes the traceless part with respect to g. Linearizing in (X, v) at $(X, v, \theta) = (0, 0, 0)$, we find

$$\begin{aligned}\mathcal{N}_0'(Y,a) &= \frac{d}{d\epsilon}\Big(\delta_g\big[\overbrace{\varphi_{\epsilon Y,1}^*(g)}^{\circ}\big] + \sum_i (\epsilon a_i)\omega_i\Big)\Big|_{\epsilon=0} \\ &= \Big(\delta_g[\overbrace{\mathcal{L}_g Y^\flat}^{\circ}] + \sum_i a_i\omega_i\Big) \\ &= \Big(\Box Y^\flat + \sum_i a_i\omega_i\Big),\end{aligned}$$

where $Y^\flat$ is the dual one-form to Y. The adjoint map $(\mathcal{N}_0')^* : C^{m+2,\alpha}(T^*M) \to C^{m,\alpha}(TM) \times \mathbb{R}^\kappa$ is given by

$$(\mathcal{N}_0')^*(\eta) = \Big((\Box\eta)^\sharp, \int_M \langle \eta, \omega_i\rangle \, dV_g\Big), \tag{3.19}$$

where $(\Box\eta)^\sharp$ is the vector field dual to $\Box\eta$.

If η is in the kernel of the adjoint, the first equation implies that η is a conformal Killing form, while the second implies that η is orthogonal (in L^2) to the space of conformal Killing forms. It follows that $\eta = 0$, so the map $\mathcal{N}_0'$ is surjective.

Omitting a few technical details for simplicity, applying an infinite-dimensional version of the implicit function theorem (which will be discussed in detail below in Lecture 4), given $\theta_1 \in C^{\ell+1,\alpha}(S^2(T^*M))$ small enough we can solve the equation $\mathcal{N}_{\theta_1} = 0$; i.e., there is a vector field $X \in C^{\ell+2,\alpha}(TM)$, and a $v \in \mathbb{R}^\kappa$, such that

$$\delta_g[\overbrace{\varphi^* g_1}^{\circ}] + \sum_i v_i\omega_i = 0, \tag{3.20}$$

where $\varphi = \varphi_{X,1}$. Letting $\tilde{\theta} = \varphi^* g_1 - g$, then $\tilde{\theta}$ satisfies

$$\delta_g[\overset{\circ}{\tilde{\theta}}] + \sum_i v_i\omega_i = 0, \tag{3.21}$$

Pairing with ω_j, for $j = 1 \dots \kappa$, and integrating by parts, we see that $v_j = 0$, and we are done. $\square$

EXERCISE 3.7. Verify the above formula (3.19) for $(\mathcal{N}_0')^*$.

EXERCISE 3.8. By adding a scaling factor to the map $\mathcal{N}$, modify the above argument to show that we can find a constant c (depending upon g_1), and find

$$\tilde{\theta} \equiv e^c \varphi^* g_1 - g, \tag{3.22}$$

so that in addition to the traceless part of $\tilde{\theta}$ being TT, $\tilde{\theta}$ also satisfies

$$\int_M tr_g \tilde{\theta} \, dV_g = 0.$$

That is, we can also "gauge away" the scale-invariance of the functional. Equivalently, we can look at a slice of unit-volume metrics modulo diffeomorphism.

REMARK 3.9. The reason this is called an "infinitesimal" version of the Slice Theorem is because the full Ebin-Palais Slice Theorem constructs a local slice for the action of the diffeomorphism group, see [**Ebi68**]. The main difficulty is that the natural action of the diffeomorphism group on the space of Riemannian metrics is not differentiable as a mapping of Banach spaces (with say Sobolev or Hölder norms). It is however differentiable as a mapping of ILH spaces, see [**Omo70, Koi78**]. For the purposes of these lectures, we will content ourselves with the infinitesimal version, and will not go into details about the full slice theorem

PROOF OF THEOREM 3.5. Combining the above discussions, given any g_1 sufficiently near g, we can write

$$\varphi^* g_1 = g + \tilde{\theta}, \tag{3.23}$$

with $\tilde{\theta} = f \cdot g + z$ with $\int_M f dV_g = 0$, and z is TT. Then

$$\begin{aligned}
\tilde{\mathcal{E}}(g_1) &= \tilde{\mathcal{E}}(\varphi^* g_1) \text{ (from diffeomorphism invariance)} \\
&= \tilde{\mathcal{E}}(g + \tilde{\theta}) \\
&= \tilde{\mathcal{E}}(g) + \tilde{\mathcal{E}}'_g(\tilde{\theta}) + \tilde{\mathcal{E}}''_g(f \cdot g + z, f \cdot g + z) + \text{remainder} \\
&= \tilde{\mathcal{E}}(g) + \tilde{\mathcal{E}}''_g(f \cdot g, f \cdot g) + \tilde{\mathcal{E}}''_g(z, z) + \text{remainder.}
\end{aligned}$$

□

4. Saddle point structure and the smooth Yamabe invariant

We have seen that the functional $\tilde{\mathcal{E}}$ is minimizing in the conformal directions, but maximizing (modulo a finite-dimensional subspace) in the TT directions. So an Einstein metric is always a saddle point for $\mathcal{E}$. This suggests defining the following min-max type invariant.

First, we minimize in the conformal direction:

$$Y(M, [g]) = \inf_{\tilde{g} \in [g]} \tilde{\mathcal{E}}(g).$$

This is called the *conformal Yamabe invariant.*

The min-max invariant is then defined by

$$Y(M) = \sup_{g \in \mathcal{M}} Y(M, [g]),$$

which we will call the *smooth Yamabe invariant* of M, also known as the *σ-invariant* of M. This was defined independently by Osamu Kobayashi [**Kob87**] and Richard Schoen [**Sch89**].

4.1. Some known cases

We will not focus on smooth Yamabe invariants in this lecture, but only list a few known cases:

- $Y(S^n) = Y(S^n, [g_S]) = n(n-1)Vol(S^n)^{\frac{2}{n}}$.
- $Y(S^1 \times S^{n-1}) = Y(S^n, [g_S])$, proved by Schoen [**Sch89**].
- $Y(\mathbb{RP}^3) = Y(\mathbb{RP}^3, [g_S])$, proved by Bray-Neves [**BN04**].
- If (M^3, g_H) is compact hyperbolic, then $Y(M^3) = Y(M^3, [g_H])$. This follows from Perelman's work, see [**AIL07**].

- $Y(\mathbb{CP}^2) = Y(\mathbb{CP}^2, [g_{\mathrm{FS}}]) = 12\pi\sqrt{2}$, where g_{FS} is the Fubini-Study metric, proved by LeBrun [**LeB97**], see also [**GL98**].

There are many other cases for which the Yamabe invariant is known, but we do not list them here. We note that an effective tool in dimension four is Seiberg-Witten Theory, see [**LeB99, Sun09**]. Also, there are also many known estimates on Yamabe invariants (see for example [**ADH13, Pet00**]), but there is not a single known example of a compact manifold M with positive Yamabe invariant which has been shown to satisfy $0 < Y(M) < Y(S^n)$ in dimensions $n \geq 5$.

4.2. Some unknown cases

It is a very difficult problem to determine Yamabe invariants in general. Here are a few prominent unknown cases:

- What is $Y(S^n/\Gamma)$, where S^n/Γ is a spherical space form with $|\Gamma| > 1$? Is it achieved by the round metric? The only known case is the Bray-Neves result listed above.
- What is $Y(\mathbb{CP}^2 \# \mathbb{CP}^2)$? The only result known is due to O. Kobayashi [**Kob87**]:
$$Y(\mathbb{CP}^2 \# \mathbb{CP}^2) \geq Y(\mathbb{CP}^2).$$
- What is $Y(\mathbb{CP}^2 \# \overline{\mathbb{CP}}^2)$? Again, the only result known is
$$Y(\mathbb{CP}^2 \# \overline{\mathbb{CP}}^2) \geq Y(\mathbb{CP}^2).$$
- What is $Y(S^2 \times S^2)$? The only known result is that
$$Y(S^2 \times S^2) > Y(S^2 \times S^2, g_{S^2} + g_{S^2})$$
(strict inequality). This follows from Exercise 1.14 and a result of Böhm-Wang-Ziller that CSC metrics sufficiently near an Einstein metric are also Yamabe minimizers in their conformal class [**BWZ04**, Theorem C].

We end by noting there are relatively few theorems giving conditions for the uniqueness of a Yamabe metric. There is Obata's Theorem 2.6, and the result of [**BWZ04**] mentioned above; also see [**dLPZ12**] and [**Kat94**].

LECTURE 4

The moduli space of Einstein metrics

1. Moduli space of Einstein metrics

Next, given an Einstein metric g with $Ric(g) = \lambda \cdot g$, we would like understand the space of solutions of the equation

$$Ric(\tilde{g}) = \lambda \cdot \tilde{g} \tag{4.1}$$

with $\tilde{g}$ near g. This will be infinite-dimensional since $\varphi^* g$ will also be a solution for any diffeomorphism $\varphi : M \to M$. Therefore, we need to look at the space of Einstein metrics modulo diffeomorphism. Our goal is to prove:

THEOREM 4.1. *Assume g is Einstein with $Ric(g) = \lambda \cdot g$ and $\lambda < 0$. Then the space of Einstein metrics near g modulo diffeomorphism is locally isomorphic to the zero set of a map*

$$\Psi : H^1_E \to H^1_E, \tag{4.2}$$

where

$$H^1_E = \{h \in S^2(T^*M) : \delta_g h = 0, tr_g h = 0, \Delta h + 2Rm * h = 0\}, \tag{4.3}$$

where $Rm * h$ is the operator defined above in (1.27).

Elements in the space H^1_E are called *infinitesimal Einstein deformations.*

1.1. Ellipticity

The diffeomorphism invariance also means that the above equation cannot be elliptic. Indeed, differentiating

$$Ric(\varphi_t^* g) = \varphi_t^*(Ric(g)) \tag{4.4}$$

yields

$$Ric'(\mathcal{L}_X g) = \mathcal{L}_X(Ric(g)) = \lambda \cdot \mathcal{L}_X g. \tag{4.5}$$

EXERCISE 4.2. Show that this implies that the symbol of Ric' is not elliptic.

We will next describe a procedure called "gauging" which shows in effect, that the diffeomorphism directions are the only obstruction to ellipticity. This is somewhat analogous to the "Coulomb gauge" in electrodynamics.

1.2. A gauge choice

Recall from above, that at an Einstein metric satisfying $Ric(g) = \lambda \cdot g$, the linearized Ricci tensor is given by

(4.6)

$$(Ric')_{ij} = \frac{1}{2}\Big(- \Delta h_{ij} + \nabla_i(\delta h)_j + \nabla_j(\delta h)_i - \nabla_i \nabla_j (trh) - 2R_{iljp}h^{lp} + 2\lambda h_{ij}\Big).$$

Define the operator

$$\beta_g h = \delta_g h - \frac{1}{2} d(tr_g h). \tag{4.7}$$

EXERCISE 4.3. *Show that*

$$\frac{1}{2}\mathcal{L}\beta_g h = \frac{1}{2}\Big(\nabla_i(\delta h)_j + \nabla_j(\delta h)_i - \nabla_i\nabla_j(trh)\Big). \tag{4.8}$$

Combining the above expressions,

$$(Ric' - \frac{1}{2}\mathcal{L}\beta_g)h = \frac{1}{2}\Big(- \Delta h - 2Rm * h + 2\lambda h\Big). \tag{4.9}$$

2. The nonlinear map

Given $\theta \in C^{2,\alpha}(S^2T^*M)$, consider the map

$$P_g : C^{2,\alpha}(S^2(T^*M)) \to C^{0,\alpha}(S^2(T^*M)) \tag{4.10}$$

defined by

$$P_g(\theta) = Ric(g+\theta) - \lambda\cdot(g+\theta) - \frac{1}{2}\mathcal{L}_{g+\theta}\beta_g\theta. \tag{4.11}$$

PROPOSITION 4.4. *The operator P_g is elliptic.*

PROOF. This is immediate: from the above, the linearized operator at $\theta = 0$ is

$$P'_g h = \frac{1}{2}\Big(- \Delta h - 2Rm * h\Big), \tag{4.12}$$

which is clearly elliptic. □

We next see that zeroes of P_g are in fact Einstein metrics.

PROPOSITION 4.5. *Assume that $\lambda < 0$. If $\theta \in C^{3,\alpha}$ is sufficiently near g and satisfies $P_g(\theta) = 0$, then $Ric(g+\theta) = \lambda(g+\theta)$, and $\theta \in C^\infty$.*

PROOF. Apply the operator $\beta_{g+\theta}$ to the equation $P_g(\theta) = 0$ to obtain

$$\beta_{g+\theta}\mathcal{L}_{g+\theta}\beta_g\theta = 0 \tag{4.13}$$

A computation shows that this is equivalently (exercise):

$$(\Delta_{g+\theta} + Ric(g+\theta))(\beta_g\theta) = 0. \tag{4.14}$$

Since θ is sufficiently small in $C^{2,\alpha}$ norm, and $Ric(g)$ is strictly negative definite, then $Ric_{g+\theta}$ is also strictly negative definite. Pairing with $\beta_g\theta$ and integrating by parts then shows that $\beta_g\theta = 0$. □

EXERCISE 4.6. Prove the regularity statement in Proposition 4.5. (Hint: Letting $\tilde{g} = g + \theta$, in harmonic coordinates, the Ricci tensor can be written in the form

$$Ric_{kl}(\tilde{g}) = -\frac{1}{2}\tilde{g}^{ij}\partial^2_{ij}\tilde{g}_{kl} + Q_{kl}(\partial\tilde{g},\tilde{g}) \tag{4.15}$$

where $Q(\partial\tilde{g},\tilde{g})$ is an expression that is quadratic in $\partial\tilde{g}$, polynomial in $\tilde{g}$ and has $\sqrt{|\tilde{g}|}$ in its denominator. Use a bootstrap argument in these coordinates.)

EXERCISE 4.7. Show that we only need to assume that $\theta \in C^{2,\alpha}$. (Hint: instead of differentiating in the first step, integrate by parts.)

Next, we have a converse up to diffeomorphism: Einstein metrics near to g can be gauged to yield zeroes of P_g.

PROPOSITION 4.8. *If $\tilde{g}$ is an Einstein metric near g with Einstein constant λ, then there exists a diffeomorphism $\varphi : M \to M$ such that $\tilde{\theta} = \varphi^*\tilde{g} - g$ satisfies $P_g(\tilde{\theta}) = 0$.*

The proof uses a modified (infinitesimal) Ebin-Palais slice theorem using the Bianchi gauge:

LEMMA 4.9. *For each metric g_1 in a sufficiently small $C^{\ell+1,\alpha}$-neighborhood of g ($\ell \geq 1$), there is a $C^{\ell+2,\alpha}$-diffeomorphism $\varphi : M \to M$ such that*

$$\tilde{\theta} \equiv \varphi^* g_1 - g \tag{4.16}$$

satisfies

$$\beta_g(\tilde{\theta}) = 0 \tag{4.17}$$

PROOF. The proof is almost identical to that of Theorem 3.6, and is omitted. □

PROOF OF PROPOSITION 4.8. If $\tilde{g}$ is Einstein then $\varphi^*\tilde{g}$ is also Einstein. Since

$$\beta_g(\tilde{\theta}) = \beta_g(\varphi^*\tilde{g} - g) = 0, \tag{4.18}$$

we obviously obtain a zero of P_g. □

3. Structure of nonlinear terms

Let us write

$$P_g(\theta) = P_g(0) + P_g'(\theta) + Q_g(\theta). \tag{4.19}$$

The following proposition is crucial, and shows that the nonlinear term is manageable.

PROPOSITION 4.10. *For θ_1, θ_2 sufficiently small, there exists a constant C so that*

$$\|Q_g(\theta_1) - Q_g(\theta_2)\|_{C^{0,\alpha}} \leq C(\|\theta_1\|_{C^{2,\alpha}} + \|\theta_2\|_{C^{2,\alpha}})\|\theta_1 - \theta_2\|_{C^{2,\alpha}}. \tag{4.20}$$

PROOF. The proof is left as an exercise, with a few hints. First, show that

$$\Gamma(g+h)^k_{ij} = \Gamma(g)^k_{ij} + \frac{1}{2}(g+h)^{km}\left\{\nabla_j h_{im} + \nabla_i h_{jm} - \nabla_m h_{ij}\right\}. \tag{4.21}$$

In shorthand, we can therefore write the covariant derivative of any tensor T as

$$\nabla_{g+h}T = \nabla_g T + (g+h)^{-1} * \nabla_g h * T, \tag{4.22}$$

where the notation $*$ denotes taking various contractions of the tensors involved (the exact indices contracted do not matter for the conclusion).

Next, for any metric $\tilde{g}$, the curvature tensor can be written in shorthand as

$$Rm_{\tilde{g}} = \nabla_{\tilde{g}}\Gamma_{\tilde{g}}. \tag{4.23}$$

Using (4.22), show that this implies an expansion of the form

$$Rm(g+h) = Rm(g) + (g+h)^{-1} * \nabla^2 h + (g+h)^{-2} * \nabla h * \nabla h. \tag{4.24}$$

Contract with $(g+h)^{-1}$ to get $Ric(g+h)$ and then use the formula

$$(g+h)^{-1} = g^{-1} - g^{-1}(g+h)^{-1}h. \tag{4.25}$$

to pull out the terms in the linearization, and (4.20) will then follow from the resulting expression for Q_g. □

4. Existence of the Kuranishi map

The following is the main tool used to construct the map Ψ, see for example [**Biq13a**, Lemma 8.3].

LEMMA 4.11. *Let $H : E \to F$ be a smooth map between Banach spaces. Define $Q = H - H(0) - H'(0)$. Assume that there are positive constants C_1, s_0, C_2 so that the following are satisfied:*

- *(1) The nonlinear term Q satisfies*
$$\|Q(x) - Q(y)\|_F \leq C_1(\|x\|_E + \|y\|_E)\|x - y\|_E$$
for every $x, y \in B_E(0, s_0)$.
- *(2) The linearized operator at 0, $H'(0) : E \to F$ is an isomorphism with inverse bounded by C_2.*

If s and $\|H(0)\|_F$ are sufficiently small (depending upon C_1, s_0, C_2), then there is a unique solution $x \in B_E(0, s)$ of the equation $H(x) = 0$.

OUTLINE OF PROOF. The equation $H(x) = 0$ expands to

$$H(0) + H'(0)(x) + Q(x) = 0. \tag{4.26}$$

If we let $x = Gy$, where G is the inverse of $H'(0)$, then we have

$$H(0) + y + Q(Gy) = 0, \tag{4.27}$$

or

$$y = -H(0) - Q(Gy). \tag{4.28}$$

In other words, y is a fixed point of the mapping

$$T : y \mapsto -H(0) - Q(Gy). \tag{4.29}$$

With the assumptions in the lemma, it follows that T is a contraction mapping, so a fixed point exists by the standard fixed point theorem ($T^n y_0$ converges to a unique fixed point for any y_0 sufficiently small). □

To prove Theorem 4.1, we next construct the map

$$\Psi : H^1_E \to H^1_E, \tag{4.30}$$

whose zero set is locally isomorphic to the zero set of P. Consider $H = \Pi \circ P$, where Π is projection to the orthogonal complement of H^1_E. The differential of this map is now surjective. Choose any complement K to the space H^1_E, and restrict the mapping to this complement. Equivalently, let G be any right inverse, i.e., $H'(0)G = Id$, and let K be the image of G. Given a kernel element $x_0 \in H^1_E$, the equation $H(x_0 + Gy) = 0$ expands to

$$H(0) + H'(0)(x_0 + Gy) + Q(x_0 + Gy) = 0. \tag{4.31}$$

We therefore need to find a fixed point of the map

$$T_{x_0} : y \mapsto -H(0) - Q(x_0 + Gy), \tag{4.32}$$

and the proof is the same as before.

To finish the proof of Theorem 4.1, we need to identify the kernel of the linearized operator.

PROPOSITION 4.12. *If $\lambda < 0$, then $Ker(P'_g)$ consists exactly of transverse-traceless tensors satisfying*

$$\Delta h + 2Rm * h = 0. \tag{4.33}$$

PROOF. If $P'(h) = 0$, then h is smooth by elliptic regularity, Also,

$$P'h = Ric'(h) - \lambda h - \frac{1}{2}\mathcal{L}_g \beta_g h. \tag{4.34}$$

Applying β_g to this equation, yields $\beta_g \mathcal{L}_g \beta_g h = 0$, so $\beta_g h = 0$ by the above argument. Taking a trace, we find that

$$\Delta(tr_g(h)) + 2\lambda \cdot tr_g(h) = 0, \tag{4.35}$$

so $tr_g(h) = 0$ since $\lambda < 0$. □

5. Rigidity of Einstein metrics

In general it is quite difficult to construct the map Ψ explicitly, but one of the easiest consequences of the above discussion is the following (see [**Koi78**]):

COROLLARY 4.13. *If $Ric(g) = \lambda \cdot g$ with $\lambda < 0$, and $H^1_E = \{0\}$ then g is rigid (isolated as an Einstein metric). That is, if g_t is a path of Einstein metrics passing through g, all with Einstein constant $\lambda < 0$, then there exist a path of diffeomorphisms $\varphi_t : M \to M$ such that $g_t = \varphi_t^* g$.*

We next discuss a few cases where it is known that $H^1_E = 0$. For a longer list, see [**Koi78, Koi80, Koi82**].

5.1. The negative case

In the hyperbolic case, in Lecture 1 we proved that for $n \geq 3$, $H^1_E = \{0\}$, so hyperbolic manifolds are locally rigid as Einstein metrics. In fact, something much stronger is true in dimension four:

THEOREM 4.14 (Besson-Courtois-Gallot [**BCG95**]). *If (M^4, g) is compact and hyperbolic, then g is the unique Einstein metric on M, up to scaling.*

This is proved using completely different methods than we have discussed here (using the notion of volume entropy), which we will not have time to go into. This is a generalization of Mostow rigidity, which says that hyperbolic metrics are determined up to scaling by homotopy type in dimensions $n \geq 3$ [**Mos68**]. An analogous rigidity result was proved for complex hyperbolic 4-manifolds using Seiberg-Witten Theory in [**LeB95**].

Einstein metrics with negative sectional curvature are also locally rigid:

EXERCISE 4.15. Show that any Einstein metric with negative sectional curvature is rigid, that is, $H^1_E = \{0\}$.

In the case of $n = 2$, we saw in Lecture 1 that elements of H^1_E are Codazzi tensors. Using some elementary Riemann surface theory, it is possible to identify these with the space of real parts of holomorphic quadratic differentials, see for example [**EE69**]. By the Riemann-Roch Theorem, this space is of real dimension $6\ell - 6$, where ℓ is the genus for genus $\ell \geq 2$ [**Don11**].

In the case of $n \geq 3$, Codazzi tensors do not give infinitesimal Einstein deformations of a hyperbolic metric. However, they do yield infinitesimal deformations of the locally conformally flat structure [**Laf83**]. Hyperbolic metrics admitting such deformations are called *bendable*. For examples, see [**JM87**].

5.2. The positive case

It is possible to modify the above construction so that it works also in the positive Einstein case, but we leave this as an exercise. The main difference is that the gauge term should be chosen differently. We will just mention two issues that arise.

- A positive Einstein metric can admit a nontrivial group of isometries (identity component). Letting $\mathrm{Isom}(M, g)$ denote the isometry group, we note that $\mathrm{Isom}(M, g)$ acts on the space of symmetric tensors, and therefore, by linearizing at the identity transformation, so does the space of Killing fields $\mathfrak{K}$, which is the Lie algebra of $\mathrm{Isom}(M, g)$. Taking this action into account, the end result is that the map Ψ is equivariant with respect to the isometry group, and the actual moduli space is locally described by $\Psi^{-1}(0)/\mathfrak{K}$, rather than just $\Psi^{-1}(0)$.
- In the case of the sphere, we run into the problem of first nontrivial eigenfunctions yielding pure trace kernel. However, these can also be "gauged away" since they arise as tangents to conformal diffeomorphisms. See for example [**GV11**, Section 6] for details.

Next, we discuss a few known rigid examples in the positive case:

- Any spherical space form S^n/Γ with the round metric g_S. In this case, we saw in Lecture 1 that the linearized operator obviously has trivial kernel, see Corollary 1.10.
- $S^2 \times S^2$ with the product metric $g_1 + g_2$. In this case, we saw in Exercise 1.14 that the first two eigenvalues of of the linearized operator are -1 and 1, thus 0 does not occur as an eigenvalue.
- $(\mathbb{CP}^n, g_{\mathrm{FS}})$, where g_{FS} is the Fubini-Study metric. We will not have time to prove this case in these lectures, since the nicest proof involves the theory of deformations of Kähler-Einstein metrics, and needs a considerable amount of background in complex geometry.

It is remarked that there are examples of positive Einstein metrics admitting nontrivial infinitesimal Einstein deformations, yet which are rigid as Einstein metrics, for example $S^2 \times \mathbb{CP}^{2\ell}$ [**Koi82**]. This shows that, in general, determining the Kuranishi is not an easy problem.

5.3. The zero case

We saw that, in the case of a flat metric, H^1 consists of parallel sections. The Kuranishi map turns out to be identically zero in this case, since all of these are "integrable", corresponding to deformations of the flat structure.

EXERCISE 4.16. Determine the moduli space of flat structures on a 2-torus. (Hint: a flat structure is equivalent to a choice of lattice in $\mathbb{R}^2$.)

Another special class of Ricci-flat metrics are Calabi-Yau metrics, which are Kähler Ricci-flat metrics. It is known that Calabi-Yau metrics are *unobstructed*:

THEOREM 4.17 (Bogomolov-Tian). *For a Calabi-Yau metric (X, g), the Kuranishi map $\Psi \equiv 0$. That is, every infinitesimal Einstein deformation integrates to an actual deformation.*

We will not discuss the proof, since it involves a considerable amount of complex geometry, see [**Bog78, Tia87**].

We also mention that Dai-Wei-Wang have proved stability results for manifolds admitting a parallel spinor [**DWW05**].

LECTURE 5

Quadratic curvature functionals

1. Quadratic curvature functionals

First recall that the curvature tensor admits the orthogonal decomposition

$$Rm = W + \frac{1}{n-2} E \owedge g + \frac{R}{2n(n-1)} g \owedge g, \tag{5.1}$$

where

$$E = Ric - \frac{R}{n} g \tag{5.2}$$

is the *traceless Ricci tensor*. The $\owedge$ symbol is the Kulkarni-Nomizu product, which takes 2 symmetric $(0,2)$ tensors and produces a $(0,4)$ tensor with the same algebraic symmetries of the curvature tensor, and is defined by

$$\begin{aligned} A \owedge B(X,Y,Z,W) =& A(X,Z)B(Y,W) - A(Y,Z)B(X,W) \\ & - A(X,W)B(Y,Z) + A(Y,W)B(X,Z). \end{aligned}$$

The tensor W occurring in (5.1) is called the Weyl tensor (use (5.1) to *define* the Weyl tensor), and is the part of the curvature tensor which lies in the kernel of the Ricci contraction map. An important property of the Weyl tensor is given by:

EXERCISE 5.1. The Weyl tensor, viewed as a $(1,3)$-tensor with components ${W_{ijk}}^l$, is pointwise conformally invariant. That is, if $\tilde{g} = f \cdot g$ where f is a strictly positive function, then $W(\tilde{g}) = W(g)$.

We will now turn our attention to functionals on the space of Riemannian metrics $\mathcal{M}$ which are quadratic in the curvature; see [**Bes08, Bla00, Smo05**] for surveys.

A basis for the space of quadratic curvature functionals is

$$\mathcal{W}(g) = \int_M |W_g|^2 dV_g, \quad \rho(g) = \int_M |Ric_g|^2 dV_g, \quad \mathcal{S}(g) = \int_M R_g^2 dV_g. \tag{5.3}$$

Let us now restrict the rest of this lecture to dimension four. In this dimension, for M compact without boundary, the Chern-Gauss-Bonnet formula states that

$$32\pi^2 \chi(M) = \int_M |W_g|^2 dV_g - 2\int_M |Ric_g|^2 dV_g + \frac{2}{3}\int_M R_g^2 dV_g, \tag{5.4}$$

where $\chi(M)$ is the Euler characteristic of M. This implies that any one of the functionals in (5.3) can be written as a linear combination of the other two (plus a topological term).

REMARK 5.2. We are using the tensor norm on $|W|^2$, that is $|W|^2 = W^{ijkl}W_{ijkl}$, where all indices are summed from 1 to 4. This differs from the norm of W as a mapping on 2-forms by a factor of 4, that is

$$|W|^2 = 4\|\widehat{W}\|^2 \tag{5.5}$$

(see below for the definition of $\widehat{W}$).

We next present the Euler-Lagrange equations of these functionals:

PROPOSITION 5.3 (Berger [**Ber70**]). *The Euler-Lagrange equations of the functionals in* (5.3) *are given by*

$$(\nabla \mathcal{W})_{ij} = -4\Big(\nabla^k\nabla^l W_{ikjl} + \frac{1}{2}R^{kl}W_{ikjl}\Big), \tag{5.6}$$

$$(\nabla \rho)_{ij} = -\Delta(Ric)_{ij} - 2R_{ikjl}R^{kl} + \nabla_i\nabla_j R - \frac{1}{2}(\Delta R)g_{ij} + \frac{1}{2}|Ric|^2 g_{ij}, \tag{5.7}$$

$$(\nabla \mathcal{S})_{ij} = 2\nabla_i\nabla_j R - 2(\Delta R)g_{ij} - 2RR_{ij} + \frac{1}{2}R^2 g_{ij}. \tag{5.8}$$

OUTLINE OF PROOF. Let $g(t)$ be a path of metrics such that $g(0) = g$ and $g'(0) = h$. Using the formula for the derivative of the inverse of a matrix

$$(g^{pq})' = -g^{pk}h_{kl}g^{lq}, \tag{5.9}$$

the formula for the derivative of the volume element (1.16), and the formula for the linearization of the Ricci tensor (1.25), equation (5.7) follows upon integrating by parts. Next, recalling the formula for the linearization of the scalar curvature (1.14), the formula (5.8) follows similarly.

Finally, instead of computing the linearization of the Weyl tensor directly, use the Chern-Gauss-Bonnet formula (5.4) to express the Euler-Lagrange equations of $\mathcal{W}$ as a linear combination of the Euler-Lagrange equations of the other two functionals. Note that the formula obtained in this way shows that $\nabla\mathcal{W}$ depends only upon the Ricci tensor (and it covariant derivatives). Use the Bianchi identities to show these are equivalent to the form (5.6). □

We point out some obvious critical metrics:

PROPOSITION 5.4. *Any Einstein metric is critical for all of the functionals* $\mathcal{W}, \rho$, *and* $\mathcal{S}$. *Also, any scalar-flat metric is critical for* $\mathcal{S}$.

PROOF. These statements follow easily from the Euler-Lagrange equations computed above in Proposition 5.3. □

We note that the functional $\mathcal{S}$ has been deeply studied in Kähler geometry; critical points of this functional when restricted to a Kähler class are known as extremal Kähler metrics, see [**Cal82, Cal85**]. The Euler-Lagrange equations of the restricted functional are that the gradient of the scalar curvature is the real part of a holomorphic vector field. In particular, constant scalar curvature Kähler metrics are extremal.

Also, the functional $\mathcal{W}$ has been studied in depth (see for example [**Gur98**]), and it was introduced by Rudolf Bach in 1921 [**Bac21**]. Thus the Euler-Lagrange tensor $\nabla\mathcal{W}$ is known as the *Bach tensor.* Conformal invariance of the functional $\mathcal{W}$ implies that the Bach tensor is also conformally invariant.

2. Curvature in dimension four

If (M^4, g) is oriented, the Hodge star operator on Λ^2 satisfies $*^2 = I$. The space of 2-forms then decomposes into

$$\Lambda^2 = \Lambda^2_+ \oplus \Lambda^2_-, \tag{5.10}$$

the $+1$ and -1 eigenspaces of the Hodge star operator, respectively. Note that $dim_{\mathbb{R}}(\Lambda^2) = 6$, and $dim_{\mathbb{R}}(\Lambda^2_\pm) = 3$. Elements of Λ^2_+ are called *self dual* 2-forms, and elements of Λ^2_- are called *anti-self-dual* 2-forms

We fix an oriented orthonormal basis $\{e_1, e_2, e_3, e_4\}$ and denote the dual basis by $\{e^1, e^2, e^3, e^4\}$. Define

$$\begin{aligned}\omega_1^\pm &= e^1 \wedge e^2 \pm e^3 \wedge e^4,\\ \omega_2^\pm &= e^1 \wedge e^3 \pm e^4 \wedge e^2,\\ \omega_3^\pm &= e^1 \wedge e^4 \pm e^2 \wedge e^3.\end{aligned}$$

Note that $*\omega_i^\pm = \pm\omega_i^\pm$, and $\frac{1}{\sqrt{2}}\omega_i^\pm$ is an orthonormal basis of $\Lambda^2_\pm$.

In dimension 4 there is the special coincidence that the curvature operator acts on 2-forms, and the space of 2-forms decomposes as above. Recall from above that full curvature tensor decomposes as

$$Rm = W + \frac{1}{2} E \owedge g + \frac{R}{24} g \owedge g. \tag{5.11}$$

Consider the curvature tensor as a mapping on 2-forms defined by

$$\widehat{Rm}(\omega) = \frac{1}{4} \sum_{i,j,k,l} R_{ijkl}\omega_{kl} e^i \wedge e^j, \tag{5.12}$$

where

$$\omega = \frac{1}{2} \sum_{i,j} \omega_{ij} e^i \wedge e^j. \tag{5.13}$$

We call this mapping the *curvature operator*, which has a corresponding decomposition, see [**ST69**]:

$$\widehat{Rm} = \left(\begin{array}{c|c} \widehat{W}^+ + \frac{R}{12} I & \widehat{E} \\ \hline \widehat{E} & \widehat{W}^- + \frac{R}{12} I \end{array} \right). \tag{5.14}$$

The operators $\widehat{W}^+$ and $\widehat{W}^-$ are traceless as endomorphisms of Λ^2_+ and Λ^2_-, respectively.

EXERCISE 5.5. Prove the decomposition (5.14). (Hint: this is equivalent to saying that $\widehat{W}$ commutes with $*$ and $\mathcal{E}$ anti-commutes with $*$, where $\widehat{W}$ and $\widehat{E}$ are the operators on 2-forms corresponding to W and $\frac{1}{2} E \owedge g$, respectively.)

A beautiful theorem relates the L^2-norms of the tensors $W^\pm$ to the topology of the manifold, and is called the Hirzebruch Signature Theorem.

THEOREM 5.6 (Hirzebruch [**Hir95**]). *Let (M^4, g) be compact and oriented. Then*

$$48\pi^2\tau(M) = \int_M |W_g^+|^2 dV_g - \int_M |W_g^-|^2 dV_g,$$

where $\tau = b_2^+ - b_2^-$ is the signature of M.

3. Einstein metrics in dimension four

From Proposition 5.3, it is not hard to see that any Einstein metric is critical for all three functionals in (5.3). One of the only known obstructions to the existence of Einstein metrics is the following inequality which is called the Hitchin-Thorpe Inequality.

THEOREM 5.7 (Hitchin-Thorpe). *If (M^4, g) is Einstein and oriented, then*

$$2\chi(M) \geq 3|\tau(M)|, \tag{5.15}$$

with equality if and only if g is flat or finitely covered by a $K3$ surface with a Ricci-flat metric.

PROOF. This inequality follows from the Chern-Gauss-Bonnet Formula and Hirzebruch Signature Theorem, this was first noted in [**Tho69**], and the equality case was characterized by Hitchin in [**Hit74**].

We give an outline of the equality case: if equality holds, then one concludes that $R = 0$ and that either $W^+ = 0$ or $W^- = 0$. Reversing orientation if necessary, we may assume that $W^+ = 0$. If g is flat, then we are done, so assume that W^- does not vanish identically. Since g is assumed to be Einstein, then the entire top half of (5.14) vanishes. This says that the bundle $\Lambda^2_+(T^*M)$ is flat. Since $W^- \not\equiv 0$, the Chern-Gauss-Bonnet theorem implies that $\chi(M) > 0$. If $b_1(M)$ (the first Betti number) were non-zero, then from Hodge Theory, there would exists a non-trivial harmonic 1-form α. But since $Ric \equiv 0$, by the Bochner formula on 1-forms (3.5), α would be parallel. The dual vector field would be a non-zero vector field on M, contradicting the fact that $\chi(M) > 0$ (from the Poincaré-Hopf Index Theorem for vector fields). So we conclude that $b_1(M) = 0$. By the Cheeger-Gromoll splitting theorem, we can then conclude that $\pi_1(M)$ is finite [**CG72**], so we just consider the universal cover $\tilde{M}$ of M. Since the bundle $\Lambda^2_+(T^*\tilde{M})$ is flat and $\tilde{M}$ is simply-connected, it must be trivial, and consequently the holonomy can be reduced to SU(2), which implies that g is Kähler with vanishing first Chern class, and must therefore be a K3 surface. □

EXERCISE 5.8. (i) Show that if $k\#\mathbb{CP}^2$ admits an Einstein metric then $k \leq 3$. (ii) Show that if $\mathbb{CP}^2\#k\overline{\mathbb{CP}}^2$ admits an Einstein metric then $k \leq 8$.

We also mention there are improvements of (5.15) using Seiberg-Witten Theory, see for example [**LeB99**].

Next, we will list some examples of Einstein metrics in dimension 4. The only known compact examples with positive Einstein constant in dimension four:

- S^4 or $\mathbb{RP}^4$ with the round metric.
- $S^2 \times S^2$ with the product metric, its orientable $\mathbb{Z}/2\mathbb{Z}$ quotient $G(2,4)$, $\mathbb{RP}^2 \times \mathbb{RP}^2$ with the product metric, and $S^2 \times \mathbb{RP}^2$ with the product metric.

- $\mathbb{CP}^2$ with the Fubini-Study metric.
- $\mathbb{CP}^2\#\overline{\mathbb{CP}}^2$ with the Page metric, an explicit U(2)-invariant Einstein metric, see [**Pag78a**]. This admits a non-orientable quotient $\mathbb{CP}^2\#\mathbb{RP}^4$.
- $\mathbb{CP}^2\#2\overline{\mathbb{CP}}^2$ with the Chen-LeBrun-Weber metric [**CLW08**]. This metric is conformal to an extremal Kähler metric. See Section 4 in Lecture 9 below for a more discussion of this metric.
- $\mathbb{CP}^2\#k\overline{\mathbb{CP}}^2$, $k = 3, \dots, 8$, admits positive Kähler-Einstein metrics (Tian-Yau [**TY87**], Tian [**Tia90**]).

It is an interesting problem to find other topological manifolds admitting positive Einstein metrics, and also to possibly find other Einstein metrics on the manifolds listed above. For example, it is unknown whether S^4 admits an Einstein metric which is not of constant curvature. However, if one exists, it is known that its Yamabe energy cannot be too large.

THEOREM 5.9 (Gursky [**Gur00**]). *Suppose S^4 admits a positive Einstein metric g which is not isometric to the standard round metric, normalized so that $Ric(g) = 3g$. Then*

$$Vol(g) < \frac{8}{9}\pi^2 = \frac{1}{3}Vol(g_S). \tag{5.16}$$

The only known compact examples with zero Einstein constant in dimension four:

- K3 surface with Calabi-Yau Ricci-flat metric [**Yau78**], and its quotients.
- Flat metrics.

There are many more examples of Einstein metrics with negative Einstein constant. Of course, any hyperbolic manifold is an example. Complex hyperbolic manifolds are another interesting class with negative Einstein constant [**LeB95**]. Any Kähler manifold with $c_1 < 0$ carries an Einstein metric by Aubin-Yau [**Aub78, Yau78**]. There are in fact many such manifolds, for example, any non-singular hypersurface in $\mathbb{CP}^3$ of degree $d > 4$ satisfies $c_1 < 0$. The case $d = 4$ is the K3 surface which has $c_1 = 0$, and carries the Calabi-Yau Ricci-flat metric mentioned above. Also, see [**And10**] for a nice survey and other examples.

3.1. Higher dimensions

We note that, in dimensions $n > 4$, there is no known topological obstruction to the existence of an Einstein metric. There are in fact quite a large number of known examples of Einstein metrics in higher dimensions (too many to list here). It could be the case that every compact manifold of dimension $n > 4$ admits an Einstein metric.

4. Optimal metrics

Another interesting class of metrics are called optimal metrics, and are defined to be those metrics which *globally* minimize the functional

$$\mathcal{R}(g) = \int_M |Rm_g|^2 dV_g. \tag{5.17}$$

Using the formula

$$|Rm_g|^2 = |W_g|^2 + 2|Ric_g|^2 - \frac{1}{3}R_g^2, \tag{5.18}$$

the Chern-Gauss-Bonnet formula (5.4) may be written

$$\mathcal{R}(g) = 32\pi^2\chi(M) + 4\int_M |E_g|^2 dV_g, \tag{5.19}$$

which shows immediately that Einstein metrics are necessarily optimal.

We will not go into much more details about optimal metrics in general, but just make a few remarks taken from [**LeB04**]:

- There are optimal metrics which are not Einstein.
- There are compact 4-manifolds which do not admit optimal metrics.
- There are topological 4-manifolds which admit an optimal metric for some smooth structure, but do not admit any optimal metric for a different smooth structure.

We may also write

$$\mathcal{R}(g) = -32\pi^2(\chi(M) + 3\tau(M)) + \int_M \left(\frac{R_g^2}{3} + 4|W_g^+|^2\right) dV_g. \tag{5.20}$$

Thus we see that another class of optimal metrics are those with $W^+ \equiv 0$ and $R = 0$, these are called scalar-flat anti-self-dual metrics, which we will discuss in more detail next.

REMARK 5.10. On a related note, we mention that the functionals ρ and $\mathcal{S}$ are known to be globally minimized in dimension four at a negative Kähler-Einstein metric. This is proved in [**LeB01**] using Seiberg-Witten theory, along with many other interesting results regarding minimal volumes.

EXERCISE 5.11. (i) Prove (5.20) using the Hirzebruch Signature Theorem 5.6. (ii) Show that

$$32\pi^2(\chi(M) - 3\tau(M)) = \int_M \left(-|W_g^+|^2 + \frac{1}{6}R_g^2 + 3|W_g^-|^2 - 2|E_g|^2\right) dV_g. \tag{5.21}$$

If g is Kähler then the first two terms cancel (see (6.26) below), and one is left with

$$32\pi^2(\chi(M) - 3\tau(M)) = \int_M \left(3|W_g^-|^2 - 2|E_g|^2\right) dV_g \tag{5.22}$$

(see [**Der83**]). Show that this implies that g_{FS} is the unique Kähler-Einstein metric on $\mathbb{CP}^2$, up to scaling.

5. Anti-self-dual or self-dual metrics

In dimension 4, the curvature condition

$W^+ = 0$ is called anti-self-dual (ASD),

$W^- = 0$ is called self-dual (SD).

Since the Weyl tensor is conformally invariant and the Hodge star operator on 2-forms in dimension 4 is also conformally invariant, it follows that either condition is conformally invariant.

PROPOSITION 5.12. *If (M^4, g_0) compact and self-dual, then g_0 is a global minimizer of the functional $\mathcal{W}$, so is necessarily a critical metric for $\mathcal{W}$.*

PROOF. For any metric g on M, using the Hirzebruch Signature Theorem, we have

$$\begin{aligned}\mathcal{W}(g) = \int_M |W_g|^2 \, dV_g &= \int_M |W_g^+|^2 \, dV_g + \int_M |W_g^-|^2 \, dV_g \\ &= 48\pi^2\tau(M) + 2\int_M |W_g^-|^2 dV_g \geq 48\pi^2\tau(M),\end{aligned}$$

with equality if and only if g is self-dual. □

The only obvious topological obstruction to existence of a self-dual or anti-self-dual metric comes from the Hirzebruch Signature Theorem.

PROPOSITION 5.13. *If (M, g) is self-dual (anti-self-dual) then $\tau \geq 0$ ($\tau \leq 0$) with equality if and only if g is locally conformally flat.*

One can get a stronger restriction if one assumes the scalar curvature is positive:

PROPOSITION 5.14. *If (M, g) is self-dual (anti-self-dual) and $R > 0$, then $b_2^- = 0$ ($b_2^+ = 0$).*

PROOF. On 2-forms in dimension 4, the Weitzenböck formula is

$$(\Delta_H \omega^\pm)_{ij} = -(\Delta\omega^\pm)_{ij} - \sum_{l,m} W^\pm_{lmij}\omega^\pm_{lm} + \frac{R}{3}\omega^\pm_{ij}. \tag{5.23}$$

The result follows by choosing $\omega^\pm$ be harmonic, pairing (5.23) with $\omega^\pm$, and integrating. □

LECTURE 6

Anti-self-dual metrics

1. Deformation theory of anti-self-dual metrics

There are now a wealth of examples of anti-self-dual metrics on 4-manifolds. But before we get into specific examples, let us consider the deformation theory of these metrics. Let (M, g) denote an anti-self-dual 4-manifold (the deformation theory of self-dual metrics is identical upon reversing orientation). The anti-self-dual deformation complex is given by

$$\Gamma(T^*M) \xrightarrow{\mathcal{K}_g} \Gamma(S_0^2(T^*M)) \xrightarrow{\mathcal{D}} \Gamma(S_0^2(\Lambda_+^2)), \tag{6.1}$$

where $\mathcal{K}_g$ is the conformal Killing operator defined by

$$(\mathcal{K}_g(\omega))_{ij} = \nabla_i \omega_j + \nabla_j \omega_i - \frac{1}{2}(\delta\omega) g_{ij}, \tag{6.2}$$

with $\delta\omega = \nabla^i \omega_i$, $S_0^2(T^*M)$ denotes traceless symmetric tensors, and $\mathcal{D} = (\mathcal{W}^+)'_g$ is the linearized self-dual Weyl curvature operator. This complex is elliptic [**KK92**].

The cohomology groups of this complex are given as follows:

$$H_c^0(M, g) = \{\kappa \in T^*M \mid \mathcal{K}_g \kappa = 0\}. \tag{6.3}$$

Letting $\mathcal{D}_g$ denote $(W^+)'_g$, we have

$$H_c^1(M, g) = \{h \in S_0^2(T^*M) \mid \mathcal{D}_g h = 0,\ \delta_g h = 0\}. \tag{6.4}$$

Finally,

$$H_c^2(M, g) = \{Z \in S_0^2(\Lambda_+^2) \mid \mathcal{D}_g^* Z = 0\}, \tag{6.5}$$

where $\mathcal{D}_g^*$ is the formal L^2-adjoint of $\mathcal{D}_g$.

If M is a compact manifold then there is a formula for the index depending only upon topological quantities. The analytical index is given by

$$Ind(M, g) = \dim(H_c^0(M, g)) - \dim(H_c^1(M, g)) + \dim(H_c^2(M, g)). \tag{6.6}$$

The index is given in terms of topology via the Atiyah-Singer index theorem:

THEOREM 6.1 (I.M. Singer). *If (M, g) is compact and anti-self-dual, then*

$$Ind(M, g) = \frac{1}{2}(15\chi(M) + 29\tau(M)). \tag{6.7}$$

where $\chi(M)$ is the Euler characteristic and $\tau(M)$ is the signature of M

This was first computed by I.M. Singer in 1978, see also [**LeB92**, page 303] and [**KK92**].

The cohomology groups of the complex (6.1) yield information about the local structure of the moduli space of anti-self-dual conformal classes, which we discuss next (see also [**Ito93, KK92**]). Letting $\mathrm{Conf}(M,g)$ denote the conformal automorphism group, we note that $\mathrm{Conf}(M,g)$ acts on the space of symmetric tensors, and therefore, by linearizing at the identity transformation, so does $H^0_c(M,g)$, since the space of conformal Killing Fields is the Lie algebra of $\mathrm{Conf}(M,g)$.

THEOREM 6.2. *If (M,g) is anti-self-dual, then there is a map*

$$\Psi : H^1_c(M,g) \to H^2_c(M,g) \tag{6.8}$$

called the Kuranishi map *which is equivariant with respect to the action of the conformal group. The moduli space of anti-self-dual conformal structures near g (anti-self-dual metrics modulo diffeomorphism and conformal equivalence) is locally isomorphic $\Psi^{-1}(0)/H^0_c(M,g)$.*

PROOF. The proof is similar to the Einstein case considered above, and we only give a brief outline here. If g is anti-self-dual, for $\theta \in S^2_0(T^*M)$, consider the map

$$P_g : \Gamma(S^2_0(T^*M)) \to \Gamma(S^2_0(\Lambda^2_+(T^*M))) \times \Gamma(T^*M) \tag{6.9}$$

defined by

$$P_g(\theta) = \{\Pi_g(W^+(g+\theta)), \delta_g\theta\}, \tag{6.10}$$

where Π_g is projection onto $S^2_0(\Lambda^2_+)$ with respect to g. The linearized operator at $\theta = 0$ is given by

$$P'_g(h) = (\mathcal{D}_g h, \delta_g h). \tag{6.11}$$

This is a mixed-order elliptic operator in the sense of Douglis-Nirenberg [**DN55**]. The kernel is isomorphic to H^1_c, while the cokernel is isomorphic to $H^2_c \oplus H^0_c$. The gauging result in Theorem 3.6 can be refined to allow conformal deformations, which then allows one to gauge to be transverse-traceless, see for example [**GV11**, Theorem 2.11], or [**FM77**]. The result then follows from the implicit function theorem in Lemma 4.11. □

In general, it is a difficult problem to compute the Kuranishi map, but we have the obvious corollary:

COROLLARY 6.3. *If (M,g) is anti-self-dual and satisfies*

$$H^0_c(M,g) = \{0\} \text{ and } H^2_c(M,g) = \{0\}, \tag{6.12}$$

then the moduli space of anti-self-dual conformal structures near g is a smooth finite-dimensional manifold of dimension $\dim(H^1_c(M,g))$.

2. Weitzenböck formulas

For (M,g) Einstein, with $Ric = \lambda \cdot g$, define the *Lichnerowicz Laplacian* by

$$\Delta_L h_{ij} = \Delta h_{ij} + 2R_{ipjq}h^{pq} - 2\lambda \cdot h_{ij}. \tag{6.13}$$

We next have the following Weitzenböck formulae.

THEOREM 6.4 (Kobayashi [**Kob85**], Itoh [**Ito95**]). *If (M,g) is compact and self-dual Einstein with $Ric = \lambda \cdot g$, then*

$$\mathcal{D}^*\mathcal{D}h = \frac{1}{2}\Big(\Delta_L + 2\lambda\Big)\Big(\Delta_L + \frac{4}{3}\lambda\Big)h \tag{6.14}$$

$$\mathcal{D}\mathcal{D}^*Z = \frac{1}{12}(3\Delta - 8\lambda)(\Delta - 2\lambda)Z. \tag{6.15}$$

PROOF. One uses the formula for the adjoint operator

$$(\mathcal{D}^* Z)_{ij} = 2(\nabla^k \nabla^l + \frac{1}{2} R^{kl}) Z_{ikjl}. \tag{6.16}$$

The computations are lengthy, so are left as an exercise. □

REMARK 6.5. We note that the gradient of $\mathcal{W}$ may also be written as

$$\begin{aligned} \nabla \mathcal{W} &= -8(\nabla^k \nabla^l + \frac{1}{2} R^{kl}) W^+_{ikjl} \\ &= -8(\nabla^k \nabla^l + \frac{1}{2} R^{kl}) W^-_{ikjl}. \end{aligned} \tag{6.17}$$

The details are left as an exercise, see [**Ito95**]. It follows that if g is any anti-self-dual metric, then

$$(\nabla \mathcal{W})'(h) = -4\mathcal{D}^* \mathcal{D} h. \tag{6.18}$$

EXERCISE 6.6. Compute $H^i_c(M, g)$ for the following examples:

- S^4 with the round metric g_S. This is locally conformally flat, so obviously anti-self-dual.
- $\mathbb{CP}^2$ with the Fubini-Study metric g_{FS}. This is self-dual with respect to the complex orientation. To see this, the tensor W^- must be in $S^2_0(\Lambda^{1,1}_0)$ (see (6.22) below). The isometry group of g_{FS} contains $\mathrm{SU}(3)$, and the stabilizer of a point contains $\mathrm{SU}(2)$. It is not hard to see that $\mathrm{SU}(2)$ acts on $\Lambda^{1,1}_0$ as the standard representation of $\mathrm{SO}(3) = \mathrm{SU}(2)/\mathbb{Z}_2$. Consequently, the only tensor in $S^2_0(\Lambda^{1,1}_0)$ invariant under $\mathrm{SU}(2)$ is the zero tensor, so $W^- \equiv 0$.
- $S^1 \times S^3$ with the product metric g (this is not Einstein, so you cannot directly use the Weitzenböck formulas. But it is locally conformally flat). What is the dimension of the moduli space near g?

We mention the following conjecture.

CONJECTURE 6.7 (I.M. Singer). *If (M, g) is anti-self-dual and $R > 0$, then*

$$H^2_c(M, g) = 0. \tag{6.19}$$

In the Einstein case, this follows easily from the Weitzenböck formula (6.15). However, Hitchin proved that the only smooth positive ASD Einstein metrics are (S^4, g_S) or $(\mathbb{CP}^2, g_{\mathrm{FS}})$, see [**Bes08**, Theorem 13.30].

A wealth of examples of anti-self-dual metrics have been found since this conjecture was made, and all of the ones with positive scalar curvature have turned out to have $H^2_c(M, g) = 0$. But despite all of the evidence, a proof of this conjecture remains elusive.

3. Calabi-Yau metric on K3 surface

The K3 surface is defined to be a nondegenerate quartic surface in $\mathbb{CP}^3$, that is

$$K3 = \{[z_0, z_1, z_2, z_3] \in \mathbb{CP}^3 \mid z_0^4 + z_1^4 + z_2^4 + z_3^4 = 0\}.$$

The topology of K3 is described by: $\pi_1(K3) = \{e\}$, $b_2 = 22$, $b_2^+ = 3$, $b_2^- = 19$.

Since $c_1(K3) = 0$, by Yau's solution of the Calabi conjecture [**Yau78**], K3 admits a Ricci-flat Kähler metric g_Y.

PROPOSITION 6.8. *$(K3, g_Y)$ is anti-self-dual with respect to the complex orientation.*

PROOF. To see this we use that fact that for any Kähler metric, W^+ is entirely determined by the scalar curvature. In fact,

$$\widehat{W}^+ = \frac{R}{12}(3\omega \odot \omega - I), \tag{6.20}$$

where ω is the Kähler form. To see this, one proves the following: for a Kähler surface

$$\Lambda^2_+ = \mathbb{R} \cdot \omega \oplus (\Lambda^{2,0} \oplus \Lambda^{0,2}) \tag{6.21}$$

$$\Lambda^2_- = \Lambda^{1,1}_0. \tag{6.22}$$

Next, use the fact that if (M, g, J) is Kähler, then

$$Rm(X, Y, Z, W) = Rm(JX, JY, Z, W) = Rm(X, Y, JZ, JW), \tag{6.23}$$

$$Ric(X, Y) = Ric(JX, JY). \tag{6.24}$$

This implies that the curvature operator

$$\widehat{R} \in S^2(\Lambda^{1,1}). \tag{6.25}$$

Choose an ONB so that the Kähler form is given by ω_1^+. Since ω_+^2 and ω_+^3 are in $\Lambda^{2,0} \oplus \Lambda^{0,2}$, which is orthogonal to the space of $(1,1)$-forms, they must be annihilated by $\widehat{W}^+ + \frac{R}{12}I$. The formula (6.20) then follows since $\widehat{W}^+$ is traceless. □

REMARK 6.9. Note that (6.20) implies that

$$|W_g^+|^2 = \frac{1}{6}R_g^2 \tag{6.26}$$

for any Kähler metric g.

EXERCISE 6.10. Using Theorem 6.1 and the Weitzenböck formulas in Theorem 6.4, show that:

- $\dim(H_c^0(K3, g_Y)) = 0$.
- $\dim(H_c^1(K3, g_Y)) = 57$.
- $\dim(H_c^2(K3, g_Y)) = 5$.

In fact, using the isomorphism $S_0^2(T^*M) = \Lambda^2_+ \otimes \Lambda^2_-$, it can be shown that $H_c^1(K3, g_Y)$ has a basis

$$\{\omega_I \otimes \omega_j^-, \omega_J \otimes \omega_j^-, \omega_K \otimes \omega_j^-\},$$

where $\omega_I, \omega_J, \omega_K$ are a basis of the space of self-dual harmonic 2-forms (these are actually Kähler forms for complex structures I, J, K), and $\{\omega_j^-, j = 1, \ldots, 19\}$ is a basis of the space of anti-self-dual harmonic 2-forms.

Furthermore, by the Weitzenböck formula, $H_c^1 = H_E^1$ (infinitesimal Einstein deformations) and the moduli space is exactly 57-dimensional; the Kuranishi map $\Psi \equiv 0$ by the Bogomolov-Tian Theorem 4.17.

4. Twistor methods

No discussion of anti-self-dual metrics in dimension four can be complete without mentioning twistor theory, which was first proposed by Penrose in the Lorentzian case. This was then studied in depth by Atiyah-Hitchin-Singer in the Riemannian case [**AHS78**].

Given any oriented Riemannian 4-manifold (M, g), one may consider the bundle of unit-norm self-dual 2-forms:

$$S^2 \to \mathcal{Z}^+(M) \to M. \tag{6.27}$$

This has a "tautological" almost complex structure defined as follows. At any point in $\mathcal{Z}^+(M)$, the horizontal space carries a tautological complex structure, while the vertical space has the complex structure of S^2. With the correct choice of orientation on the fiber, it was shown in [**AHS78**] that this almost complex structure is integrable if and only if the metric g is anti-self-dual. Thus to any anti-self-dual four-manifold, one can associate a compact complex 3-fold, and techniques from complex geometry may be used. We only mention that the cohomology groups $H^i_c(M,g) \otimes \mathbb{C}$ are isomorphic to $H^i(\mathcal{Z}^+(M), \Theta)$, the Kuranishi cohomology groups corresponding to the deformations of complex structure. The Kuranishi map defined above is exactly the Kuranishi map for this complex 3-fold.

In [**Poo86**] and [**Poo92**], Poon found examples of self-dual conformal classes on the connected sums $\mathbb{CP}^2\#\mathbb{CP}^2$ and $3\#\mathbb{CP}^2$ using techniques from algebraic geometry. In [**LeB91**], LeBrun gave a more explicit construction of U(1)-invariant self-dual conformal classes on $n\#\mathbb{CP}^2$ for any n, and identified the twistor spaces of these metrics. Joyce constructed a large class of toric ASD metrics on $n\#\mathbb{CP}^2$ in [**Joy95**], and these were classified by Fujiki [**Fuj00**]. Rollin-Singer constructed scalar-flat Kähler metrics on $\mathbb{CP}^2\#10\overline{\mathbb{CP}}^2$ in [**RS05**]. Honda has constructed many new examples, see for example [**Hon06, Hon07**]. There are now so many known examples that we cannot possibly list all of them here.

5. Gluing theorems for anti-self-dual metrics

A very successful technique for producing new examples of ASD metrics is given by the following.

THEOREM 6.11 (Donaldson-Friedman, Floer). *If (M_1, g_1) and $(M_2.g_2)$ are anti-self-dual and $H^2_c(M_i, g_i) = \{0\}$ for $i = 1, 2$, then there exist anti-self-dual metrics on the connected sum $M_1\#M_2$.*

Donaldson-Friedman proved Theorem 6.11 in the case of smooth manifolds using twistor theory, together with methods from the deformation theory of singular complex 3-folds [**DF89**]. LeBrun-Singer generalized this proof to the case of orbifolds with $\mathbb{Z}/2\mathbb{Z}$-orbifold points [**LS94**]. In [**Flo91**], Floer gave an analytic proof for the case of the connected sum of n copies of $\mathbb{CP}^2$. The strategy of his proof is to delete points from the summands, and conformally change the metrics to become asymptotically cylindrical. The metrics are then pasted together by very long cylindrical regions in between. An analysis of the indicial roots of the linearized problem on the cylinder together with a fixed point theorem as in Lemma 4.11, then allowed Floer to perturb to an exact solution.

There are also many interesting examples of ASD orbifolds, (see for example [**CS06, LeB04, LV12, Wri11**] and the references therein), and it is also an interesting problem to glue together orbifold metrics with complementary singularities to produce smooth examples. We mention that Kovalev-Singer presented a generalization of Floer's argument which works also in the orbifold case [**KS01**], but see also [**AV12b, RS05, LM08**] for some clarifications.

We end this lecture by mentioning Taubes' stable existence claim for anti-self-dual metrics: for any compact, oriented, smooth 4-manifold M, the manifold $M\#n\overline{\mathbb{CP}}^2$ carries an anti-self-dual metric for some n, see [**Tau92**].

LECTURE 7

Rigidity and stability for quadratic functionals

1. Strict local minimization

We saw that critical points of the Einstein-Hilbert functional in general have a saddle-point structure. However, critical points for certain quadratic functionals have a nicer local variational structure. For example, one result we will discuss in this lecture is the following. Define the functional

$$\mathcal{F}_\tau(g) = \int_M |Ric_g|^2 dV_g + \tau \int_M R_g^2 dV_g. \tag{7.1}$$

THEOREM 7.1 (Gursky-Viaclovsky [**GV11**]). *On S^4, the round metric g_S is a strict local minimizer (modulo diffeomorphisms and scaling) for the functional $\mathcal{F}_\tau$ provided that*

$$-\frac{1}{3} < \tau < \frac{1}{6}. \tag{7.2}$$

Many other stability results are given in [**GV11**]; various results are proved for hyperbolic metrics, complex projective spaces, products of spheres, and Ricci-flat metrics. But for simplicity, we will only concentrate on the case of the spherical metric in this lecture.

1.1. The Jacobi operator

Just as in the case of the Einstein-Hilbert functional, the second variation is orthogonal with respect to the splitting

$$S^2(T^*M) = \{f \cdot g\} \oplus \{\mathcal{L}(\alpha)\} \oplus \{\delta h = 0, tr_g(h) = 0\}. \tag{7.3}$$

Therefore, if h is any symmetric 2-tensor, then it decomposes as

$$h = f \cdot g + \mathcal{L}\alpha + z, \tag{7.4}$$

where z is TT. Then

$$\mathcal{F}''_\tau(h, h) = \mathcal{F}''_\tau(f \cdot g, f \cdot g) + \mathcal{F}''_\tau(z, z). \tag{7.5}$$

Consequently, to check the second variation, we only need to consider conformal variations and TT variations separately.

As mentioned above, Einstein metrics are indeed critical for $\mathcal{F}_\tau$. Let us write the second variation at an Einstein metric as

$$\mathcal{F}''_\tau(h_1, h_2) = \int_M \langle h_1, J h_2 \rangle dV_g. \tag{7.6}$$

The Jacobi operator J is given explicitly in the following for TT tensors for any Einstein metric.

PROPOSITION 7.2 ([**GV11**]). *If g is Einstein with $Ric(g) = \lambda \cdot g$ and h is TT, then the Jacobi operator of $\mathcal{F}_\tau$ is*

$$Jh = \frac{1}{2}\Big(\Delta_L + 2\lambda\Big)\Big(\Delta_L + 4\Big(1 + 2\tau\Big)\lambda\Big)h, \tag{7.7}$$

where

$$\Delta_L h_{ij} = \Delta h_{ij} + 2R_{ipjq}h^{pq} - 2\lambda \cdot h_{ij}. \tag{7.8}$$

The proof of this is a long computation, and will not be presented here.

For conformal variations we have the following.

PROPOSITION 7.3 ([**GV11**]). *If g is Einstein with $Ric(g) = \lambda \cdot g$ and $h = fg$, then*

$$tr_g(Jf) = \frac{4 + 12\tau}{2}(3\Delta + 4\lambda)\Delta f. \tag{7.9}$$

Again, the proof is a long computation, and will not be presented here.

1.2. The case of the round sphere

We will now restrict to the case of (S^4, g_S). In this case, the Lichnerowicz Laplacian on TT-tensors is

$$\Delta_L h = \Delta h - 8h. \tag{7.10}$$

PROPOSITION 7.4. *The least eigenvalue of the rough Laplacian on TT-tensors is 8.*

PROOF. The proof is left as an exercise, with the following hint: use the inequality

$$\int_{S^4} |\nabla_i h_{jk} + \nabla_j h_{ki} + \nabla_k h_{ij}|^2 dV_{g_S} \geq 0. \tag{7.11}$$

□

Consequently, the least eigenvalue of the Lichnerowicz Laplacian on TT-tensors is 16. Proposition 7.2 then implies that if

$$\tau < \frac{1}{6}, \tag{7.12}$$

then the Jacobi operator is positive definite when restricted to TT-tensors. This results in the upper bound in (7.2).

Proposition 7.3 implies that the Jacobi operator is non-negative in conformal directions for

$$-\frac{1}{3} < \tau, \tag{7.13}$$

with the zero eigenvalues given by $h = f \cdot g$, where f is a lowest nontrivial eigenfunction of Δ (by Lichnerowicz' Theorem). This results in the lower bound in (7.2).

To summarize, we have shown that on (S^4, g_S), the second variation is strictly positive on TT-tensors, and strictly positive in conformal directions (except for lowest nontrivial eigenfunction directions) in the range

$$-\frac{1}{3} < \tau < \frac{1}{6}. \tag{7.14}$$

We next need to integrate this result to make a conclusion about the actual behavior of the functional in a neighborhood of the spherical metric. To this end, using a modification of the Ebin-Palais slicing, we can ignore Lie derivative directions (as before), and we can also ignore the conformal zero eigentensors using conformal diffeomorphisms, so the functional is in fact strictly locally minimized at the spherical metric modulo diffeomorphisms. For details, we refer the reader to [**GV11**, Section 6].

Notice that for $\tau = -(1/4)$, the functional is $\int |E|^2$, so is obviously strictly minimized for this τ, but our improvement of the range of τ for minimization has an interesting application, which we will discuss next.

1.3. A reverse Bishop's inequality

The classical Bishop's inequality implies that if (M^4, g) is a closed manifold with $Ric(g) \geq Ric(S^4, g_S) = 3g$, then the volume satisfies $Vol(g) \leq Vol(S^4, g_S)$, and equality holds only if (M, g) is isometric to (S^4, g_S). An interesting consequence of strict local minimization for $\tau = 0$ is that, locally, a "reverse Bishop's inequality" holds.

COROLLARY 7.5 ([**GV11**]). *On (S^4, g_S), there exists a neighborhood U of g_S in the $C^{2,\alpha}$-norm such that if $\tilde{g} \in U$ with*

$$Ric(\tilde{g}) \leq 3\tilde{g}, \tag{7.15}$$

then

$$Vol(\tilde{g}) \geq Vol(g_S), \tag{7.16}$$

with equality if and only if $\tilde{g} = \phi^* g_S$ for some diffeomorphism $\phi : M \to M$.

There remain some very interesting questions:

- What is the largest neighborhood U for which this holds?
- Is the functional $\int_{S^4} |Ric_g|^2 dV_g$ globally minimized at g_S?

2. Local description of the moduli space

We will next discuss a way to describe local structure of the moduli space of solutions using a procedure analogous to that for the Einstein equations which we discussed above in Lecture 4. We again consider the functional $\mathcal{F}_\tau$, and denote the Euler-Lagrange equations by

$$\nabla \mathcal{F}_\tau = 0. \tag{7.17}$$

Due to diffeomorphism invariance, the linearization of (7.17) is not elliptic, so we have to make a gauge choice. We will work in transverse-traceless gauge, so define the operator

$$\beta_g h = \delta_g h - \frac{1}{4} d(tr_g h). \tag{7.18}$$

Also, since the functional is scale invariant, we will be interested in the space

$$\overline{S}^2_0(T^*M) = \Big\{ h \in S^2(T^*M) \ \Big| \ \int_M (tr_g h) dV_g = 0 \Big\}. \tag{7.19}$$

Recall that $\mathfrak{K}$ denotes the Lie algebra of Killing vector fields.

THEOREM 7.6. *Assume g is critical for $\mathcal{F}_\tau$ with $\tau \neq -1/3$. Then the space of critical metrics near g modulo diffeomorphism and scaling is locally isomorphic to $\Psi^{-1}(0)/\mathfrak{K}$, where Ψ is a smooth mapping*

$$\Psi : H^1_\tau \to H^1_\tau, \tag{7.20}$$

with

$$H^1_\tau = \{h \in C^\infty(\overline{S}^2_0(T^*M)) \mid (\nabla\mathcal{F}_\tau)'_g h = 0,\ \beta_g h = 0\}.$$

Consequently, if g is infinitesimally rigid, then g is rigid.

REMARK 7.7. For $\tau = -1/3$, the functional is equivalent to the L^2-norm of the Weyl tensor, so is conformally invariant. The above result then holds if one restricts to the space of pointwise traceless tensors, and one considers the moduli space of conformal classes near g. For details, see [**GV11**, Section 2.3].

The first step to prove Theorem 7.6 is to construct a nonlinear mapping whose zeroes correspond to the moduli space (locally). Assume $\mathcal{U} \subset S^2(T^*M)$ is a neighborhood of the zero section, sufficiently small so that $\theta \in \mathcal{U}_0 \Rightarrow \tilde{g} = g + \theta$ is a metric. We define the map

$$P_g : \mathcal{U} \to \overline{S}^2_0(T^*M), \tag{7.21}$$

by

$$P_g(\theta) = \nabla\mathcal{F}(g+\theta) + \frac{1}{2}\mathcal{K}_{g+\theta}[\beta_g\mathcal{K}_g\beta_g\theta]. \tag{7.22}$$

We have the following analogue of Proposition 4.5.

PROPOSITION 7.8. *If $\tau \neq -1/3$, then the linearized operator of P_g at g is elliptic. Furthermore, if $P_g(\theta) = 0$, and $\theta \in C^{4,\alpha}$ is sufficiently small for some $0 < \alpha < 1$, then $B^t(g+\theta) = 0$ and $\theta \in C^\infty$.*

PROOF. The proof involves an integration-by-parts argument. It is crucial that the equations are variational (since $\mathcal{F}_\tau$ is the functional), so $\delta\nabla\mathcal{F}_\tau = 0$. This is equivalent to diffeomorphism invariance of $\mathcal{F}_\tau$. The proof is similar to that of Proposition 4.5. □

We also have an analogue of Proposition 4.8.

PROPOSITION 7.9. *If $g_1 = g + \theta_1$ is a critical metric in a sufficiently small $C^{k+1,\alpha}$-neighborhood of g ($k \geq 3$), then there exists a $C^{k+2,\alpha}$-diffeomorphism $\phi : M \to M$ and a constant c such that*

$$e^c\phi^* g_1 = g + \tilde{\theta} \tag{7.23}$$

with

$$P_g(\tilde{\theta}) = 0 \tag{7.24}$$

and

$$\int_M tr_g\tilde{\theta}\, dV_g = 0. \tag{7.25}$$

PROOF. The proof is a gauging argument using an infinitesimal Ebin-Palais gauging as done above in the proof of Theorem 3.6, the details are similar and are omitted. □

We also require a proposition analogous to Proposition 4.10, which shows that the nonlinear terms are under control.

PROPOSITION 7.10. *There exists a constant C such that if we write*

$$P_g(h) = P_g(0) + S_g h + Q_g(h), \tag{7.26}$$

then for $h_1, h_2 \in C^{4,\alpha}$ of sufficiently small norm,

$$\|Q_g(h_1) - Q_g(h_2)\|_{C^\alpha} \leq C(\|h_1\|_{C^{4,\alpha}} + \|h_2\|_{C^{4,\alpha}}) \cdot \|h_1 - h_2\|_{C^{4,\alpha}}. \tag{7.27}$$

Since the equation is fourth order, the proof is involved, and we refer the reader to [**GV11**, Lemma 2.13] for the details. We then follow the same procedure as in Lecture 4 to construct the Kuranishi map, using the implicit function theorem. The details are left to the reader.

To finish the proof, we note that the gauge term is also carefully chosen so that solutions of the linearized equation must be in the transverse-traceless gauge. That is, if $(P_g)'h = 0$ then we have separately,

$$(\nabla \mathcal{F}_t)'(h) = 0 \text{ and } \delta \overset{\circ}{h} = 0, \tag{7.28}$$

which is an analogue of Proposition 4.12.

3. Some rigidity results

For h transverse-traceless, recall from above that the linearized operator of ∇F_τ at an Einstein metric is given by

$$(\nabla F_\tau)'h = \frac{1}{2}\Big(\Delta_L + \frac{1}{2}R\Big)\Big(\Delta_L + \Big(1 + 2\tau\Big)R\Big)h. \tag{7.29}$$

- This formula was previously obtained for the linearized Bach tensor, which is the case of $\tau = -1/3$ by O. Kobayashi [**Kob85**].
- Recall that infinitesimal Einstein deformations are given by TT kernel of the operator $\Delta_L + \frac{1}{2}R$, which we studied in Lecture 4. These deformations are still present, but note there is now the possibility of non-Einstein deformations.

Also, recall that for $h = fg$, we have

$$tr_g((\nabla F_\tau)'h) = 2(1 + 3\tau)(3\Delta + R)(\Delta f). \tag{7.30}$$

The rigidity question is then reduced to a separate analysis of the eigenvalues of Δ_L on transverse-traceless tensors, and of Δ on functions. Such an analysis of the eigenvalues of these operators yields the following rigidity theorems. Let H^1_τ denote the space of transverse-traceless kernel of the linearized operator. In the case of the Fubini-Study metric, we have

THEOREM 7.11 ([**GV11**]). *On $(\mathbb{CP}^2, g_{\mathrm{FS}})$, $H^1_\tau = \{0\}$ provided that $\tau < 1/6$.*

In the case of the product metric, we have

THEOREM 7.12 ([**GV11**]). *On $(S^2 \times S^2, g_{S^2 \times S^2})$, $H^1_\tau = \{0\}$ provided that $\tau < 0$ and $\tau \neq -1/2$. If $\tau = -1/2$, then H^1_τ is one-dimensional and spanned by the element $g_1 - g_2$.*

In particular, in the case $\tau = -1/2$, this gives an example of a deformation of critical metrics for this functional which is not an Einstein deformation.

Many other rigidity results are presented in [**GV11**], which we will not go into detail here. We have stated the above two results because these rigidity results will play a crucial rôle in the gluing construction which will be discussed in Lecture 9.

4. Other dimensions

We restricted the above discussion to dimension 4 for simplicity, since quadratic functionals are scale-invariant in that dimension. In dimensions other than four, if $\mathcal{F}$ denotes a quadratic curvature functional, then the volume-normalized functional

$$\tilde{\mathcal{F}}[g] = Vol(g)^{\frac{4}{n}-1}\mathcal{F}[g]. \tag{7.31}$$

is scale-invariant. Many of the results stated above also have analogues in other dimensions. For example, Corollary 7.5 has an analogue in higher dimensions:

THEOREM 7.13 ([**GV11**]). *Let (M, g) be a sphere, space form, or complex projective space, normalized so that $Ric(g) = (n-1)g$. Then there exists a $C^{2,\alpha}$-neighborhood U of g such that if $\tilde{g} \in U$ with $Ric(\tilde{g}) \leq (n-1)\tilde{g}$, then $Vol(\tilde{g}) \geq Vol(g)$ with equality if and only if $\tilde{g} = \phi^* g$ for some diffeomorphism $\phi : M \to M$.*

For the proof, and for other examples of rigidity and stability of Einstein metrics for quadratic curvature functionals, we refer the reader to [**GV11**].

We mention that quadratic functionals in dimension 3 were considered by Anderson in [**And97, And01**]. Also, critical points of the functional $\tilde{\mathcal{F}}_{\sigma_2}$ (defined above in (2.21)) were studied in dimension 3 in [**GV01**]. For rigidity results involving other types of functionals see [**MØ13**].

LECTURE 8

ALE metrics and orbifold limits

1. Ricci-flat ALE metrics

In order to understand limits of Einstein metrics, one one first have some understanding of a class of complete non-compact metrics, which are defined as follows.

DEFINITION 8.1. A complete Riemannian manifold (X^4, g) is called *asymptotically locally Euclidean* or *ALE* of order τ if there exists a finite subgroup $\Gamma \subset \mathrm{SO}(4)$ acting freely on S^3 and a diffeomorphism $\psi : X \setminus K \to (\mathbb{R}^4 \setminus B(0,R))/\Gamma$ where K is a compact subset of X, and such that under this identification,

$$(\psi_* g)_{ij} = \delta_{ij} + O(\rho^{-\tau}), \tag{8.1}$$

$$\partial^{|k|}(\psi_* g)_{ij} = O(\rho^{-\tau-k}), \tag{8.2}$$

for any partial derivative of order k, as $r \to \infty$, where ρ is the distance to some fixed basepoint.

Note that this definition is really just the same as Definition 2.7, with the addition of a group at infinity:

1.1. Eguchi-Hanson metric

We next recall the Eguchi-Hanson metric, which was the first example of a non-trivial Ricci-flat ALE space [**EH79**]. It is given by

$$g_{\mathrm{EH}} = \frac{dr^2}{1 - r^{-4}} + r^2\Big[\sigma_1^2 + \sigma_2^2 + (1 - r^{-4})\sigma_3^2\Big], \tag{8.3}$$

where r is a radial coordinate, and $\{\sigma_1, \sigma_2, \sigma_3\}$ is a left-invariant coframe on S^3 (viewed as the Lie group $\mathrm{SU}(2)$). This has an apparent singularity at $r = 1$, so redefine the radial coordinate to be $\hat{r}^2 = r^2 - 1$, and attach a $\mathbb{CP}^1$ at $\hat{r} = 0$. After taking a quotient by $\mathbb{Z}_2$, the metric then extends smoothly over the added $\mathbb{CP}^1$, is Ricci-flat, ALE at infinity of order 4, and is diffeomorphic to $\mathcal{O}(-2) \cong T^*S^2$. This space is hyperkähler, that is, there are three independent complex structures denoted by I, J, and K satisfying the quaternion relations

$$I^2 = J^2 = K^2 = IJK = -1. \tag{8.4}$$

Denote the corresponding Kähler forms by $\omega_I, \omega_J, \omega_K$, which are 3 linearly independent harmonic self-dual 2-forms.

Since this space is non-compact, let us look at $H^1_{E,-}(X,g)$, which we define to be decaying infinitesimal Einstein deformations, that is, those $h \in S^2(T^*X)$ satisfying all the conditions in (4.3), and also which satisfy $h = O(r^{-\epsilon})$ for some $\epsilon > 0$. Using the isomorphism $S^2_0(T^*M) = \Lambda^2_+ \otimes \Lambda^2_-$, it follows from the discussions in Lecture 6 that $H^1_{E,-}(X,g)$ has a basis

$$\{\omega_I \otimes \omega^-, \omega_J \otimes \omega^-, \omega_K \otimes \omega^-\}, \tag{8.5}$$

where ω^- is a non-trivial L^2 harmonic 2-form, see [**Pag78b**]. Consequently,

$$\dim(H^1_{E,-}(X,g)) = 3. \tag{8.6}$$

However, the Eguchi-Hanson metric is known to be rigid (up to scaling) as an Einstein ALE metric. This means that these infinitesimal deformations do not integrate to non-trivial Einstein deformations. It turns out that these elements can in fact be written as Lie derivatives, and can be understood as gluing parameters for a certain gluing problem which we will discuss in more detail below.

1.2. Hyperkähler ALE metrics

After the Eguchi-Hanson metric was found, Gibbons-Hawking wrote down a metric ansatz depending on the choice of n monopole points in $\mathbb{R}^3$, giving an anti-self-dual ALE hyperkähler metric with group $\mathbb{Z}/n\mathbb{Z}$ at infinity, which are called multi-Eguchi-Hanson metrics [**GH78, Hit79**]. In 1989, Kronheimer then classified all hyperkähler ALE spaces in dimension 4, [**Kro89a, Kro89b**]. To describe these, we consider the following subgroups of SU(2):

- Type $A_n, n \geq 1$: Γ the cyclic group $\mathbb{Z}_{n+1}$,

$$(z_1, z_2) \mapsto (e^{2\pi ip/(n+1)}z_1, e^{-2\pi ip/(n+1)}z_2), \quad 0 \leq p \leq n. \tag{8.7}$$

  acting on $\mathbb{R}^4$, which is identified with $\mathbb{C}^2$ via the map

$$(x_1, y_1, x_2, y_2) \mapsto (x_1 + iy_1, x_2 + iy_2) = (z_1, z_2). \tag{8.8}$$

  Writing a quaternion $q \in \mathbb{H}$ as $\alpha + j\beta$ for $\alpha, \beta \in \mathbb{C}$, we can also describe the action as generated by $e^{2\pi i/(n+1)}$, acting on the left.
- Type $D_n, n \geq 3$: Γ the binary dihedral group $\mathbb{D}^*_{n-2}$ of order $4(n-2)$. This is generated by $e^{\pi i/(n-2)}$ and j, both acting on the left.
- Type $E_6 : \Gamma = \mathbb{T}^*$, the binary tetrahedral group of order 24, double cover of $A(4)$.
- Type $E_7 : \Gamma = \mathbb{O}^*$, the binary octohedral group of order 48, double cover of $S(4)$.
- Type $E_8 : \Gamma = \mathbb{I}^*$, the binary icosahedral group of order 120, double cover of $A(5)$.

More specifically, Kronheimer showed that for any of these groups Γ, there do in fact exist simply-connected hyperkähler ALE spaces with group Γ at infinity, and moreover he completely classified these as hyperkähler quotients.

We will not go into details about Kronheimer's construction, but just briefly discuss the identification of the space of decaying infinitesimal Einstein deformations, given by the argument in [**Biq13a**, Proposition 1.1]. The operator Δ_L acting on transverse-traceless tensors can be identified with the operator $d_- d_-^*$ where

$$d_- : \Omega^1 \otimes \Omega^2_+ \to \Omega^2_- \otimes \Omega^2_+ \cong \Gamma(S^2_0(T^*X)) \tag{8.9}$$

is the exterior derivative. Since Ω^2_+ has a basis of parallel sections $\{\omega_I, \omega_J, \omega_K\}$, the proposition follows since the L^2-cohomology $H^2_{(2)}(X)$ is isomorphic to the usual cohomology $H^2(X)$ [**Car98**]. Consequently, the formal dimension of the moduli space of any such metric is $3b_2^-$. Note by the Weitzenböck formulas given in Lecture 6, these deformations are equivalent to decaying infinitesimal anti-self-dual deformations, see [**Via13b**]. Kronheimer has shown that the actual dimension of the moduli space of hyperkähler metrics on these spaces is $3b_2^- - 3$, which implies that there are infinitesimal deformations which are not integrable. As in the case of the Eguchi-Hanson metric, these can also be understood as gluing parameters. The properties of these spaces are summarized in Table 8.1.

TABLE 8.1. Invariants of hyperkähler ALE spaces.

Type	Γ	$\lvert\Gamma\rvert$	b_2^-	χ
$A_n, n \geq 1$	$\mathbb{Z}_{n+1}$	$n+1$	n	$n+1$
$D_m, m \geq 3$	$\mathbb{D}^*_{m-2}$	$4(m-2)$	m	$m+1$
E_6	$\mathbb{T}^*$	24	6	7
E_7	$\mathbb{O}^*$	48	7	8
E_8	$\mathbb{I}^*$	120	8	9

To close this brief discussion of hyperkähler metrics, we note here the following interesting conjecture due to Bando, Kasue, and Nakajima [**BKN89**]:

CONJECTURE 8.2. *If (M, g) is a simply-connected Ricci-flat ALE space in dimension four, then g is hyperkähler.*

The following result, which is joint work of the author and Michael Lock, shows that the conjecture is true, provided one restricts to the diffeomorphism types of minimal resolutions of $\mathbb{C}^2/\Gamma$, or any iterated blow-up thereof.

THEOREM 8.3 ([**LV15**]). *Let $\Gamma \subset \mathrm{U}(2)$ be a finite subgroup which acts freely on S^3, and let X be diffeomorphic to the minimal resolution of $\mathbb{C}^2/\Gamma$, or any iterated blow-up thereof. If g is a Ricci-flat ALE metric on X, then $\Gamma \subset \mathrm{SU}(2)$ and g is hyperkähler.*

REMARK 8.4. There are other interesting complete Einstein metrics with different asymptotics at infinity, known as "ALF", "ALG", and "ALH" gravitational instantons. We will not have time to discuss these, and refer the reader to [**BM11, Min10**] for a nice discussion of these types of metrics. There are also many interesting examples of gravitational instantons which have non-integral volume growth exponent [**Hei12**].

2. Non-collapsed limits of Einstein metrics

The results in Lecture 4 give a local description of the moduli space near a fixed Einstein metric. One would also like to understand global properties of the moduli space of Einstein metrics, for example, what are the possible limits of sequences of Einstein metrics?

DEFINITION 8.5. A *Riemannian orbifold* (M^4, g) is a topological space which is a smooth manifold of dimension 4 with a smooth Riemannian metric away from finitely many singular points. At a singular point p, M is locally diffeomorphic to a cone $\mathcal{C}$ on S^3/Γ, where $\Gamma \subset \mathrm{SO}(4)$ is a finite subgroup acting freely on S^3. Furthermore, at such a singular point, the metric is locally the quotient of a smooth Γ-invariant metric on B^4 under the orbifold group Γ.

In the non-collapsing case, the following is known, due to Anderson, Bando-Kasue-Nakajima, and Tian.

THEOREM 8.6 ([**And89**], [**BKN89**], [**Tia90**]). *Let (M_i, g_i) sequence of Einstein manifolds of dimension 4 satisfying*

$$\int_{M_i} |Rm_{g_i}|^2 dV_{g_i} < \Lambda, \ \mathrm{diam}(g_i) < D, \ Vol(g_i) > V > 0. \tag{8.10}$$

Then for a subsequence $\{j\} \subset \{i\}$,

$$(M_j, g_j) \xrightarrow{\text{Cheeger-Gromov}} (M_\infty, g_\infty), \tag{8.11}$$

where (M_∞, g_∞) is an orbifold with finitely many singular points.

The above convergence is in the Cheeger-Gromov sense which means that the metrics converge in the Gromov-Hausdorff sense to the limit space as a metric space, but away from the singular points, the convergence is moreover smooth (after pulling back by diffeomorphisms). Rescaling such a sequence to have bounded curvature near a singular point and taking a pointed limit yields Ricci-flat ALE spaces, also called "bubbles". This bubbling description can be refined; what we just described produces a "deepest bubble". Choosing different scalings can yield a tree of ALE Ricci-flat orbifolds at a singular point, see [**Nak88**] for a nice description of this process.

2.1. K3 example

Eguchi-Hanson metrics arise as bubbles for certain sequences of Calabi-Yau metrics on K3, and was suggested by [**KT87**]:

EXAMPLE 8.7. *There exists a sequence of Ricci-flat metrics g_i on $K3$ satisfying*

$$(K3, g_i) \longrightarrow (T^4/\{\pm 1\}, g_{\text{flat}}). \tag{8.12}$$

At each of the 16 singular points, an Eguchi-Hanson metric on T^*S^2 "bubbles off".

Note that since each Eguchi-Hanson metric has 3 infinitesimal Einstein deformations, and the torus has 10 flat deformations, modulo scaling the parameter count is 57, which is in nice agreement with the count made in Section 3 of Lecture 6. See [**Pag78b**] for a nice heuristic description of this. Also see [**LS94**] for a rigorous construction of a Calabi-Yau metric by anti-self-dual gluing methods, and the note of Donaldson [**Don10**] for a rigorous construction using Kähler-Einstein techniques.

REMARK 8.8. There is another very interesting limit of Calabi-Yau metrics on K3, known as the "large complex structure limit" [**GW00**]. A sequence of these metrics collapses to a limiting 2-sphere, so the limit is not described by Theorem 8.6. Away from 24 points, the sequence collapses with bounded curvature, and gives a nice illustration of the ϵ-regularity theorem of Cheeger-Tian [**CT06**].

2.2. Desingularization

We next ask the follow question:

- Can you reverse this process? That is, can you start with an Einstein orbifold, "glue on" hyperkähler bubbles at the singular points, and resolve to a smooth Einstein metric?

The "answer" is:

- In general, the answer is "no", since this gluing problem is obstructed; there are always decaying infinitesimal Einstein deformations of non-trivial Ricci-flat ALE spaces.

In the ASD case, the relevant operator maps between different bundles, and the index is not necessarily zero. However, the index is always zero in the Einstein case, so this makes the problem much more difficult. We remark that sometimes, Einstein metrics *can* be found by gluing techniques, but only when using some extra structure. As mentioned above, Calabi-Yau metrics on K3 can be produced using ASD gluing techniques [**LS94**], or Kähler-Einstein techniques [**Don10**]. For the G_2 and $Spin(7)$ cases, see [**Joy96, Joy99**].

2.3. Biquard's Theorem

We next discuss a beautiful result which says that the answer to the above question is "yes" in a certain case. The setting is a class of complete non-compact Einstein metrics with negative Einstein constant. If M^4 is a compact manifold with boundary ∂M, then a metric g on M^4 is said to be *conformally compact* if $\tilde{g} = \rho^2 g$ has some regularity (e.g., Hölder) up to the boundary, where ρ is a defining function for the boundary which satisfies $\rho^{-1}(0) = \partial M$ and $d\rho \neq 0$ on ∂M. If $|d\rho|_{\tilde{g}} = 1$, then g limits to a hyperbolic metric as $\rho \to 0$, such a metric is called *asymptotically hyperbolic.* If it is in addition Einstein (necessarily with negative Einstein constant), then (M, g) is called *asymptotically hyperbolic Einstein,* or AHE for short. We note that there is an induced conformal class on the boundary manifold at infinity. This definition should be thought of as a generalization of the hyperbolic ball, with the conformal class of the round sphere at infinity. Biquard's result is the following.

THEOREM 8.9 (Biquard [**Biq13a**]). *Let (M^4, g) be an AHE metric with a $\mathbb{Z}/2\mathbb{Z}$ orbifold singularity at $p \in M$. If (M^4, g) is rigid (i.e., g admits no nontrivial infinitesimal Einstein deformations), then the singularity can be resolved to a AHE metric by gluing on an Eguchi-Hanson metric if and only if*

$$\det(\mathcal{R}^+)(p) = 0, \tag{8.13}$$

where $\mathcal{R}^+$ is the upper-left 3×3 block in (5.14).

As we discussed above, the Eguchi-Hanson metric admits a 3-dimensional space of infinitesimal Einstein deformations, so this gluing problem is obstructed. One of these deformations corresponds to a scaling parameter, and the other two correspond to rotations in SO(4)/U(2). Biquard is able to overcome these obstructions using the freedom to perturb the boundary conformal class of the AH Einstein metric.

Recently, Biqard has given a generalization of Theorem 8.9 to allow orbifolds with more general ADE-type singularities, see [**Biq13b**] for the precise statement of this extension.

3. B^t-flat metrics

We next return to critical metrics of quadratic curvature functionals. We will be interested in the functional

$$\mathcal{B}_t[g] = \int_M |W_g|^2 \, dV_g + t \int_M R_g^2 \, dV_g.$$

REMARK 8.10. From the Chern-Gauss-Bonnet Theorem 5.4, this is really the most general quadratic functional in dimension 4, up to scaling. We have chosen to normalize this way to take advantage of the conformal invariance of $\mathcal{W}$.

The Euler-Lagrange equations of $\mathcal{B}_t$ are given by

$$B^t \equiv B + tC = 0, \tag{8.14}$$

where B is the *Bach tensor* defined by (5.6)

$$B_{ij} \equiv -4\Big(\nabla^k\nabla^l W_{ikjl} + \frac{1}{2}R^{kl}W_{ikjl}\Big), \tag{8.15}$$

and C is the tensor defined by (5.8)

$$C_{ij} = 2\nabla_i\nabla_j R - 2(\Delta R)g_{ij} - 2RR_{ij} + \frac{1}{2}R^2 g_{ij}. \tag{8.16}$$

From conformal invariance of the functional $\mathcal{W}$, it follows that the Bach-tensor is conformally invariant. We will refer to such a critical metric as a *B^t-flat metric.* Note that any Einstein metric is critical for $\mathcal{B}_t$, but there are in fact non-Einstein B^t-flat metrics, as we shall see.

For $t \neq 0$, by taking a trace of the E-L equations, it follows that

$$\Delta R = 0. \tag{8.17}$$

If M is compact, this implies $R = constant$. Consequently, the B^t-flat condition is equivalent to

$$B = 2tR \cdot E, \tag{8.18}$$

where E denotes the traceless Ricci tensor. That is, the Bach tensor is a constant multiple of the traceless Ricci tensor, which is indeed a natural generalization of the Einstein condition.

3.1. B^t-flat ALE metrics

Of course, all of the hyperkähler ALE spaces described above are also B^t-flat, but there are many more non-Einstein examples.

A large source of examples is the following. If (M, g) is Bach-flat and has positive scalar curvature, then we can convert (M, g) into an asymptotically flat (AF) metric

$$(N, g_N) = (M \setminus \{p\}, G^2 g)$$

using the Green's function for the conformal Laplacian G based at p. Since (M, g) is Bach-flat, then (N, g_N) is also Bach-flat (from conformal invariance) and scalar-flat (since we used the Green's function). Consequently, g_N is B^t-flat for all $t \in \mathbb{R}$. This gives a large family of examples of non-trivial asymptotically flat B^t-flat metrics, in contrast to the Ricci-flat case.

By taking the sum of Green's functions based at several points, one can also obtain many examples with several ends. In the special case of the sphere, with two points, one obtains the Euclidean Schwarzschild metric

$$g = \Big(1 + \frac{m}{r^2}\Big)^2 g_0, \tag{8.19}$$

where g_0 is the Euclidean metric. This metric plays a very important rôle in the Riemannian Penrose Inequality, see for example [**Bra01, BL09**].

Another family of non-trivial examples is the following. In [**LeB88**], LeBrun presented the first known examples of scalar-flat ALE spaces of negative mass, which gave counterexamples to extending the positive mass theorem to ALE spaces. We briefly describe these as follows. Define

$$g_{\mathrm{LB}} = \frac{dr^2}{1 + Ar^{-2} + Br^{-4}} + r^2\Big[\sigma_1^2 + \sigma_2^2 + (1 + Ar^{-2} + Br^{-4})\sigma_3^2\Big], \tag{8.20}$$

where r is a radial coordinate, $\{\sigma_1, \sigma_2, \sigma_3\}$ is a left-invariant coframe on S^3, and $A = n - 2$, $B = 1 - n$. There is an apparent singularity at $r = 1$, so redefine the radial coordinate to be $\hat{r}^2 = r^2 - 1$, and attach a $\mathbb{CP}^1$ at $\hat{r} = 0$. After taking a quotient by $\mathbb{Z}_n$, with action given by the diagonal action

$$(z_1, z_2) \mapsto e^{2\pi i p/n}(z_1, z_2), \quad 0 \le p \le n - 1, \tag{8.21}$$

the metric then extends smoothly over the added $\mathbb{CP}^1$, is ALE at infinity, and is diffeomorphic to $\mathcal{O}(-n)$. The mass (as defined in (2.15)) is computed to be $-2(n-2)$, which is negative when $n > 2$. These metrics are scalar-flat Kähler (so are anti-self-dual, and thus Bach-flat), and satisfy $b^2_- = 1, \tau = -1$, and $\chi = 2$.

In 2004, Calderbank and Singer produced many examples of toric ALE anti-self-dual metrics, which are moreover scalar-flat Kähler, and have cyclic groups at infinity contained in U(2) [**CS04**]. In 2014, the author, jointly with Michael Lock, produced new examples of ALE scalar-flat Kähler metrics with group at infinity *any* finite subgroup of U(2) acting freely on S^3 [**LV14**].

4. Non-collapsed limits of B^t-flat metrics

For $t \neq 0$, the B^t-flat equation can be rewritten as

$$\Delta Ric = Rm * Ric. \tag{8.22}$$

If $t = 0$, equation (8.22) is satisfied provided one assumes in addition that g has constant scalar curvature. With slightly different geometric assumptions, a similar orbifold-compactness theorem as in the Einstein case holds for sequences of metrics satisfying (8.22):

THEOREM 8.11 (Tian-Viaclovsky [**TV05a, TV05b, TV08**]). *Let (M_i, g_i) be a sequence of 4-dimensional manifolds satisfying (8.22) and*

$$\int_{M_i} |Rm_{g_i}|^2 dV_{g_i} < \Lambda, \; Vol(B(q,s)) > Vs^4 > 0, \; b_1(M_i) < B, \tag{8.23}$$

for all $s \le diam(M)/2$. Then for a subsequence $\{j\} \subset \{i\}$,

$$(M_j, g_j) \xrightarrow{\text{Cheeger-Gromov}} (M_\infty, g_\infty), \tag{8.24}$$

where (M_∞, g_∞) is a multi-fold satisfying (8.22), with finitely many singular points.

Similar to the Einstein case, rescaling such a sequence to have bounded curvature near a singular point yields ALE metrics satisfying (8.22). An important difference with the Einstein case is that the ALE spaces can have multiple ends (this is ruled out in the Einstein case by the Cheeger-Gromoll splitting theorem). Thus singular points are more general in that multiple orbifold cones can touch at a singular point, thus the terminology "multi-fold". Another difference is that a smooth point of the limit can in fact be a singular point of convergence, this cannot happen in the Einstein case (by Bishop's volume comparison theorem).

The key point in this theorem is the following related volume growth theorem:

THEOREM 8.12 (Tian-Viaclovsky [**TV08**]). *Let (M, g) be a metric satisfying* (8.22) *on a smooth, complete four-dimensional manifold M with*

$$\int_M |\mathrm{Rm}_g|^2 dV_g \leq \Lambda, \tag{8.25}$$

for some constant Λ.

Assume that

$$\mathrm{Vol}(B(q,s)) \geq V_0 s^4, \text{ for all } q \in M, \text{ and } s \leq diam(M)/2, \tag{8.26}$$

$$b_1(M) < B_1, \tag{8.27}$$

where V_0, B_1 are constants. Then there exists a constant V_1, depending only upon V_0, Λ, B_1, such that $\mathrm{Vol}(B(p,r)) \leq V_1 \cdot r^4$, *for all $p \in M$ and $r > 0$.*

Once one has this volume growth estimate, the proof of Theorem 8.11 is fairly similar to the Einstein case, see [**TV05b**]. The key part of the proof of Theorem 8.12 is the volume growth theorem in [**TV05a**], which depends on a Sobolev constant bound. Subsequently, using a point-picking argument, it was shown in [**TV08**] that the Sobolev constant bound can be replaced with a lower volume growth bound, which is the version we stated here. But we note that if a Sobolev constant bound is assumed, then the assumption on b_1 is not necessary [**TV08**].

4.1. Desingularization questions

It is natural to ask the same question that we asked in the Einstein case:

- Can you reverse this process? That is, start with an critical orbifold, "glue on" critical ALE metrics at the singular points, and resolve to a smooth critical metric?

The "answer" is still:

- In general, the answer is "no", because this gluing problem is obstructed; there are always decaying infinitesimal deformations of non-trivial B^t-flat ALE metrics.

However, in Lecture 10 we will discuss a recent theorem which says that the answer is "yes" in certain cases.

LECTURE 9

Regularity and volume growth

In this Lecture, we will discuss some of the main ideas involved in the proofs of Theorems 8.11 and 8.12. Also, we will give an outline of the proof of the existence of an Einstein metric on $\mathbb{CP}^2\#2\overline{\mathbb{CP}}^2$ due to Chen-LeBrun-Weber [**CLW08**].

1. Local regularity

We consider any system of the type

$$\Delta Ric = Rm * Ric. \tag{9.1}$$

Any Riemannian metric satisfies

$$\Delta Rm = L(\nabla^2 Ric) + Rm * Rm, \tag{9.2}$$

where $L(\nabla^2 Ric)$ denotes a linear expression in second derivatives of the Ricci tensor, and $Rm * Rm$ denotes a term which is quadratic in the curvature tensor (see [**Ham82**, Lemma 7.2]).

For a compact 4-manifold (M, g), we define the Sobolev constant C_S as the best constant C_S so that for all $f \in C^{0,1}(M)$ (Lipschitz) we have

$$\|f\|_{L^4} \leq C_S \|\nabla f\|_{L^2} + Vol^{-1/4}\|f\|_{L^2}. \tag{9.3}$$

If (X, g) is a complete, noncompact 4-manifold, the Sobolev constant C_S is defined as the best constant C_S so that for all $f \in C_c^{0,1}(X)$ (Lipschitz with compact support), we have

$$\|f\|_{L^4} \leq C_S \|\nabla f\|_{L^2}. \tag{9.4}$$

The following local regularity theorem is known as an "ϵ-regularity" theorem:

THEOREM 9.1 (Tian-Viaclovsky [**TV05a**]). *Assume that (9.1) is satisfied, let $r < diam(X)/2$, and $B(p, r)$ be a geodesic ball around the point p, and $k \geq 0$. Then there exist constants ϵ_0, C_k (depending upon C_S) so that if*

$$\|Rm\|_{L^2(B(p,r))} = \left\{ \int_{B(p,r)} |Rm|^2 dV_g \right\}^{1/2} \leq \epsilon_0,$$

then

$$\sup_{B(p,r/2)} |\nabla^k Rm| \leq \frac{C_k}{r^{2+k}} \left\{ \int_{B(p,r)} |Rm|^2 dV_g \right\}^{1/2} \leq \frac{C_k \epsilon_0}{r^{2+k}}.$$

In the case of harmonic curvature, $\delta Rm = 0$, one has an equation on the full curvature tensor

$$\Delta Rm = Rm * Rm. \tag{9.5}$$

In this case, the result follow by a Moser iteration procedure, for details we refer to [**Aku94, And89, Nak88, Tia90**]. This is a generalization of an ϵ-regularity theorem of Uhlenbeck [**Uhl82a, Uhl82b**]. We also note that this theorem was extended to extremal Kähler metrics by Chen-Weber [**CW11**]. Also, for Einstein metrics, dependence on the Sobolev constant was removed in [**CT06**].

Even though second derivatives of the Ricci tensor occur in (9.2), overall the principal symbol of the system (9.1) and (9.2) is in triangular form. The equations (9.1) and (9.2), when viewed as an elliptic system, together with the bound on the Sobolev constant, are the key to the proof of theorem 9.1, which is an involved iteration procedure, and we will refer the reader to [**TV05a**] for details.

In relation to Theorem 8.11, this ϵ-regularity result is the key to proving that there are only finitely many points at which the curvature can blow-up. This is because each such point must account for at least ϵ_0 of the L^2-norm of curvature, which is assumed to be bounded for the sequence. Consequently, at strictly positive distance away from the singular points, the curvature is bounded, and a subsequential limiting space with finitely many singular points can be obtained using fundamental ideas of Cheeger [**Che70**] and Gromov [**Gro07**]. The subsequence will converge to the limit in the Gromov-Hausdorff sense, but the convergence away from the singular points is much stronger in the following sense. Define $\Omega_{\delta,j} \subset M_j$ to be the set of points with distance to the singular set bounded below by $\delta > 0$. For j large, these subsets all be diffeomorphic, and after pulling-back by diffeomorphisms to a fixed manifold, the metrics converge in any $C^{k,\alpha}$-norm in coordinate charts as $j \to \infty$. To say more about the structure of the singularities, we need an upper volume growth estimate, which we discuss next.

2. Volume growth estimate

We emphasize that, in the Einstein case, an upper volume growth estimate on balls follows from Bishop's volume comparison theorem [**BC64**]. For metrics satisfying a system of the form (9.1), it is much more difficult to obtain an upper volume growth estimate since we are not assuming any pointwise Ricci curvature bound. The following is the key result:

THEOREM 9.2 (Tian-Viaclovsky [**TV05a**]). *Let (X, g) be a complete, noncompact, n-dimensional Riemannian manifold with base point p. Assume that there exists a constant $C_1 > 0$ so that*

$$Vol(B(q,s)) \geq C_1 s^n, \tag{9.6}$$

for any $q \in X$, and all $s \geq 0$. Assume furthermore that as $r \to \infty$,

$$\sup_{S(r)} |Rm_g| = o(r^{-2}), \tag{9.7}$$

where $S(r)$ denotes the sphere of radius r centered at p. If $b_1(X) < \infty$, then (X, g) has finitely many ends, and there exists a constant C_2 (depending on g) so that

$$Vol(B(p,r)) \leq C_2 r^n. \tag{9.8}$$

Furthermore, each end is ALE of order 0.

OUTLINE OF PROOF. The entire proof of this theorem is over 20 pages long; we only give an extremely rough outline containing the main ideas. For $s > 1$, consider a sequence of dyadic annuli $A(s^i, s^{i+1})$. We can assume that there is a subsequence $\{j\} \subset \{i\}$ so that

$$\mathcal{H}^{n-1}(S(s^{j+1})) \geq (1-\eta_j)\mathcal{H}^{n-1}(S(s^j))s^{n-1} \tag{9.9}$$

for some sequence $\eta_j \to 0$ as $j \to \infty$. Otherwise, this would contradict the lower volume growth assumption. Letting $A_j = A(s^j, s^{j+1})$, we show that as $j \to \infty$,

$$\frac{1}{Vol(A_j)} \int_{A_j} |\Delta r^2 - 2n| dV_g \to 0. \tag{9.10}$$

The proof of this is a long computation and uses the coarea formula, we will omit the details.

If we rescale the annuli to unit size, that is, let

$$(\tilde{A}_j, \tilde{g}_j) = (A_j, s^{-2j}g), \tag{9.11}$$

the curvature decay estimate (9.7) implies that

$$|Rm(\tilde{g}_j)| \to 0 \tag{9.12}$$

as $j \to \infty$, so the metric $\tilde{g}_j$ is limiting to a flat metric.

Note that if $Vol(\tilde{A}_j) < C$ for some constant C, then (9.10) would imply that $\Delta\tilde{r}^2 \to 2n$ as $j \to \infty$, which implies that the rescaled distance function is limiting to the Euclidean distance function, so the sequence of rescaled annuli would converge to a Euclidean annulus. However, we do not yet know that the volumes of the rescaled annuli are bounded. To deal with this, we use a contradiction argument. If the rescaled volumes are not bounded, then we show it is possible to divide the annuli into finitely many regions with large but bounded volume, and prove that there is always at least one "nice" connected, non-collapsed region. This chopping procedure is one of the most delicate parts of the proof. We then apply the above rescaling argument to the sequence of "nice" regions, and show that these regions converge to portions of Euclidean annuli. Since Euclidean annuli "close up", it follows that the entire annular regions are in fact converging to Euclidean annuli. This contradiction proves that the rescaled annuli have bounded volume. Since this can be done for any $s > 1$, the upper volume estimate (9.8) follows. It then follows that all tangent cones at infinity are Euclidean cones, from which we conclude that the metric is ALE of order 0.

A important technical point arises with the above argument. In general, annuli might have many connected boundary components, and a sequence of connected components of annuli with the inner boundaries having more than 1 connected component would cause a problem. However, the assumption on the first Betti number ensures that this situation cannot arise. □

Note that Theorem 9.2 is a result for noncompact spaces, but this result does in fact imply the volume growth result stated in Theorem 8.12. This is done by a contradiction argument, see [**TV05b, TV08**] for the details.

3. ALE order and removable singularity theorems

The upper volume growth estimate implies that all tangent cones of the limit space are ALE of order 0 which implies that the limit space has C^0-orbifold singularities. That is, after passing to a local cover as in Definition 8.5, the metric only has an extension to a C^0-metric in a neighborhood of the origin. Another important ingredient in the proof of Theorem 8.11 is therefore to prove that the singularities of the limit are *smooth* orbifold points, that is, after passing to a local cover, the metric can be extended to a C^∞-metric over the origin. A closely related problem is to obtain the optimal ALE order of the spaces which bubble out, which we will discuss next.

A crucial result in the Ricci-flat case was obtained by Cheeger-Tian: if (M^n, g) is Ricci-flat ALE of order 0, there exists a change of coordinates at infinity so that (M^n, g) is ALE of order n, where n is the dimension [**CT94**]. This generalized and extended the work of Bando-Kasue-Nakajima [**BKN89**], who employed improved Kato inequalities together with a Moser iteration argument. The Cheeger-Tian method has the advantage of finding the *optimal* order of curvature decay, without relying on Kato inequalities.

In the case of anti-self-dual scalar-flat metrics, or scalar-flat metrics with harmonic curvature, it was proved in [**TV05a**] that such spaces are ALE of order τ for any $\tau < 2$, using the technique of Kato inequalities. Subsequently, this was generalized to Bach-flat metrics and metrics with harmonic curvature in dimension 4 in [**Str10**], using the Cheeger-Tian technique. This technique was generalized to obstruction-flat metrics in any dimension in [**AV12a**], which is a generalization of the Bach-flat condition in dimension four, see [**Gra09**].

The method in [**AV12a**] applies to much more general systems than just the obstruction tensors, and works in any dimension $n \geq 3$. Given two tensor fields A, B, the notation $A * B$ will mean a linear combination of contractions of $A \otimes B$ yielding a symmetric 2-tensor. The main result is:

THEOREM 9.3 (Ache-Viaclovsky [**AV12a**]). *Let $k = 1$ if $n = 3$, or $1 \leq k \leq \frac{n}{2} - 1$ if $n \geq 4$. Assume that (M, g) is scalar-flat, ALE of order 0, and satisfies*

$$\Delta_g^k Ric = \sum_{j=2}^{k+1} \sum_{\alpha_1 + \ldots + \alpha_j = 2(k+1) - 2j} \nabla_g^{\alpha_1} Rm * \ldots * \nabla_g^{\alpha_j} Rm. \tag{9.13}$$

Then (M, g) is ALE of order $n - 2k$.

For $k = 1$, this is simply

$$\Delta Ric = Rm * Rm. \tag{9.14}$$

We emphasize that this is more general than (9.1), since the right hand side is allowed to be quadratic in the full curvature tensor. This is satisfied in particular by scalar-flat Kähler metrics and metrics with harmonic curvature in any dimension, and also anti-self-dual metrics in dimension 4. These special cases were previously considered in [**Che09**] using improved Kato inequalities and a Moser iteration technique. The optimal decay for scalar-flat anti-self-dual ALE metrics was previously considered in [**CLW08**, Proposition 13]. The case of extremal Kähler ALE metrics was considered in [**CW11**]. As mentioned above, the cases of Bach-flat metrics and metrics with harmonic curvature in dimension 4 were considered in [**Str10**].

The main idea of the proof of Theorem 9.3 is based on the method of Cheeger-Tian from [**CT94**], and is roughly to show that the optimal ALE decay rate is determined by the rates of decaying solutions of the gauged, linearized equation on a Euclidean cone. This step uses a fundamental technique of Leon Simon called the *Three Annulus Lemma* [**Sim85**]. An analysis of the decay rates of solutions of the gauged linearized equation, together with an estimate on the nonlinear terms in the equation, then yields Theorem 9.3.

The same technique also yields a removable singularity theorem for higher-order systems:

THEOREM 9.4 (Ache-Viaclovsky [**AV12a**]). *Let $k=1$ if $n=3$, or $1 \leq k \leq \frac{n}{2}-1$ if $n \geq 4$. Assume that $(B_\rho(0) \setminus \{0\}, g)$ has constant scalar curvature and satisfies*

$$\Delta_g^k Ric = \sum_{j=2}^{k+1} \sum_{\alpha_1+\ldots+\alpha_j=2(k+1)-2j} \nabla_g^{\alpha_1} Rm * \ldots * \nabla_g^{\alpha_j} Rm. \tag{9.15}$$

If the origin is a C^0-orbifold point for g, then the metric extends to a smooth orbifold metric in $B_\rho(0)$.

In particular, this says that the limit space in Theorem 8.11 is a smooth multi-fold.

4. Chen-LeBrun-Weber metric

The volume growth theorem was used in a fundamental way in [**CLW08**] in order to obtain an Einstein metric on $\mathbb{CP}^2\#2\overline{\mathbb{CP}}^2$, in this section we will give a brief overview of the proof.

THEOREM 9.5 (Chen-LeBrun-Weber [**CLW08**]). *There exists a positive Einstein metric on $M = \mathbb{CP}^2\#2\overline{\mathbb{CP}}^2$.*

OUTLINE OF PROOF. The first step is to consider Kähler classes which are invariant under a torus action, and bilaterally symmetric. That is, only Kähler classes from which the (-1) curves in the blow-up have the same area are considered. The space of such Kähler classes is 1-dimensional, and one can parametrize these classes by the area of the (-1) curves; call this parameter x. Chen-Lebrun-Weber then consider the functional

$$\mathcal{A}([\omega]) = \frac{(c_1 \cdot [\omega])^2}{[\omega]^2} - \frac{1}{32\pi^2}\mathcal{F}(\xi, [\omega]), \tag{9.16}$$

where $\mathcal{F}(\xi, [\omega])$ is the Futaki invariant, with ξ the extremal vector field of the class $[\omega]$, see [**Fut83, Fut88, FM93**]. It is next observed that the graph of $\mathcal{A}$ as a function of x has a strict local minimum at a certain positive value of x, call this value x_0. If one can prove that an extremal Kähler metric g_{x_0} exists in this class corresponding to x_0, then $\mathcal{S}$ would have a critical point at g_{x_0} when restricted to the set of toric, bilaterally symmetric Kähler classes on M. From (6.26), the functional $\mathcal{W}$ would also have such a critical point.

To proceed further, we need to understand the structure of the Bach tensor for a Kähler metric. The Bach tensor is a symmetric tensor, and since we have a complex structure, we can consider the tensors B^+ and B^-, the J-invariant and J anti-invariant parts of B, respectively. A computation shows that

$$B^+ = -\frac{2}{3}\left(RE + 2(\nabla^2 R)_0^+\right), \tag{9.17}$$

$$B^- = \frac{2}{3}(\nabla^2 R)_0^-, \tag{9.18}$$

where $(\nabla^2 R)_0^+$ and $(\nabla^2 R)_0^-$ denote the J-invariant and J anti-invariant parts of the traceless Hessian of the scalar curvature, respectively, see [**Der83, ACG03**]. This implies the following:

PROPOSITION 9.6 (Derdzinski [**Der83**]). *If (M, g, J) is Kähler and Bach-flat, then (M, g, J) is extremal and the metric $\tilde{g} = R_g^{-2} g$ is Einstein near any point with $R_g \neq 0$.*

PROOF. This follows from equations (9.17) and (9.18) by noting that the condition $(\nabla^2 R)_0^- = 0$ is exactly the condition for g to be extremal Kähler [**Cal82**], and the second claim follows from the conformal transformation formula (2.7). □

If the extremal metric g_{x_0} exists, since this metric is critical for $\mathcal{W}$ restricted to the space of invariant Kähler classes, (9.17) and (9.18) show that the Bach tensor can be viewed is a harmonic $(1,1)$-form and can therefore be used as a Kähler variation. Consequently, g_{x_0} would be Bach-flat. It turns out that this metric must have strictly positive scalar curvature, so Proposition 9.6 yields the desired conformally Einstein metric.

The main part of the proof is therefore to show that the extremal Kähler metric g_{x_0} exists. To show this, a continuity argument is used. For x small, extremal Kähler metrics are known to exist in these Kähler classes by work of Arezzo-Pacard-Singer [**APS11**]; this was an extension of the gluing result of Arezzo-Pacard for constant scalar curvature Kähler metrics [**AP06, AP09**]. The set of Kähler classes admitting extremal Kähler metrics is known to be open by LeBrun-Simanca [**LS93**]. A compactness theorem is used to show that the set of x for which there exists an extremal Kähler metric is also closed for $x \leq x_0$. Consequently, from connectedness of the interval $(0, x_0]$, an extremal Kähler metric exists at x_0.

To show the compactness, an extension of Theorem 8.11 to extremal Kähler metrics is used [**CW11**]. We note that the main part of [**CW11**] is to extend the ϵ-regularity result in Theorem 9.1 to the class of extremal Kähler metrics; the volume growth result in Theorem 9.2 is still crucial in order to obtain the compactness theorem. Given a sequence of extremal Kähler metrics g_{x_i} for $x_i \to x \leq x_0$ as $i \to \infty$, an orbifold limit can be obtained provided that the Sobolev constant can be controlled, which is proved in [**CLW08**, Section 5]. If the curvatures of this sequence were not bounded, then a nontrivial Kähler scalar-flat ALE space must bubble off at some point. However, a detailed analysis of possible bubbles, employing the toric and bilateral symmetries, shows that non-trivial bubbles can be ruled out, thereby proving compactness. □

LECTURE 10

A gluing theorem for B^t-flat metrics

1. Existence of critical metrics

Let us begin by stating the main result:

THEOREM 10.1 (Gursky-Viaclovsky [**GV13**]). *A B^t-flat metric exists on the manifolds in the table for some t near the indicated value of t_0.*

TABLE 10.1. Simply-connected examples with one bubble

Topology of connected sum	Value(s) of t_0
$\mathbb{CP}^2\#\overline{\mathbb{CP}}^2$	$-1/3$
$S^2\times S^2\#\overline{\mathbb{CP}}^2 = \mathbb{CP}^2\#2\overline{\mathbb{CP}}^2$	$-1/3$, $-(9m_1)^{-1}$
$S^2\times S^2\#S^2\times S^2$	$-2(9m_1)^{-1}$

The constant m_1 is the mass of the Green's function metric of the product metric $S^2\times S^2$, defined in (2.15).

We make some remarks:

- $M = \mathbb{CP}^2\#\overline{\mathbb{CP}}^2$ admits an $U(2)$-invariant Einstein metric called the "Page metric" [**Pag78a**]. M does not admit any Kähler-Einstein metric, but the Page metric is conformal to an extremal Kähler metric.
- $M = \mathbb{CP}^2\#2\overline{\mathbb{CP}}^2$ admits a toric invariant Einstein metric called "Chen-LeBrun-Weber metric" described in the previous lecture [**CLW08**]. Again, M does not admit any Kähler-Einstein metric, but the Chen-LeBrun-Weber metric is conformal to an extremal Kähler metric.
- $M = S^2\times S^2\#S^2\times S^2$ does not admit any Kähler metric, it does not even admit an almost complex structure. Our metric is the first known example of a "canonical" metric on this manifold.

1.1. The approximate metric

The critical metrics in Theorem 10.1 are found by perturbing from an "approximate" critical metric. We describe this construction next.

- Let (Z, g_Z) and (Y, g_Y) be Einstein manifolds, and assume that g_Y has positive scalar curvature.
- Choose basepoints $z_0\in Z$ and $y_0\in Y$.
- Convert (Y, g_Y) into an asymptotically flat (AF) metric

$$(N, g_N) = (Y\setminus\{y_0\}, G^2 g_Y)$$

using the Green's function for the conformal Laplacian based at y_0. Since (M, g) is Bach-flat, then (N, g_N) is also Bach-flat (from conformal invariance) and scalar-flat (since we used the Green's function). Consequently, g_N is B^t-flat for all $t \in \mathbb{R}$.

- Let $a > 0$ be small, and consider $Z \setminus B(z_0, a)$. Scale the compact metric to $(Z, \tilde{g} = a^{-4} g_Z)$. Attach this metric to the metric $(N \setminus B(a^{-1}), g_N)$ using cutoff functions near the boundary, to obtain a smooth metric on the connected sum $Z \# \overline{Y}$.

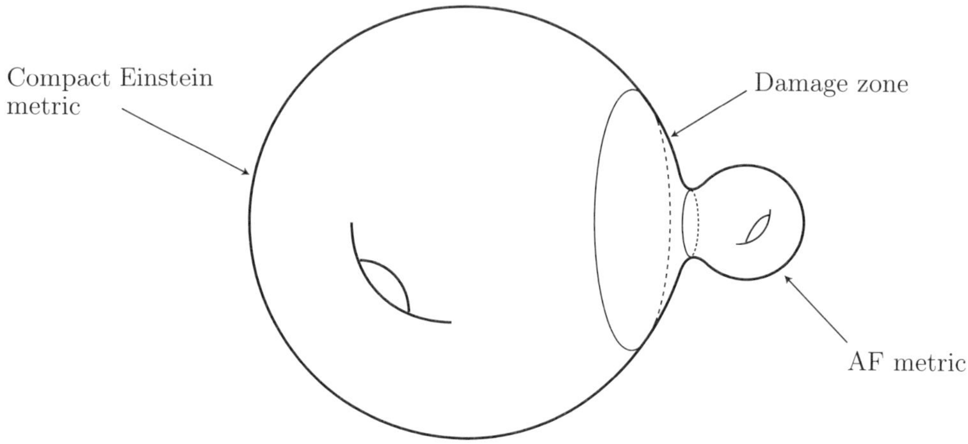

FIGURE 10.1. The approximate metric.

Since both g_Z and g_N are B^t-flat, this metric is an "approximate" B^t-flat metric, with vanishing B^t tensor away from the "damage zone", where cutoff functions were used.

2. Lyapunov-Schmidt reduction

In general, there are several degrees of freedom in this approximate metric.

- The scaling parameter a (1-dimensional).
- Rotational freedom when attaching (6-dimensional).
- Freedom to move the base points of either factor (8-dimensional).

There are a total of 15 gluing parameters, which yield a 15-dimensional space of "approximate" kernel of the linearized operator. Using a Lyapunov-Schmidt reduction argument, one can reduce the problem to that of finding a zero of the *Kuranishi map*

$$\Psi : U \subset \mathbb{R}^{15} \to \mathbb{R}^{15}. \tag{10.1}$$

- It is crucial to use certain weighted norms to find a bounded right inverse for the linearized operator.
- This 15-dimensional problem is too difficult in general. We will take advantage of various symmetries in order to reduce to only 1 free parameter: the scaling parameter a.

2.1. Technical theorem

The leading term of the Kuranishi map corresponding to the scaling parameter is given by:

THEOREM 10.2 ([**GV13**]). *As $a \to 0$, then for any $\epsilon > 0$,*

$$\Psi_1 = \Big(\frac{2}{3} W(z_0) \circledast W(y_0) + 4tR(z_0)\mathrm{mass}(g_N)\Big)\omega_3 a^4 + O(a^{6-\epsilon}), \tag{10.2}$$

where $\omega_3 = Vol(S^3)$, and the product of the Weyl tensors is given by

$$W(z_0) \circledast W(y_0) = \sum_{ijkl} W_{ijkl}(z_0)(W_{ijkl}(y_0) + W_{ilkj}(y_0)), \tag{10.3}$$

where $W_{ijkl}(\cdot)$ denotes the components of the Weyl tensor in a normal coordinate system at the corresponding point.

We note that the product $\circledast$ depends upon the coordinate systems chosen, and therefore in general depends upon a rotation parameter, and obviously on the base points of the gluing.

3. The building blocks

We next discuss the "building blocks" of the gluing construction.

3.1. The Fubini-Study metric

We let $(\mathbb{CP}^2, g_{\mathrm{FS}})$ denote the Fubini-Study metric, scaled so that $Ric = 6g$. We consider the following group actions.

Torus action:

$$[z_0, z_1, z_2] \mapsto [z_0, e^{i\theta_1} z_1, e^{i\theta_2} z_2]. \tag{10.4}$$

Flip symmetry:

$$[z_0, z_1, z_2] \mapsto [z_0, z_2, z_1]. \tag{10.5}$$

The Green's function metric of the Fubini-Study metric $\hat{g}_{FS}$ is also known as the Burns metric, and is completely explicit (in fact, it is the case $n = 1$ of (8.20)) with mass given by

$$\mathrm{mass}(\hat{g}_{FS}) = 2. \tag{10.6}$$

3.2. The product metric

The next building block is $(S^2 \times S^2, g_{S^2 \times S^2})$, the product of 2-dimensional spheres of Gaussian curvature 1, with $Ric = g$. We consider the following group actions.

Torus action:

(10.7) Product of rotations fixing north and south poles.

Flip symmetry:

$$(p_1, p_2) \mapsto (p_2, p_1). \tag{10.8}$$

The Green's function metric $\hat{g}_{S^2 \times S^2}$ of the product metric does not seem to have a known explicit description. We will denote

$$m_1 = \mathrm{mass}(\hat{g}_{S^2 \times S^2}). \tag{10.9}$$

By the positive mass theorem of Schoen-Yau, $m_1 > 0$. Note that since $S^2 \times S^2$ is spin, this also follows from Witten's proof of the positive mass theorem. The value of m_1 has recently been determined:

THEOREM 10.3 ([**Via13c**]). *The mass* $m_1 \sim .5872$ *and may be written as an explicit infinite sum.*

The explicit formula for the mass is lengthy, and will not be written here. We just note that this implies that $(-9m_1)^{-1} \sim= -.1892$ so as a corollary we see that the manifold $S^2 \times S^2 \# \overline{\mathbb{CP}}^2$ admits a B^t-flat metrics for at least two different values of t. Note that it is shown in [**GV13**] that the metrics obtained in Theorem 10.1 are not Einstein, so these metrics are distinct from the Chen-LeBrun-Weber metric.

4. Remarks on the proof

We first remark that:

- We impose the toric symmetry and "flip" symmetry in order to reduce the number of free parameters to 1 (only the scaling parameter). That is, we perform an equivariant gluing.
- The special value of t_0 is computed by
$$\frac{2}{3}W(z_0) \circledast W(y_0) + 4t_0 R(z_0)\text{mass}(g_N) = 0.$$
This choice of t_0 makes the leading term of the Kuranishi map vanish, and is furthermore a *nondegenerate* zero (if $R(z_0) > 0$; $\text{mass}(g_N) > 0$ by the positive mass theorem).

We next outline the spaces used in the construction of the naïve approximate metric:

- (i) $\mathbb{CP}^2 \# \overline{\mathbb{CP}}^2$; the Fubini-Study metric with a Burns metric attached at one fixed point. This case admits a $U(2)$-action.
- (ii) $S^2 \times S^2 \# \overline{\mathbb{CP}}^2 = \mathbb{CP}^2 \# 2\overline{\mathbb{CP}}^2$; the product metric on $S^2 \times S^2$ with a Burns metric attached at one fixed point. Alternatively, we can view this as the Fubini-Study metric on $\mathbb{CP}^2$, with a Green's function $S^2 \times S^2$ metric attached at one fixed point. For this topology, we will therefore construct two different critical metrics. Both of these will have toric symmetry plus invariance under the flip symmetry.
- (iii) $S^2 \times S^2 \# S^2 \times S^2$; the product metric on $S^2 \times S^2$ with a Green's function $S^2 \times S^2$ metric attached at one fixed point. This metric is toric and flip-symmetric.

We note that an equivariant gluing is carried out – the metrics obtained in Theorem 10.1 retain the indicated symmetries.

By imposing other discrete symmetries, we can perform the gluing operation with more than one bubble. For example, we can find critical metrics on $\mathbb{CP}^2 \# 3\overline{\mathbb{CP}}^2$ $3\# S^2 \times S^2$, $\mathbb{CP}^2 \# 3(S^2 \times S^2)$, $S^2 \times S^2 \# 4\overline{\mathbb{CP}}^2$, and $5\# S^2 \times S^2$ (see [**GV13**, Table 1.2]).

The product metric on $S^2 \times S^2$ admits the quotient $S^2 \times S^2/\mathbb{Z}_2$ where $\mathbb{Z}_2$ acts by the antipodal map on both factors. It is well-known that this quotient is diffeomorphic to $G(2,4)$, the Grassmannian of 2-planes in $\mathbb{R}^4$, see for example [**ST69**]. Another quotient is $\mathbb{RP}^2 \times \mathbb{RP}^2$. The product metric descends to an

Einstein metric on both of these quotients. We can also use these quotient spaces as building blocks to obtain non-simply connected examples. We do not list all of the examples here, but just note that we find a critical metric on $G(2,4)\#G(2,4)$, which has infinite fundamental group, and therefore does not admit any positive Einstein metric by Myers' Theorem. For the complete list of non-simply-connected examples, see [**GV13**, Table 1.3, Table B.1].

4.1. Ellipticity and gauging

The B^t-flat equations are not elliptic due to diffeomorphism invariance. A gauging procedure analogous to the Bianchi gauge is used. This was already discussed above in Lecture 7, with the following note. From (5.4), we can write

$$\mathcal{F}_\tau = 16\pi^2\chi(M) + \frac{1}{2}\mathcal{B}_{2(\tau+\frac{1}{3})}. \tag{10.10}$$

Taking gradients, we obtain the relation

$$\nabla\mathcal{B}_t = 2\nabla\mathcal{F}_{\frac{t}{2}-\frac{1}{3}}. \tag{10.11}$$

It follows from the formula for P that the linearized operator is given by

$$S^t h = (B' + tC')h + \mathcal{K}_g\delta_g\mathcal{K}_g\delta_g\overset{\circ}{h}, \tag{10.12}$$

where B' and C' are the linearizations of B and C respectively. Therefore the discussion in Lecture 7 applies, with τ replaced by $\frac{t}{2}-\frac{1}{3}$.

4.2. Rigidity

The rigidity results we need were discussed above. We mention here the resulting linearized operator as we change $\tau \to t$. For h transverse-traceless (TT), the linearized operator at an Einstein metric is given by

$$S^t h = \Big(\Delta_L + \frac{1}{2}R\Big)\Big(\Delta_L + \Big(\frac{1}{3}+t\Big)R\Big)h. \tag{10.13}$$

Next, for $h = fg$, we have

$$tr_g(S^t h) = 6t(3\Delta + R)(\Delta f). \tag{10.14}$$

The above rigidity results, Theorems 7.11 and 7.12, are then as follows (stated in terms of t instead of τ).

THEOREM 10.4 ([**GV11**]). *On $(\mathbb{CP}^2, g_{\mathrm{FS}})$, $H^1_t = \{0\}$ provided that $t < 1$.*

For the case of the product metric:

THEOREM 10.5 ([**GV11**]). *On $(S^2\times S^2, g_{S^2\times S^2})$, $H^1_t = \{0\}$ provided that $t < 2/3$ and $t \neq -1/3$. If $t = -1/3$, then H^1_t is one-dimensional and spanned by the element $g_1 - g_2$.*

In relation to Theorem 10.1, the positive mass theorem says that $t_0 < 0$, so luckily we are in the rigidity range of the factors. Consequently, there is no cokernel arising from deformations of the building blocks.

4.3. Refined approximate metric

The approximate metric described above is not good enough. It must be improved by matching up leading terms of the metrics by solving certain auxiliary linear equations, so that the cutoff function disappears from the leading term. This step is inspired by the work of Biquard mentioned above. Let (Z, g_Z) be the compact metric. In Riemannian normal coordinates,

$$(g_Z)_{ij}(z) = \delta_{ij} - \frac{1}{3}R_{ikjl}(z_0)z^k z^l + O^{(4)}(|z|^4)_{ij} \tag{10.15}$$

as $z \to z_0$.

Let (N, g_N) be the Green's function metric of (Y, g_Y), then we have

$$(g_N)_{ij}(x) = \delta_{ij} - \frac{1}{3}R_{ikjl}(y_0)\frac{x^k x^l}{|x|^4} + 2A\frac{1}{|x|^2}\delta_{ij} + O^{(4)}(|x|^{-4+\epsilon}) \tag{10.16}$$

as $|x| \to \infty$, for any $\epsilon > 0$. Note that the constant A is given by

$$\text{mass}(g_N) = 12A - R(y_0)/12. \tag{10.17}$$

We consider $a^{-4}g_Z$ and let $z = a^2 x$, then we have

$$a^{-4}(g_Z)_{ij}(x) = \delta_{ij} - a^4\frac{1}{3}R_{ikjl}(z_0)x^k x^l + \cdots . \tag{10.18}$$

Note that the second order terms do not agree. One needs to construct new metrics on the factors so that these terms agree. This is done by solving the linearized equation on each factor with prescribed leading term the second order term of the *other* metric. We will describe this procedure next.

4.4. The obstruction

On (N, g_N), one would like to solve

$$\begin{aligned} S^t\tilde{h} &= 0 \\ \tilde{h} &= -a^4\frac{1}{3}R_{ikjl}(z_0)x^k x^l + O(|x|^\epsilon), \end{aligned} \tag{10.19}$$

as $x \to \infty$. However, it turns out this equation is obstructed, so there is not necessarily a solution. However, using some Fredholm theory in weighted spaces, it turns out that one *can* solve the modified equation

$$S^t\tilde{h} = \lambda \cdot k_1,$$

where k_1 pairs nontrivially with the decaying cokernel o_1 on the AF space (N, g_N). (There is also a considerable amount of work involved in order to prove that the space of decaying cokernel is 1-dimensional; the symmetries are crucial for this.)

- A similar procedure is carried out on the compact factor, except this is unobstructed (since the compact factor is rigid), so this does not contribute to the leading term of the Kuranishi map.

The leading term is the computed by the following. Pairing with the cokernel element o_1,

$$\begin{aligned} \lambda &= \lim_{r\to\infty}\int_{B(r)}\langle S\tilde{h}, o_1\rangle \\ &= \lim_{r\to\infty}(\text{ spherical boundary integrals}). \end{aligned}$$

This limit can be computed explicitly using the expansion of the cokernel element

$$(o_1)_{ij} = \frac{2}{3} W_{ikjl}(y_0) \frac{x^k x^l}{|x|^4} + f g_{ij} + O(|x|^{-4+\epsilon}) \tag{10.20}$$

as $x \to \infty$, where f satisfies

$$\Delta f = -\frac{1}{3} \langle Ric, o_1 \rangle, \tag{10.21}$$

together with the expansion

$$\tilde{h} = -a^4 \frac{1}{3} R_{ikjl}(z_0) x^k x^l + O(|x|^\epsilon), \tag{10.22}$$

as $x \to \infty$.

The complete computation is very lengthy, and we refer to [**GV13**] for the details.

4.5. Final remarks

The proof shows that there is a dichotomy. Either

- (i) there is a critical metric at exactly the critical t_0, in which case there would necessarily be a 1-dimensional moduli space of solutions for this fixed t_0, or
- (ii) for each value of the gluing parameter a sufficiently small, there will be a critical metric for a corresponding value of $t_0 = t_0(a)$. The dependence of t_0 on a will depend on the next term in the expansion of the Kuranishi map.

Bibliography

[AB03] Kazuo Akutagawa and Boris Botvinnik, *Yamabe metrics on cylindrical manifolds*, Geom. Funct. Anal. **13** (2003), no. 2, 259–333. MR1982146 (2004e:53051)

[AB04] Kazuo Akutagawa and Boris Botvinnik, *The Yamabe invariants of orbifolds and cylindrical manifolds, and L^2-harmonic spinors*, J. Reine Angew. Math. **574** (2004), 121–146. MR2099112 (2005g:53055)

[ACG03] V. Apostolov, D. M. J. Calderbank, and P. Gauduchon, *The geometry of weakly self-dual Kähler surfaces*, Compositio Math. **135** (2003), no. 3, 279–322. MR1956815 (2004f:53045)

[ADH13] Bernd Ammann, Mattias Dahl, and Emmanuel Humbert, *Smooth Yamabe invariant and surgery*, J. Differential Geom. **94** (2013), no. 1, 1–58. MR3031859

[AHS78] M. F. Atiyah, N. J. Hitchin, and I. M. Singer, *Self-duality in four-dimensional Riemannian geometry*, Proc. Roy. Soc. London Ser. A **362** (1978), no. 1711, 425–461. MR506229 (80d:53023)

[AIL07] Kazuo Akutagawa, Masashi Ishida, and Claude LeBrun, *Perelman's invariant, Ricci flow, and the Yamabe invariants of smooth manifolds*, Arch. Math. (Basel) **88** (2007), no. 1, 71–76. MR2289603 (2007j:53070)

[Aku94] Kazuo Akutagawa, *Yamabe metrics of positive scalar curvature and conformally flat manifolds*, Differential Geom. Appl. **4** (1994), no. 3, 239–258. MR95i:58046

[Aku12] Kazuo Akutagawa, *Computations of the orbifold Yamabe invariant*, Math. Z. **271** (2012), no. 3-4, 611–625. MR2945574

[And89] Michael T. Anderson, *Ricci curvature bounds and Einstein metrics on compact manifolds*, J. Amer. Math. Soc. **2** (1989), no. 3, 455–490. MR90g:53052

[And97] Michael T. Anderson, *Extrema of curvature functionals on the space of metrics on 3-manifolds*, Calc. Var. Partial Differential Equations **5** (1997), no. 3, 199–269. MR1438146 (98a:58041)

[And01] Michael T. Anderson, *Extrema of curvature functionals on the space of metrics on 3-manifolds. II*, Calc. Var. Partial Differential Equations **12** (2001), no. 1, 1–58. MR1808106 (2002a:58012)

[And10] Michael T. Anderson, *A survey of Einstein metrics on 4-manifolds*, Handbook of geometric analysis, No. 3, Adv. Lect. Math. (ALM), vol. 14, Int. Press, Somerville, MA, 2010, pp. 1–39. MR2743446 (2012a:53071)

[AP06] Claudio Arezzo and Frank Pacard, *Blowing up and desingularizing constant scalar curvature Kähler manifolds*, Acta Math. **196** (2006), no. 2, 179–228. MR2275832 (2007i:32018)

[AP09] Claudio Arezzo and Frank Pacard, *Blowing up Kähler manifolds with constant scalar curvature. II*, Ann. of Math. (2) **170** (2009), no. 2, 685–738. MR2552105 (2010m:32025)

[APS11] Claudio Arezzo, Frank Pacard, and Michael Singer, *Extremal metrics on blowups*, Duke Math. J. **157** (2011), no. 1, 1–51. MR2783927 (2012k:32024)

[Aub76] Thierry Aubin, *Équations différentielles non linéaires et problème de Yamabe concernant la courbure scalaire*, J. Math. Pures Appl. (9) **55** (1976), no. 3, 269–296.

[Aub78] Thierry Aubin, *Équations du type Monge-Ampère sur les variétés kählériennes compactes*, Bull. Sci. Math. (2) **102** (1978), no. 1, 63–95. MR494932 (81d:53047)

[Aub98] Thierry Aubin, *Some nonlinear problems in Riemannian geometry*, Springer-Verlag, Berlin, 1998.

[AV12a] Antonio G. Ache and Jeff A. Viaclovsky, *Obstruction-flat asymptotically locally Euclidean metrics*, Geom. Funct. Anal. **22** (2012), no. 4, 832–877.

[AV12b] Antonio G. Ache and Jeff A. Viaclovsky, *Asymptotics of the self-dual deformation complex*, J. Geom. Anal. **25** (2015), no. 2, 951–1000. MR3319957

[Bac21] R. Bach, *Zur weylschen relativitatstheorie und der weylschen erweiterung des krummungstensorbegriffs*, Math. Zeit. **9** (1921), 112–135.

[Bar86] Robert Bartnik, *The mass of an asymptotically flat manifold*, Comm. Pure Appl. Math. **39** (1986), no. 5, 661–693. MR849427 (88b:58144)

[BC64] Richard L. Bishop and Richard J. Crittenden, *Geometry of manifolds*, Academic Press, New York, 1964. MR29#6401

[BCG95] G. Besson, G. Courtois, and S. Gallot, *Entropies et rigidités des espaces localement symétriques de courbure strictement négative*, Geom. Funct. Anal. **5** (1995), no. 5, 731–799. MR1354289 (96i:58136)

[BE69] M. Berger and D. Ebin, *Some decompositions of the space of symmetric tensors on a Riemannian manifold*, J. Differential Geometry **3** (1969), 379–392. MR0266084 (42 #993)

[Ber70] Marcel Berger, *Quelques formules de variation pour une structure riemannienne*, Ann. Sci. École Norm. Sup. (4) **3** (1970), 285–294. MR0278238 (43 #3969)

[Bes08] Arthur L. Besse, *Einstein manifolds*, Classics in Mathematics, Springer-Verlag, Berlin, 2008, Reprint of the 1987 edition. MR2371700 (2008k:53084)

[BG08] Thomas P. Branson and A. Rod Gover, *Origins, applications and generalisations of the Q-curvature*, Acta Appl. Math. **102** (2008), no. 2-3, 131–146. MR2407527 (2010a:53025)

[Biq13a] Olivier Biquard, *Désingularisation de métriques d'Einstein. I*, Invent. Math. **192** (2013), no. 1, 197–252. MR3032330

[Biq13b] Olivier Biquard, *Désingularisation de métriques d'Einstein. II*, arXiv.org:1311.0956, 2013.

[BKN89] Shigetoshi Bando, Atsushi Kasue, and Hiraku Nakajima, *On a construction of coordinates at infinity on manifolds with fast curvature decay and maximal volume growth*, Invent. Math. **97** (1989), no. 2, 313–349. MR90c:53098

[BL09] Hubert L. Bray and Dan A. Lee, *On the Riemannian Penrose inequality in dimensions less than eight*, Duke Math. J. **148** (2009), no. 1, 81–106. MR2515101 (2010f:53051)

[Bla00] David E. Blair, *Spaces of metrics and curvature functionals*, Handbook of differential geometry, Vol. I, North-Holland, Amsterdam, 2000, pp. 153–185.

[BM09] Simon Brendle and Fernando C. Marques, *Blow-up phenomena for the Yamabe equation. II*, J. Differential Geom. **81** (2009), no. 2, 225–250. MR2472174 (2010k:53050)

[BM11] Olivier Biquard and Vincent Minerbe, *A Kummer construction for gravitational instantons*, Comm. Math. Phys. **308** (2011), no. 3, 773–794. MR2855540 (2012m:53101)

[BN04] Hubert L. Bray and André Neves, *Classification of prime 3-manifolds with Yamabe invariant greater than* $\mathbb{RP}^3$, Ann. of Math. (2) **159** (2004), no. 1, 407–424. MR2052359 (2005f:53048a)

[Bog78] F. A. Bogomolov, *Hamiltonian Kählerian manifolds*, Dokl. Akad. Nauk SSSR **243** (1978), no. 5, 1101–1104.

[Bra01] Hubert L. Bray, *Proof of the Riemannian Penrose inequality using the positive mass theorem*, J. Differential Geom. **59** (2001), no. 2, 177–267. MR1908823 (2004j:53046)

[Bre05] Simon Brendle, *Convergence of the Yamabe flow for arbitrary initial energy*, J. Differential Geom. **69** (2005), no. 2, 217–278. MR2168505 (2006e:53119)

[Bre07] Simon Brendle, *Convergence of the Yamabe flow in dimension 6 and higher*, Invent. Math. **170** (2007), no. 3, 541–576. MR2357502 (2008k:53136)

[Bre08] Simon Brendle, *Blow-up phenomena for the Yamabe equation*, J. Amer. Math. Soc. **21** (2008), no. 4, 951–979. MR2425176 (2009m:53084)

[BWZ04] C. Böhm, M. Wang, and W. Ziller, *A variational approach for compact homogeneous Einstein manifolds*, Geom. Funct. Anal. **14** (2004), no. 4, 681–733. MR2084976 (2005g:53074)

[Cal82] Eugenio Calabi, *Extremal Kähler metrics*, Seminar on Differential Geometry, Ann. of Math. Stud., vol. 102, Princeton Univ. Press, Princeton, N.J., 1982, pp. 259–290. MR645743 (83i:53088)

[Cal85] Eugenio Calabi, *Extremal Kähler metrics. II*, Differential geometry and complex analysis, Springer, Berlin, 1985, pp. 95–114. MR780039 (86h:53067)

[Car98] Gilles Carron, *Une suite exacte en L^2-cohomologie*, Duke Math. J. **95** (1998), no. 2, 343–372.

[CG72] Jeff Cheeger and Detlef Gromoll, *The splitting theorem for manifolds of nonnegative Ricci curvature*, J. Differential Geometry **6** (1971/72), 119–128. MR0303460 (46 #2597)

[CGY02a] Sun-Yung A. Chang, Matthew J. Gursky, and Paul Yang, *An a priori estimate for a fully nonlinear equation on four-manifolds*, J. Anal. Math. **87** (2002), 151–186, Dedicated to the memory of Thomas H. Wolff. MR1945280 (2003k:53036)

[CGY02b] Sun-Yung A. Chang, Matthew J. Gursky, and Paul C. Yang, *An equation of Monge-Ampère type in conformal geometry, and four-manifolds of positive Ricci curvature*, Ann. of Math. (2) **155** (2002), no. 3, 709–787. MR1923964 (2003j:53048)

[Che70] Jeff Cheeger, *Finiteness theorems for Riemannian manifolds*, Amer. J. Math. **92** (1970), 61–74. MR0263092 (41 #7697)

[Che09] Szu-Yu Sophie Chen, *Optimal curvature decays on asymptotically locally euclidean manifolds*, arXiv.org:0911.5538, 2009.

[CLW08] Xiuxiong Chen, Claude Lebrun, and Brian Weber, *On conformally Kähler, Einstein manifolds*, J. Amer. Math. Soc. **21** (2008), no. 4, 1137–1168.

[CS04] David M. J. Calderbank and Michael A. Singer, *Einstein metrics and complex singularities*, Invent. Math. **156** (2004), no. 2, 405–443.

[CS06] David M. J. Calderbank and Michael A. Singer, *Toric self-dual Einstein metrics on compact orbifolds*, Duke Math. J. **133** (2006), no. 2, 237–258. MR2225692 (2007g:53041)

[CT94] Jeff Cheeger and Gang Tian, *On the cone structure at infinity of Ricci flat manifolds with Euclidean volume growth and quadratic curvature decay*, Invent. Math. **118** (1994), no. 3, 493–571.

[CT06] Jeff Cheeger and Gang Tian, *Curvature and injectivity radius estimates for Einstein 4-manifolds*, J. Amer. Math. Soc. **19** (2006), no. 2, 487–525 (electronic). MR2188134 (2006i:53042)

[CW11] Xiuxiong Chen and Brian Weber, *Moduli spaces of critical Riemannian metrics with $L^{\frac{n}{2}}$ norm curvature bounds*, Adv. Math. **226** (2011), no. 2, 1307–1330.

[Der83] Andrzej Derdziński, *Self-dual Kähler manifolds and Einstein manifolds of dimension four*, Compositio Math. **49** (1983), no. 3, 405–433. MR707181 (84h:53060)

[DF89] S. Donaldson and R. Friedman, *Connected sums of self-dual manifolds and deformations of singular spaces*, Nonlinearity **2** (1989), no. 2, 197–239.

[dLPZ12] L. L. de Lima, P. Piccione, and M. Zedda, *A note on the uniqueness of solutions for the Yamabe problem*, Proc. Amer. Math. Soc. **140** (2012), no. 12, 4351–4357. MR2957225

[DM08] Zindine Djadli and Andrea Malchiodi, *Existence of conformal metrics with constant Q-curvature*, Ann. of Math. (2) **168** (2008), no. 3, 813–858. MR2456884 (2009h:53074)

[DN55] Avron Douglis and Louis Nirenberg, *Interior estimates for elliptic systems of partial differential equations*, Comm. Pure Appl. Math. **8** (1955), 503–538.

[Don10] Simon Donaldson, *Calabi-Yau metrics on Kummer surfaces as a model glueing problem*, arXiv:1007.421, 2010.

[Don11] Simon Donaldson, *Riemann surfaces*, Oxford Graduate Texts in Mathematics, vol. 22, Oxford University Press, Oxford, 2011. MR2856237

[DWW05] Xianzhe Dai, Xiaodong Wang, and Guofang Wei, *On the stability of Riemannian manifold with parallel spinors*, Invent. Math. **161** (2005), no. 1, 151–176.

[Ebi68] David G. Ebin, *On the space of Riemannian metrics*, Bull. Amer. Math. Soc. **74** (1968), 1001–1003. MR0231410 (37 #6965)

[EE69] Clifford J. Earle and James Eells, *A fibre bundle description of Teichmüller theory*, J. Differential Geometry **3** (1969), 19–43. MR0276999 (43 #2737a)

[EH79] Tohru Eguchi and Andrew J. Hanson, *Self-dual solutions to Euclidean gravity*, Ann. Physics **120** (1979), no. 1, 82–106.

[Flo91] Andreas Floer, *Self-dual conformal structures on $l\mathbb{C}\mathrm{P}^2$*, J. Differential Geom. **33** (1991), no. 2, 551–573.

[FM77] Arthur E. Fischer and Jerrold E. Marsden, *The manifold of conformally equivalent metrics*, Canad. J. Math. **29** (1977), no. 1, 193–209. MR0445537 (56 #3877)

[FM93] Akito Futaki and Toshiki Mabuchi, *Uniqueness and periodicity of extremal Kähler vector fields*, Proceedings of GARC Workshop on Geometry and Topology '93 (Seoul,

1993), Lecture Notes Ser., vol. 18, Seoul Nat. Univ., Seoul, 1993, pp. 217–239. MR1270938 (95c:32028)

[Fuj00] Akira Fujiki, *Compact self-dual manifolds with torus actions*, J. Differential Geom. **55** (2000), no. 2, 229–324. MR1847312 (2002k:57085)

[Fut83] A. Futaki, *An obstruction to the existence of Einstein Kähler metrics*, Invent. Math. **73** (1983), no. 3, 437–443. MR718940 (84j:53072)

[Fut88] Akito Futaki, *Kähler-Einstein metrics and integral invariants*, Lecture Notes in Mathematics, vol. 1314, Springer-Verlag, Berlin, 1988. MR947341 (90a:53053)

[GH78] G. W. Gibbons and S. W. Hawking, *Gravitational multi-instantons*, Physics Letters B **78** (1978), no. 4, 430–432.

[GL98] M. J. Gursky and C. LeBrun, *Yamabe invariants and* Spin^c *structures*, Geom. Funct. Anal. **8** (1998), no. 6, 965–977. MR1664788 (99k:53078)

[Gra09] C. Robin Graham, *Extended obstruction tensors and renormalized volume coefficients*, Adv. Math. **220** (2009), no. 6, 1956–1985. MR2493186 (2010e:53060)

[Gro07] Misha Gromov, *Metric structures for Riemannian and non-Riemannian spaces*, english ed., Modern Birkhäuser Classics, Birkhäuser Boston, Inc., Boston, MA, 2007, Based on the 1981 French original, With appendices by M. Katz, P. Pansu and S. Semmes, Translated from the French by Sean Michael Bates. MR2307192 (2007k:53049)

[Gur98] Matthew J. Gursky, *The Weyl functional, de Rham cohomology, and Kähler-Einstein metrics*, Ann. of Math. (2) **148** (1998), no. 1, 315–337. MR1652920 (99i:58036)

[Gur00] Matthew J. Gursky, *Four-manifolds with* $\delta W^+ = 0$ *and Einstein constants of the sphere*, Math. Ann. **318** (2000), no. 3, 417–431. MR1800764 (2002c:53074)

[GV01] Matthew J. Gursky and Jeff A. Viaclovsky, *A new variational characterization of three-dimensional space forms*, Invent. Math. **145** (2001), no. 2, 251–278. MR1872547 (2002j:53039)

[GV03] Matthew J. Gursky and Jeff A. Viaclovsky, *A fully nonlinear equation on four-manifolds with positive scalar curvature*, J. Differential Geom. **63** (2003), no. 1, 131–154. MR2015262 (2004h:53052)

[GV07] Matthew J. Gursky and Jeff A. Viaclovsky, *Prescribing symmetric functions of the eigenvalues of the Ricci tensor*, Ann. of Math. (2) **166** (2007), no. 2, 475–531. MR2373147 (2008k:53068)

[GV11] Matthew J. Gursky and Jeff A. Viaclovsky, *Rigidity and stability of Einstein metrics for quadratic curvature functionals*, J. Reine Angew. Math. **700** (2015), 37–91. MR3318510

[GV13] Matthew J. Gursky and Jeff A. Viaclovsky, *Critical metrics on connected sums of Einstein four-manifolds*, to appear in Adv. Math., arXiv.org:1303.0827, 2013.

[GW00] Mark Gross and P. M. H. Wilson, *Large complex structure limits of* $K3$ *surfaces*, J. Differential Geom. **55** (2000), no. 3, 475–546. MR1863732 (2003a:32042)

[GW03] Pengfei Guan and Guofang Wang, *A fully nonlinear conformal flow on locally conformally flat manifolds*, J. Reine Angew. Math. **557** (2003), 219–238. MR1978409 (2004e:53101)

[GW06] Yuxin Ge and Guofang Wang, *On a fully nonlinear Yamabe problem*, Ann. Sci. École Norm. Sup. (4) **39** (2006), no. 4, 569–598. MR2290138 (2007k:53040)

[Ham82] Richard S. Hamilton, *Three-manifolds with positive Ricci curvature*, J. Differential Geom. **17** (1982), no. 2, 255–306. MR84a:53050

[Hei12] Hans-Joachim Hein, *Gravitational instantons from rational elliptic surfaces*, J. Amer. Math. Soc. **25** (2012), no. 2, 355–393. MR2869021

[Hir95] Friedrich Hirzebruch, *Topological methods in algebraic geometry*, Classics in Mathematics, Springer-Verlag, Berlin, 1995, Translated from the German and Appendix One by R. L. E. Schwarzenberger, With a preface to the third English edition by the author and Schwarzenberger, Appendix Two by A. Borel, Reprint of the 1978 edition. MR1335917 (96c:57002)

[Hit74] Nigel Hitchin, *Compact four-dimensional Einstein manifolds*, J. Differential Geometry **9** (1974), 435–441.

[Hit79] N. J. Hitchin, *Polygons and gravitons*, Math. Proc. Cambridge Philos. Soc. **85** (1979), no. 3, 465–476.

[Hon06] Nobuhiro Honda, *Non-Moishezon twistor spaces of* $4\mathbf{CP}^2$ *with non-trivial automorphism group*, Trans. Amer. Math. Soc. **358** (2006), no. 5, 1897–1920 (electronic). MR2197434 (2006j:32020)

[Hon07] Nobuhiro Honda, *Self-dual metrics and twenty-eight bitangents*, J. Differential Geom. **75** (2007), no. 2, 175–258. MR2286821 (2008c:32029)

[Ito93] Mitsuhiro Itoh, *Moduli of half conformally flat structures*, Math. Ann. **296** (1993), no. 4, 687–708.

[Ito95] Mitsuhiro Itoh, *The Weitzenböck formula for the Bach operator*, Nagoya Math. J. **137** (1995), 149–181.

[JM87] Dennis Johnson and John J. Millson, *Deformation spaces associated to compact hyperbolic manifolds*, Discrete groups in geometry and analysis (New Haven, Conn., 1984), Progr. Math., vol. 67, Birkhäuser Boston, Boston, MA, 1987, pp. 48–106.

[Joy95] Dominic D. Joyce, *Explicit construction of self-dual 4-manifolds*, Duke Math. J. **77** (1995), no. 3, 519–552.

[Joy96] Dominic D. Joyce, *Compact Riemannian 7-manifolds with holonomy* G_2. *I, II*, J. Differential Geom. **43** (1996), no. 2, 291–328, 329–375. MR1424428 (97m:53084)

[Joy99] Dominic Joyce, *A new construction of compact 8-manifolds with holonomy* Spin(7), J. Differential Geom. **53** (1999), no. 1, 89–130. MR1776092 (2002a:53063)

[Kat94] Shin Kato, *Examples of non-Einstein Yamabe metrics with positive scalar curvature*, Tokyo J. Math. **17** (1994), no. 1, 187–189. MR1279579 (95f:53078)

[KK92] A. D. King and D. Kotschick, *The deformation theory of anti-self-dual conformal structures*, Math. Ann. **294** (1992), no. 4, 591–609.

[KMS09] M. A. Khuri, F. C. Marques, and R. M. Schoen, *A compactness theorem for the Yamabe problem*, J. Differential Geom. **81** (2009), no. 1, 143–196. MR2477893 (2010e:53065)

[Kob85] Osamu Kobayashi, *On a conformally invariant functional of the space of Riemannian metrics*, J. Math. Soc. Japan **37** (1985), no. 3, 373–389.

[Kob87] Osamu Kobayashi, *Scalar curvature of a metric with unit volume*, Math. Ann. **279** (1987), no. 2, 253–265. MR919505 (89a:53048)

[Koi78] Norihito Koiso, *Nondeformability of Einstein metrics*, Osaka J. Math. **15** (1978), no. 2, 419–433. MR504300 (81h:53046)

[Koi80] Norihito Koiso, *Rigidity and stability of Einstein metrics—the case of compact symmetric spaces*, Osaka J. Math. **17** (1980), no. 1, 51–73.

[Koi82] Norihito Koiso, *Rigidity and infinitesimal deformability of Einstein metrics*, Osaka J. Math. **19** (1982), no. 3, 643–668.

[Kro89a] P. B. Kronheimer, *The construction of ALE spaces as hyper-Kähler quotients*, J. Differential Geom. **29** (1989), no. 3, 665–683.

[Kro89b] P. B. Kronheimer, *A Torelli-type theorem for gravitational instantons*, J. Differential Geom. **29** (1989), no. 3, 685–697.

[KS01] A. Kovalev and M. Singer, *Gluing theorems for complete anti-self-dual spaces*, Geom. Funct. Anal. **11** (2001), no. 6, 1229–1281.

[KT87] Ryoichi Kobayashi and Andrey N. Todorov, *Polarized period map for generalized* $K3$ *surfaces and the moduli of Einstein metrics*, Tohoku Math. J. (2) **39** (1987), no. 3, 341–363. MR902574 (89g:32029)

[Küh88] Wolfgang Kühnel, *Conformal transformations between Einstein spaces*, Conformal geometry (Bonn, 1985/1986), Aspects Math., E12, Vieweg, Braunschweig, 1988, pp. 105–146. MR979791 (90b:53055)

[Laf83] Jacques Lafontaine, *Modules de structures conformes plates et cohomologie de groupes discrets*, C. R. Acad. Sci. Paris Sér. I Math. **297** (1983), no. 13, 655–658.

[LeB88] Claude LeBrun, *Counter-examples to the generalized positive action conjecture*, Comm. Math. Phys. **118** (1988), no. 4, 591–596.

[LeB91] Claude LeBrun, *Explicit self-dual metrics on* $\mathbf{CP}_2\#\cdots\#\mathbf{CP}_2$, J. Differential Geom. **34** (1991), no. 1, 223–253.

[LeB92] Claude LeBrun, *Twistors, Kähler manifolds, and bimeromorphic geometry. I*, J. Amer. Math. Soc. **5** (1992), no. 2, 289–316.

[LeB95] Claude LeBrun, *Einstein metrics and Mostow rigidity*, Math. Res. Lett. **2** (1995), no. 1, 1–8. MR1312972 (95m:53067)

[LeB97] Claude LeBrun, *Yamabe constants and the perturbed Seiberg-Witten equations*, Comm. Anal. Geom. **5** (1997), no. 3, 535–553. MR1487727 (98j:58032)

[LeB99] Claude LeBrun, *Einstein metrics and the Yamabe problem*, Trends in mathematical physics (Knoxville, TN, 1998), AMS/IP Stud. Adv. Math., vol. 13, Amer. Math. Soc., Providence, RI, 1999, pp. 353–376. MR1708770 (2000f:53057)

[LeB01] Claude LeBrun, *Ricci curvature, minimal volumes, and Seiberg-Witten theory*, Invent. Math. **145** (2001), no. 2, 279–316.

[LeB04] Claude LeBrun, *Curvature functionals, optimal metrics, and the differential topology of 4-manifolds*, Different faces of geometry, Int. Math. Ser. (N. Y.), vol. 3, Kluwer/Plenum, New York, 2004, pp. 199–256. MR2102997 (2005h:53055)

[Lic58] André Lichnerowicz, *Géométrie des groupes de transformations*, Travaux et Recherches Mathématiques, III. Dunod, Paris, 1958. MR0124009 (23 #A1329)

[LL03] Aobing Li and Yanyan Li, *On some conformally invariant fully nonlinear equations*, Comm. Pure Appl. Math. **56** (2003), no. 10, 1416–1464. MR1988895 (2004e:35072)

[LM08] Claude LeBrun and Bernard Maskit, *On optimal 4-dimensional metrics*, J. Geom. Anal. **18** (2008), no. 2, 537–564.

[LP87] John M. Lee and Thomas H. Parker, *The Yamabe problem*, Bull. Amer. Math. Soc. (N.S.) **17** (1987), no. 1, 37–91.

[LS93] Claude LeBrun and Santiago R. Simanca, *On the Kähler classes of extremal metrics*, Geometry and global analysis (Sendai, 1993), Tohoku Univ., Sendai, 1993, pp. 255–271. MR1361191 (96h:58037)

[LS94] Claude LeBrun and Michael Singer, *A Kummer-type construction of self-dual 4-manifolds*, Math. Ann. **300** (1994), no. 1, 165–180.

[LV12] Michael T. Lock and Jeff A. Viaclovsky, *Anti-self-dual orbifolds with cyclic quotient singularities*, J. Eur. Math. Soc. (JEMS) **17** (2015), 2805–2842. MR3420523

[LV14] Michael T. Lock and Jeff A. Viaclovsky, *A smörgåsbord of scalar-flat Kähler ALE surfaces*, to appear in Crelle's Journal, arXiv.org:1410.6461, 2014.

[LV15] Michael T. Lock and Jeff A. Viaclovsky, *Quotient singularities, eta invariants, and self-dual metrics*, to appear in Geom. Topol., arXiv.org:1501.03234, 2015.

[Min10] Vincent Minerbe, *On the asymptotic geometry of gravitational instantons*, Ann. Sci. Éc. Norm. Supér. (4) **43** (2010), no. 6, 883–924. MR2778451

[MØ13] Niels Martin Møller and Bent Ørsted, *Rigidity of conformal functionals on spheres*, International Mathematics Research Notices (2013), 1–38.

[Mos68] G. D. Mostow, *Quasi-conformal mappings in n-space and the rigidity of hyperbolic space forms*, Inst. Hautes Études Sci. Publ. Math. (1968), no. 34, 53–104. MR0236383 (38 #4679)

[Nak88] Hiraku Nakajima, *Hausdorff convergence of Einstein 4-manifolds*, J. Fac. Sci. Univ. Tokyo Sect. IA Math. **35** (1988), no. 2, 411–424. MR90e:53063

[Oba62] Morio Obata, *Certain conditions for a Riemannian manifold to be isometric with a sphere*, J. Math. Soc. Japan **14** (1962), 333–340. MR0142086 (25 #5479)

[Oba72] Morio Obata, *The conjectures on conformal transformations of Riemannian manifolds*, J. Differential Geometry **6** (1971/72), 247–258. MR0303464 (46 #2601)

[Omo70] Hideki Omori, *On the group of diffeomorphisms on a compact manifold*, Global Analysis (Proc. Sympos. Pure Math., Vol. XV, Berkeley, Calif., 1968), Amer. Math. Soc., Providence, R.I., 1970, pp. 167–183. MR0271983 (42 #6864)

[Pag78a] Don Page, *A compact rotating gravitational instanton*, Physics Letters B **79** (1978), no. 3, 235 – 238.

[Pag78b] Don N. Page, *A physical picture of the $K3$ gravitational instanton*, Physics Letters B **80** (1978), no. 1-2, 55 – 57.

[Pet00] Jimmy Petean, *The Yamabe invariant of simply connected manifolds*, J. Reine Angew. Math. **523** (2000), 225–231. MR1762961 (2001g:53075)

[Poo86] Y. Sun Poon, *Compact self-dual manifolds with positive scalar curvature*, J. Differential Geom. **24** (1986), no. 1, 97–132.

[Poo92] Y. Sun Poon, *On the algebraic structure of twistor spaces*, J. Differential Geom. **36** (1992), no. 2, 451–491.

[RS05] Yann Rollin and Michael Singer, *Non-minimal scalar-flat Kähler surfaces and parabolic stability*, Invent. Math. **162** (2005), no. 2, 235–270. MR2199006 (2007b:32035)

[Sch84] Richard M. Schoen, *Conformal deformation of a Riemannian metric to constant scalar curvature*, J. Differential Geom. **20** (1984), no. 2, 479–495.

[Sch89] Richard M. Schoen, *Variational theory for the total scalar curvature functional for Riemannian metrics and related topics*, Topics in calculus of variations (Montecatini Terme, 1987), Lecture Notes in Math., vol. 1365, Springer, Berlin, 1989, pp. 120–154.

[Sim85] Leon Simon, *Isolated singularities of extrema of geometric variational problems*, Harmonic mappings and minimal immersions (Montecatini, 1984), Lecture Notes in Math., vol. 1161, Springer, Berlin, 1985, pp. 206–277.

[Smo05] N. K. Smolentsev, *Spaces of Riemannian metrics*, Sovrem. Mat. Prilozh. (2005), no. 31, Geometriya, 69–147. MR2464555 (2010a:58015)

[ST69] I. M. Singer and J. A. Thorpe, *The curvature of 4-dimensional Einstein spaces*, Global Analysis (Papers in Honor of K. Kodaira), Univ. Tokyo Press, Tokyo, 1969, pp. 355–365.

[Str10] Jeffrey Streets, *Asymptotic curvature decay and removal of singularities of Bach-flat metrics*, Trans. Amer. Math. Soc. **362** (2010), no. 3, 1301–1324.

[STW07] Wei-Min Sheng, Neil S. Trudinger, and Xu-Jia Wang, *The Yamabe problem for higher order curvatures*, J. Differential Geom. **77** (2007), no. 3, 515–553. MR2362323 (2008i:53048)

[Sun09] Chanyoung Sung, *Surgery, Yamabe invariant, and Seiberg-Witten theory*, J. Geom. Phys. **59** (2009), no. 2, 246–255. MR2492194 (2010a:53054)

[SY79] Richard M. Schoen and Shing Tung Yau, *On the proof of the positive mass conjecture in general relativity*, Comm. Math. Phys. **65** (1979), no. 1, 45–76.

[SY81] Richard M. Schoen and Shing Tung Yau, *Proof of the positive mass theorem. II*, Comm. Math. Phys. **79** (1981), no. 2, 231–260.

[SY88] Richard M. Schoen and Shing Tung Yau, *Conformally flat manifolds, Kleinian groups and scalar curvature*, Inventiones Mathematicae **92** (1988), no. 1, 47–71.

[Tau92] Clifford Henry Taubes, *The existence of anti-self-dual conformal structures*, J. Differential Geom. **36** (1992), no. 1, 163–253. MR1168984 (93j:53063)

[Tho69] John A. Thorpe, *Some remarks on the Gauss-Bonnet integral*, J. Math. Mech. **18** (1969), 779–786.

[Tia87] Gang Tian, *Smoothness of the universal deformation space of compact Calabi-Yau manifolds and its Petersson-Weil metric*, Mathematical aspects of string theory (San Diego, Calif., 1986), Adv. Ser. Math. Phys., vol. 1, World Sci. Publishing, Singapore, 1987, pp. 629–646.

[Tia90] G. Tian, *On Calabi's conjecture for complex surfaces with positive first Chern class*, Invent. Math. **101** (1990), no. 1, 101–172. MR1055713 (91d:32042)

[Tru68] Neil S. Trudinger, *Remarks concerning the conformal deformation of Riemannian structures on compact manifolds*, Ann. Scuola Norm. Sup. Pisa (3) **22** (1968), 265–274.

[TV05a] Gang Tian and Jeff Viaclovsky, *Bach-flat asymptotically locally Euclidean metrics*, Invent. Math. **160** (2005), no. 2, 357–415.

[TV05b] Gang Tian and Jeff Viaclovsky, *Moduli spaces of critical Riemannian metrics in dimension four*, Adv. Math. **196** (2005), no. 2, 346–372.

[TV08] Gang Tian and Jeff Viaclovsky, *Volume growth, curvature decay, and critical metrics*, Comment. Math. Helv. **83** (2008), no. 4, 889–911.

[TY87] Gang Tian and Shing-Tung Yau, *Kähler-Einstein metrics on complex surfaces with* $C_1 > 0$, Comm. Math. Phys. **112** (1987), no. 1, 175–203. MR904143 (88k:32070)

[Uhl82a] Karen K. Uhlenbeck, *Connections with L^p bounds on curvature*, Comm. Math. Phys. **83** (1982), no. 1, 31–42. MR83e:53035

[Uhl82b] Karen K. Uhlenbeck, *Removable singularities in Yang-Mills fields*, Comm. Math. Phys. **83** (1982), no. 1, 11–29. MR83e:53034

[Via00] Jeff A. Viaclovsky, *Conformal geometry, contact geometry, and the calculus of variations*, Duke Math. J. **101** (2000), no. 2, 283–316. MR1738176 (2001b:53038)

[Via06] Jeff Viaclovsky, *Conformal geometry and fully nonlinear equations*, Inspired by S. S. Chern, Nankai Tracts Math., vol. 11, World Sci. Publ., Hackensack, NJ, 2006, pp. 435–460. MR2313345 (2008c:53030)

[Via10] Jeff A. Viaclovsky, *Monopole metrics and the orbifold Yamabe problem*, Ann. Inst. Fourier (Grenoble) **60** (2010), no. 7, 2503–2543 (2011). MR2866998 (2012m:53078)

[Via13a] Jeff A. Viaclovsky, *Einstein metrics and Yamabe invariants of weighted projective spaces*, Tohoku Math. J. (2) **65** (2013), no. 2, 297–311. MR3079291

[Via13b] Jeff A. Viaclovsky, *An index theorem on anti-self-dual orbifolds*, Int. Math. Res. Not. IMRN (2013), no. 17, 3911–3930. MR3096914

[Via13c] Jeff A. Viaclovsky, *The mass of the product of spheres*, Comm. Math. Phys. **335** (2015), no. 1, 17–41. MR3314498

[Wri11] Dominic Wright, *Compact anti-self-dual orbifolds with torus actions*, Selecta Math. (N.S.) **17** (2011), no. 2, 223–280. MR2803843 (2012j:53054)

[Yam60] Hidehiko Yamabe, *On a deformation of Riemannian structures on compact manifolds*, Osaka Math. J. **12** (1960), 21–37.

[Yau78] Shing Tung Yau, *On the Ricci curvature of a compact Kähler manifold and the complex Monge-Ampère equation. I*, Comm. Pure Appl. Math. **31** (1978), no. 3, 339–411. MR480350 (81d:53045)

Min-max theory and a proof of the Willmore conjecture

Fernando C. Marques and André Neves

IAS/Park City Mathematics Series
Volume 22, 2013

Min-max theory and a proof of the Willmore conjecture

Fernando C. Marques and André Neves

Organization

Introduction

The purpose of this lecture series is to present the main ideas involved in our solution [**13**] of the Willmore conjecture. Our proof is based on the min-max theory of minimal surfaces developed more than thirty years ago by Almgren and Pitts. We also present the ideas of joint work of the authors with I. Agol [**1**], in which we prove a conjecture of Freedman, He and Wang about the energy of two-component links in three-space. We thank Rafael Montezuma for taking excellent notes during the lectures and for typing them up in a clear and organized way.

Given a closed surface Σ of a fixed topological type, we would like to determine what is its best immersion in $\mathbb{R}^3$. If the genus is zero it is natural to think that the solution is a round sphere, but if the genus is one or higher the answer is not so clear and we need to give the problem a precise mathematical sense. We will accomplish this goal by defining a natural energy that we can minimize in the class of surfaces in three-space with a given topological type.

Let $\Sigma \subset \mathbb{R}^3$ be a closed, immersed, smooth surface of genus g. The local geometry of Σ is described by its principal curvatures k_1 and k_2, from which we can define

- the mean curvature $H = \frac{1}{2}(k_1 + k_2)$,
- the Gauss curvature $K = k_1 k_2$.

Since we would like our energy to have the property of scale-invariance, it is natural to consider integrands that are quadratic in the principal curvatures. By the Gauss-Bonnet Theorem the integral of the Gauss curvature K is a topological invariant:

$$\int_\Sigma K d\Sigma = 2\pi\chi(\Sigma),$$

Princeton University, Department of Mathematics, Fine Hall, Princeton, NJ 08544
E-mail address: coda@impa.br
Imperial College London, Huxley Building, 180 Queen's Gate, London SW7 2RH United Kingdom
E-mail address: a.neves@imperial.ac.uk

therefore it does not distinguish between the infinitely many curved shapes a surface of genus g can assume in $\mathbb{R}^3$. Fortunately there is something else we can do, namely we can integrate the square of the mean curvature. This is precisely the definition of the Willmore energy of Σ:

$$\mathcal{W}(\Sigma) = \int_\Sigma H^2 d\Sigma.$$

This functional is remarkably symmetric. Not only is it invariant under ambient isometries and scalings, it is also invariant under inversions hence it is preserved by any conformal transformation of $\mathbb{R}^3$. This fact was known back in the 1920s by Blaschke [**5**] and Thomsen [**22**].

Indeed,

$$\int_\Sigma H^2 d\Sigma = \int_\Sigma \Big\{K + \frac{(k_1-k_2)^2}{4}\Big\} d\Sigma = 2\pi\chi(\Sigma) + \frac{1}{2}\int_\Sigma |\mathring{A}|^2 d\Sigma,$$

where $\mathring{A}$ is the trace-free second fundamental form. It is an exercise to check that the expression $|\mathring{A}|^2 d\Sigma$ is pointwise conformally invariant. Hence $W(F(\Sigma)) = W(\Sigma)$ if F is conformal.

The idea that the round sphere is the best possible immersion of a two-sphere in three-space is compatible with the fact that the round shape is the absolute minimizer of the Willmore energy. This is the content of a theorem of Willmore:

Theorem
(Willmore 1960's): *If $\Sigma \subset \mathbb{R}^3$ is a closed surface, then*

$$\mathcal{W}(\Sigma) \geq 4\pi$$

and equality holds precisely when Σ is a round sphere.

Motivated by this result, Willmore asked himself what should be the minimum energy if we fix a genus $g > 0$. He made the following conjecture:

Willmore Conjecture
([**24**], 1965): *If $\Sigma \subset \mathbb{R}^3$ is a torus, then $\mathcal{W}(\Sigma) \geq 2\pi^2$.*

EXAMPLE 0.1. Let $\Sigma_{\sqrt{2}}$ be the torus obtained by rotation of a circle with center at distance $\sqrt{2}$ of the axis of revolution and radius 1. Then

$$\mathcal{W}(\Sigma_{\sqrt{2}}) = 2\pi^2.$$

Willmore proved that this torus minimizes the energy among the class of tori of revolution with a circular section. Somehow he had the intuition that the best torus of all should be an anchor ring.

A critical point Σ for the functional $\mathcal{W}$ is called a *Willmore surface.* The Euler-Lagrange equation for this variational problem is of higher order:

$$\Delta H + 2(H^2 - K)H = 0,$$

where K denotes the Gauss curvature. The rich analysis associated to this equation together with its most subtle aspects is a subject of the lectures delivered by Tristan Rivière [**17**] in this same 23nd Annual PCMI Summer Session.

The torus $\Sigma_{\sqrt{2}}$ is actually very special. In order to see this we need to recall that the stereographic projection $\pi : S^3 \setminus \{p\} \to \mathbb{R}^3$, $p \in S^3$, is a conformal transformation. Since the Willmore energy is conformally invariant we can try to formulate an equivalent problem for surfaces in the three-sphere S^3 instead. If $\Sigma \subset S^3 \setminus \{p\}$, we can calculate the energy of the projection $\tilde{\Sigma} = \pi(\Sigma) \subset \mathbb{R}^3$:

$$\int_{\tilde{\Sigma}} \tilde{H}^2 d\tilde{\Sigma} = \int_{\Sigma} (1 + H^2)\, d\Sigma,$$

where H now denotes the mean curvature of Σ with respect to the spherical geometry.

This suggests:

DEFINITION 0.2. For $\Sigma \subset S^3$, the Willmore energy is defined to be

$$\mathcal{W}(\Sigma) = \int_{\Sigma} (1 + H^2)\, d\Sigma.$$

One advantage of considering the problem for surfaces in S^3 is that a relation with the area functional becomes apparent. It follows immediately from the above definition that for $\Sigma \subset S^3$ one always has $\mathcal{W}(\Sigma) \geq \text{area}(\Sigma)$, and $\mathcal{W}(\Sigma) = \text{area}(\Sigma)$ if and only if Σ is a minimal surface.

There are infinitely many known closed minimal surfaces in S^3 (see Lawson [**11**]), the simplest two being:

- the equator (great sphere) $S^2_1(0)$ with area 4π,
- the Clifford torus $\hat{\Sigma} = S^1(1/\sqrt{2}) \times S^1(1/\sqrt{2}) \subset S^3$ with area $2\pi^2$.

It is no coincidence that if we choose the right stereographic projection we will have $\pi(\hat{\Sigma}) = \Sigma_{\sqrt{2}}$.

The Willmore conjecture has a long history and we refer the reader to [**13**] for an account. We mention a result of particular relevance to our approach:

Theorem
(Li and Yau' 1982, [**12**]): *If $F : \Sigma \to S^3$ is an immersion and there exists $p \in S^3$ such that $\#F^{-1}(p) = k$, then $\mathcal{W}(\Sigma) \geq 4\pi k$. In particular, if Σ is not embedded then $\mathcal{W}(\Sigma) \geq 8\pi$.*

A proof can be given by exploiting the conformal invariance of the functional in S^3. Fix $v \in \mathbb{R}^4$ and consider the vector field $x \mapsto v^T(x)$, given by tangential projection of v on T_xS^3. This is a conformal killing vector field in S^3 that generates a flow ϕ_t of conformal transformations called centered dilations (these will play an important role later). By conformal invariance of the Willmore energy, we have

$$\mathcal{W}(\Sigma) = \mathcal{W}(\phi_t(\Sigma)) = \int_{\phi_t(\Sigma)} (1 + H^2_{\phi_t(\Sigma)}) d\phi_t(\Sigma) \geq \text{area}(\phi_t(\Sigma)).$$

If we choose v so that $-v \in \Sigma$ and if there are k sheets of Σ passing through $-v$, then as we let $t \to \infty$ the surfaces $\phi_t(\Sigma)$ will converge to the union of k great spheres. Hence $\mathcal{W}(\Sigma) \geq 4\pi k$ by the inequality above.

Because of the result of Li and Yau, we may assume the surface is embedded. In [**13**] we proved the following theorem:

THEOREM A. *Let $\Sigma \subset S^3$ be a closed, embedded smooth surface with genus $g \geq 1$. Then $\mathcal{W}(\Sigma) \geq 2\pi^2$, and $\mathcal{W}(\Sigma) = 2\pi^2$ if and only if Σ is a conformal image of the Clifford torus.*

The above statement implies the Willmore conjecture is true. In order to prove it we first considered the particular case of minimal surfaces:

THEOREM B. *Let* $\Sigma \subset S^3$ *be a closed, embedded smooth surface with genus* $g \geq 1$. *If* Σ *is minimal then* $\text{area}(\Sigma) \geq 2\pi^2$, *and* $\text{area}(\Sigma) = 2\pi^2$ *if and only if* Σ *is the image of the Clifford torus under an ambient isometry.*

Our proof is based on min-max theory for the area functional. Min-max methods are useful to detect critical points of functions, especially those of higher Morse index. Perhaps the simplest example of the method is given by the Mountain Pass Theorem. Given a closed submanifold $S \subset \mathbb{R}^n$ and a differentiable function $f : S \to \mathbb{R}$ with two strict minimum points p and q, connected by a path $\gamma_0 : [0,1] \to S$, $\gamma_0(0) = p, \gamma_0(1) = q$, one can consider the homotopy class Π of γ_0 relative to $\{0,1\}$ and define the min-max invariant to be

$$c = \inf_{\gamma \in \Pi} \sup_{t \in [0,1]} f(\gamma(t)).$$

One can then prove that there exists a critical point $x_0 \in S$ of f, with index at most one, such that $f(x_0) = c$. In general, to detect index k critical points one needs to perform the min-max procedure over k-parameter families.

Of course the critical points of the area functional are precisely the minimal surfaces. The equator in S^3 as a minimal surface has Morse index one, and in fact can be produced by doing min-max over one-dimensional sweepouts of S^3.

Our work was motivated by the following similar question:

Key Question:
Is it possible to produce the Clifford torus as a min-max minimal surface?

Since the Morse index of the Clifford torus is five, we need to first find natural 5-parameter families. Our proof starts by constructing, for each embedded closed surface $\Sigma \subset S^3$, a *canonical family* of surfaces $\Sigma_{(v,t)} \subset S^3$, where $(v,t) \in B^4 \times (-\pi, \pi)$, with the properties that:

- $\Sigma_{(0,0)} = \Sigma$,
- $\text{area}(\Sigma_{(v,t)}) \leq \mathcal{W}(\Sigma)$ for every $(v,t) \in B^4 \times (-\pi,\pi)$.

Each surface $\Sigma_{(v,t)}$ is an equidistant surface of some conformal image of Σ, and the area estimate follows by conformal invariance of the Willmore energy and a calculation of Ros [**18**] (see also Heintze-Karcher [**9**]). The next step is to understand the geometric and topological properties of the canonical family, especially the behavior as we let the parameter (v,t) converge to the boundary of the parameter space $\partial\left(\overline{B}^4 \times [-\pi,\pi]\right)$. We prove that:

- for any sequence $\{(v_i,t_i)\}$ that converges to $\partial\left(\overline{B}^4 \times [-\pi,\pi]\right)$, a subsequence of $\{\Sigma_{(v_i,t_i)}\}$ converges to some round sphere (possibly trivial) in S^3.

An important ingredient in our approach is a theorem of Urbano:

Theorem
(Urbano' 1990): *If* $\Sigma \subset S^3$ *is a closed minimal surface with genus* $g \geq 0$ *and such that* $\text{index}(\Sigma) \leq 5$, *then* Σ *is either a Clifford torus (of index* 5*) or a great sphere (of index* 1*).*

Finally, we use the min-max theory of Almgren and Pitts. This uses ideas of Geometric Measure Theory and is general enough to apply to our setting (that of the canonical family). This will allow us to answer the Key Question affirmatively and in the process find a solution to the Willmore conjecture.

Part 1. Canonical Family and Degree Calculation

In this section we define the canonical family and derive a crucial topological identity used in the proof. We begin by introducing some notation:

- $\Sigma \subset S^3$ denotes a closed surface,
- $B^4 \subset \mathbb{R}^4$ is the open unit ball, with $\partial B^4 = S^3$,
- $B^4_R(Q) = \{x \in \mathbb{R}^4 : |x - Q| < R\}$,
- $B_r(q) = \{x \in S^3 : d(x,q) < r\}$, where d is the spherical distance.

For each $v \in B^4$, we consider the conformal map $F_v : S^3 \to S^3$ defined by

$$F_v(x) = \frac{1-|v|^2}{|x-v|^2}(x-v) - v.$$

We have that $F_0 = \text{id}$, and if $v \neq 0$ then F_v is a centered dilation that fixes $\pm v/|v|$. Note that if $v \to p \in S^3$, then $F_v(x) \to -p$ for every $x \in S^3$, $x \neq p$.

Let $S^3 \setminus \Sigma = A \cup A^*$, A and A^* the connected components, and let N be the unit normal vector to Σ pointing into A^*. We consider the following images under F_v:

$$A_v = F_v(A), \quad A^*_v = F_v(A^*) \quad \text{and} \quad \Sigma_v = F_v(\Sigma) = \partial A_v.$$

The unit normal vector to Σ_v is given by $N_v = DF_v(N)/|DF_v(N)|$. We also consider the signed distance function to Σ_v in S^3:

$$d_v(x) = \begin{cases} d(x,\Sigma_v) & \text{if } x \notin A_v \\ -d(x,\Sigma_v) & \text{if } x \in A_v. \end{cases}$$

This is a Lipschitz function that is smooth near Σ_v.

DEFINITION 0.3 (Canonical Family). For $v \in B^4$ and $t \in (-\pi,\pi)$, define

$$\Sigma_{(v,t)} = \partial A_{(v,t)},$$

where $A_{(v,t)} = \{x \in S^3 : d_v(x) < t\}$ (an open subset of S^3 with finite perimeter).

REMARK 1. If $\psi_{(v,t)} : \Sigma_v \to S^3$ is given by $\psi_{(v,t)}(y) = exp^{\perp}_y(tN_v(y))$, then $\psi_{(v,t)}(y) = \cos(t)y + \sin(t)N_v(y)$. Since the distance of any point to Σ_v can be realized by a minimizing geodesic, we have that

$$\Sigma_{(v,t)} \subset \psi_{(v,t)}(\{\text{Jac } \psi_{(v,t)}(y) \geq 0\}).$$

Area Estimate

THEOREM 0.4 (Ros, [**18**]). *Let Σ_t, $t \in (-\pi,\pi)$, be an equidistant surface of an embedded closed surface $\Sigma \subset S^3$. Then*

$$\text{area}(\Sigma_t) \leq \mathcal{W}(\Sigma). \tag{1}$$

Moreover, if Σ is not a geodesic sphere and

$$\text{area}(\Sigma_t) = \mathcal{W}(\Sigma),$$

then $t = 0$ and Σ is a minimal surface.

PROOF. Consider $\{e_1, e_2\} \subset T_y\Sigma$ an orthonormal basis such that $DN|_y(e_i) = -k_i e_i, i = 1, 2$, where k_i are the principal curvatures of Σ. Let $\psi_t(y) = \cos(t)y + \sin(t)N(y)$ for $y \in \Sigma$, as in the previous remark. Since $D\psi_t|_y(e_i) = (\cos(t) - k_i \sin(t))e_i$, we have

$$\begin{aligned}\text{Jac } \psi_t &= (\cos(t) - k_1 \sin(t))(\cos(t) - k_2 \sin(t)) \\ &= \cos^2(t) - (k_1 + k_2)\cos(t)\sin(t) + k_1 k_2 \sin^2(t).\end{aligned}$$

By a simple calculation, this gives

$$\text{Jac } \psi_t = (1 + H^2) - (\sin(t) + H\cos(t))^2 - \frac{(k_1 - k_2)^2}{4}\sin^2(t).$$

Using this and the area formula we obtain

$$\begin{aligned}\text{area}(\Sigma_t) &\leq \text{area}\big(\psi_t(\{\text{Jac } \psi_t(p) \geq 0\})\big) \\ &\leq \int_{\{\text{Jac } \psi_t \geq 0\}} (\text{Jac } \psi_t)\, d\Sigma \\ &\leq \int_{\{\text{Jac } \psi_t \geq 0\}} (1 + H^2)\, d\Sigma - \sin^2(t) \int_{\{\text{Jac } \psi_t \geq 0\}} \frac{(k_1 - k_2)^2}{4}\, d\Sigma \\ &\leq \int_\Sigma (1 + H^2)\, d\Sigma - \sin^2(t) \int_{\{\text{Jac } \psi_t \geq 0\}} \frac{(k_1 - k_2)^2}{4}\, d\Sigma \\ &= \mathcal{W}(\Sigma) - \frac{\sin^2(t)}{2} \int_{\{\text{Jac } \psi_t \geq 0\}} |\mathring{A}|^2 d\Sigma \\ &\leq \mathcal{W}(\Sigma).\end{aligned}$$

If Σ is not a geodesic sphere then $\int_\Sigma |\mathring{A}|^2 d\Sigma > 0$. Therefore, if we also have $\text{area}(\Sigma_t) = \mathcal{W}(\Sigma)$ then the above chain of inequalities implies that $\{\text{Jac } \psi_t \geq 0\} = \Sigma$ and $t = 0$. But in this case we have $\text{area}(\Sigma) = \mathcal{W}(\Sigma)$ and hence Σ must be a minimal surface. $\square$

If we apply the estimate of Ros to our canonical family and use the conformal invariance of the Willmore energy we get:

THEOREM 0.5. *For every $(v, t) \in B^4 \times (-\pi, \pi)$, we have*

$$\text{area}(\Sigma_{(v,t)}) \leq \mathcal{W}(\Sigma). \tag{2}$$

Moreover, if Σ is not a geodesic sphere and

$$\text{area}(\Sigma_{(v,t)}) = \mathcal{W}(\Sigma),$$

then $t = 0$ and Σ_v is a minimal surface.

Boundary behavior

Next we must analyze what happens to the surfaces $\Sigma_{(v,t)}$ as $v \to S^3$. This is important because before applying min-max theory we need to have a well-defined (and continuous) family of surfaces up to the boundary of the parameter space. The notation for symmetric difference here is, as usual, $X \Delta Y = (X \setminus Y) \cup (Y \setminus X)$. We will think of two open sets (of finite perimeter) X and Y as being close to each other if $\text{vol}(X \Delta Y)$ is small. This is to say that ∂X and ∂Y, as integral two-currents, are close in the flat topology (see [**14**], [**20**]).

Suppose we have sequences $v_n \in B^4$ and $t_n \in (-\pi, \pi)$ such that $(v_n, t_n) \to (v, t) \in \overline{B}^4 \times [-\pi, \pi]$. The case $v \in \overline{B}^4 \setminus \Sigma$ is not so difficult to analyze:

CASE 1. *If* $v \in B^4$, *then* $\Sigma_{v_n} \to \Sigma_v$ *smoothly and*

$$vol\ (A_{(v_n,t_n)} \Delta A_{(v,t)}) \to 0.$$

CASE 2. *If* $v \in A$, *then* $\Sigma_{v_n} \to -v$ *in Hausdorff distance, and*

$$vol\ (A_{(v_n,t_n)} \Delta B_{\pi+t}(v)) \to 0.$$

CASE 3. *If* $v \in A^*$, *then* $\Sigma_{v_n} \to -v$ *in Hausdorff distance and*

$$vol\ (A_{(v_n,t_n)} \Delta B_t(-v)) \to 0.$$

The subtle case is: $v_n \to v = p \in \Sigma$. For example, if $v_n = (1 - 1/n)v$ one can check that $A_{v_n} \to B_{\pi/2}(-N)$ in the sense that

$$\text{vol}\ (A_{v_n} \Delta B_{\pi/2}(-N)) \to 0.$$

This at first sight seems to suggest some discontinuity because for nearby points in S^3 (outside Σ) the limit of A_{v_n} always has either full or zero volume, according to the previously analyzed cases. The explanation is that there are limits (always geodesic balls) of any intermediate volume if we vary the angle at which v_n converges to $p \in \Sigma$. In fact, the limit of $A_{v_n} = F_{v_n}(A)$ will be determined by p and the angle of convergence.

In order to explain that, let us first introduce some notation. Let

$$D^2_+(r) = \{s = (s_1, s_2) \in \mathbb{R}^2 : |s| < r, s_1 \geq 0\},$$

and consider $\varepsilon > 0$ sufficiently small so that $\Lambda : \Sigma \times D^2_+(3\varepsilon) \to \overline{B}^4$ given by

$$\Lambda(p, s) = (1 - s_1)(\cos(s_2)p + \sin(s_2)N(p))$$

is a diffeomorphism onto a tubular neighbourhood of Σ in $\overline{B}^4$. We set $\Omega_r = \Lambda(\Sigma \times D^2_+(r))$.

If $v_n \to v = p \in \Sigma$, we can write $v_n = \Lambda(p_n, (s_{n1}, s_{n2}))$, with $p_n \to p$ and $|s_n| \to 0$. Let $B_{p_n} = B^4_{\sqrt{2}}(-N(p_n)) \cap S^3$. The idea is to show that the limit of $A_{v_n} = F_{v_n}(A)$ is the same limit of $F_{v_n}(B_{p_n})$. We must learn to compute the conformal images of balls and sectors in S^3.

Conformal images of balls and sectors

Let $p, N \in S^3$ with $\langle p, N\rangle = 0$, and let $r > 0$. In the appendix of [**13**] we study the conformal images of sectors of S^3 of the following form:

$$\Delta(p, N, r) = S^3 \setminus \{B_r(\cos(r)p + \sin(r)N) \cup B_r(\cos(r)p - \sin(r)N)\},$$

and of balls

$$B_{\pi/2}(-N) = B^4_{\sqrt{2}}(-N) \cap S^3.$$

Suppose $v = (1 - s)(\cos(t)p + \sin(t)N)$, with $0 < s \leq \varepsilon_0$ and $|t| \leq \varepsilon_0$, where $\varepsilon_0 > 0$ is sufficient small. Let

$$\overline{Q} = -\frac{t/s}{\sqrt{1 + (t/s)^2}}p - \frac{1}{\sqrt{1 + (t/s)^2}}N \in S^3,$$

and

$$\overline{R} = \sqrt{2\left(1 - \frac{t/s}{\sqrt{1 + (t/s)^2}}\right)}.$$

We have:

PROPOSITION 0.6. *There exists a universal constant $C > 0$ so that:*

(i) $$B^4_{\overline{R}-C\sqrt{|(s,t)|}}(\overline{Q}) \cap S^3 \subset F_v(B^4_{\sqrt{2}}(-N) \cap S^3) \subset B^4_{\overline{R}+C\sqrt{|(s,t)|}}(\overline{Q}) \cap S^3$$

and

(ii) $$F_v(\Delta(p,N,r)) \subset \overline{B}^4_{\overline{R}+C\sqrt{|(s,t)|}}(\overline{Q}) \setminus \overline{B}^4_{\overline{R}-C\sqrt{|(s,t)|}}(\overline{Q}).$$

Let $v_n \to v = p \in \Sigma$, with $v_n = \Lambda(p_n, (s_{n1}, s_{n2}))$, $p_n \to p$ and $|s_n| \to 0$, and let $B_{p_n} = B^4_{\sqrt{2}}(-N(p_n)) \cap S^3$. Suppose also that there is an angle of convergence:

$$\lim_{n\to\infty} \frac{s_{n2}}{s_{n1}} = k \in [-\infty, \infty].$$

One can always choose $r_0 > 0$ sufficiently small such that

$$B_{r_0}\left(\cos(r_0)p_n - \sin(r_0)N(p_n)\right) \subset A,$$

and

$$\overline{A} \subset S^3 \setminus B_{r_0}\left(\cos(r_0)p_n + \sin(r_0)N(p_n)\right).$$

for all n. This implies $A \Delta B_{p_n} \subset \Delta(p_n, N(p_n), r_0)$, and thus

$$F_{v_n}(A\Delta B_{p_n}) \subset F_{v_n}(\Delta(p_n, N(p_n), r_0)).$$

Using the above proposition, we have

$$F_{v_n}(A)\Delta F_{v_n}(B_{p_n}) \subset \overline{B}^4_{\overline{R}_n+C\sqrt{|s_n|}}(\overline{Q}_n) \setminus \overline{B}^4_{\overline{R}_n-C\sqrt{|s_n|}}(\overline{Q}_n),$$

where

$$\overline{Q}_n \to -\frac{k}{\sqrt{1+k^2}}p - \frac{1}{\sqrt{1+k^2}}N(p) \in S^3$$

and

$$\overline{R}_n \to \sqrt{2\left(1 - \frac{k}{\sqrt{1+k^2}}\right)}.$$

Then $\mathrm{vol}(F_{v_n}(A)\Delta F_{v_n}(B_{p_n})) \to 0$. But $F_{v_n}(B_{p_n})$ converges if we have that $s_{n2}/s_{n1} \to k \in [-\infty, \infty]$, by part (i) of the last proposition. Summarizing, we have:

PROPOSITION 0.7. *Consider a sequence $(v_n, t_n) \in B^4 \times (-\pi, \pi)$ converging to $(v,t) \in \overline{B}^4 \times [-\pi, \pi]$.*

(i) *If $v \in B^4$ then*

$$\lim_{n\to\infty} vol\left(A_{(v_n,t_n)} \Delta A_{(v,t)}\right) = 0.$$

(ii) *If $v \in A$ then*

$$\lim_{n\to\infty} vol\left(A_{(v_n,t_n)} \Delta B_{\pi+t}(v)\right) = 0$$

and, given any $\delta > 0$,

$\Sigma_{(v_n,t_n)} \subset \overline{B}_{\pi+t+\delta}(v) \setminus B_{\pi+t-\delta}(v)$ *for all n sufficiently large.*

(iii) *If $v \in A^*$ then*

$$\lim_{n\to\infty} vol\left(A_{(v_n,t_n)} \Delta B_t(-v)\right) = 0$$

and, given any $\delta > 0$,

$\Sigma_{(v_n,t_n)} \subset \overline{B}_{t+\delta}(-v) \setminus B_{t-\delta}(-v)$ *for all n sufficiently large.*

(iv) *If* $v = p \in \Sigma$ *and*

$$v_n = \Lambda(p_n, (s_{n1}, s_{n2})) \text{ with } \lim_{n\to\infty} \frac{s_{n2}}{s_{n1}} = k \in [-\infty, \infty],$$

then

$$\lim_{n\to\infty} vol\left(A_{(v_n,t_n)} \Delta B_{\overline{r}_k+t}(\overline{Q}_{p,k})\right) = 0$$

and, given any $\delta > 0$,

$$\Sigma_{(v_n,t_n)} \subset \overline{B}_{\overline{r}_k+t+\delta}(\overline{Q}_{p,k}) \setminus B_{\overline{r}_k+t-\delta}(\overline{Q}_{p,k}) \text{ for all } n \text{ sufficiently large.}$$

REMARK 2. In Section 12 of [**13**], we also prove that the canonical family $\Sigma_{(v,t)} = \partial A_{(v,t)}$ has the property of *no concentration of area (or mass)*. This means that for every $\delta > 0$, there exists $r > 0$ such that

$$\text{area}(\Sigma_{(v,t)} \cap B_r(q)) \leq \delta \quad \text{for every } q \in S^3 \text{ and } (v,t) \in B^4 \times (-\pi, \pi).$$

Reparametrizing Map $T : \overline{B}^4 \to \overline{B}^4$

Based on the previous proposition, we will reparametrize the canonical family in such a way that it can be continuously extended to the whole $\overline{B}^4 \times [-\pi, \pi]$. Let $\phi : [0, 3\varepsilon] \to [0,1]$ be a smooth function such that

- $\phi([0,\varepsilon]) = 0$,
- ϕ is strictly increasing in $[\varepsilon, 2\varepsilon]$,
- $\phi([2\varepsilon, 3\varepsilon]) = 1$.

Define $T : \overline{B}^4 \to \overline{B}^4$ by

$$T(v) = \begin{cases} v & \text{if } v \in \overline{B}^4 \setminus \Omega_{3\varepsilon} \\ \Lambda(p, \phi(|s|)s) & \text{if } v = \Lambda(p,s) \in \Omega_{3\varepsilon}. \end{cases}$$

The map T is continuous and collapses Ω_ε onto Σ by preserving the ratio s_{n2}/s_{n1}. Moreover, $T : B^4 \setminus \overline{\Omega}_\epsilon \to B^4$ is a homeomorphism.

For each $v \in B^4 \setminus \overline{\Omega}_\varepsilon$, put $C(v,t) = \Sigma_{(T(v),t)}$. Using the results in the previous section we know that as $v \to \partial(B^4 \setminus \overline{\Omega}_\varepsilon)$, the surface $C(v,t)$ converges to some geodesic sphere (round sphere). We extend C to Ω_ε by making it constant along the curves $s_1 \mapsto \Lambda(p, s_1, s_2)$. By construction, we have that

$$C(v,t) \text{ is continuous (in volume sense) in } \overline{B}^4 \times [-\pi, \pi]$$

and that

$$C(v,t) = \partial B_{\overline{r}(v)+t}(\overline{Q}(v))$$

for $v \in \partial B^4$, where the *center map* $\overline{Q} : S^3 \to S^3$ is given by

$$\overline{Q}(v) = \begin{cases} -T(v) & \text{if } v \in A^* \setminus \overline{\Omega}_\epsilon \\ T(v) & \text{if } v \in A \setminus \overline{\Omega}_\epsilon \\ -\frac{s}{\varepsilon}p - \frac{\sqrt{\varepsilon^2 - s^2}}{\varepsilon} N(p) & \text{if } v = \cos(s)p + \sin(s)N(p), s \in [-\varepsilon, \varepsilon], \end{cases}$$

and $\overline{r} : S^3 \to [0, \pi]$ is the function given by

$$\overline{r}(v) = \begin{cases} 0 & \text{if } v \in A^* \setminus \overline{\Omega}_\varepsilon, \\ \pi & \text{if } v \in A \setminus \overline{\Omega}_\varepsilon, \\ \frac{\pi}{2} - \arctan\left(\frac{s_2}{\sqrt{\varepsilon^2 - s_2^2}}\right) & \text{if } v = \Lambda(p,s) \in S^3 \cap \overline{\Omega}_\varepsilon. \end{cases}$$

Degree calculation

The main topological ingredient is the discovery that the genus of the original surface Σ can be read off the topological properties of the canonical family at the boundary:

THEOREM 0.8. $\deg(\overline{Q}) = \text{genus}(\Sigma)$.

PROOF. Let dV and $d\text{vol}$ be the standard volume forms of S^3 and $\mathbb{R}^4$, respectively. First notice that $\overline{Q}$ is piecewise smooth. We calculate its degree by integrating $\overline{Q}^*(dV)$ on S^3, and using the formula

$$\int_{S^3} \overline{Q}^*(dV) = \deg(\overline{Q}) \int_{S^3} dV.$$

In order to proceed with the calculation, we divide S^3 in three domains, namely $A^* \setminus \Omega_\varepsilon$, $A \setminus \Omega_\varepsilon$ and $\Omega_\epsilon \cap S^3$.

Then

$$\int_{A^*\setminus\Omega_\varepsilon} \overline{Q}^*(dV) = \int_{A^*\setminus\Omega_\varepsilon} (-T)^*(dV) = \int_{A^*} dV = \text{vol}(A^*).$$

Analogously,

$$\int_{A\setminus\Omega_\varepsilon} \overline{Q}^*(dV) = \text{vol}(A).$$

Now we let $G : \Sigma \times [-\varepsilon, \varepsilon] \to S^3 \cap \Omega_\varepsilon$ be the diffeomorphism given by $G(p,t) = \cos(t)p + \sin(t)N(p)$. If $Q = \overline{Q} \circ G : \Sigma \times [-\varepsilon, \varepsilon] \to S^3$, then

$$Q(p,t) = -\frac{t}{\varepsilon}p - \frac{\sqrt{\varepsilon^2 - t^2}}{\varepsilon}N(p).$$

Using this reparametrization, the third integral can be calculated as

$$\int_{\Omega_\varepsilon \cap S^3} \overline{Q}^*(dV) = \int_{\Sigma\times[-\varepsilon,\varepsilon]} Q^*(dV).$$

If $\{e_1, e_2\} \subset T_p\Sigma$ denotes an orthonormal basis with $DN|_p(e_i) = -k_i e_i$, where k_i are the principal curvatures, we have

$$\begin{aligned} Q^*(dV)(e_1, e_2, \partial_t) &= dV(DQ(e_1), DQ(e_2), DQ(\partial_t)) \\ &= d\text{vol}(DQ(e_1), DQ(e_2), DQ(\partial_t), Q) \\ &= \Big(-\frac{t}{\varepsilon} + \frac{\sqrt{\varepsilon^2 - t^2}}{\varepsilon}k_1\Big)\Big(-\frac{t}{\varepsilon} + \frac{\sqrt{\varepsilon^2 - t^2}}{\varepsilon}k_2\Big)\frac{(-1)}{\sqrt{\varepsilon^2 - t^2}}. \end{aligned}$$

The Gauss curvature of Σ satisfies $K = 1 + k_1 k_2$ by the Gauss equation, hence

$$\begin{aligned} \int_{\Sigma\times[-\varepsilon,\varepsilon]} Q^*(dV) &= -\iint \frac{1}{\varepsilon^2}\Big(k_1 k_2\sqrt{\varepsilon^2 - t^2} - (k_1 + k_2)t + \frac{t^2}{\sqrt{\varepsilon^2 - t^2}}\Big)dt d\Sigma \\ &= -\frac{\pi}{2}\int_\Sigma (K-1)d\Sigma - \frac{\pi}{2}\int_\Sigma d\Sigma \\ &= -\frac{\pi}{2}\int_\Sigma K d\Sigma. \end{aligned}$$

By the Gauss-Bonnet Theorem,

$$\int_{\Sigma\times[-\varepsilon,\varepsilon]} Q^*(dV) = \pi^2(2g - 2).$$

Adding up the results and using that $\mathrm{vol}(S^3) = 2\pi^2$ (!), we get

$$\int_{S^3} \overline{Q}^*(dV) = \mathrm{vol}(S^3) + \pi^2(2g-2) = 2\pi^2 g = g\int_{S^3} dV.$$

This finishes the proof. □

Part 2. Min-Max Theory and Ruling Out Great Spheres

The idea is to perform the min-max procedure for the area functional over topologically nontrivial families of cycles (in the sense of Geometric Measure Theory). In the 1960s, Almgren [**2, 3**] devised a very general min-max theory that applied to any ambient dimension and codimension, and to any number of parameters. He was able to prove the existence of a possibly singular minimal variety (a stationary integral varifold). In 1981, Pitts [**16**] improved the theory of Almgren and proved regularity in the case of hypersurfaces, provided the ambient dimension $n \leq 5$. Pitts proved this by showing that the stationary integral varifold of Almgren can be chosen to satisfy the an almost minimizing property. Then Schoen and Simon [**19**] extended the regularity to higher dimensions, allowing the possibility of singular sets of codimension 7.

In particular, one gets the following result:

THEOREM 0.9. *Every compact Riemannian manifold (M^n, g), $n \leq 7$, contains a smooth embedded minimal hypersurface $\Sigma \subset M$.*

The minimal hypersurface Σ can be obtained by applying the min-max technique to the class of sweepouts of M, topologically nontrivial one-parameter families of hypersurfaces. These are the families of surfaces $\{\Sigma_t\}_{t\in[0,1]}$ such that we can write $\Sigma_t = \partial\Omega_t$ with Ω_t varying continuously (in volume sense), where $\Omega_0 = 0$ and $\Omega_1 = M$. The family of level sets $\{x \in M : f(x) = t\}$ of a Morse function $f : M \to [0,1]$ is an example of a sweepout.

For technical reasons the theory of Almgren and Pitts is formulated discretely, namely the space of parameters is discretized. Before describing this approach it is useful to have an idealized version that applies to families defined in the n-dimensional cube and that helps clarifying the main ideas involved in the proofs.

Let us restrict ourselves to the case of three-dimensional ambient spaces for simplicity. We denote by $Z_2(M^3)$ the space of integral 2-cycles in M (integral 2-currents with zero boundary). We would like to run min-max with maps $\Phi : I^n \to Z_2(M^3)$ continuous in the flat topology, and defined on the n-dimensional cube $I^n = [0,1]^n$. If Π is the homotopy class of Φ relative to the boundary ∂I^n, we would define the width of Π by

$$L(\Pi) = \inf\{L(\Phi') : \Phi' \in \Pi\}, \tag{3}$$

where $L(\Phi') = \sup\{M(\Phi'(x)) : x \in I^n\}$. Here $M(T)$ denotes the mass of the current T. Any sequence $\{\phi_i\}_i \subset \Pi$ with $L(\phi_i) \to L(\Pi)$ is called an *optimal sequence.*

The prototypical min-max theorem is:

Min-max Theorem:

If $L(\Pi) > \sup\{M(\Phi(x)) : x \in \partial I^n\}$, then there exists a smooth embedded minimal surface Σ (possibly disconnected, with integer multiplicities) in M so that $L(\Pi) =$ area(Σ). Moreover, if $\{\phi_i\}_i \subset \Pi$ is an optimal sequence then we can choose Σ so that for a subsequence $\{j\} \subset \{i\}$ and some $x_j \in I^n$ we have that Σ is the limit of $\phi_j(x_j)$ in varifold sense.

EXAMPLE 0.10. Given an orientable compact Riemannian three-manifold M, we can choose a Morse function $f : M \to [0,1]$ and define $\Phi(t) = \{x \in M : f(x) = t\}$, $t \in [0,1]$. Let Π_1 denote the homotopy class (relative to $\{0,1\}$) of Φ. Since for any $\Phi' \in \Pi_1$, $\Phi'(t) = \partial\Omega'(t)$, there exists $t_0 \in [0,1]$ such that $M(\Omega'(t_0)) = \text{vol}(M)/2$, the isoperimetric inequality tells us that $L(\Pi_1) > 0$. By the Min-max Theorem there will be a closed embedded minimal surface $\Sigma \subset M$ such that $L(\Pi_1) = \text{area}(\Sigma)$. If M is the unit three-sphere we can take a standard foliation by round spheres: $\Phi(t) = \{x \in S^3 : x_4 = 1 - 2t\}$, $t \in [0,1]$, to conclude that $L(\Pi_1) \leq 4\pi$. But the area of any closed minimal surface in S^3 is at least 4π, with equality only if the surface is a great sphere. Therefore the Almgren-Pitts min-max minimal surface of S^3 is the equator with area $L(\Pi_1) = 4\pi$.

Almgren-Pitts min-max theory

We briefly describe the Almgren-Pitts min-max theory by comparing it with the idealized theory mentioned above. We need the following notation:

- $I(1,k)$ denotes the cell complex on I^1 whose 1-cells and 0-cells are, respectively,

$$[0,3^{-k}], [3^{-k}, 2\cdot 3^{-k}], \ldots, [1-3^{-k}, 1], \text{ and } [0], [3^{-k}], \ldots, [1-3^{-k}], [1].$$

- $I(n,k)$ is the n-dimensional cell complex on I^n given by

$$I(n,k) = I(1,k) \otimes \ldots \otimes I(1,k) \ (n \text{ times}).$$

- $I(n,k)_0$ is the subset of all 0-cells, or vertices, of $I(n,k)$.
- $I(n,k)_1$ is the subset of all 1-cells of $I(n,k)$.

In the Almgren-Pitts min-max theory one replaces a continuous map defined on the whole I^n by a sequence of maps ϕ_i defined in finer and finer grids $I(n,k_i)_0$ ($k_i \to \infty$). Appropriate discretized notions of homotopy are also provided. The continuity in the flat topology is replaced by the notion of fineness in the mass norm (a much stronger topology). This consists in requiring that for some $\delta_i \to 0$ we have $M(\phi_i(x) - \phi_i(y)) < \delta_i$, for all x, y such that $[x,y] \in I(n,k_i)_1$. The advantage of using the mass norm $M(S,T) = M(S-T)$ is that it can easily be localized, making it ideal for area comparisons, cut and paste arguments, and thus, regularity theory. The other advantage is that the mass functional is continuous in the M-norm, but not in the flat topology. Because this is a much stronger topology, some serious interpolation work is needed to adapt families like $\{x_4 = 1-2t\}$ in S^3 to this setting. The property of no concentration of area (see Remark 2) is very useful in this technical part. There are other treatments to min-max theory, such as [**21, 6**], but those require stronger regularity assumptions and do not apply to the families we are interested in.

The diagram below explains how to translate some usual notions to the Almgren-Pitts discrete setting:

Continuous	**Discrete**
$\Phi : I^n \to Z_2(M)$	$S = \{\phi_i\}_i, \phi_i : I(n,k_i)_0 \to Z_2(M)$
Φ continuous in the flat topology	$M(\phi_i(x) - \phi_i(y)) < \delta_i, \forall [x,y] \in I(n,k_i)_1$
$L(\Phi) = \sup\{M(\Phi(x)) : x \in I^n\}$	$L(S) = \lim_i \sup\{M(\phi_i(x)) : x \in I(n,k_i)_0\}$
$\Phi \simeq \Phi'$ homotopy	$\phi_i \simeq_{\delta_i} \phi'_i$ and $\delta_i \to 0$
$L(\Pi) = \{L(\Phi') : \Phi' \in \Pi\}$	$L(\Pi) = \{L(S') : S' \in \Pi\}$

Ruling out great spheres (Topological Argument).
Since $\overline{B}^4 \times [-\pi, \pi]$ is homeomorphic to the 5-cube, we can reparametrize once again the min-max family $C : \overline{B}^4 \times [-\pi, \pi] \to Z_2(S^3)$ to get a map $\Phi : I^5 \to Z_2(S^3)$ as stated below. We conveniently choose the reparametrization so that at the boundary of I^5 the great spheres are achieved at $\partial I^4 \times \{1/2\}$.

THEOREM 0.11. *Let $\Sigma \subset S^3$ be an embedded, closed, genus g surface. The map $\Phi : I^5 \to Z_2(S^3)$ satisfies the following:*

(i) *Φ is continuous in the flat topology,*
(ii) *$\Phi(I^4 \times 0) = \Phi(I^4 \times 1) = 0$,*
(iii) *for every $x \in \partial I^4$, $\{\Phi(x,t)\}_{t\in[0,1]}$ is the standard sweepout of S^3 by oriented geodesic spheres centered at $Q(x) \in S^3$,*
(iv) *$\Phi(x, 1/2) = \partial B_{\pi/2}(Q(x))$ for every $x \in \partial I^4$,*
(v) *the degree of the center map $Q : \partial I^4 \to S^3$ is equal to g.*

Moreover,

$$\sup\{M(\Phi(x)) : x \in I^5\} < \mathcal{W}(\Sigma).$$

We will explain the topological argument assuming the idealized min-max theory mentioned before. Note that in [**13**] we do everything in the discrete setting of Almgren and Pitts.

Consider Π the homotopy class of Φ relative to ∂I^5. Of course we have

$$\sup\{M(\Phi(x)) : x \in \partial I^5\} = 4\pi.$$

The next theorem implies that the homotopy class Π is topologically nontrivial when $g \geq 1$, in the sense that the critical point produced out of min-max in Π cannot be a great sphere.

THEOREM 0.12. *If $g \geq 1$, then $L(\Pi) > 4\pi$.*

Before proving that, we introduce some notation. We denote by $\mathcal{V}_2(S^3)$ the closure, in the weak topology, of the space of 2-dimensional rectifiable varifolds with support contained in S^3. The topology of $\mathcal{V}_2(S^3)$ can be metrized by the F-metric $F(\cdot,\cdot)$. We define $\mathcal{T} \subset \mathcal{V}_2(S^3)$ to be the space of unoriented great spheres. The map that associates to each element in $\mathcal{T}$ the pair of centers $\{q, -q\}$ of the corresponding great sphere induces a homeomorphism between $\mathcal{T}$ and $\mathbb{RP}^3$.

Let $|T|$ denote the varifold in $\mathcal{V}_2(S^3)$ obtained by ignoring the orientation of the integral 2-current $T \in Z_2(S^3)$. We consider the map

$$|\Phi| : \partial I^4 \times \{1/2\} \to \mathbb{RP}^3 \simeq \mathcal{T}$$

given by $|\Phi|(x, 1/2) = |\Phi(x, 1/2)| = \{Q(x), -Q(x)\}$. The degree calculation gives

$$|\Phi|_*(\partial I^4 \times \{1/2\}) = 2g \in H_3(\mathbb{RP}^3, \mathbb{Z}).$$

PROOF OF THEOREM 0.12. Suppose, by contradiction, that $L(\Pi) = 4\pi$. Then there exists a sequence $\{\phi_i\}_i \subset \Pi$ so that

$$\sup_{x\in I^5} M(\phi_i(x)) \leq 4\pi + \frac{1}{i}.$$

Recall that $\phi_i = \Phi$ on ∂I^5. Let $\varepsilon_0 > 0$ small (to be chosen later) and take $\delta > 0$ such that

$$y = (x,t) \in J_\delta = \partial I^4 \times [1/2 - \delta, 1/2 + \delta] \Rightarrow F(|\Phi(y)|, |\Phi(x, 1/2)|) \leq \varepsilon_0.$$

Then consider $0 < \varepsilon_1 \leq \varepsilon_0$ with the property that

$$y \in \partial I^4 \times I, \mathbf{F}(|\Phi(y)|, \mathcal{T}) \leq \varepsilon_1 \Rightarrow y \in J_\delta.$$

If we could find a 4-dimensional manifold R so that $\partial R = \partial I^4 \times \{1/2\}$ and a continuous function $f : R \to \mathbb{RP}^3$ with $f|_{\partial R} = |\Phi|$, then we would have

$$|\Phi|_*(\partial I^4 \times \{1/2\}) = f_*(\partial R) = \partial f_*(R) = 0 \text{ in } H_3(\mathbb{RP}^3, \mathbb{Z}).$$

This would be a contradiction if $g \geq 1$.

Here is how we proceed. We first construct a sequence of 4-dimensional $R(i) \subset I^5$ with support$(\partial R(i)) \subset \partial I^4 \times I$. These submanifolds are constructed so that $|\phi_i(x)|$ is sufficient close to $\mathcal{T}$ for any $x \in R(i)$, when i is large. We begin with

$$\overline{A}(i) = \{x \in I^5 : \mathbf{F}(|\phi_i(x)|, \mathcal{T}) \geq \varepsilon_1\},$$

$$A(i) \subset \overline{A}(i) \text{ the connected component of } I^4 \times 0.$$

REMARK 3. In the actual proof ([**13**]) the regions $\overline{A}(i)$ and $A(i)$ are constructed as unions of a collection of cubes of a subdivision of I^5. We use singular homology to make things rigorous.

CLAIM 1. *$A(i) \cap (I^4 \times \{1\}) = \emptyset$ if i is sufficiently large.*

PROOF. If not, up to a subsequence, we can find $\gamma_i : [0, 1] \to I^5$ with $\gamma_i(0) \in I^4 \times 0$, $\gamma_i(1) \in I^4 \times 1$ and $\gamma_i([0, 1]) \subset A(i)$. Since $\phi_i \in \Pi$ for every i, we have $\phi_i \circ \gamma_i \in \Pi_1$, the class of homotopically nontrivial one-parameter sweepouts of S^3. We know that $L(\Pi_1) = 4\pi$, the lowest area a closed minimal surface can have in the unit three-sphere. Then

$$4\pi = L(\Pi_1) \leq L(\phi_i \circ \gamma_i) \leq 4\pi + 1/i.$$

This implies that the sequence $\{\phi_i \circ \gamma_i\}_i$ is optimal in Π_1. By the Min-max Theorem, and after passing to a subsequence, we find a smooth embedded minimal surface $\Sigma \subset S^3$ and a sequence $x_i \in I$ so that

$$\text{area}(\Sigma) = 4\pi \text{ and } F(|(\phi_i \circ \gamma_i)(x_i)|, \Sigma) \to 0.$$

The only minimal surfaces in S^3 with area equal to 4π are the great spheres, hence $\phi_i(\gamma_i(x_i))$ gets arbitrarily close to $\mathcal{T}$, in a clear contradiction with the fact that $\gamma_i(x_i) \in A(i)$. □

In particular, $A(i) \cap \partial I^5 \subset (\partial I^4 \times I) \cup (I^4 \times 0)$. Define

$$R(i) = (\partial A(i)) \cap \text{int}(I^5).$$

Therefore $F(|\phi_i(x)|, \mathcal{T}) = \varepsilon_1$ for every $x \in R(i)$ and, if i is large enough, $\partial R(i) \subset J_\delta$. Take $C(i) = \partial A(i) \cap (\partial I^4 \times I)$. Because $\partial(\partial A(i)) = 0$ and $\partial A(i) = R(i) \cup C(i) \cup (I^4 \times 0)$, we have

$$\partial C(i) = \partial R(i) \cup \partial(I^4 \times 0)$$

and this implies

$$[\partial R(i)] = [\partial I^4 \times \{1/2\}] \text{ in } H_3(J_\delta, \mathbb{Z}).$$

We still have to deal with the facts that

(1) surfaces along $R(i)$ are close to $\mathcal{T}$, but not exactly in $\mathcal{T}$,
(2) the cycles $\partial R(i)$ and $\partial I^4 \times \{1/2\}$ are homologous and close to each other, but not necessarily equal.

We need a projection argument to conclude.

DEFINITION 0.13. Let $\widehat{\Phi} : \partial I^4 \times I \to \mathbb{RP}^3 \simeq \mathcal{T}$ be the continuous map given by $\widehat{\Phi}(x,t) = |\Phi(x,1/2)|$. In particular,

$$\widehat{\Phi}_*(\partial I^4 \times \{1/2\}) = 2g$$

in $H_3(\mathbb{RP}^3, \mathbb{Z})$.

CLAIM 2. *There exists a continuous map $f_i : R(i) \to \mathcal{T}$ with $f_i|_{\partial R(i)} = \widehat{\Phi}|_{\partial R(i)}$.*

PROOF. We can choose a triangulation Δ of $R(i)$ so that if $[x,y] \in \Delta_1$, $x, y \in \Delta_0$, then

$$F(|\phi_i(x)|, |\phi_i(y)|) \leq \varepsilon_0.$$

First define $f_i : \Delta_0 \to \mathcal{T}$ by putting $f_i(x) = \widehat{\Phi}(x)$ if $x \in \partial R(i)$, and $f_i(x) = y$ if $x \notin \partial R(i)$, where we choose $y \in \mathcal{T}$ so that $\mathbf{F}(|\phi_i(x)|, y) = \mathbf{F}(|\phi_i(x)|, \mathcal{T})$. If $x, y \in \Delta_0$ are adjacent, the triangle inequality gives

$$F(f_i(x), f_i(y)) \leq 3\varepsilon_0.$$

If ε_0 is sufficient small (depending only on the size of convex balls in $\mathbb{RP}^3$ with standard geometry), the map f_i can be extended to the 1-skeleton of $|\Delta| = R(i)$ by keeping $f_i|_{\partial R(i)} = \widehat{\Phi}|_{\partial R(i)}$. By choosing ε_0 even smaller to start with, we can make sure that we can extend inductively from the $(i-1)$-skeleton to the i-skeleton, keeping $f_i|_{\partial R(i)} = \widehat{\Phi}|_{\partial R(i)}$, until we get to the 4-skeleton. Then we have our $f_i : R(i) \to \mathcal{T}$. □

Now we have

$$\widehat{\Phi}_*[\partial R(i)] = (f_i)_*[\partial R(i)] = [\partial (f_i)_*(R(i))] = 0 \text{ in } H_3(\mathbb{RP}^3, \mathbb{Z}).$$

On the other hand, the degree calculation and the fact that

$$[\partial R(i)] = [\partial I^4 \times \{1/2\}] \text{ in } H_3(\partial I^4 \times I, \mathbb{Z})$$

gives

$$\widehat{\Phi}_*[\partial R(i)] = \widehat{\Phi}_*[\partial I^4 \times \{1/2\}] = 2g \text{ in } H_3(\mathbb{RP}^3, \mathbb{Z}).$$

Contradiction if $g \geq 1$. Then $L(\Pi) > 4\pi$. □

Part 3. Proof of the Main Theorems and the Energy of Links

First we prove the statement about minimal surfaces:

THEOREM B. *Let $\Sigma \subset S^3$ be a closed, embedded smooth surface with genus $g \geq 1$. If Σ is minimal then* area$(\Sigma) \geq 2\pi^2$, *and* area$(\Sigma) = 2\pi^2$ *if and only if Σ is the image of the Clifford torus under an ambient isometry.*

REMARK 4. The general index bound is missing in current min-max theory: the fact that if Σ is produced by min-max with k parameters, one should have index$(\Sigma) \leq k$. Therefore we will exploit the extra structure coming from the canonical family.

PROOF. Consider

$$\mathcal{F}_1 = \{S \subset S^3 : S \text{ is an embedded, closed minimal surface of genus } g \geq 1\}.$$

By standard arguments in Geometric Measure Theorey (see appendix of [**13**]) there exists $\Sigma \in \mathcal{F}_1$ satisfying

$$\text{area}(\Sigma) = \inf_{S \in \mathcal{F}_1} \text{area}(S). \tag{4}$$

CLAIM 3. *Let* $\Sigma \in \mathcal{F}_1$ *be so that* (4) *holds. Then* $\text{index}(\Sigma) \leq 5$.

If Σ_t is a variation of the minimal surface $\Sigma = \Sigma_0$, the formula for the second variation of area gives

$$\frac{d^2}{dt^2}\text{area}(\Sigma_t)|_{t=0} = -\int_\Sigma \psi L\psi d\Sigma,$$

where $L = \Delta + |A|^2 + Ric(N, N)$ is the Jacobi operator of Σ, ψN is the variational vector field and N is the unit normal vector to Σ. Recall that we constructed a canonical family $\Sigma_{(v,t)}$, $(v, t) \in \overline{B}^4 \times [-\pi, \pi]$, satisfying the area estimate

$$\text{area}(\Sigma_{(v,t)}) \leq \mathcal{W}(\Sigma) = \text{area}(\Sigma)$$

(because here Σ is a minimal surface).

Consider the variational vector fields of $\Sigma_{(v,t)}$ at $(v = 0, t = 0)$:

$$\psi_i = \langle N, e_i\rangle, \quad i = 1, 2, 3, 4 \text{ and } \psi_5 = 1,$$

where e_i are the canonical directions in $\mathbb{R}^4$. Let $\mathrm{E} = \text{span}\{\psi_1, \ldots, \psi_5\}$.

We can prove that $(0, 0)$ is a nondegenerate local maximum point of $(v, t) \mapsto \text{area}(\Sigma_{(v,t)})$:

LEMMA 0.14. *We have*

$$-\int_\Sigma \psi L\psi d\Sigma < 0 \;\forall\, \psi \in \mathrm{E} \setminus \{0\},$$

unless Σ *is totally geodesic.*

PROOF. Let $\delta > 0$ be small, and let

$$f(v, t) = \text{area}(\Sigma_{(v,t)}), \quad (v, t) \in B^4_\delta(0) \times (-\delta, \delta).$$

Since Σ is minimal, we have $f(0, 0) = \text{area}(\Sigma) = \mathcal{W}(\Sigma)$ and $Df(0, 0) = 0$. We also know that $f(v, t) \leq f(0, 0)$ for every $(v, t) \in B^4_\delta(0) \times (-\delta, \delta)$. Hence $D^2 f(0, 0) \leq 0$ and this means that

$$-\int_\Sigma \psi L\psi \, d\Sigma \leq 0 \quad \text{for every } \psi \in E.$$

Suppose the lemma is not true. We could find $\phi \in E \setminus \{0\}$ such that

$$-\int_\Sigma \phi L\psi \, d\Sigma = -\int_\Sigma \psi L\phi \, d\Sigma = 0 \quad \text{for every } \psi \in E.$$

Hence

$$\int_\Sigma \phi \psi_i \, d\Sigma = 0 \quad \text{for every } 1 \leq i \leq 4, \quad \text{and} \int_\Sigma L\phi \, d\Sigma = 0. \tag{5}$$

This implies, since $\psi_5 = 1 \in E$, the existence of $c \in \mathbb{R}$ such that

$$1 = c\phi + \psi, \quad \text{where} \quad \psi = \sum_{i=1}^{4} a_i\psi_i.$$

Hence, because ψ is an eigenfunction of L with $L\psi - 2\psi = 0$, we have

$$\int_\Sigma (|A|^2 + 2) \, d\Sigma = \int_\Sigma L(1) \, d\Sigma = \int_\Sigma (cL\phi + L\psi) \, d\Sigma = 2\int_\Sigma \psi \, d\Sigma.$$

On the other hand, we also have $1 = c^2\phi^2 + 2c\phi\psi + \psi^2$. If we integrate over Σ, we obtain from (5):

$$\text{area}(\Sigma) = \int_\Sigma (c^2\phi^2 + 2c\phi\psi + \psi^2)\,d\Sigma \geq \int_\Sigma \psi^2\,d\Sigma = \int_\Sigma \psi(1 - c\phi)\,d\Sigma$$
$$= \int_\Sigma \psi\,d\Sigma.$$

Hence

$$2\,\text{area}(\Sigma) \leq \int_\Sigma (|A|^2 + 2)\,d\Sigma = 2\int_\Sigma \psi\,d\Sigma \leq 2\,\text{area}(\Sigma).$$

This implies $A = 0$ and so Σ is a great sphere. This contradicts our assumption that $\Sigma \in \mathcal{F}_1$. □

Suppose, by contradiction, that index$(\Sigma) \geq 6$. The idea is to construct a comparison map

$$\Phi' : I^5 \to \mathcal{Z}_2(S^3)$$

which coincides with Φ, the map given by Theorem 0.11, in a neighborhood of ∂I^5 and such that

$$\sup_{x\in I^5} M(\Phi'(x)) < \text{area}(\Sigma).$$

If index$(\Sigma) \geq 6$, then there exists $\varphi \in C^\infty(\Sigma)$ with

$$-\int_\Sigma \varphi L\varphi d\Sigma < 0 \quad \text{and} \quad -\int_\Sigma \varphi L\psi_i d\Sigma = 0, \text{ for all } i = 1, \ldots, 5.$$

Choose a vector field X in S^3 such that $X = \varphi N$ on Σ, and let Γ_s be its flow. Define $f(v,t,s) = \text{area}(\Gamma_s(\Sigma_{(v,t)}))$. Then

$$f(0,0,0) = \text{area}(\Sigma), \quad Df(0,0,0) = 0, \quad D^2f(0,0,0) < 0,$$

hence f has a strict local maximum at the origin. Replace $\Sigma_{(v,t)}$ with

$$\Sigma'_{(v,t)} = \Gamma_{\beta(v,t)}(\Sigma_{(v,t)}),$$

where the function β is chosen to satisfy $0 \leq \beta \leq \delta$, $\beta(v,t) = 0$ if $|v| + |t| \geq \delta$ and $\beta(v,t) = \delta$ if $|v| + |t| \leq \delta/2$, where $\delta > 0$ is a sufficiently small number. Then area$(\Sigma'_{(v,t)}) < \Sigma$ if $|v| + |t| < \delta$ and $\Sigma'_{(v,t)} = \Sigma_{(v,t)}$ if $|v| + |t| \geq \delta$.

Therefore

$$\text{area}(\Sigma'_{(v,t)}) < \text{area}(\Sigma) \quad \forall \quad (v,t) \in B^4 \times (-\pi, \pi).$$

This is because the equality area$(\Sigma_{(v,t)}) = \mathcal{W}(\Sigma) = \text{area}(\Sigma)$ can only happen if $t = 0$ and Σ_v is minimal. But when Σ is minimal we have the formula area$(\Sigma_v) = \text{area}(\Sigma) - 4\int \frac{\langle v, N(x)\rangle^2}{|x-v|^4} d\Sigma$. Hence, if we also had that Σ_v is minimal, $v \in B^4$, then area$(\Sigma_v) = \mathcal{W}(\Sigma_v) = \mathcal{W}(\Sigma) = \text{area}(\Sigma)$, and the formula would imply that $\langle v, N\rangle = 0$ along Σ. This only happens for $v \neq 0$ if Σ is invariant under the flow of $x \in S^3 \mapsto v^T(x) \in T_xS^3$, i.e., if Σ is a great sphere containing $\{v, -v\}$.

We can reparametrize the new family $\Sigma'_{(v,t)}$ just as we did for $\Sigma_{(v,t)}$ to get families $C'(v,t)$ and $\Phi' : I^5 \to Z_2(S^3)$ with the same basic properties and

$$\sup_{x\in I^5} M(\Phi'(x)) \leq \sup\{\text{area}(\Sigma'_{(v,t)}) : (v,t) \in \overline{B}^4 \times [-\pi, \pi]\} < \text{area}(\Sigma).$$

By applying min-max theory to Π', the homotopy class of Φ' relative to ∂I^5, we obtain a closed minimal surface Σ' so that $\text{area}(\Sigma') = L(\Pi')$. Since the boundary values of Φ and Φ' coincide, Theorem 0.12 applies and we get $L(\Pi') > 4\pi$. Hence

$$4\pi < \text{area}(\Sigma') = L(\Pi') < \text{area}(\Sigma) \leq 2\pi^2 < 8\pi,$$

where the $2\pi^2$ estimate comes from the fact that the Clifford torus is a surface in $\mathcal{F}_1$. This implies the multiplicity of Σ' is 1, otherwise we would have $\text{area}(\Sigma') \geq 8\pi$. Also, we must have $\text{genus}(\Sigma') \geq 1$, because by a theorem of Almgren [**4**] the only minimal surfaces of genus $g = 0$ in S^3 are the great spheres. Contradiction, since this implies $\Sigma' \in \mathcal{F}_1$ but $\text{area}(\Sigma') < \text{area}(\Sigma)$. Therefore $\text{index}(\Sigma) \leq 5$.

By Urbano's theorem, Σ is a Clifford torus and Theorem B is proved. □

Now we are ready to prove our main result:

THEOREM A. *Let $\Sigma \subset S^3$ be a closed, embedded smooth surface with genus $g \geq 1$. Then $\mathcal{W}(\Sigma) \geq 2\pi^2$, and $\mathcal{W}(\Sigma) = 2\pi^2$ if and only if Σ is a conformal image of the Clifford torus.*

PROOF. Let $\Phi : I^5 \to Z_2(S^3)$ be the min-max family constructed from Σ as before, and let Π be its homotopy class relative to ∂I^5. By applying min-max theory to Π, we obtain a minimal surface $\widehat{\Sigma}$ with $\text{area}(\widehat{\Sigma}) = L(\Pi)$. Again, from Theorem 0.12, we get

$$4\pi < \text{area}(\widehat{\Sigma}) = L(\Pi) \leq W(\Sigma).$$

If $\widehat{\Sigma}$ has multiplicity greater than 1, then $\text{area}(\widehat{\Sigma}) \geq 8\pi > 2\pi^2$. If not, the fact that $\text{area}(\widehat{\Sigma}) > 4\pi$ and Almgren's theorem imply $\text{genus}(\widehat{\Sigma}) \geq 1$. In any case, $\text{area}(\widehat{\Sigma}) \geq 2\pi^2$ and therefore

$$W(\Sigma) \geq 2\pi^2.$$

For the rigidity part, suppose $W(\Sigma) = 2\pi^2$. In this case $\text{area}(\Sigma_{(v,t)}) \leq W(\Sigma) = 2\pi^2$. This implies that for some $(v_0, t_0) \in B^4 \times (-\pi, \pi)$ we have $\text{area}(\Sigma_{(v_0,t_0)}) = W(\Sigma)$. Otherwise we would be able to run the min-max process again and produce a non-spherical minimal surface with area strictly below $2\pi^2$, contradicting Theorem B.

But $\text{area}(\Sigma_{(v_0,t_0)}) = W(\Sigma)$ implies $t_0 = 0$ and that Σ_{v_0} is a minimal surface. Then $\text{area}(\Sigma_{v_0}) = W(\Sigma_{v_0}) = W(\Sigma) = 2\pi^2$. By the rigidity part of Theorem B, $\Sigma_{v_0} = F_{v_0}(\Sigma)$ must be a Clifford torus, with $F_{v_0} \in \text{Conf}(S^3)$. This finishes the proof of Theorem A. □

By analyzing what we have done, we can answer affirmatively the key question posed in the introduction. The following corollary presents the Clifford torus as the min-max surface of a 5-dimensional family in S^3.

Corollary
Let $\widehat{\Pi}$ be the homotopy class (relative to ∂I^5) of the canonical family $\widehat{\Phi} : I^5 \to \mathcal{Z}_2(S^3)$ associated with the Clifford torus $\widehat{\Sigma} = S^1(\frac{1}{\sqrt{2}}) \times S^1(\frac{1}{\sqrt{2}}) \subset S^3$. Then

$$\mathbf{L}(\widehat{\Pi}) = \text{area}(\widehat{\Sigma}) = 2\pi^2.$$

Energy of links

Finally we would like to describe another application of these ideas, this time to the problem of determining the best 2-component link in three-space. A 2-component link in $\mathbb{R}^3$ is a pair (γ_1, γ_2) of rectifiable curves $\gamma_i : S^1 \to \mathbb{R}^3$, $i = 1, 2$, such that $\gamma_1(S^1) \cap \gamma_2(S^1) = \emptyset$. The *Möbius cross energy* of the link (γ_1, γ_2) is defined to be

$$E(\gamma_1, \gamma_2) = \int_{S^1 \times S^1} \frac{|\gamma_1'(s)||\gamma_2'(t)|}{|\gamma_1(s) - \gamma_2(t)|^2}\, ds\, dt.$$

This energy was introduced in [**7**] (see also [**15**] for knot energies) and has the property, like the Willmore energy of surfaces, of conformal invariance.

It was conjectured by Freedman-He-Wang (1994) that the Möbius energy of any nontrivial link (those that cannot be isotoped to the union of the boundaries of two disjoint disks) should be at least $2\pi^2$. The equality is attained by the stereographic projection of the so-called *standard Hopf link*:

$$\hat{\gamma}_1(s) = (\cos s, \sin s, 0, 0) \in S^3 \quad \text{and} \quad \hat{\gamma}_2(t) = (0, 0, \cos t, \sin t) \in S^3.$$

It follows from a result of He that it suffices to prove the conjecture for links (γ_1, γ_2) that have linking number $\operatorname{lk}(\gamma_1, \gamma_2) = \pm 1$. This is what I. Agol and the present authors prove in [**1**]:

Theorem

([**1**]) *Let $\gamma_i : S^1 \to \mathbb{R}^3$, $i = 1, 2$, be a 2-component link in $\mathbb{R}^3$ with $|\operatorname{lk}(\gamma_1, \gamma_2)| = 1$. Then $E(\gamma_1, \gamma_2) \geq 2\pi^2$.*

Moreover, if $E(\gamma_1, \gamma_2) = 2\pi^2$ then there exists a conformal map $F : \mathbb{R}^4 \to \mathbb{R}^4$ such that $(F \circ \gamma_1, F \circ \gamma_2)$ describes the standard Hopf link up to orientation.

We prove this result by constructing a *canonical family* for the problem: given a link (γ_1, γ_2) in $S^3 \subset \mathbb{R}^4$ (the conformal invariance of the energy extends to any Euclidean space [**10**]) we construct a 5-parameter family of surfaces in S^3 with the same basic properties of the canonical family for the Willmore problem, and such that the area of any surface in the family is bounded above by the Möbius energy of the link. The fact that $|\operatorname{lk}(\gamma_1, \gamma_2)| = 1$ implies the center map $Q : S^3 \to S^3$ associated with the family satisfies $|\deg(Q)| = 1$. Therefore the topological argument applies and we conclude the existence of at least one surface in the family with area greater than or equal to $2\pi^2$. This establishes the inequality $E(\gamma_1, \gamma_2) \geq 2\pi^2$, and after some extra work one can also prove the rigidity part.

In what follows we sketch the construction. We start with a definition. The *Gauss map* of (γ_1, γ_2) in $\mathbb{R}^4$, denoted by $g = G(\gamma_1, \gamma_2)$, is the Lipschitz map $g : S^1 \times S^1 \to S^3$ defined by

$$g(s, t) = \frac{\gamma_1(s) - \gamma_2(t)}{|\gamma_1(s) - \gamma_2(t)|}.$$

Note that we can reparametrize the curves γ_1, γ_2 so that they are Lipschitz and parametrized proportionally to the arc length. The starting point is the observation that

$$|\operatorname{Jac} g|(s, t) \leq \frac{|\gamma_1'(s)||\gamma_2'(t)|}{|\gamma_1(s) - \gamma_2(t)|^2},$$

hence

$$\operatorname{area}(g(S^1 \times S^1)) \leq E(\gamma_1, \gamma_2).$$

Like in the Willmore problem, we can start playing with conformal transformations. Given $v \in \mathbb{R}^4$, we define the conformal map

$$F_v : \mathbb{R}^4 \setminus \{v\} \to \mathbb{R}^4, \quad F_v(x) = \frac{x - v}{|x - v|^2}.$$

If $v \in B^4$, we can compute

$$F_v(S_1^3(0)) = S^3_{\frac{1}{1-|v|^2}}(c(v)) \quad \text{where} \quad c(v) = \frac{v}{1 - |v|^2}.$$

Hence, if g_v denotes the Gauss map of the link $(F_v \circ \gamma_1, F_v \circ \gamma_2)$, then it follows from the previous observation and the conformal invariance of the energy that

$$\text{area}(g_v(S^1 \times S^1)) \leq E(F_v \circ \gamma_1, F_v \circ \gamma_2) = E(\gamma_1, \gamma_2).$$

for every $v \in B^4$. This already gives us a four-dimensional family of surfaces (parametrized tori) in S^3 whose areas are bounded above by $E(\gamma_1, \gamma_2)$. As in the Willmore problem, we need to introduce a fifth parameter.

Given $w \in \mathbb{R}^4$ and $\lambda \in \mathbb{R}$, we set $D_{w,\lambda}(x) = \lambda(x - w) + w$, where $x \in \mathbb{R}^4$. Finally, given $v \in \overline{B}^4$ and $z \in (0, 1)$, we also define

$$b(v, z) = \frac{(2z - 1)}{(1 - |v|^2 + z)(1 - z)}$$

and

$$a(v, z) = 1 + (1 - |v|^2)b(v, z) = 1 + \frac{(1 - |v|^2)(2z - 1)}{(1 - |v|^2 + z)(1 - z)}.$$

For each $v \in B^4$ fixed, $z \to a(v, z)$ is a nondecreasing parametrization of $(0, +\infty)$.

If $\gamma_1(S^1) \cup \gamma_2(S^1) \subset S^3$, we define the canonical five-dimensional family of surfaces associated to (γ_1, γ_2) to be

$$C(v, z) = g_{(v,z)\#}(S^1 \times S^1) \in \mathcal{Z}_2(S^3),$$

for $(v, z) \in B^4 \times (0, 1)$, where $g_{(v,z)} : S^1 \times S^1 \to S^3$ is the Gauss map $g_{(v,z)} = G(F_v \circ \gamma_1, D_{c(v),a(v,z)} \circ F_v \circ \gamma_2)$:

$$g_{(v,z)}(s, t) = \frac{(F_v \circ \gamma_1)(s) - (D_{c(v),a(v,z)} \circ F_v \circ \gamma_2)(t)}{|(F_v \circ \gamma_1)(s) - (D_{c(v),a(v,z)} \circ F_v \circ \gamma_2)(t)|}.$$

The notation $f_\#(T)$ is used to indicate the push-forward of the current T under the Lipschitz map f. Note that both curves $F_v \circ \gamma_1$ and $D_{c(v),a(v,z)} \circ F_v \circ \gamma_2$ are contained in spheres centered at $c(v)$, and that $g_{(v,1/2)} = G(F_v \circ \gamma_1, F_v \circ \gamma_2)$ since $a(v, 1/2) = 1$.

The fact that $\text{area}(C(v, z)) \leq E(\gamma_1, \gamma_2)$ for every $(v, z) \in B^4 \times (0, 1)$ follows from the conformal invariance of the energy and the estimates below:

$$\begin{aligned}
|\text{Jac}\, g_{(v,z)}|(s, t) &\leq \frac{|(F_v \circ \gamma_1)'(s)||(D_{c(v),a(v,z)} \circ F_v \circ \gamma_2)'(t)|}{|F_v \circ \gamma_1(s) - D_{c(v),a(v,z)} \circ F_v \circ \gamma_2(t)|^2} \\
&\leq \frac{a(v, z)|(F_v \circ \gamma_1)'(s)||(F_v \circ \gamma_2)'(t)|}{a(v, z)|F_v \circ \gamma_1(s) - F_v \circ \gamma_2(t)|^2 + b(v, z)^2} \\
&\leq \frac{|(F_v \circ \gamma_1)'(s)||(F_v \circ \gamma_2)'(t)|}{|F_v \circ \gamma_1(s) - F_v \circ \gamma_2(t)|^2}.
\end{aligned}$$

The map C admits an extension $C : \overline{B}^4 \times [0, 1] \to \mathcal{Z}_2(S^3)$, and its boundary behavior is summarized in the following result:

Proposition

There exists a constant $c > 0$ *such that for every* $p \in S^3$ *we have*

(i) $C(p, 1/2) = -\mathrm{lk}(\gamma_1, \gamma_2) \cdot \partial B_{\pi/2}(p)$,

(ii) $\mathrm{supp}(C(p, z)) \subset \overline{B}_{\pi/2}(p) \setminus B_{r(z)}(p)$ *if* $z \in [1/2, 1]$,

(iii) $\mathrm{supp}(C(p, z)) \subset \overline{B}_{\pi/2}(-p) \setminus B_{\pi - r(z)}(-p)$ *if* $z \in [0, 1/2]$,

where

$$r(z) = \cos^{-1}\left(\frac{b(z)}{\sqrt{|b(z)|^2 + c^2}}\right) \in [0, \pi] \quad \textit{and} \quad b(z) = \frac{2z - 1}{z(1 - z)}.$$

REMARK 5. Notice that $r(0) = \pi$, $r(1/2) = \pi/2$, and $r(1) = 0$.

For instance, given $p \in S^3 \setminus (\gamma_1(S^1) \cup \gamma_2(S^1))$ we have that the link

$$(F_p \circ \gamma_1, F_p \circ \gamma_2)$$

is contained in the affine hyperplane

$$P_{(p,1/2)} = \{x \in \mathbb{R}^4 \, : \, \langle x, p \rangle = -1/2\}.$$

This implies that

$$\mathrm{supp}\, C(p, 1/2) \subset G(F_p \circ \gamma_1, F_p \circ \gamma_2)(S^1 \times S^1) \subset \partial B_{\pi/2}(-p),$$

and so by the Constancy Theorem [**20**, Theorem 26.27] we have that

$$C(p, 1/2) = k \cdot \partial B_{\pi/2}(-p)$$

for some integer k.

In order to determine the value of k, let P be the oriented hyperplane with normal vector $p \in S^3$ and let ω_P be its volume form. Let ω, ω_{S^3}, $\omega_{\mathbb{R}^4}$ denote, respectively, the volume form of $\partial B_{\pi/2}(-p) \subset S^3$ (the exterior unit normal is p), S^3, and $\mathbb{R}^4$. Note that

$$\partial B_{\pi/2}(-p) = \{x \in \mathbb{R}^4 : \langle x, p \rangle = 0\} \cap S^3 \subset P.$$

If $g = G(\tilde{\gamma}_1, \tilde{\gamma}_2)$, with $\tilde{\gamma}_i = F_p \circ \gamma_i$, then

$$\begin{aligned} g^*\omega\left(\frac{\partial}{\partial s}, \frac{\partial}{\partial t}\right) &= \omega\left(\frac{\partial g}{\partial s}, \frac{\partial g}{\partial t}\right) = \omega_{S^3}\left(\frac{\partial g}{\partial s}, \frac{\partial g}{\partial t}, p\right) \\ &= \omega_{\mathbb{R}^4}\left(\frac{\partial g}{\partial s}, \frac{\partial g}{\partial t}, p, g\right) = -\omega_{\mathbb{R}^4}\left(\frac{\partial g}{\partial s}, \frac{\partial g}{\partial t}, g, p\right) \\ &= -\omega_P\left(\frac{\partial g}{\partial s}, \frac{\partial g}{\partial t}, g\right) = -\det\left(\frac{\partial g}{\partial s}, \frac{\partial g}{\partial t}, g\right) \\ &= \frac{\det(\tilde{\gamma}_1', \tilde{\gamma}_2', \tilde{\gamma}_1 - \tilde{\gamma}_2)}{|\tilde{\gamma}_1 - \tilde{\gamma}_2|^3}. \end{aligned}$$

Since

$$\mathrm{lk}(\tilde{\gamma}_1, \tilde{\gamma}_2) = \int_{S^1 \times S^1} \frac{\det(\tilde{\gamma}_1', \tilde{\gamma}_2', \tilde{\gamma}_1 - \tilde{\gamma}_2)}{|\tilde{\gamma}_1 - \tilde{\gamma}_2|^3} ds\, dt$$

by the Gauss formula, and $\mathrm{lk}(\tilde{\gamma}_1, \tilde{\gamma}_2) = \mathrm{lk}(\gamma_1, \gamma_2)$ we have that

$$k = \frac{1}{4\pi} C(p, 1/2)(\omega) = \frac{1}{4\pi} \int_{S^1 \times S^1} g^*\omega = \mathrm{lk}(\gamma_1, \gamma_2).$$

By continuity of C, we also have $C(p,1/2) = -\mathrm{lk}(\gamma_1,\gamma_2)\cdot\partial B_{\pi/2}(p)$ for $p \in \gamma_1(S^1)\cup \gamma_2(S^1)$, proving item (i).

The boundary values of C are not yet round spheres, but by the previous proposition they are contained in hemispheres that vary continuously. The idea now is to apply area-decreasing deformations, taking place inside these hemispheres, so that at the end of the deformations the surface gets identified with a round sphere.

Let $p \in S^3$, $\lambda \in [0,\pi/2]$. For $t \in [0,1]$, we define a retraction map

$$R_{(p,\lambda,t)} : \overline{B}_{\pi/2}(p) \setminus B_\lambda(p) \to \overline{B}_{\pi/2}(p) \setminus B_\lambda(p)$$

by

$$R_{(p,\lambda,t)}(x) = \exp_p\left(\left((1-t) + t\frac{\lambda}{d(p,x)}\right)\exp_p^{-1}(x)\right).$$

Notice that $R_{(p,\pi/2,t)} : \partial B_{\pi/2}(p) \to \partial B_{\pi/2}(p)$ is the identity map for every $p \in S^3$ and $t \in [0,1]$. The $R_{(p,\lambda,t)}$ are length-decreasing maps that satisfy

- $R_{(p,\lambda,0)}(x) = x$ for all $x \in \overline{B}_{\pi/2}(p) \setminus B_\lambda(p)$,
- $R_{(p,\lambda,1)}\left(\overline{B}_{\pi/2}(p) \setminus B_\lambda(p)\right) \subset \partial B_\lambda(p)$.

We now define $\tilde{C} : \overline{B}^4_2(0) \times [0,1] \to \mathcal{Z}_2(S^3)$ by

$$\tilde{C}(v,t) = \begin{cases} C(v,t) & \text{if } v \in \overline{B}^4_1(0) \text{ and } t \in [0,1], \\ R(-\frac{v}{|v|}, \pi - r(t), |v|-1)_\# C(\frac{v}{|v|}, t) & \text{if } v \in \overline{B}^4_2(0) \setminus \overline{B}^4_1(0) \text{ and } t \in [0,1/2], \\ R(\frac{v}{|v|}, r(t), |v|-1)_\# C(\frac{v}{|v|}, t) & \text{if } v \in \overline{B}^4_2(0) \setminus \overline{B}^4_1(0) \text{ and } t \in [1/2,1]. \end{cases}$$

If $v \in \partial\overline{B}^4_2(0)$, then $\mathrm{supp}(\tilde{C}(v,t)) \subset \partial B_{r(t)}(\frac{v}{|v|})$. By the Constancy Theorem there exists $k(v,t) \in \mathbb{Z}$ such that

$$\tilde{C}(v,t) = k(v,t)\cdot \partial B_{r(t)}(\frac{v}{|v|}).$$

Since $\tilde{C}(v,1/2) = -\mathrm{lk}(\gamma_1,\gamma_2)\cdot\partial B_{\pi/2}(\frac{v}{|v|})$, it follows by continuity in the flat topology that $k(v,t) = -\mathrm{lk}(\gamma_1,\gamma_2)$ for every $v \in \partial\overline{B}^4_2(0)$ and $t \in [0,1]$.

After an appropriate reparametrization, we get:

Theorem

Let (γ_1,γ_2) be a 2-component link in S^3 with $\mathrm{lk}(\gamma_1,\gamma_2) = -1$. *The map*

$$\Phi : I^5 \to \mathcal{Z}_2(S^3),$$

obtained by reparametrization of $\tilde{C}$, is continuous in the flat topology and satisfies the following properties:

(1) $\Phi(x,0) = \Phi(x,1) = 0$ *for any* $x \in I^4$,
(2) $\Phi(x,t) = \partial B_{r(t)}\left(\frac{f(x)}{|f(x)|}\right)$ *for every* $(x,t) \in \partial I^4 \times I$,
(3) $\Phi(x,1/2) = \partial B_{\pi/2}(\frac{f(x)}{|f(x)|})$ *for any* $x \in \partial I^4$,
(4) $\sup\{M(\Phi(x)) : x \in I^5\} \le E(\gamma_1,\gamma_2)$,
(5) *the center map* $Q : \partial I^4 \to S^3$, *given by* $Q(x) = \frac{f(x)}{|f(x)|}$, *satisfies* $\deg(Q) = 1$.

We can consider the homotopy class Π relative to ∂I^5 of the family $\Phi : I^5 \to \mathcal{Z}_2(S^3)$. By the previous theory, since $\deg(Q) \neq 0$, we conclude that $L(\Pi) \geq 2\pi^2$. Therefore there exists $y \in I^5$ such that $\text{area}(\Phi(y)) \geq 2\pi^2$. Since $\text{area}(\Phi(x)) \leq E(\gamma_1, \gamma_2)$ for every $x \in I^5$, we get $E(\gamma_1, \gamma_2) \geq 2\pi^2$. The case $\text{lk}(\gamma_1, \gamma_2) = 1$ also follows by a change in orientation. This proves the Freedman-He-Wang conjecture. The rigidity part follows by exploiting the fact that one must have

$$\sup\{\text{area}(C(v, z)) : (v, z) \in B^4 \times (0, 1)\} = E(\gamma_1, \gamma_2)$$

if $E(\gamma_1, \gamma_2) = 2\pi^2$.

Bibliography

[1] I. Agol, F. C. Marques and A. Neves, *Min-max theory and the energy of links*, arXiv:1205.0825 (2012)

[2] F. Almgren, *The homotopy groups of the integral cycle groups.* Topology (1962), 257–299. MR0146835 (26:4355)

[3] F. Almgren, *The theory of varifolds.* Mimeographed notes, Princeton (1965).

[4] F. Almgren, *Some interior regularity theorems for minimal surfaces and an extension of Bernstein's theorem.* Ann. of Math. (2) 84 (1966), 277–292. MR0200816 (34:702)

[5] W. Blaschke, *Vorlesungen Über Differentialgeometrie III,* Berlin: Springer (1929).

[6] T. Colding and C. De Lellis, *The min-max construction of minimal surfaces,* Surveys in Differential Geometry VIII , International Press, (2003), 75–107. MR2039986 (2005a:53008)

[7] M. Freedman, Z-X. He, Z. Wang, *Möbius energy of knots and unknots,* Ann. of Math. (2) 139 (1994), no. 1, 1–50. MR1259363 (94j:58038)

[8] Zheng-Xu He, *On the minimizers of the Möbius cross energy of links*, Experiment. Math. **11** (2002), no. 2, 244–248. MR1959266 (2003k:58016)

[9] E. Heintze and H. Karcher, *A general comparison theorem with applications to volume estimates for submanifolds,* Annales scientifiques de l'E.N.S. 11 (1978), 451–470. MR533065 (80i:53026)

[10] D. Kim and R. Kusner, *Torus knots extremizing the Möbius energy,* Experiment. Math. Volume 2, Issue 1 (1993), 1-9. MR1246479 (94j:58039)

[11] B. Lawson, *Complete minimal surfaces in S^3,* Ann. of Math. (2) 92 (1970), 335–374. MR0270280 (42:5170)

[12] P. Li and S-T. Yau, *A new conformal invariant and its applications to the Willmore conjecture and the first eigenvalue of compact surfaces,* Invent. Math. 69 (1982), 269–291. MR674407 (84f:53049)

[13] F. C. Marques and A. Neves, *Min-max theory and the Willmore conjecture*, Annals of Math. (2) 179 (2014), 683–782. MR3152944

[14] F. Morgan, *Geometric measure theory: a beginner's guide, third edition,* Academic Press, 2000. MR1775760 (2001j:49001)

[15] J. O'Hara, *Energy of a knot,* Topology 30 (1991), 241–247. MR1098918 (92c:58017)

[16] J. Pitts, *Existence and regularity of minimal surfaces on Riemannian manifolds,* Mathematical Notes 27, Princeton University Press, Princeton, (1981). MR626027 (83e:49079)

[17] T. Rivière, *Weak immersions of surfaces with finite total curvature*, Lectures at the 23nd Annual PCMI Summer Session (2013).

[18] A. Ros, *The Willmore conjecture in the real projective space,* Math. Res. Lett. 6 (1999), 487–493. MR1739208 (2001a:53016)

[19] R. Schoen and L. Simon, *Regularity of stable minimal hypersurfaces.* Comm. Pure Appl. Math. 34 (1981), 741–797. MR634285 (82k:49054)

[20] L. Simon, *Lectures on geometric measure theory,* Proceedings of the Centre for Mathematical Analysis, Australian National University, Canberra, (1983). vii+272 pp. MR756417 (87a:49001)

[21] F. Smith, *On the existence of embedded minimal 2-spheres in the 3-sphere, endowed with an arbitrary Riemannian metric*, supervisor L. Simon, University of Melbourne (1982).

[22] G. Thomsen, *Über Konforme Geometrie, I: Grundlagen der Konformen Flächentheorie,* Abh. Math. Sem. Hamburg (1923), 31–56.

[23] F. Urbano, *Minimal surfaces with low index in the three-dimensional sphere,* Proc. Amer. Math. Soc. 108 (1990), 989–992. MR1007516 (90h:53073)

[24] T. J. Willmore, *Note on embedded surfaces,* An. Sti. Univ."Al. I. Cuza" Iasi Sect. I a Mat. (N.S.) 11B (1965) 493–496. MR0202066 (34:1940)

Weak immersions of surfaces with L^2-bounded second fundamental form

Tristan Riviére

IAS/Park City Mathematics Series
Volume 22, 2013

Weak immersions of surfaces with L^2-bounded second fundamental form

Tristan Rivière

The first appearance of the *Willmore functional* is in the work of Sophie Germain on elastic surfaces. Following the main lines of J. Bernoulli and Euler's studies of the mechanics of sticks in the first half of the eighteenth century, Germain formulated what she called the *fundamental hypothesis*: *at one point of the surface the elastic force which counterbalances the external forces is proportional to the sum of the principal curvature at this point*, i.e., what we now call the *Mean curvature*. Although Germain's work was controversial at the time and was criticized by contemporaries including Simeon Poisson, the *Willmore energy* was finally established on physical grounds by G. Kirchhoff in 1850 [**Ki**] as the free energy of an elastic membrane modulo the addition of a null Lagrangian (see for instance [**LaLi**]).

The *Willmore energy* was probably first considered in geometry in the early twentieth century by Wilhelm Blaschke in his efforts to merge *minimal surface theory* with *conformal invariance*. He proved that the Willmore energy of a closed surface remains invariant under conformal transformations of the ambient space. As proved quite recently [**MoN**], this Lagrangian is the unique one possessing this property modulo the addition of a multiple of the Gauss curvature.

Because of its simplicity and fundamental nature, the Willmore energy appears in many areas of science and technology. Beyond *non-linear elasticity* and *conformal geometry*, one sees it for instance in *general relativity*, where the *Willmore energy* is the main term in the so called *Hawking Mass*. It arises also in *cell biology* as the main term of the free elastic energy of lipid bilayer membranes and is called the *Helfrich Energy* there. The *Willmore Energy* in its umbilic form also arises in the design of multifocal optical elements (see for instance [**KaRu**]), etc.

For many years, the only surfaces known to be critical for the Willmore Lagrangian were the minimal surfaces and their images under conformal transformation. This perhaps explains why there was essentially no work on the variational theory of the Willmore functional during the several decades after Blaschke's seminal work. In the short but decisive paper [**Wi**] in 1965, Tom Willmore reintroduced this Lagrangian, which now bears his name. He formulated there what is now called the *Willmore Conjecture*, which asserts that the Willmore energy of a compact surface of non-zero genus in three-dimensional Euclidian space is at least $2\pi^2$, and this minimal energy is only achieved by the stereographic projection of the *Clifford Torus* in S^3 (and its conformal transformations). This famous conjecture was proved in 2011 by Fernando Codá Marques and André Neves in [**MN**] and their

Swiss Federal Institute of Technology Zurich, Department of Mathematics, 8092 Zurich, Switzerland

E-mail address: tristan.riviere@math.ethz.ch

proof is the subject of one of the mini-course given at the present Park City Summer School 2013.

In the present series of five lectures, we develop some fundamental tools in the analytic study of the Willmore Lagrangian motivated by the following set of questions:

i) Does there exists a minimizer of the Willmore functional among all smooth immersions of a fixed surface Σ^2? If there is, can one estimate the energy and special properties of such a minimizer?
ii) Does there exist a minimizer of the Willmore functional among a more restricted class of immersions, for example, considering only the conformal immersions relative to a fixed conformal class c on Σ, or among all immersions of Σ into $\mathbb{R}^3$ which enclose a domain of given volume and which have a fixed area?
iii) How stable is the Willmore equation? In other words, does a sequence of "approximately Willmore" surfaces, e.g. Palais-Smale sequences for the Willmore energy, necessarily converge to a Willmore surface?
iv) Can one produce Willmore Surfaces using *min-max arguments* ? More specifically can one apply fundamental variational principles such as *Ekeland's variational Principles* or the *Mountain pass lemma* to the Willmore functional?

Question i) was first considered by Leon Simon in [**Sim86**] and completely solved when Σ is the torus T^2 in [**Sim93**]. Later, in [**BK03**], a condition shown in [**Sim93**] to be sufficient to guarantee the existence of minimizing Willmore surfaces was proved to hold for every oriented closed surface. Simon's approach has two main features. First, it is an *ambient approach*: namely, surfaces are viewed as subsets[1] of the ambient Euclidian space. The second is the use of *biharmonic approximation*, where almost-minimizing surfaces are "improved" by replacing small pieces by *biharmonic graphs*. The main drawback of this strategy is that it seems to be only well-suited for the study of minimizing surfaces, but not for questions of the form iii) and iv) for non-minimizing critical points. Simon's *ambient approach* was also used in [**KS**] to give partial answers to question ii).

In 2010, the author introduced the notion of *weak immersions* in order to provide a suitable framework in which general variations of the Willmore Lagrangian (see [**Riv10**]) are well posed. The goal of this mini-course is to present fundamental properties of this approach, which turns out to be a fundamental tool for addressing all of the questions i)...iv) . To illustrate this, we explain in the last lecture how to use these properties to give a new proof of Simon's existence result. More completely, we address in this mini-course the following questions:

a) Does a weak immersion define a smooth conformal structure ?
b) What happens to a sequence of weak immersions of a given surface Σ^2 for which the Willmore energy is uniformly bounded? Does it convergence in some sense to a weak immersion ?
c) Is there a weak formulation of the Willmore equation which is compatible with the notion of weak immersion ?

---

[1] More precisely, this approaches considers immersed surfaces as varifolds, i.e., Radon measures in the Grassman manifold of 2-planes of Euclidian space corresponding to the tangent bundle of these surfaces immersed into the tangent bundle the euclidian ambient space.

d) Are weak immersions which are solutions to the Willmore equation necessarily smooth?

The five lectures of this mini-course correspond to a presentation of the 2006 and 2010 papers [**Riv06**] and [**Riv10**], published respectively in 2008 and 2014. Since that time, weak immersions have played a crucial role in answering questions i)–iv) see [**Riv10**], [**Riv15**], [**KMR**] and [**Riv16**].

LECTURE 1

Notations and fundamental results on the differential geometry of surfaces

1. Notations

For a smooth map $\vec{e}\colon D^2 \to \mathbb{R}^m$, denote

$$\nabla\vec{e} := \begin{pmatrix} \partial_{x_1}\vec{e} \\ \partial_{x_2}\vec{e} \end{pmatrix}. \tag{1.1}$$

The associated 1-form is denoted by $d\vec{e}$ and in local coordinates given by

$$d\vec{e} = \partial_{x_1}\vec{e}\, dx_1 + \partial_{x_2}\vec{e}\, dx_2.$$

Further, define

$$\nabla^{\perp}\vec{e} := \begin{pmatrix} -\partial_{x_2}\vec{e} \\ \partial_{x_1}\vec{e} \end{pmatrix}. \tag{1.2}$$

It corresponds to the 1-form $*d\vec{e}$, where $*$ is the Hodge operator from $\wedge^1\mathbb{R}^2$ to $\wedge^1\mathbb{R}^2$. Indeed, in local coordinates we have

$$*d\vec{e} = \partial_{x_1}\vec{e} * dx_1 + \partial_{x_2}\vec{e} * dx_2 = -\partial_{x_2}\vec{e}\, dx_1 + \partial_{x_1}\vec{e}\, dx_2.$$

For smooth maps $\vec{e}, \vec{f}\colon D^2 \to \mathbb{R}^m$, denote

$$\langle\vec{e}, \nabla\vec{f}\rangle := \begin{pmatrix} \langle\vec{e}, \partial_{x_1}\vec{f}\rangle \\ \langle\vec{e}, \partial_{x_2}\vec{f}\rangle \end{pmatrix}, \qquad \langle\vec{e}, \nabla^{\perp}\vec{f}\rangle := \begin{pmatrix} -\langle\vec{e}, \partial_{x_2}\vec{f}\rangle \\ \langle\vec{e}, \partial_{x_1}\vec{f}\rangle \end{pmatrix}, \tag{1.3}$$

$$\vec{e}\times\nabla\vec{f} := \begin{pmatrix} \vec{e}\times\partial_{x_1}\vec{f} \\ \vec{e}\times\partial_{x_2}\vec{f} \end{pmatrix}, \qquad \vec{e}\times\nabla^{\perp}\vec{f} := \begin{pmatrix} -\vec{e}\times\partial_{x_2}\vec{f} \\ \vec{e}\times\partial_{x_1}\vec{f} \end{pmatrix}, \tag{1.4}$$

where $\langle\cdot,\cdot\rangle$ is the inner product of $\mathbb{R}^m$ and we use the $\times$-operation, if $m = 3$. Similarly, for $\lambda\colon D^2 \to \mathbb{R}$ and $\vec{e}\colon D^2 \to \mathbb{R}^m$ we write

$$\nabla\lambda\cdot\vec{e} := \begin{pmatrix} \partial_{x_1}\lambda\,\vec{e} \\ \partial_{x_2}\lambda\,\vec{e} \end{pmatrix}, \quad \nabla^{\perp}\lambda\cdot\vec{e} := \begin{pmatrix} -\partial_{x_2}\lambda\,\vec{e} \\ \partial_{x_1}\lambda\,\vec{e} \end{pmatrix}. \tag{1.5}$$

As notation suggests, we write for instance

$$\langle\nabla^{\perp}\vec{e}, \nabla\vec{f}\rangle = \langle\partial_{x_1}\vec{e}, \partial_{x_2}\vec{f}\rangle - \langle\partial_{x_2}\vec{e}, \partial_{x_1}\vec{f}\rangle, \tag{1.6}$$

and if $m = 3$,

$$\nabla\vec{e}\times\nabla^{\perp}\vec{f} = -\partial_{x_1}\vec{e}\times\partial_{x_2}\vec{f} + \partial_{x_2}\vec{e}\times\partial_{x_1}\vec{f}.$$

Naturally, denote

$$|\nabla\vec{e}|^2 := \langle\nabla\vec{e}, \nabla\vec{e}\rangle.$$

Similarly,

$$\nabla\lambda\cdot\nabla\vec{e} := \partial_{x_1}\lambda\,\partial_{x_1}\vec{e}+\partial_{x_2}\lambda\,\partial_{x_2}\vec{e}, \tag{1.7}$$

$$\nabla^{\perp}\lambda\cdot\nabla\vec{e} := -\partial_{x_2}\lambda\,\partial_{x_1}\vec{e}+\partial_{x_1}\lambda\,\partial_{x_2}\vec{e}. \tag{1.8}$$

Applying the divergence operator *div* to a vector field

$$\vec{X}=\begin{pmatrix}\vec{X}_1\\ \vec{X}_2\end{pmatrix}$$

with components $\vec{X}_i\colon D^2\to\mathbb{R}^m$ (for instance $\vec{X}$ of the form (1.1), (1.2), (1.3), (1.4), (1.5)) is to be understood as

$$div\,\vec{X}=\partial_{x_1}\vec{X}_1+\partial_{x_2}\vec{X}_2.$$

Note that

$$div\langle\vec{e},\nabla^{\perp}\vec{f}\rangle=\langle\nabla\vec{e},\nabla^{\perp}\vec{f}\rangle=-\langle\nabla^{\perp}\vec{e},\nabla\vec{f}\rangle \tag{1.9}$$

and for $m=3$,

$$\begin{aligned} div\left[\vec{e}\times\nabla^{\perp}\vec{f}\right] &= -\partial_{x_1}\vec{e}\times\partial_{x_2}\vec{f}+\partial_{x_2}\vec{e}\times\partial_{x_1}\vec{f}\\ &=\nabla\vec{f}\times\nabla^{\perp}\vec{e}. \end{aligned} \tag{1.10}$$

Similarly, we use the above introduced notation for 1-forms. For instance, $\langle\vec{e},d\vec{f}\rangle$ is the 1-form, which is in coordinates given by

$$\langle\vec{e},d\vec{f}\rangle:=\langle\vec{e},\partial_{x_1}\vec{f}\rangle\,dx_1+\langle\vec{e},\partial_{x_2}\vec{f}\rangle\,dx_2$$

and thus the associated 1-form to the vector on the left hand side of (1.3). Finally, define the 2-form

$$\langle d\vec{e},d\vec{f}\rangle:=d\langle\vec{e},d\vec{f}\rangle, \tag{1.11}$$

which in local coordinates is given by

$$\langle d\vec{e},d\vec{f}\rangle=\Big(\langle\partial_{x_1}\vec{e},\partial_{x_2}\vec{f}\rangle-\langle\partial_{x_2}\vec{e},\partial_{x_1}\vec{f}\rangle\Big)\,dx_1\wedge dx_2. \tag{1.12}$$

Observe that the Jacobian occurring in (1.12) is (1.6).

2. Immersions and their geometry

A smooth map $\vec{\Phi}\colon M\to N$ between two smooth manifolds M and N is called *immersion* if $d\vec{\Phi}_p\colon T_pM\to T_{f(p)}N$ is injective for all $p\in M$.

In this course, Σ denotes a smooth 2-*dimensional closed oriented manifold* and we usually consider immersions (later understood in a weak sense) from Σ into $\mathbb{R}^m$.

2.1. Submanifolds of $\mathbb{R}^3$

We start to consider a particularly easy class of immersions, namely the class of 2-dimensional submanifolds of $\mathbb{R}^3$ (the immersion being provided by the inclusion map).

Let S be a 2-dimensional submanifold of $\mathbb{R}^3$. We assume S to be oriented and we denote by $\vec{n}$ the associated *Gauss map*: the unit normal giving this orientation.

The *first fundamental form* is the induced metric on S that we denote by g: For any $p\in S$ and $\vec{X},\vec{Y}\in T_pS$, it is given by

$$g_p(\vec{X},\vec{Y}):=\langle\vec{X},\vec{Y}\rangle,$$

where $\langle\cdot,\cdot\rangle$ is the canonical inner product in $\mathbb{R}^3$. The *volume form* associated to g on S is locally given by

$$dvol_g := \sqrt{det(g(\partial_{x_i},\partial_{x_j}))}\; dx_1 \wedge dx_2,$$

where (x_1, x_2) are arbitrary local positive coordinates.[1]

The *second fundamental form* at $p \in S$ is the bilinear map which assigns to a pair of vectors $\vec{X}, \vec{Y}$ in T_pS an orthogonal vector to T_pS that we shall denote by $\vec{\mathbb{I}}(\vec{X},\vec{Y})$. This normal vector expresses how much the Gauss map varies along these directions $\vec{X}$ and $\vec{Y}$. Precisely, it is given by

$$\begin{array}{rccc} \vec{\mathbb{I}}_p\colon & T_pS \times T_pS & \longrightarrow & N_pS \\ & (\vec{X},\vec{Y}) & \mapsto & -\langle d\vec{n}_p(\vec{X}),\vec{Y}\rangle\, \vec{n}(p). \end{array} \tag{1.13}$$

Extending smoothly $\vec{X}$ and $\vec{Y}$ first locally on S and then in a neighborhood of p in $\mathbb{R}^3$, one has, using $\langle\vec{n},\vec{Y}\rangle = 0$ on S,

$$\begin{aligned} \vec{\mathbb{I}}(\vec{X},\vec{Y}) &= \langle\vec{n}, d\vec{Y}(\vec{X})\rangle\, \vec{n} \quad = d\vec{Y}(X) - \nabla_{\vec{X}}\vec{Y} \\ &= \overline{\nabla}_{\vec{X}}\vec{Y} - \nabla_{\vec{X}}\vec{Y}. \end{aligned} \tag{1.14}$$

Here, ∇ is the Levi-Civita connection on S generated by g, and we have $\nabla_{\vec{X}}\vec{Y} = \pi_T(d\vec{Y}(X))$, where π_T is the orthogonal projection onto TS. $\overline{\nabla}$ is the Levi-Civita connection associated to the flat metric and is simply given by $\overline{\nabla}_{\vec{X}}\vec{Y} = d\vec{Y}(X)$.

An elementary but fundamental property of the second fundamental form says that it is symmetric[2]. It can then be diagonalized in an orthonormal basis and the two eigenvalues κ_1 and κ_2 are called the *principal curvatures* of the surface at p. The *mean curvature* is then given by

$$H := \frac{\kappa_1 + \kappa_2}{2}$$

and the *mean curvature vector* by

$$\vec{H} := H\,\vec{n} = \frac{1}{2}\,\mathrm{tr}(g^{-1}\,\vec{\mathbb{I}}) = \frac{1}{2}\sum_{i,j=1}^{2} g^{ij}\,\vec{\mathbb{I}}(\partial_{x_i},\partial_{x_j}), \tag{1.15}$$

where (x_1, x_2) are arbitrary local coordinates. Here, $(g^{ij})_{ij}$ denotes the inverse matrix to $(g_{ij}) := (g(\partial_{x_i},\partial_{x_j}))$. In particular, if $(\vec{e}_1,\vec{e}_2)$ is an orthonormal basis of T_pS, (1.15) becomes

$$\vec{H} = \frac{\vec{\mathbb{I}}(\vec{e}_1,\vec{e}_2) + \vec{\mathbb{I}}(\vec{e}_2,\vec{e}_2)}{2}. \tag{1.16}$$

[1] Local coordinates, denoted by (x_1, x_2), are provided by a diffeomorphism x from an open set in $\mathbb{R}^2$ onto an open set in Σ. For any point q in this open set of S we shall denote by $x_i(q)$ the canonical coordinates in $\mathbb{R}^2$ of $x^{-1}(q)$. Finally ∂_{x_i} is the vector-field on S given by $\partial x/\partial x_i$.

[2] This can be seen combining equation (1.14) with the fact that the two Levi-Civita connections ∇ and $\overline{\nabla}$ are torsion-free, i.e. we have

$$\nabla_{\vec{X}}\vec{Y} - \nabla_{\vec{Y}}\vec{X} = [X, Y]$$

and

$$\overline{\nabla}_{\vec{X}}\vec{Y} - \overline{\nabla}_{\vec{Y}}\vec{X} = [X, Y].$$

The Gauss curvature is given by

$$K := \frac{det\left(\langle\vec{n},\vec{\mathbb{I}}(\partial_{x_i},\partial_{x_j})\rangle\right)}{det(g_{ij})} = \kappa_1\ \kappa_2. \tag{1.17}$$

The *Willmore functional* of the surface Σ is defined by

$$W(S) = \int_S |\vec{H}|^2\ dvol_g = \frac{1}{4}\int_S |\kappa_1+\kappa_2|^2\ dvol_g.$$

The *Gauss-Bonnet theorem*[3] asserts that the integral of $K\ dvol_g$ is proportional to a topological invariant of S: $\chi(S)$, the Euler characteristic of S. Precisely, one has

$$\begin{aligned}\int_S K\ dvol_g = \int_S \kappa_1\ \kappa_2\ dvol_g &= 2\pi\ \chi(S)\\ &= 4\pi\ (1-g(S)),\end{aligned} \tag{1.18}$$

where $g(S)$ denotes the genus of S. Combining the definition of W and this last identity one obtains[4]

$$\begin{aligned}W(S)-\pi\ \chi(S) &= \frac{1}{4}\int_S(\kappa_1^2+\kappa_2^2)\ dvol_g\\ &= \frac{1}{4}\int_S|\vec{\mathbb{I}}|^2_g\ dvol_g = \frac{1}{4}\int_S |d\vec{n}|^2_g\ dvol_g.\end{aligned} \tag{1.19}$$

Hence modulo the addition of a topological term, the Willmore energy corresponds to the *Sobolev homogeneous $\dot{H}^1-$energy of the Gauss map* for the induced metric g.

2.2. Immersions of an abstract surface into $\mathbb{R}^m$.

Let us now consider the general case that $\vec{\Phi}\colon\Sigma\to\mathbb{R}^m$ is a immersion of an abstract 2-dimensional oriented closed manifold Σ into $\mathbb{R}^m$.

The *first fundamental form* associated to the immersion is the metric $g := \vec{\Phi}^*g_{\mathbb{R}^m}$ induced by $\vec{\Phi}$, where $g_{\mathbb{R}^m}$ is the canonical metric on $\mathbb{R}^m$: For all $p\in\Sigma$ and $X,Y\in T_p\Sigma$,

$$g_p(X,Y) := \langle d\vec{\Phi}_p(X), d\vec{\Phi}_p(Y)\rangle$$

where $\langle\cdot,\cdot\rangle$ is the canonical inner product in $\mathbb{R}^m$. The *volume form* associated to g on Σ is locally given by

$$dvol_g := \sqrt{det(g(\partial_{x_i},\partial_{x_j}))}\ dx_1\wedge dx_2,$$

where (x_1,x_2) are arbitrary local positive coordinates.

We shall denote by $\vec{e}$ the map which to a point in Σ assigns the oriented 2-plane[5], given by the push-forward by $\vec{\Phi}$ of the oriented tangent space $T_p\Sigma$. Using a

---

[3]See for instance **[dC76]**

[4]The last identity comes from the fact that at a point p, taking an orthonormal basis $(\vec{e}_1,\vec{e}_2)$ of T_pS, one has:

$$|d\vec{n}|^2_g = \sum_{i,j=1}^2 \langle d\vec{n}(\vec{e_i}),\vec{e}_j\rangle^2 = \sum_{i,j=1}^2 |\vec{\mathbb{I}}(\vec{e}_i,\vec{e}_j)|^2 = |\vec{\mathbb{I}}|^2,$$

since $\langle d\vec{n},\vec{n}\rangle = 0$.

[5]We denote by $\tilde{G}_p(\mathbb{R}^m)$, the Grassman space of oriented p-planes in $\mathbb{R}^m$ that we interpret as the space of unit simple p-vectors in $\mathbb{R}^m$ which is included in the Grassmann algebra $\wedge^p\mathbb{R}^m$.

positive orthonormal basis $(\vec{e}_1, \vec{e}_2)$ of $\vec{\Phi}_* T_p\Sigma$, an explicit expression of $\vec{e}$ is given by

$$\vec{e} = \vec{e}_1 \wedge \vec{e}_2 .$$

With these notations the *Gauss map* which to every point p assigns the oriented $(m-2)$-orthogonal plane to $\vec{\Phi}_* T_p\Sigma$ is given by

$$\vec{n} = \star\vec{e} = \vec{n}_1 \wedge \cdots \wedge \vec{n}_{m-2} .$$

Here, $\star$ is the Hodge operator[6] from $\wedge^2\mathbb{R}^m$ to $\wedge^{m-2}\mathbb{R}^m$ and $(\vec{n}_1, \dots, \vec{n}_{m-2})$ is a positive orthonormal basis of the oriented normal plane to $\vec{\Phi}_* T_p\Sigma$. (Consequently, $(\vec{e}_1, \vec{e}_2, \vec{n}_1, \cdots, \vec{n}_{m-2})$ is an orthonormal basis of $\mathbb{R}^m$.)

We shall denote by $\pi_{\vec{n}}$ the orthogonal projection onto the $(m-2)$-plane at p given by $\vec{n}(p)$.

The *second fundamental form* associated to the immersion $\vec{\Phi}$ is the following map:

$$\begin{aligned} \vec{\mathbb{I}}_p \colon T_pS \times T_pS &\longrightarrow (\vec{\Phi}_* T_p\Sigma)^\perp \\ (\vec{X}, \vec{Y}) &\mapsto \pi_{\vec{n}}\left(d^2\vec{\Phi}(X, Y)\right), \end{aligned}$$

where X and Y are extended smoothly into local smooth vector fields around p. One easily verifies that, although $d^2\vec{\Phi}(X,Y)$ might depend on these extensions, $\pi_{\vec{n}}(d^2\vec{\Phi}(X,Y))$ does not and we have then defined a tensor.

Let $\vec{X} := d\vec{\Phi}(X)$ and $\vec{Y} := d\vec{\Phi}(Y)$. Denote also by π_T the orthogonal projection onto $\vec{\Phi}_* T_q\Sigma$.

$$\begin{aligned} \pi_{\vec{n}}\left(d^2\vec{\Phi}(X,Y)\right) &= d(d\vec{\Phi}(X))(Y) - \pi_T\left(d(d\vec{\Phi}(X))(Y)\right) \\ &= d\vec{X}(Y) - \nabla_Y X \qquad (1.20) \\ &= \overline{\nabla}_{\vec{Y}}\vec{X} - \nabla_Y X, \end{aligned}$$

where $\overline{\nabla}$ is the Levi-Civita connection in $\mathbb{R}^m$ for the canonical metric and ∇ is the Levi-Civita connection on $T\Sigma$ induced by the metric g. Here again, as in the 3-d case in the previous subsection, from the fact that Levi-Civita connections are torsion-free we can deduce the symmetry of the second fundamental form.

Similarly to the 3-d case, the *mean curvature vector*[7] is given by

$$\vec{H} := \frac{1}{2}\,\mathrm{tr}(g^{-1}\,\vec{\mathbb{I}}) = \frac{1}{2}\sum_{i,j=1}^{2} g^{ij}\,\vec{\mathbb{I}}(\partial_{x_i}, \partial_{x_j}), \qquad (1.21)$$

[6] The Hodge operator on $\mathbb{R}^m$ is the linear map from $\wedge^p\mathbb{R}^m$ to $\wedge^{m-p}\mathbb{R}^m$ which to a $p-$vector α assigns the $(m-p)$-vector $\star\alpha$ on $\mathbb{R}^m$, which is characterized by the following property: for any p-vector β in $\wedge^p\mathbb{R}^m$,

$$\beta \wedge \star\alpha = \langle \beta, \alpha \rangle\, \varepsilon_1 \wedge \cdots \wedge \varepsilon_m ,$$

where $(\varepsilon_1, \cdots, \varepsilon_m)$ is the canonical orthonormal basis of $\mathbb{R}^m$ and $\langle \cdot, \cdot \rangle$ is the canonical inner product on $\wedge^p\mathbb{R}^m$.

[7] observe that the notion of mean curvature H does not make sense any more in codimension larger than 1 unless a normal direction is given.

where (x_1, x_2) are arbitrary local coordinates in Σ and $(g^{ij})_{ij}$ is the inverse matrix to $(g(\partial_{x_i}, \partial_{x_j}))$.

We can now give the general formulation of the *Willmore energy* of an immersion $\vec{\Phi}$ in $\mathbb{R}^m$ of an abstract surface Σ:

$$W(\vec{\Phi}) := \int_\Sigma |\vec{H}|^2 \, dvol_g.$$

A fundamental theorem by Gauss gives an expression of the intrinsic Gauss curvature in terms of the second fundamental form of any immersion of the surface in $\mathbb{R}^m$. Precisely this theorem says (see theorem 2.5 chapter 6 of [**dC92**])

$$K = \langle \vec{\mathbb{I}}(e_1, e_1), \vec{\mathbb{I}}(e_2, e_2) \rangle - \langle \vec{\mathbb{I}}(e_1, e_2), \vec{\mathbb{I}}(e_1, e_2) \rangle, \tag{1.22}$$

where (e_1, e_2) is an arbitrary orthonormal basis of $T_p\Sigma$.

Note that in particular, the Gauss curvature is controlled by the second fundamental form as follows:

$$\begin{aligned} |K| &\le |\vec{\mathbb{I}}(\vec{e}_1, \vec{e}_1)||\vec{\mathbb{I}}(\vec{e}_2, \vec{e}_2)| + |\vec{\mathbb{I}}(\vec{e}_1, \vec{e}_2)|^2 \\ &\le \frac{1}{2} \left(|\vec{\mathbb{I}}(\vec{e}_1, \vec{e}_1)|^2 + |\vec{\mathbb{I}}(\vec{e}_2, \vec{e}_2)|^2 + 2|\vec{\mathbb{I}}(\vec{e}_1, \vec{e}_2)|^2 \right) \\ &= \frac{1}{2} |\vec{\mathbb{I}}|_g^2. \end{aligned} \tag{1.23}$$

From identity (1.22) we furthermore deduce easily

$$|\vec{\mathbb{I}}|_g^2 = 4|\vec{H}|^2 - 2K. \tag{1.24}$$

Hence, using Gauss-Bonnet theorem, we obtain the following expression of the Willmore energy of an immersion into $\mathbb{R}^m$ of an arbitrary closed surface:

$$W(\vec{\Phi}) = \frac{1}{4} \int_\Sigma |\vec{\mathbb{I}}|_g^2 \, dvol_g + \pi \, \chi(\Sigma). \tag{1.25}$$

Let us take a local *normal frame* around $p \in \Sigma$: a smooth map $(\vec{n}_1, \cdots, \vec{n}_{m-2})$ from a neighborhood $U \subset \Sigma$ into $(S^{m-1})^{m-2}$ such that for any point $q \in U$, $(\vec{n}_1(q), \cdots, \vec{n}_{m-2}(q))$ realizes a positive orthonormal basis of $(\vec{\Phi}_* T_q\Sigma)^\perp$. Then

$$\pi_{\vec{n}} \left(d^2\vec{\Phi}(X, Y) \right) = \sum_{\alpha=1}^{m-2} \langle d^2\vec{\Phi}(X, Y), \vec{n}_\alpha \rangle \, \vec{n}_\alpha,$$

from which we deduce the following expression, which is the natural extension of (1.13):

$$\vec{\mathbb{I}}(X, Y) = -\sum_{\alpha=1}^{m-2} \langle d\vec{n}_\alpha(X), \vec{Y} \rangle \, \vec{n}_\alpha, \tag{1.26}$$

where we denote $\vec{Y} := d\vec{\Phi}(Y)$. Let (e_1, e_2) be an orthonormal basis of $T_p\Sigma$. Then the previous expression of the second fundamental form implies

$$|\vec{\mathbb{I}}|_g^2 = \sum_{i,j=1}^{2} \sum_{\alpha=1}^{m-2} |\langle d\vec{n}_\alpha(e_i), \vec{e}_j \rangle|^2 = \sum_{i=1}^{2} \sum_{\alpha=1}^{m-2} |\langle d\vec{n}_\alpha, \vec{e}_i \rangle|^2. \tag{1.27}$$

Observe that

$$
\begin{aligned}
d\vec{n} &= \sum_{\alpha=1}^{m-2} (-1)^{\alpha-1} d\vec{n}_\alpha \wedge_{\beta\neq\alpha} \vec{n}_\beta \\
&= \sum_{i=1}^{2}\sum_{\alpha=1}^{m-2} (-1)^{\alpha-1} \langle d\vec{n}_\alpha, \vec{e}_i\rangle \; \vec{e}_i \wedge_{\beta\neq\alpha} \vec{n}_\beta .
\end{aligned}
\tag{1.28}
$$

$(\vec{e}_i \wedge_{\beta\neq\alpha} \vec{n}_\beta)$ for $\alpha = 1\ldots, m-2$ and $i = 1, 2$ realizes a linearly independent family of $2(m-2)$ orthonormal vectors in $\wedge^{m-2}\mathbb{R}^m$. Hence

$$
|d\vec{n}|_g^2 = \sum_{i=1}^{2}\sum_{\alpha=1}^{m-2} |\langle d\vec{n}_\alpha, \vec{e}_i\rangle|^2 = |\vec{\mathbb{I}}|_g^2 . \tag{1.29}
$$

Combining (1.25) and (1.29), we obtain

$$
W(\vec{\Phi}) = \frac{1}{4}\int_\Sigma |d\vec{n}|_g^2 \; dvol_g + \pi\; \chi(\Sigma), \tag{1.30}
$$

which generalizes identity (1.19) to arbitrary immersions of closed 2-dimensional surfaces.

The *negative Laplace-Beltrami operator* Δ_g associated to the metric $g = g_{\vec{\Phi}}$, acting on a smooth function $f\colon \Sigma \to \mathbb{R}$, is in local coordinates given by

$$
\Delta_g f = \frac{1}{\sqrt{\det(g_{ij})}}\; \partial_{x_i}\left(g^{ij}\sqrt{\det(g_{ij})}\partial_{x_j} f\right), \tag{1.31}
$$

where we use the Einstein summation convention.

2.3. Immersions into a Riemannian manifold

We now want to have a short glance at a more general framework, namely immersions mapping into a Riemannian manifold:

Let $\vec{\Phi}$ be an immersion from a 2-dimensional closed, oriented manifold Σ into an arbitrary oriented Riemannian manifold $(M^m, \overline{g})$ of dimension $m \geq 3$.

The first fundamental form g of $\vec{\Phi}$ is the pull-back of the metric $\overline{g}$ by $\vec{\Phi}$: For all $p \in \Sigma$ and $X, Y \in T_p\Sigma$, it is given by

$$
g_p(X,Y) := \overline{g}\left(d\vec{\Phi}_p(X), d\vec{\Phi}_p(Y)\right).
$$

For defining the second fundamental form one formally replaces the exterior differential d with the Levi-Civita connection $\overline{\nabla}$ of $(M, \overline{g})$. Precisely, the second fundamental form associated to the immersion $\vec{\Phi}$ at a point $p \in \Sigma$ is the following map:

$$
\begin{aligned}
\vec{\mathbb{I}}_p : T_p\Sigma \times T_p\Sigma &\longrightarrow (\vec{\Phi}_* T_p\Sigma)^\perp \\
(X, Y) &\mapsto \pi_{\vec{n}}\left(\overline{\nabla}_{\vec{Y}}(d\vec{\Phi}(X))\right),
\end{aligned}
$$

where $\pi_{\vec{n}}$ denotes the orthogonal projection from $T_{\vec{\Phi}(p)}M$ onto $(\vec{\Phi}_* T_p\Sigma)^\perp$, the space orthogonal to $\vec{\Phi}_*(T_p\Sigma)$ with respect to the metric $\overline{g}$. As before we use the notation $\vec{Y} = d\vec{\Phi}(Y)$.

As before, the *mean curvature vector* is given by

$$\vec{H} = \frac{1}{2}\,\mathrm{tr}(g^{-1}\vec{\mathbb{I}}) = \frac{1}{2}\sum_{i,j=1}^{2} g^{ij}\,\vec{\mathbb{I}}(\partial_{x_i},\partial_{x_j}), \tag{1.32}$$

where we are using local coordinates (x_1, x_2) on Σ. The Willmore energy of $\vec{\Phi}$ is defined as

$$W(\vec{\Phi}) := \int_\Sigma |\vec{H}|^2\, dvol_g.$$

3. Conformal invariance of the Willmore Energy

DEFINITION 1.1. *i) Two metrics g and h on a smooth manifold M are said to be* conformal *(or conformally equivalent), if there exists a smooth function $\mu\colon M \to \mathbb{R}$ such that*

$$h = e^{2\mu} g.$$

ii) An immersion $\vec{\Phi}\colon M \to N$ between two Riemannian manifolds (M, g) and (N, k) is called conformal *if the pull-back metric $h := \vec{\Phi}^* k$ is conformally equivalent to g on M.*

Example.

i) Let $U \subset \mathbb{C}$ be an open subset and $\vec{\Phi}\colon U \to \mathbb{C}$ a diffeomorphism onto its image. $\vec{\Phi}$ is conformal if and only if it is either holomorphic (if it is orientation-preserving) or antiholomorphic (if it is orientation-reversing). In particular, the space of conformal transformations of $\mathbb{C}$ is infinite-dimensional.

ii) The group of conformal transformations in dimension $m \geq 3$ reduces to a finite-dimensional group. Precisely, Liouville's theorem gives that any conformal diffeomorphism $\vec{\Phi}$ from $U \subset \mathbb{R}^m$ into $\mathbb{R}^m$ is of the form

$$\vec{\Phi}(x) = a + \alpha\,\frac{A(x - x_0)}{|x - x_0|^{\varepsilon}},$$

for some $a \in \mathbb{R}^m$, $x_0 \in \mathbb{R}^m \setminus U$, $\alpha \in \mathbb{R}$, an orthogonal matrix A and $\varepsilon \in \{0, 2\}$.

iii) The space of conformal maps from 2D into 3D is infinite-dimensional. It contains non-analytic mappings.

The conformal invariance of the Willmore energy was known since the work of Blaschke [**Bla29**] in 3 dimensions, in the general case it is a consequence of the following theorem due to Bang Yen Chen [**Che74**].

THEOREM 1.2. *Let $\vec{\Phi}$ be the immersion of a 2-dimensional oriented manifold Σ into a Riemannian manifold (M, g).*

Let h be a conformally equivalent metric to g, given by

$$h := e^{2\mu} g.$$

We denote by $\vec{H}^g$ and $\vec{H}^h$ the mean curvature vectors of the immersion $\vec{\Phi}$ in (M, g) and (M, h) respectively. We also denote by K^g and K^h the scalar curvatures of $(\Sigma, \vec{\Phi}^* g)$ and $(\Sigma, \vec{\Phi}^* h)$ respectively. Furthermore, $\overline{K}^g$ and $\overline{K}^h$ denote the sectional *curvatures of the subspace $\vec{\Phi}_* T_p\Sigma$ in the manifold (M,g) and (M,h) respectively. With the previous notations the following pointwise identity holds:*

$$e^{2\mu\circ\vec{\Phi}}\left(|\vec{H}^h|_h^2 - K^h + \overline{K}^h\right) = |\vec{H}^g|_g^2 - K^g + \overline{K}^g. \tag{1.33}$$

Proof of theorem 1.2. See [**Riv**].

The conformal invariance of the Willmore energy is a corollary of theorem 1.2.

COROLLARY 1.3. *Let Σ be a smooth 2-dimensional closed oriented manifold and let $\vec{\Phi}$ be an immersion of Σ into an oriented Riemannian manifold (M^m, g).*

Let Ψ be a positive conformal diffeomorphism from (M^m, g) into another oriented Riemannian manifold (N^m, k).

Then the following equality holds:

$$W(\vec{\Phi}) + \int_\Sigma \overline{K}^g \, dvol_{\vec{\Phi}^* g} = W(\Psi\circ\vec{\Phi}) + \int_\Sigma \overline{K}^k \, dvol_{(\Psi\circ\vec{\Phi})^* k}, \tag{1.34}$$

where $\overline{K}^g$ (resp. $\overline{K}^k$) is the sectional curvature of the 2-plane $\vec{\Phi}_* T\Sigma$ in (M^m, g) (resp. of the 2-plane $\Psi_*\vec{\Phi}_* T\Sigma^2$ in (N^m, k)).

Proof of corollary 1.3. By definition Ψ realizes an isometry between $(M, \Psi^* k)$ and (N, k). Let $\mu\colon M \to \mathbb{R}$ such that

$$e^{2\mu} g = \Psi^* k.$$

We can apply the previous theorem with $h = \Psi^* k$ and obtain

$$\left[|\vec{H}^{\vec{\Phi}^* g}|^2 - K^{\vec{\Phi}^* g} + \overline{K}^g\right] = e^{2\mu\circ\vec{\Phi}}\left[|\vec{H}^{(\Psi\circ\vec{\Phi})^* k}|^2 - K^{(\Psi\circ\vec{\Phi})^* k} + \overline{K}^k\right].$$

Furthermore, we have

$$dvol_{\vec{\Phi}^* g} = e^{-2\mu\circ\vec{\Phi}} \, dvol_{(\Psi\circ\vec{\Phi})^* k}.$$

Hence combining the two last facts gives the following pointwise identity everywhere on Σ:

$$\begin{aligned} &\left[|\vec{H}^{\vec{\Phi}^* g}|^2 - K^{\vec{\Phi}^* g} + \overline{K}^g\right] dvol_{\vec{\Phi}^* g} \\ &\quad = \left[|\vec{H}^{(\Psi\circ\vec{\Phi})^* k}|^2 - K^{(\Psi\circ\vec{\Phi})^* k} + \overline{K}^k\right] dvol_{(\Psi\circ\vec{\Phi})^* k}, \end{aligned} \tag{1.35}$$

(1.34) is obtained by integrating (1.35) over Σ, the scalar curvature terms canceling each other on both sides of the identity due to Gauss-Bonnet theorem. □

Example.

i) Let $\vec{\Phi}\colon \Sigma \to \mathbb{R}^m$ be an immersion into $\mathbb{R}^m$. Define Ψ to be the inverse of the stereographic projection

$$\pi\colon S^m \setminus \{(0,\dots,0,1)\} \to \mathbb{R}^m.$$

Recall that π is conformal, thus applying Corollary 1.3 gives

$$\int_\Sigma |\vec{H}^{\vec{\Phi}^* g_{\mathbb{R}^m}}|^2 dvol_{\vec{\Phi}^* g_{\mathbb{R}^m}} = \int_\Sigma \left(|\vec{H}^{\Psi\circ\vec{\Phi})^* k}|^2 + 1\right) dvol_{(\Psi\circ\vec{\Phi})^* k},$$

where k denotes the round metric on S^m.

ii) For $a \in \mathbb{R}^m$, consider the inversion i_a at a, which for $x \in \mathbb{R}^m \setminus \{a\}$ is given by

$$i_a(x) = \frac{x-a}{|x-a|^2}.$$

i_a, restricted to $\mathbb{R}^m \setminus B_\delta(a)$, is a conformal diffeomorphism.

Given a smooth immersion $\vec{\Phi}\colon \Sigma \to \mathbb{R}^m$ with $a \notin \vec{\Phi}(\Sigma)$, we have thus

$$W(i_a \circ \vec{\Phi}) = W(\vec{\Phi}),$$

due to Corollary 1.3.

If $a \in \vec{\Phi}(\Sigma)$, then the situation is different. In fact, in this case we have

$$W(i_a \circ \vec{\Phi}) = W(\vec{\Phi}) + 4\pi \cdot \mathrm{Card}\{\vec{\Phi}^{-1}(a)\}.$$

4. Two-dimensional geometry in isothermal charts

Definition 1.4. *Let $\vec{\Phi}\colon \Sigma \to \mathbb{R}^m$ be an immersion of Σ. A chart $\psi\colon D^2 \to \Sigma$ is called* isothermal *or* conformal *for $\vec{\Phi}$, if*

$$\left\{\begin{array}{ll} \langle \partial_{x_1}(\vec{\Phi}\circ\psi), \partial_{x_2}(\vec{\Phi}\circ\psi)\rangle = 0 & \text{in } D^2 \\ |\partial_{x_1}(\vec{\Phi}\circ\psi)| = |\partial_{x_2}(\vec{\Phi}\circ\psi)| & \text{in } D^2. \end{array}\right. \tag{1.36}$$

Here, $\langle \partial_{x_1}(\vec{\Phi}\circ\psi), \partial_{x_2}(\vec{\Phi}\circ\psi)\rangle$ denotes the usual inner product in $\mathbb{R}^m$.

Note that a different way of formulating (1.36) would be to say that the map $\vec{\Phi}\circ\psi\colon D^2 \to \mathbb{R}^m$ is conformal (cf. Definition 1.1 ii)).

In isothermal charts many objects defined in subsection 2 take an easier form, which we want to explore now.

Let $\psi\colon D^2 \to \Sigma$ be an isothermal chart for the immersion $\vec{\Phi}\colon \Sigma \to \mathbb{R}^m$.

The first fundamental form in the coordinates provided by ψ is

$$\psi^* g_{\vec{\Phi}} = e^{2\lambda}(dx_1^2 + dx_2^2),$$

where $e^\lambda = |\partial_{x_1}(\vec{\Phi}\circ\psi)| = |\partial_{x_2}(\vec{\Phi}\circ\psi)|$. The volume element is given by

$$dvol_g := e^{2\lambda}\, dx_1 \wedge dx_2 \quad .$$

Moreover, for the second fundamental form, we have

$$\vec{\mathbb{I}}_{ij} := \vec{\mathbb{I}}(\partial_{x_i}, \partial_{x_j}) = \pi_{\vec{n}}\left(\partial_{x_j}\partial_{x_i}(\vec{\Phi}\circ\psi)\right). \tag{1.37}$$

Note that

$$\langle\partial^2_{x_1}(\vec{\Phi}\circ\psi),\partial_{x_1}(\vec{\Phi}\circ\psi)\rangle$$

$$=\partial_{x_1}\underbrace{\langle\partial_{x_1}(\vec{\Phi}\circ\psi),\partial_{x_1}(\vec{\Phi}\circ\psi)\rangle}_{=e^{2\lambda}}-\langle\partial_{x_1}(\vec{\Phi}\circ\psi),\partial^2_{x_1}(\vec{\Phi}\circ\psi)\rangle$$

$$=\frac{1}{2}\partial_{x_1}\left(e^{2\lambda}\right)=\frac{1}{2}\partial_{x_1}\langle\partial_{x_2}(\vec{\Phi}\circ\psi),\partial_{x_2}(\vec{\Phi}\circ\psi)\rangle \tag{1.38}$$

$$=\langle\partial_{x_1}\partial_{x_2}(\vec{\Phi}\circ\psi),\partial_{x_2}(\vec{\Phi}\circ\psi)\rangle$$

$$=\partial_{x_2}\underbrace{\langle\partial_{x_1}(\vec{\Phi}\circ\psi),\partial_{x_2}(\vec{\Phi}\circ\psi)\rangle}_{=0}-\langle\partial^2_{x_2}(\vec{\Phi}\circ\psi),\partial_{x_1}(\vec{\Phi}\circ\psi)\rangle$$

$$=-\langle\partial^2_{x_2}(\vec{\Phi}\circ\psi),\partial_{x_1}(\vec{\Phi}\circ\psi)\rangle.$$

Similarly, one obtains that

$$\langle\partial^2_{x_1}(\vec{\Phi}\circ\psi),\partial_{x_2}(\vec{\Phi}\circ\psi)\rangle=-\langle\partial^2_{x_2}(\vec{\Phi}\circ\psi),\partial_{x_2}(\vec{\Phi}\circ\psi)\rangle. \tag{1.39}$$

(1.37), (1.38) and (1.39) together imply that

$$\vec{\mathbb{I}}_{11}+\vec{\mathbb{I}}_{22}=\pi_{\vec{n}}\left(\Delta(\vec{\Phi}\circ\psi)\right)=\Delta(\vec{\Phi}\circ\psi), \tag{1.40}$$

where $\Delta=\partial^2_{x_1}+\partial^2_{x_2}$ denotes the negative flat Laplacian.

From expression (1.31) of the intrinsic negative Laplace-Beltrami operator Δ_g in coordinates the following relation follows immediately:

$$\Delta_g=e^{-2\lambda}\Delta. \tag{1.41}$$

Using (1.40) and (1.41), the mean curvature vector takes the following form:

$$\vec{H}=\frac{1}{2}\operatorname{tr}(g^{-1}\,\vec{\mathbb{I}})=\frac{e^{-2\lambda}}{2}\sum_{i,j=1}^{2}\delta^{ij}\,\vec{\mathbb{I}}_{ij}=\frac{e^{-2\lambda}}{2}\Delta(\vec{\Phi}\circ\psi) \tag{1.42}$$

$$=\frac{1}{2}\Delta_g(\vec{\Phi}\circ\psi). \tag{1.43}$$

The Willmore functional has therefore the expression

$$W(\vec{\Phi})=\frac{1}{4}\int_\Sigma|\Delta_g\vec{\Phi}|^2\,dvol_g. \tag{1.44}$$

Induced orthonormal moving frame. Given an isothermal chart $\psi\colon D^2\to\Sigma$ for an immersion $\vec{\Phi}\colon\Sigma\to\mathbb{R}^m$, we can look at the frame

$$(\vec{e}_1,\vec{e}_2)=e^{-\lambda}\,(\partial_{x_1}(\vec{\Phi}\circ\psi),\partial_{x_2}(\vec{\Phi}\circ\psi)).$$

This realizes a *tangent orthonormal moving frame*, i.e. a mapping from D^2 to $\vec{\Phi}_*(TD^2\times TD^2)$ such that at every $(x_1,x_2)\in D^2$, the pair $(\vec{e}_1,\vec{e}_2)(x_1,x_2)$ realizes a positive orthonormal basis of $\vec{\Phi}_*(T_{(x_1,x_2)}D^2)$.

A simple computation, similarly as the one in (1.38) shows

$$\langle \vec{e}_1, \nabla \vec{e}_2 \rangle = -\nabla^\perp \lambda. \tag{1.45}$$

In particular it follows

$$div \langle \vec{e}_1, \nabla \vec{e}_2 \rangle = 0. \tag{1.46}$$

Note that this identity can be writen independently of the parametrization as

$$d^{*_g} \langle \vec{e}_1, d\vec{e}_2 \rangle = 0, \tag{1.47}$$

which one refers to as the *Coulomb condition.* A frame $(\vec{e}_1, \vec{e}_2)$ satisfying (1.47) is called a *Coulomb frame.*

The following identity of 2-forms on D^2 holds:

$$d\langle \vec{e}_1, d\vec{e}_2 \rangle = K\ dvol_g, \tag{1.48}$$

where K is the Gauss curvature of (D^2, g).

To see this, let e_i be the vector field on D^2 given by $d\vec{\Phi}(e_i) = \vec{e}_i$ for $i = 1, 2$. Further, denote

$$D_{e_i} \vec{e}_j := \pi_{\vec{n}}(d\vec{e}_j(e_i)).$$

Then, using Cartan's formula[8], we obtain

$$\begin{aligned}
& d\langle \vec{e}_1, d\vec{e}_2 \rangle (e_1, e_2) \\
&= d\big(\langle \vec{e}_1, d\vec{e}_2(e_2) \rangle\big)(e_1) - d\big(\langle \vec{e}_1, d\vec{e}_2(e_1) \rangle\big)(e_2) - \langle \vec{e}_1, d\vec{e}_2([e_1, e_2]) \rangle \\
&= \langle d\vec{e}_1(e_1), d\vec{e}_2(e_2) \rangle - \langle d\vec{e}_1(e_2), d\vec{e}_2(e_1) \rangle \\
&\quad \underbrace{+ \langle \vec{e}_1, d(d\vec{e}_2(e_2))(e_1) \rangle - \langle \vec{e}_1, d\left(d\vec{e}_2(e_1)\right)(e_2) \rangle - \langle \vec{e}_1, d\vec{e}_2([e_1, e_2]) \rangle}_{= \langle \vec{e}_1, \vec{R}(\vec{e}_1, \vec{e}_2, \vec{e}_2) \rangle = 0} \\
&\overset{(*)}{=} \langle D_{e_1} \vec{e}_1, D_{e_2} \vec{e}_2 \rangle - \langle D_{e_2} \vec{e}_1, D_{e_1} \vec{e}_2 \rangle \\
&= \langle \vec{\mathbb{I}}(e_1, e_1), \vec{\mathbb{I}}(e_2, e_2) \rangle - \left| \vec{\mathbb{I}}(e_1, e_2) \right|^2 \\
&= K. \qquad (1.50)
\end{aligned}$$

In the last identity we have made use of the Gauss theorem (1.22). By $\vec{R}$ we denote the Riemannian $(3, 1)$-curvature tensor of the exterior differential d, the Levi-Civita connection associated to $(\mathbb{R}^m, \langle \cdot, \cdot \rangle)$.

In $(*)$ we used that

$$\langle d\vec{e}_1(e_1), d\vec{e}_2(e_2) \rangle = \langle D_{e_1} \vec{e}_1, D_{e_2} \vec{e}_2 \rangle + \langle \nabla_{e_1} \vec{e}_1, \nabla_{e_2} \vec{e}_2 \rangle,$$

---

[8]The Cartan formula for the exterior differential of a 1-form α on a differentiable manifolfd M^m says that for any pair of vector fields X, Y on this manifold the following identity holds:

$$d\alpha(X, Y) = d(\alpha(Y))(X) - d(\alpha(X))(Y) - \alpha([X, Y]), \tag{1.49}$$

see Corollary 1.122 chapter I of [**GHL04**].

where the second summand vanishes, since $\nabla_{e_1}\vec{e}_1$ ($\nabla_{e_2}\vec{e}_2$ resp.) is oriented along $\vec{e}_2$ ($\vec{e}_1$ resp.):

$$\langle\nabla_{e_i}\vec{e}_i,\vec{e}_i\rangle = e_i(\underbrace{\langle\vec{e}_i,\vec{e}_i\rangle}_{\equiv 1}) - \langle\nabla_{e_i}\vec{e}_i,\vec{e}_i\rangle = 0,$$

for $i=1,2$. Similarly, one shows that

$$\langle\nabla_{e_1}\vec{e}_2,\vec{e}_2\rangle = \langle\nabla_{e_2}\vec{e}_1,\vec{e}_1\rangle = 0$$

such that

$$\langle\nabla_{e_2}\vec{e}_1,\nabla_{e_1}\vec{e}_2\rangle = 0.$$

From (1.50), the claimed identity in (1.48) follows since

$$d\langle\vec{e}_1,d\vec{e}_2\rangle = d\langle\vec{e}_1,d\vec{e}_2\rangle(e_1,e_2)\; e_1^*\wedge e_2^*$$

and $e_1^*\wedge e_2^* = dvol_g$.

Combining (1.45) and (1.48) gives

$$-\Delta\lambda = -*d*d\lambda = *d\langle\vec{e}_1,d\vec{e}_2\rangle = e^{2\lambda}K,$$

where Δ denotes the negative flat Laplacian. Recall from Section 1 that $*d\lambda$ is the associated 1-form to $\nabla^\perp\lambda$.

The previous identity is the well-known expression of the Gauss curvature in isothermal coordinates in terms of the conformal factor λ, called *Liouville equation*. We have proven the following lemma.

LEMMA 1.5. *Let $\vec{\Phi}\colon D^2\to\mathbb{R}^m$ be a conformal immersion with*

$$|\partial_{x_1}\vec{\Phi}| = |\partial_{x_2}\vec{\Phi}| = e^\lambda.$$

Then

$$-\Delta\lambda = e^{2\lambda}K, \tag{1.51}$$

where K is the Gauss curvature of $(D^2,\vec{\Phi}^*g_{\mathbb{R}^m})$ and Δ is the negative flat Laplacian on D^2.

One easily obtains the following generalization.

THEOREM 1.6. *Let $\vec{\Phi}\colon\Sigma\to\mathbb{R}^m$ be a smooth immersion and $g:=\vec{\Phi}^*g_{\mathbb{R}^m}$ the induced first fundamental form. Let h be a conformally equivalent metric on Σ satisfying*

$$g = e^{2\alpha}h.$$

Then

$$-\Delta_h\alpha = e^{2\alpha}K_g - K_h, \tag{1.52}$$

where K_g and K_h are the Gauss curvatures of (Σ,g) and (Σ,h) respectively.

PROOF OF THEOREM 1.6. In the next subsection we will show that there exist isothermal coordinates for $\vec{\Phi}$, i.e. locally g is of the form

$$g = e^{2\lambda}(dx_1^2+dx_2^2). \tag{1.53}$$

Hence, we have

$$h = e^{2\sigma}(dx_1^2+dx_2^2), \tag{1.54}$$

where $\sigma:=\lambda-\alpha$.

Then we can apply Lemma 1.5 to (1.53) and (1.54) respectively and obtain

$$K_g = -e^{-2\lambda}\Delta\lambda \tag{1.55}$$

and

$$K_h = -e^{-2(\lambda-\alpha)}\Delta(\lambda-\alpha) \tag{1.56}$$

Combining (1.55) and (1.56) yields

$$K_h = e^{2\alpha}\left(K_g + e^{-2\lambda}\Delta\alpha\right) = e^{2\alpha}K_g + \Delta_h\alpha,$$

where we used (1.41) in the last identity. This finishes the proof. □

Finally, we want to take advantage of the fact that for a conformal immersion $\vec{\Phi}$ from D^2 into $\mathbb{R}^3$, $(\vec{e}_1, \vec{e}_2, \vec{n})$ is an orthonormal frame of $\vec{\Phi}^*(\mathbb{R}^3)$, where $\vec{e}_i := e^{-\lambda}\,\partial_{x_i}\vec{\Phi}$ and $\vec{n}$ is the Gauss map of $\vec{\Phi}$. The following two lemmas dealing with conformal immersions into the 3-dimensional space are easy consequences of this.

LEMMA 1.7. *Let $\vec{\Phi}\colon D^2 \to \mathbb{R}^3$ be a conformal immersion. Then*

$$K\,\vec{n} = -\frac{e^{-2\lambda}}{2}\,\nabla\vec{n}\times\nabla^{\perp}\vec{n} = -\frac{e^{-2\lambda}}{2}\,div\,[\vec{n}\times\nabla^{\perp}\vec{n}], \tag{1.57}$$

where K is the Gauss curvature of $\vec{\Phi}$.

PROOF. Note that

$$\nabla\vec{n}\times\nabla^{\perp}\vec{n} = div\,[\vec{n}\times\nabla^{\perp}\vec{n}] = -2\,\partial_{x_1}\vec{n}\times\partial_{x_2}\vec{n}.$$

Since

$$\langle\partial_{x_1}\vec{n},\vec{n}\rangle = \langle\partial_{x_2}\vec{n},\vec{n}\rangle = 0,$$

for proving (1.57), we have to show that

$$e^{2\lambda}K = |\partial_{x_1}\vec{n}\times\partial_{x_2}\vec{n}|. \tag{1.58}$$

For the Gauss curvature, we can write

$$K := \frac{det\left(\langle\vec{n},\vec{\mathbb{I}}(\partial_{x_i},\partial_{x_j})\rangle\right)}{det(g_{ij})} = e^{-2\lambda}\,det\left(-\langle\partial_{x_i}\vec{n},\vec{e}_j\rangle\right), \tag{1.59}$$

where $\vec{e}_i := e^{-\lambda}\,\partial_{x_i}\vec{\Phi}$. The parallelogram spanned by $\partial_{x_1}\vec{n}$ and $\partial_{x_2}\vec{n}$ is contained in the tangent bundle, of which $(\vec{e}_1,\vec{e}_2)$ is an orthonormal frame, hence we have

$$|\partial_{x_1}\vec{n}\times\partial_{x_2}\vec{n}| = det\left(\langle\partial_{x_i}\vec{n},\vec{e}_j\rangle\right).$$

This, together with (1.59), implies (1.58) and thus the result. □

LEMMA 1.8. *Let $\vec{\Phi}\colon D^2 \to \mathbb{R}^3$ be a conformal immersion. Then the following identity holds:*

$$-2H\;\nabla\vec{\Phi} = \nabla\vec{n} + \vec{n}\times\nabla^{\perp}\vec{n}. \tag{1.60}$$

PROOF. Consider the tangent frame $(\vec{e}_1,\vec{e}_2)$, given by

$$\vec{e}_i := e^{-\lambda}\,\partial_{x_i}\vec{\Phi},$$

where $e^{\lambda} = |\partial_{x_1}\vec{\Phi}| = |\partial_{x_2}\vec{\Phi}|$. The oriented Gauss map $\vec{n}$ is given by

$$\vec{n} = \vec{e}_1\times\vec{e}_2.$$

We have

$$\begin{aligned}\langle \vec{e}_1, \vec{n} \times \nabla^\perp \vec{n} \rangle &= -\langle \nabla^\perp \vec{n}, \vec{e}_2 \rangle, \\ \langle \vec{e}_2, \vec{n} \times \nabla^\perp \vec{n} \rangle &= \langle \nabla^\perp \vec{n}, \vec{e}_1 \rangle.\end{aligned}$$

From this we deduce

$$\left\{\begin{aligned} -\vec{n} \times \partial_{x_2}\vec{n} &= \langle \partial_{x_2}\vec{n}, \vec{e}_2 \rangle \, \vec{e}_1 - \langle \partial_{x_2}\vec{n}, \vec{e}_1 \rangle \, \vec{e}_2, \\ \vec{n} \times \partial_{x_1}\vec{n} &= -\langle \partial_{x_1}\vec{n}, \vec{e}_2 \rangle \, \vec{e}_1 + \langle \partial_{x_1}\vec{n}, \vec{e}_1 \rangle \, \vec{e}_2. \end{aligned}\right.$$

Thus,

$$\left\{\begin{aligned} \partial_{x_1}\vec{n} - \vec{n} \times \partial_{x_2}\vec{n} &= [\langle \partial_{x_2}\vec{n}, \vec{e}_2 \rangle + \langle \partial_{x_1}\vec{n}, \vec{e}_1 \rangle] \, \vec{e}_1, \\ \partial_{x_2}\vec{n} + \vec{n} \times \partial_{x_1}\vec{n} &= [\langle \partial_{x_2}\vec{n}, \vec{e}_2 \rangle + \langle \partial_{x_1}\vec{n}, \vec{e}_1 \rangle] \, \vec{e}_2. \end{aligned}\right.$$

Since

$$H = -\frac{1}{2} e^{-\lambda} \, [\langle \partial_{x_1}\vec{n}, \vec{e}_1 \rangle + \langle \partial_{x_2}\vec{n}, \vec{e}_2 \rangle],$$

we deduce (1.60) and Lemma 1.8 is proven. □

5. Existence of isothermal coordinates and the Chern moving frame method

In this subsection we want to prove the existence of isothermal coordinates for an immersion $\vec{\Phi}$ of Σ into $\mathbb{R}^m$. More precisely, we show that for any $p \in \Sigma$ there exists a neighborhood U of p and a diffeomorphism $\psi\colon D^2 \to U$ such that ψ is an isothermal chart for $\vec{\Phi}$, i.e. satisfying (1.36).

By picking any chart $\varphi\colon D^2 \to U$ around p, it suffices to prove the following theorem, which can then be applied to $\vec{\Phi} \circ \varphi$.

THEOREM 1.9. *Let $\vec{\Phi}$ be an immersion of the disc D^2 into $\mathbb{R}^m$. Then there exists a diffeomorphism ψ of D^2 such that the map $\vec{\Phi} \circ \psi\colon D^2 \to \mathbb{R}^m$ is conformal, i.e. satisfies* (1.36).

To obtain this result we will make use of the Riemann mapping theorem (see for instance [**Rud87**], chapter 14), which we want to recall now.

THEOREM 1.10. *If $U \subsetneq \mathbb{C}$ is non-empty, open and simply connected, then there exists a biholomorphic mapping f from U onto the open unit disk $D^2 = \{z \in \mathbb{C} : |z| < 1\}$.*

In the last subsection we have seen how any isothermal chart generates a Coulomb frame in the tangent bundle. S.S. Chern observed that this is in fact an exact matching, the converse being also true: starting from a Coulomb frame one can generate isothermal coordinates. This is our strategy for the proof of Theorem 1.9.

PROOF OF THEOREM 1.9. Let $\vec{\Phi}\colon D^2 \to \mathbb{R}^m$ be an immersion. We are looking for a *Coulomb tangent orthonormal moving frame* $(\vec{e}_1, \vec{e}_2)$: a tangent moving frame from D^2 to $\vec{\Phi}_*(TD^2 \times TD^2)$ such that the Coulomb condition (1.47) is satisfied.

Note that the passage from an arbitrary tangent orthonormal moving frame $(\vec{f}_1, \vec{f}_2)$ to another one $(\vec{e}_1, \vec{e}_2)$ is realized through a change of gauge which corresponds to the action of an $SO(2)$ rotation $e^{i\theta}$ on the tangent space $\vec{\Phi}_*(T_{(x_1,x_2)}D^2)$:

$$\vec{e}_1 + i\vec{e}_2 = e^{i\theta}\ (\vec{f}_1 + i\vec{f}_2). \tag{1.61}$$

One shows that the following gauge change formula holds:

$$\langle \vec{e}_1, d\vec{e}_2\rangle = \langle \vec{f}_1, d\vec{f}_2\rangle + d\theta. \tag{1.62}$$

Thus, construcing a Coulomb tangent orthonormal moving frame is the same as finding the change of gauge $\theta \in W^{1,2}(D^2)$ which minimizes the energy

$$\int_{D^2} \left| d\theta + \langle \vec{f}_1, d\vec{f}_2\rangle \right|_g^2\ dvol_g. \tag{1.63}$$

In fact, there exists a unique minimum satisfying

$$\left\{\begin{array}{rcll} d^{*_g}\left[d\theta + \langle \vec{f}_1, d\vec{f}_2\rangle\right] & = & 0 & \text{in } D^2 \\ \\ \iota^*_{\partial D^2}\left(*_g\left[d\theta + \langle \vec{f}_1, d\vec{f}_2\rangle\right]\right) & = & 0 & \text{on } \partial D^2, \end{array}\right.$$

where $\iota_{\partial D^2}$ is the canonical inclusion of ∂D^2 into $\overline{D^2}$. Denote by $(\vec{e}_1, \vec{e}_2)$ the corresponding Coulomb frame, i.e. the frame obtained by $\vec{e}_1 + i\vec{e}_2 = e^{i\theta}(\vec{f}_1 + i\vec{f}_2)$.

By Poincaré's Lemma there exists λ satisfying

$$\left\{\begin{array}{rcl} d\lambda & = & *_g\langle \vec{e}_1, d\vec{e}_2\rangle \\ \\ \displaystyle\int_{\partial D^2} \lambda & = & 0. \end{array}\right. \tag{1.64}$$

Denote moreover $e_i := d\vec{\Phi}^{-1}(\vec{e}_i)$ and let (e_1^*, e_2^*) be the dual basis to (e_1, e_2). The Cartan formula for the exterior differential of a 1-form implies

$$\begin{aligned} de_i^*(e_1, e_2) &= d(e_i^*(e_2))(e_1) - d(e_i^*(e_1))(e_2) - e_i^*([e_1, e_2]) \\ &= -e_i^*([e_1, e_2]) \\ &= -((\vec{\Phi}^{-1})^* e_i^*)([d\vec{\Phi}(e_1), d\vec{\Phi}(e_2)]) \\ &= -((\vec{\Phi}^{-1})^* e_i^*)([\vec{e}_1, \vec{e}_2]). \end{aligned} \tag{1.65}$$

The Levi-Civita connection ∇ on $\vec{\Phi}_* TD^2$ is given by $\nabla_X \vec{Y} := \pi_T(d\vec{Y}(X))$ where π_T is the orthogonal projection onto the tangent plane. Since the Levi-Civita connection is torsion-free, we have

$$[\vec{e}_1, \vec{e}_2] = \nabla_{e_1}\vec{e}_2 - \nabla_{e_2}\vec{e}_1 = \pi_T(d\vec{e}_2(e_1) - d\vec{e}_1(e_2)). \tag{1.66}$$

Since $\vec{e}_1$ and $\vec{e}_2$ have unit length, the tangential projection of $d\vec{e}_1$ (resp. $d\vec{e}_2$) are oriented along $\vec{e}_2$ (resp. $\vec{e}_1$). So we have

$$\left\{\begin{array}{l} \pi_T(d\vec{e}_2(e_1)) = \langle d\vec{e}_2, \vec{e}_1\rangle(e_1)\ \vec{e}_1 \\ \\ \pi_T(d\vec{e}_1(e_2)) = \langle d\vec{e}_1, \vec{e}_2\rangle(e_2)\ \vec{e}_2. \end{array}\right. \tag{1.67}$$

Combining (1.65), (1.66) and (1.67) gives then

$$(1.68)\qquad \begin{cases} de_1^*(e_1,e_2) = -\langle d\vec{e}_2,\vec{e}_1\rangle(e_1) \\ \\ de_2^*(e_1,e_2) = \langle d\vec{e}_1,\vec{e}_2\rangle(e_2). \end{cases}$$

Equation (1.64) gives

$$(1.69)\qquad \begin{cases} -\langle d\vec{e}_2,\vec{e}_1\rangle(e_1) = *_g d\lambda(e_1) = -d\lambda(e_2) \\ \\ \langle d\vec{e}_1,\vec{e}_2\rangle(e_2) = *_g d\lambda(e_2) = d\lambda(e_1). \end{cases}$$

Thus combining (1.68) and (1.69) yields

$$(1.70)\qquad \begin{cases} de_1^* = -d\lambda(e_2)\ e_1^*\wedge e_2^* = d\lambda\wedge e_1^* \\ \\ de_2^* = d\lambda(e_1)\ e_1^*\wedge e_2^* = d\lambda\wedge e_2^*. \end{cases}$$

We have thus finally proven that

$$(1.71)\qquad \begin{cases} d\left(e^{-\lambda}e_1^*\right) = 0 \\ \\ d\left(e^{-\lambda}e_2^*\right) = 0. \end{cases}$$

Apply Poincaré's Lemma and obtain functions ϕ_1,ϕ_2 with average 0 on the disc D^2 such that

$$(1.72)\qquad d\phi_i := e^{-\lambda}e_i^*.$$

Note that (1.72) implies that ϕ is a local homeomorphism. We can thus make the following local calculations envolving ϕ^{-1}: From (1.72) we have

$$\partial_{y_i}\phi^{-1} = e^{\lambda\circ\phi^{-1}}\,e_i$$

and consequently,

$$e^{-\lambda\circ\phi^{-1}}\ g(e_j,\partial_{y_i}\phi^{-1}) = e^{-\lambda\circ\phi^{-1}}\ e_j^*(\partial_{y_i}\phi^{-1}) = \delta_{ij}.$$

This implies

$$(1.73)\qquad g(\partial_{y_i}\phi^{-1},\partial_{y_j}\phi^{-1}) = e^{2\lambda\circ\phi^{-1}}\,\delta_{ij}$$

or in other words

$$(1.74)\qquad \langle\partial_{y_i}(\vec{\Phi}\circ\phi^{-1}),\partial_{y_j}(\vec{\Phi}\circ\phi^{-1})\rangle = e^{2\lambda\circ\phi^{-1}}\,\delta_{ij}.$$

This says in particular that there exists an open disc D^2_ε around 0 on which ϕ is a homeomorphism (and thus a diffeomorphism) and such that $\vec{\Phi}\circ\phi^{-1}\colon \phi(D^2_\varepsilon)\to\mathbb{R}^m$ is a conformal immersion. The Riemann Mapping theorem 1.10 gives the existence of a biholomorphic diffeomorphism h from D^2 into $\phi(D^2_\varepsilon)$. Thus $\vec{\Phi}\circ\phi^{-1}\circ h$ realizes a conformal immersion from D^2 onto $\vec{\Phi}(D^2_\varepsilon)$ and $\tilde{\psi} := \phi^{-1}\circ h$ is an isothermal chart.

We have been looking for a diffeomorphism ψ from D^2 to D^2 (not into D^2_ε), though, such that $\vec{\Phi}\circ\psi$ is conformal. To construct such a ψ, we will replace ϕ by its associated quasiconformal homeomorphism f, introduced in Section 6: A computation shows that (1.72) implies

$$(1.75)\qquad \langle\partial_{x_i}\phi,\partial_{x_j}\phi\rangle = 2e^{-2\lambda}g_{ij} =: h_{ij}.$$

Note that we have

$$c^{-1}(\delta_{ij}) \leq (h_{ij}) \leq c\,(\delta_{ij}). \tag{1.76}$$

We can thus apply Corollary 1.16 to ϕ and deduce that there exists a (quasiconformal) map $f\colon D^2 \to \mathbb{C}$, homeomorphic onto its image, such that $g := \phi \circ f^{-1}\colon f(D^2) \to \phi(D^2)$ is a holomorphic function. Since ϕ is a local homeomorphism and f is a homeomorphism, we have

$$\partial_z g \neq 0 \qquad \text{on } f(D^2).$$

In particular g is conformal with

$$|\partial_{y_1} g| = |\partial_{y_2} g| = \frac{1}{\sqrt{2}}|\partial_z g|.$$

Note that the following computation is again a local one, which is why we can use (1.74).

$$\begin{aligned}
&\langle \partial_{y_i}(\vec{\Phi} \circ f^{-1}), \partial_{y_j}(\vec{\Phi} \circ f^{-1})\rangle \\
&= \langle d(\vec{\Phi} \circ \phi^{-1})(\partial_{y_i} g), d(\vec{\Phi} \circ \phi^{-1})(\partial_{y_j} g)\rangle \\
&= \frac{1}{2}|\partial_z g|^2 \langle \partial_{y_i}(\vec{\Phi} \circ \phi^{-1}), \partial_{y_j}(\vec{\Phi} \circ \phi^{-1})\rangle \\
&= \frac{1}{2}|\partial_z g|^2 e^{2\lambda \circ \phi^{-1}}\, \delta_{ij}.
\end{aligned}$$

Applying the Riemann Mapping theorem 1.10 gives the existence of a biholomorphic diffeomorphism h from D^2 into $f(D^2)$. Thus $\vec{\Phi} \circ f^{-1} \circ h$ realizes a conformal immersion from D^2 onto $\vec{\Phi}(D^2)$ and $\psi := f^{-1} \circ h$ is the isothermal chart we have been looking for. □

6. Some facts on Riemann surfaces

This section collects some useful facts on Riemann surfaces that we shall need during the course. We follow [**Jos97**] and [**IT92**]. In the sequel, a *surface* denotes a 2-dimensional smooth manifold.

DEFINITION 1.11. *Let Σ be a surface. An atlas on Σ with charts $\varphi_i\colon D^2 \to U_i \subset S$ is called* conformal *if the transition maps*

$$\varphi_j^{-1} \circ \varphi_i\colon\ \varphi_i^{-1}(U_i \cap U_j) \subset \mathbb{C} \to \mathbb{C}$$

are biholomorphic. A maximal conformal atlas is called a conformal structure *on S.*

A Riemann surface *is a surface together with a conformal structure.*

Recall Example i) in Section 3 and note that Definition 1.11 provides that the transition maps of a conformal atlas are conformal and preserve orientation. In particular, a Riemann surface is oriented.

REMARK 1.12. *In other words, a Riemann surface is a complex manifold of real dimension 2. Note that any holomorphic atlas of a complex manifold M induces* *an* almost complex structure J *on* M*, that is a section in* $\operatorname{End}(TM)$ *satisfying* $J^2 = -1$*, via*

$$J\partial_{x_i} = \partial_{y_i} \quad \textit{and} \quad J\partial_{y_i} = -\partial_{x_i}. \tag{1.77}$$

Conversely, an almost complex structure J on an even-real-dimensional manifold M is not necessarily integrable*, i.e. the differentiable structure of M cannot necessarily be defined by a holomorphic atlas satisfying* (1.77)*. However, this is true in real dimension* 2*, such that any almost complex manifold of real dimension* 2 *is in fact a Riemann surface.*

Riemannian metrics and Beltrami coefficients.

THEOREM 1.13. *Let* (Σ, g) *be an oriented surface with a Riemannian metric. Then* (Σ, g) *admits a conformal structure.*

For proving Theorem 1.13, the goal is to find for each coordinate neighborhood $(U_j, (x_j, y_j))$ from an atlas $\{(U_j, (x_j, y_j))\}_j$ isothermal coordinates (u_j, v_j) for the Riemannian metric g such that it is represented as

$$\rho_j\left(du_j^2 + dv_j^2\right), \tag{1.78}$$

for ρ_j being some positive smooth function. Setting $w_j = u_j + iv_j$, the atlas $\{(U_j, w_j)\}_j$ defines then a conformal structure on Σ. Note that the Chern moving frame method, as used in the proof of Theorem 1.9, is one way to find isothermal coordinates.

We want to give the rough idea of an alternative proof for finding isothermal coordinates, which is related to *Beltrami coefficients.* If the Riemannian metric g is in local coordinates $(U, (x, y))$ given by

$$g_{11}\, dx^2 + 2g_{12}\, dxdy + g_{22}\, dy^2,$$

setting $z = x + iy$ yields the representation

$$\sigma|dz + \mu\, d\bar{z}|^2, \tag{1.79}$$

where

$$\sigma = \frac{1}{4}\left(g_{11} + g_{22} + 2\sqrt{g_{11}g_{22} - g_{12}^2}\right)$$

is a positive smooth function on U and

$$\mu = \frac{g_{11} - g_{22} + 2ig_{12}}{g_{11} + g_{22} + 2\sqrt{g_{11}g_{22} - g_{12}^2}} \tag{1.80}$$

is a complex-valued smooth function with

$$\|\mu\|_{L^\infty(U)} < 1.$$

To find an isothermal coordinate $w = u + iv$, note that such satisfies

$$\rho\left(du^2 + dv^2\right) = \rho\,|dw|^2 = \rho\,|w_z|^2\left|dz + \frac{w_{\bar{z}}}{w_z}\, d\bar{z}\right|^2, \tag{1.81}$$

using the notation from (1.78), i.e. $\rho > 0$ being a smooth function. Comparing (1.79) and (1.81), we deduce that an isothermal coordinate is an (a.e.) diffeomorphic solution w to the *Beltrami differential equation*

$$w_{\bar{z}} = \mu\, w_z, \tag{1.82}$$

where μ with $\|\mu\|_{L^\infty(U)} < 1$ is given by (1.80) and is called the *Beltrami coefficient induced by the Riemannian metric g*. The following theorem gives that (1.82) has a quasiconformal solution for any Beltrami coefficient.

DEFINITION 1.14. *A quasiconformal map $f\colon \mathbb{C} \to \mathbb{C}$ is an orientation-preserving homeomorphism of $\mathbb{C}$*

i) such that f_z and $f_{\bar{z}}$ are in $L^1_{\text{loc}}(\mathbb{C})$;
ii) for which there exists a constant $0 \leq k < 1$ such that

$$|f_{\bar{z}}| \leq k|f_z| \qquad a.e.\ on\ \mathbb{C}.$$

THEOREM 1.15 (Existence of quasiconformal mappings with complex dilation μ, [**IT92**], Theorem 4.30). *Let $\mu \in L^\infty(\mathbb{C}, \mathbb{C})$ satisfy*

$$\|\mu\|_{L^\infty(\mathbb{C})} < 1.$$

Then there exists a homeomorphism $f\colon \overline{\mathbb{C}} \to \overline{\mathbb{C}}$ which is a quasiconformal mapping of $\mathbb{C}$ with complex dilation μ, i.e.

$$f_{\bar{z}} = \mu\, f_z.$$

f is uniquely determined by the normalization conditions

$$f(0) = 0, \quad f(1) = 1, \quad f(\infty) = f(\infty).$$

We are now in the position to fill the gap we left in the proof of Theorem 1.9, using the existence of quasiconformal mappings for arbitrary Beltrami coefficients. We first make the following observation.

Given $\phi \in W^{1,\infty}(D^2, \mathbb{C})$, define

$$h_{ij} := \langle \partial_{x_i}\phi, \partial_{x_j}\phi \rangle,$$

where $\langle \cdot, \cdot \rangle$ denotes the canonical inner product in $\mathbb{R}^2$. If the symmetric 2×2-matrix h satisfies

$$c^{-1}(\delta_{ij}) \leq (h_{ij}) \leq c\,(\delta_{ij}), \tag{1.83}$$

$h = |d\phi|^2$ defines a metric. From (1.79) and (1.80) we obtain

$$|d\phi|^2 = \sigma|dz + \mu d\bar{z}|^2, \tag{1.84}$$

where

$$\frac{\partial_{\bar{z}}\phi}{\partial_z\phi} = \mu = \frac{h_{11} - h_{22} + 2ih_{12}}{h_{11} + h_{22} + 2\sqrt{h_{11}h_{22} - h_{12}^2}}. \tag{1.85}$$

Recall that μ is a Beltrami coefficient with $\|\mu\|_{L^\infty(D^2)} < 1$.

COROLLARY 1.16. *Let $\phi \in W^{1,\infty}(D^2, \mathbb{C})$ such that*

$$h_{ij} := \langle \partial_{x_i}\phi, \partial_{x_j}\phi \rangle$$

satisfies

$$c^{-1}(\delta_{ij}) \leq (h_{ij}) \leq c\,(\delta_{ij}).$$

Let μ be the associated Beltrami coefficient, given by (1.85) and f be the quasiconformal mapping with complex dilation μ (extended by 0 outside D^2), given by Theorem 1.15.

Then $\phi \circ f^{-1}\colon f(D^2) \to \phi(D^2)$ is holomorphic.

PROOF. In [**IT92**], Proposition 4.13, it is computed that the complex dilation $\mu_{f^{-1}}$ is of the form

$$\mu_{f^{-1}} = -\left(\frac{f_z}{\overline{f_z}}\mu_f\right) \circ f^{-1}.$$

Consequently,

$$\begin{aligned}
\partial_{\bar z}(\phi \circ f^{-1}) &= \partial_w \phi \circ f^{-1} \cdot (f^{-1})_{\bar z} + \partial_{\bar w}\phi \circ f^{-1} \cdot \overline{(f^{-1})_z} \\
&= \partial_w \phi \circ f^{-1} \cdot \mu_{f^{-1}}\,(f^{-1})_z + (\mu_f\,\partial_w\phi) \circ f^{-1} \cdot \overline{(f^{-1})_z} \\
&= (\mu_f\,\partial_w\phi) \circ f^{-1}\left(-\frac{f_z}{\overline{f_z}} \circ f^{-1}\,(f^{-1})_z + \overline{(f^{-1})_z}\right) \\
&= 0
\end{aligned}$$

and thus $\phi \circ f^{-1} \colon f(D^2) \to \mathbb{C}$ is holomorphic. □

The classification of Riemann surfaces.

THEOREM 1.17 (Uniformization Theorem). *Let Σ be a compact Riemann surface of genus p. Then there exists a conformal diffeomorphism*

$$\Psi\colon \Sigma \to \Sigma',$$

where Σ' is

i) a compact Riemann surface of the form $\mathbb{H}/\Gamma$, where Γ is a subgroup of $PSL(2,\mathbb{R})$ acting freely and properly discontinuously on $\mathbb{H}$, if $p \geq 2$;
ii) a compact Riemann surface $\mathbb{C}/a\mathbb{Z} + b\mathbb{Z}$, where $a, b \in \mathbb{C}$, if $p = 1$;
iii) the Riemann sphere S^2, if $p = 0$.

A corollary of Theorem 1.17 is the following.

THEOREM 1.18. *Let Σ be a compact Riemann surface, where the conformal structure is induced by the Riemannian metric g. Then there exists a conformally equivalent Riemannian metric h on Σ,*

$$g = e^{2\alpha} h,$$

with

- *constant (sectional) curvature K_h and*
- *unit volume: $vol_h(\Sigma) = 1$.*

Such h is unique.

Moreover, we have

$$K_h \begin{cases} < 0 \\ = 0 \\ > 0 \end{cases} \quad \text{if and only if } genus(\Sigma) \begin{cases} \geq 2 \\ = 1 \\ = 0 \end{cases}.$$

LECTURE 2

The space of weak immersion with L^2-bounded second fundamental form

1. Definition

Let Σ be a smooth 2-dimensional closed oriented manifold. Let g_0 be a smooth reference metric on Σ.

One defines the Sobolev spaces $W^{k,p}(\Sigma, \mathbb{R}^m)$ of measurable maps from Σ into $\mathbb{R}^m$ in the following way:

$$W^{k,p}(\Sigma, \mathbb{R}^m) = \left\{ f : \Sigma \to \mathbb{R}^m \ ; \ \sum_{l=0}^{k} \int_{\Sigma} |\nabla^l f|_{g_0}^p \, dvol_{g_0} < +\infty \right\}.$$

Since Σ is assumed to be closed, this space is independent of the choice of g_0: For any two smooth reference metrics g_0 and g_1, one can find a constant $c > 0$ such that for all $p \in \Sigma$ and $X \in T_pM$

$$c^{-1} g_0(X, X) \le g_1(X, X) \le c\, g_0(X, X). \tag{2.86}$$

We now want to define a weak notion of immersions $\vec{\Phi} : \Sigma \to \mathbb{R}^m$. We want

$$g_{\vec{\Phi}}(X, Y) = \langle d\vec{\Phi}(X), d\vec{\Phi}(Y) \rangle$$

to define an L^∞-metric on Σ.

Therefore, we require $\vec{\Phi} : \Sigma \to \mathbb{R}^m$ to be a *Lipschitz immersion*, i.e. a map satisfying

i)

$$\vec{\Phi} \in W^{1,\infty}(\Sigma, \mathbb{R}^m),$$

ii) $\exists c > 0$ s.t. $\forall p \in \Sigma,\ X \in T_pM$

$$c^{-1} g_0(X, X) \le g_{\vec{\Phi}}(X, X) \le c\, g_0(X, X). \tag{2.87}$$

Condition ii) ensures the non-degeneracy of the metric $g_{\vec{\Phi}}$. Note that this condition is again independent of the choice of g_0, due to (2.86).

For a Lipschitz immersion we can define the Gauss map as the following measurable map in $L^\infty(\Sigma)$, taking values in the Grassmannian $Gr_{m-2}(\mathbb{R}^m)$ of oriented $(m-2)$-planes in $\mathbb{R}^m$:

$$\vec{n}_{\vec{\Phi}} := \star \frac{\partial_{x_1} \vec{\Phi} \wedge \partial_{x_2} \vec{\Phi}}{|\partial_{x_1} \vec{\Phi} \wedge \partial_{x_2} \vec{\Phi}|}$$

for an arbitrary choice of local positive coordinates (x_1, x_2).

We say that a Lipschitz immersion $\vec{\Phi} : \Sigma \to \mathbb{R}^m$ has *L^2-bounded second fundamental form* if iii)

$$\vec{n} \in W^{1,2}(\Sigma, Gr_{m-2}(\mathbb{R}^m))^1.$$

In other words, condition iii) requires

$$\int_\Sigma |d\vec{n}|^2_{g_0}\ dvol_{g_0} < +\infty.$$

Summing up, we have the following definition.

DEFINITION 2.19. *The space $\mathcal{E}_\Sigma$ of* Lipschitz immersions with L^2-bounded second fundamental form *is defined as*

$$\mathcal{E}_\Sigma := \left\{\vec{\Phi}\colon \Sigma \to \mathbb{R}^m \textit{ measurable s.t. } \text{i), ii) } \textit{and } \text{iii) } \textit{hold}\right\}.$$

DEFINITION 2.20. *For $\vec{\Phi} \in \mathcal{E}_\Sigma$, denote by*

$$\mathbb{I}(\vec{\Phi}) := \int_\Sigma |\vec{\mathbb{I}}|^2_{g_{\vec{\Phi}}}\ dvol_{g_{\vec{\Phi}}} = \int_\Sigma |d\vec{n}|^2_{g_{\vec{\Phi}}}\ dvol_{g_{\vec{\Phi}}} \tag{2.88}$$

the L^2-norm of the second fundamental form of $\vec{\Phi}$.

Note that the identity in (2.88) holds by (1.29). Thus, $\mathbb{I}(\vec{\Phi})$ can also be interpreted as the Dirichlet energy of the Gauss map $\vec{n}_{\vec{\Phi}}$. Moreover, $\mathbb{I}(\vec{\Phi})$ is finite due to condition iii) in the definition of $\mathcal{E}_\Sigma$.

2. Fundamental results on integrability by compensation

2.1. Classical results from Calderon-Zygmund theory

Recall from classical Calderon-Zygmund theory the following theorem.

THEOREM 2.21. *Let $f \in L^1(D^2)$. Then there exists a unique solution in $W_0^{1,1}(D^2)$ of*

$$\begin{cases} \Delta\phi &= f \quad \textit{in } D^2 \\ \phi &= 0 \quad \textit{on } \partial D^2 \end{cases} \tag{2.89}$$

and $\phi \in W_0^{1,p}(D^2)$ for all $p < 2$, with

$$\|\nabla\phi\|_{L^p(D^2)} \leq C_p\ \|f\|_{L^1(D^2)}.$$

One might ask to which optimal space beside $\bigcap_{p<2} L^p(D^2)$ the function $\nabla\phi$ belongs to. It certainly does not belong to $L^2(D^2)$: if this was true in general, this would also hold for the Radon measure $f = \Delta \log r$ on D^2. But

$$\nabla \log r = \frac{1}{r}\partial_r \notin L^2(D^2).$$

Observe, however, that $\frac{1}{r}$ is an element of the *weak L^2-space* $L^{2,\infty}(D^2)$ of measurable functions $f\colon D^2 \to \mathbb{R}$ such that

$$|f|_{L^{2,\infty}(D^2)} = \sup_{\lambda>0}\left\{\lambda\left|\{x \in D^2;\ |f(x)| > \lambda\}\right|^{1/2}\right\} < \infty.$$

[1]The Grassman manifold $Gr_{m-2}(\mathbb{R}^m)$ can be seen as being the submanifold of the Euclidian space $\wedge^{m-2}\mathbb{R}^m$ of $(m-2)$-vectors in $\mathbb{R}^m$ made of unit simple $(m-2)$-vectors. Then one defines

$$W^{1,2}(D^2, Gr_{m-2}(\mathbb{R}^m)) := \left\{\vec{n} \in W^{1,2}(D^2, \wedge^{m-2}\mathbb{R}^m)\ ;\ \vec{n} \in Gr_{m-2}(\mathbb{R}^m) \text{ a.e.}\right\}.$$

The space $(L^{2,\infty}, |\cdot|_{L^{2,\infty}})$ is a quasi-Banach space, equivalent to a Banach space. This means that the quasi-norm $|\cdot|_{L^{2,\infty}}$ is equivalent to a norm $\|\cdot\|_{L^{2,\infty}}$. We have the continuous embedding $L^2 \subset L^{2,\infty}$.[2] Furthermore, it can be checked that, since D^2 is a bounded domain, for all $p < 2$,

$$L^p(D^2) \subset L^{2,\infty}(D^2) \tag{2.90}$$

continuously.

One shows that $(L^{2,\infty}, \|\cdot\|_{L^{2,\infty}})$ is the dual space to a space named *Lorentz space* $L^{2,1}$, which is made of measurable functions f such that

$$\int_0^{\infty} |\{x;\ |f(x)| > \lambda\}|^{1/2}\ d\lambda < \infty.$$

See [**Gra08**] for further details and more material on weak L^p and Lorentz spaces.

We have the following result.

THEOREM 2.22. *Let $f \in L^1(D^2)$. Then for the unique solution ϕ in $W_0^{1,1}(D^2)$ of*

$$\begin{cases} \Delta\phi &= f \quad \text{in } D^2 \\ \phi &= 0 \quad \text{on } \partial D^2 \end{cases}$$

we have $\nabla\phi \in L^{2,\infty}(D^2)$ and

$$\|\nabla\phi\|_{L^{2,\infty}(D^2)} \leq C\ \|f\|_{L^1(D^2)}.$$

Sketch of proof of Theorem 2.22. One checks that the gradient of the Green's function for D^2 ([**GT01**], Chapter 2.5) satisfies

$$\sup_y \||\nabla_x K|(x,y)\|_{L^{2,\infty}_x(D^2)} < \infty.$$

Since

$$\nabla\phi(x) = \int_{D^2} \nabla_x K(x,y) f(y)dy, \tag{2.91}$$

we have

$$\begin{aligned} \|\nabla\phi\|_{L^{2,\infty}(D^2)} &\leq \left\| \int_{D^2} |\nabla_x K(x,y)||f(y)|dy \right\|_{L^{2,\infty}(D^2)} \\ &\leq \int_{D^2} \|\nabla_x K(x,y)\|_{L^{2,\infty}_x(D^2)} |f(y)|dy \qquad (2.92) \\ &\leq \sup_y \|\nabla_x K\|_{L^{2,\infty}_x(D^2)}\ \|f\|_{L^1(D^2)}, \end{aligned}$$

which is the result.

From this proof one also derives immediately the following:

[2] More generally, for $p < \infty$ and $\Omega \subset \mathbb{R}^n$ open, one defines $L^{p,\infty}(\Omega)$ to be the space of measurable functions $f\colon \Omega \to \mathbb{R}$ such that

$$|f|_{L^{p,\infty}(\Omega)} := \sup_{\lambda>0} \left\{ \lambda\,|\{x \in \Omega;\ |f(x)| > \lambda\}|^{1/p} \right\} < \infty.$$

This space defines a quasi-Banach space for the quasi-norm $|\cdot|_{L^{p,\infty}(\Omega)}$ and for $1 < p < \infty$ this quasi norm is equivalent to a norm. $L^p(\Omega)$ embeds continuously in $L^{p,\infty}(\Omega)$.

THEOREM 2.23. *Let $X = (X_1, X_2)$ be a vector field in $L^1(D^2)$. Then for the solution of*

$$\begin{cases} \Delta\phi &= div X \quad in\ D^2 \\ \\ \phi &= 0 \quad\quad on\ \partial D^2, \end{cases}$$

we have $\phi \in L^{2,\infty}(D^2)$ with

$$\|\phi\|_{L^{2,\infty}(D^2)} \le C\ \|X\|_{L^1(D^2)}.$$

Indeed, we have

$$\begin{aligned} \phi(x) &= \int_{D^2} K(x,y)\ div X\ dy \\ \\ &= -\int_{D^2} \partial_{x_1} K(x,y)\ X_1 + \partial_{x_2} K(x,y)\ X_2\ dy \end{aligned}$$

and the result follows as in (2.92).

Furthermore, one can proceed in a similar manner as in the proof of Theorem 2.22 and use estimates of the gradient of the Green's function K_Σ on a closed surface Σ ([**Aub82**], Chapter 4) satisfying

$$\Delta_x K_\Sigma = \delta_{x=y} - \frac{1}{vol_{g_0}(\Sigma)},$$

to obtain the analog of Theorem 2.22 for a closed, oriented and connected surface Σ.

THEOREM 2.24. *Let Σ be a closed, oriented and connected surface. Let $f \in L^1(\Sigma)$ such that*

$$\int_\Sigma f\ dvol_{g_0} = 0.$$

Then there is a unique solution ϕ in $W^{1,1}(\Sigma)$ of

$$\Delta\phi = f \qquad in\ \mathcal{D}'(\Sigma)$$

and it satisfies

$$\|\nabla\phi\|_{L^{2,\infty}_{g_0}(\Sigma)} \le C_{g_0}\|f\|_{L^1_{g_0}(\Sigma)}.$$

One might also ask about the regularity of ϕ solving (2.89). Combining Theorem 2.21 and the Sobolev embedding

$$W_0^{1,p}(D^2) \hookrightarrow L^{p^*}(D^2), \qquad \frac{1}{p^*} = \frac{1}{p} - \frac{1}{2},$$

one deduces that

$$\phi \in \bigcap_{q<\infty} L^q(D^2).$$

This is not fully optimal: Using the embedding

$$\dot{W}^{1,(2,\infty)}(\mathbb{R}^2) \hookrightarrow \mathrm{BMO}(\mathbb{R}^2),$$

where $\mathrm{BMO}(\mathbb{R}^2)$ is the John Nierenberg space of functions of *Bounded Mean Oscillation*, one obtains that the extension of ϕ by 0 is BMO. Note that

$$\mathrm{BMO}(\mathbb{R}^2) \supsetneq L^\infty(\mathbb{R}^2),$$

where for instance $\log|x| \in \mathrm{BMO}(\mathbb{R}^2) \setminus L^\infty(\mathbb{R}^2)$.

2.2. Wente's integrability by compensation

In a famous work Henry Wente produced CMC tori immersed in $\mathbb{R}^3$ and discovered some "improvement" in the integrablility of ϕ, the solution to (2.89), as well as the one of $\nabla\phi$, when f is a Jacobian of $W^{1,2}$-functions.

THEOREM 2.25. [**Wen69**] *Let a and b be two functions in $W^{1,2}(D^2)$, and let ϕ be the unique solution in $W_0^{1,p}(D^2)$ - for $1 \le p < 2$ - of the equation*

$$\left\{ \begin{array}{lcll} \Delta\phi & - & \partial_{x_1} u\, \partial_{x_2} b - \partial_{x_1} b\, \partial_{x_2} a & \text{in } D^2 \\ \\ \phi & = & 0 & \text{on } \partial D^2. \end{array} \right. \tag{2.93}$$

Then ϕ belongs to $C^0 \cap W^{1,2}(D^2)$ and

$$\|\phi\|_{L^\infty(D^2)} + \|\nabla\phi\|_{L^2(D^2)} \le C\ \|\nabla a\|_{L^2(D^2)}\ \|\nabla b\|_{L^2(D^2)}. \tag{2.94}$$

where C is a constant independent of a and b.

PROOF. We shall first assume that a and b are smooth, so as to justify the various manipulations which we will need to perform. The conclusion of the theorem for general a and b in $W^{1,2}(D^2)$ may then be reached through a simple density argument. In this fashion, we will obtain the continuity of ϕ being the uniform limit of smooth functions.

Observe first that integration by parts and a simple application of the Cauchy-Schwarz inequality yield the estimate

$$\begin{aligned} \int_{D^2} |\nabla\phi|^2 = -\int_{D^2} \phi\, \Delta\phi \ &\le \|\phi\|_\infty\ \|\partial_x a\, \partial_y b - \partial_x b\, \partial_y a\|_1 \\ &\le 2\, \|\phi\|_\infty \|\nabla a\|_2\ \|\nabla b\|_2. \end{aligned}$$

Accordingly, if ϕ lies in $L^\infty(D^2)$, then it automatically lies in $W^{1,2}(D^2)$.

Step 1. Given two functions $\tilde{a}$ and $\tilde{b}$ in $C_0^\infty(\mathbb{C})$, which is dense in $W^{1,2}(\mathbb{C})$, we first establish the estimate (2.94) for

$$\tilde{\phi} := \frac{1}{2\pi} \log\frac{1}{r} * \left[\partial_x \tilde{a}\, \partial_y \tilde{b} - \partial_x \tilde{b}\, \partial_y \tilde{a}\right]. \tag{2.95}$$

Owing to the translation-invariance, it suffices to show that

$$|\tilde{\phi}(0)| \le C_0\ \|\nabla\tilde{a}\|_{L^2(\mathbb{C})}\ \|\nabla\tilde{b}\|_{L^2(\mathbb{C})}. \tag{2.96}$$

We have

$$\begin{aligned} \tilde{\phi}(0) &= -\frac{1}{2\pi}\int_{\mathbb{R}^2} \log r\ \partial_x\tilde{a}\, \partial_y\tilde{b} - \partial_x\tilde{b}\, \partial_y\tilde{a} \\ &= -\frac{1}{2\pi}\int_0^{2\pi}\int_0^{+\infty} \log r\ \frac{\partial}{\partial r}\left(\tilde{a}\ \frac{\partial\tilde{b}}{\partial\theta}\right) - \frac{\partial}{\partial\theta}\left(\tilde{a}\ \frac{\partial\tilde{b}}{\partial r}\right)\ dr\, d\theta \\ &= \frac{1}{2\pi}\int_0^{2\pi}\int_0^{+\infty} \tilde{a}\ \frac{\partial\tilde{b}}{\partial\theta}\ \frac{dr}{r}\ d\theta. \end{aligned}$$

Define $\overline{\tilde{a}}_r = \frac{1}{2\pi}\int_0^{2\pi} \tilde{a}(r,\theta)d\theta$, and since $\int_0^{2\pi} \frac{\partial\tilde{b}}{\partial\theta}\ d\theta = 0$, there holds

$$\tilde{\phi}(0) = \frac{1}{2\pi}\int_0^{2\pi}\int_0^{+\infty} [\tilde{a} - \overline{\tilde{a}}_r]\ \frac{\partial\tilde{b}}{\partial\theta}\ \frac{dr}{r}\ d\theta.$$

Applying successively the Cauchy-Schwarz and Poincaré inequalities on the circle S^1, we obtain

$$
\begin{aligned}
|\tilde{\phi}(0)| &\leq \frac{1}{2\pi}\int_0^{+\infty}\frac{dr}{r}\left(\int_0^{2\pi}|\tilde{a}-\bar{\tilde{a}}_r|^2\right)^{\frac{1}{2}}\left(\int_0^{2\pi}\left|\frac{\partial\tilde{b}}{\partial\theta}\right|^2\right)^{\frac{1}{2}}\\
&\leq \frac{1}{2\pi}\int_0^{+\infty}\frac{dr}{r}\left(\int_0^{2\pi}\left|\frac{\partial\tilde{a}}{\partial\theta}\right|^2\right)^{\frac{1}{2}}\left(\int_0^{2\pi}\left|\frac{\partial\tilde{b}}{\partial\theta}\right|^2\right)^{\frac{1}{2}}.
\end{aligned}
$$

Inequality (2.96) may then be inferred from the latter by applying once more the Cauchy-Schwarz inequality.

Returning to the disk D^2, we can extend a and b to $\mathbb{C}$ by setting $\tilde{a}(z) = a(\frac{z}{|z|^2})$ (and $\tilde{b}$ similarly) outside of D^2. Then

$$
\|\nabla\tilde{a}\|_{L^2(\mathbb{C})} = 2\|\nabla a\|_{L^2(D^2)}, \quad \|\nabla\tilde{b}\|_{L^2(\mathbb{C})} = 2\|\nabla b\|_{L^2(D^2)}. \tag{2.97}
$$

Let $\tilde{\phi}$ be the function in (2.95). The difference $\phi - \tilde{\phi}$ satisfies the equation

$$
\left\{
\begin{array}{rcll}
\Delta(\phi-\tilde{\phi}) &=& 0 & \text{in } D^2\\[2mm]
\phi-\tilde{\phi} &=& -\tilde{\phi} & \text{on } \partial D^2.
\end{array}
\right.
$$

The *maximum principle* and inequalities (2.96), (2.97) imply

$$
\|\phi-\tilde{\phi}\|_{L^\infty(D^2)} \leq \|\tilde{\phi}\|_{L^\infty(\partial D^2)} \leq C\|\nabla a\|_{L^2(D^2)}\ \|\nabla b\|_{L^2(D^2)}.
$$

With the triangle inequality $|\|\phi\|_\infty - \|\tilde{\phi}\|_\infty| \leq \|\phi-\tilde{\phi}\|_\infty$ and inequality (2.96) again, we reach the desired L^∞-estimate of ϕ, and therefore, by the above discussion, the theorem is proved. □

The following result, due to S. Chanillo and Y.Y. Li is a generalization of Wente's theorem 2.25.

THEOREM 2.26 ([**CL92**]). *Let a and b be two functions in $W^{1,2}(D^2)$. Let $(a^{ij})_{1\leq i,j\leq 2}$ be a 2×2 symmetric matrix-valued map in $L^\infty(D^2)$ such that there exists $C>0$ for which*

$$
C^{-1}\ |\xi|^2 \leq a^{ij}(x)\xi_i\xi_j \leq C\ |\xi|^2
$$

for all $\xi = (\xi_1,\xi_2)\in\mathbb{R}^2$ and all $x\in D^2$.

Let φ be the solution in $W^{1,p}(D^2,\mathbb{R})$ for any $1\leq p<2$ of

$$
\left\{
\begin{array}{rcll}
\partial_{x_i}\left[a^{ij}\,\partial_{x_j}\lambda\right] &=& \partial_{x_1}a\,\partial_{x_2}b-\partial_{x_2}a\,\partial_{x_1}b & \text{on } D^2\\[2mm]
\varphi &=& 0 & \text{on } \partial D^2.
\end{array}
\right. \tag{2.98}
$$

Then $\varphi\in L^\infty\cap W^{1,2}(D^2,\mathbb{R})$ and there exists $C>0$ independent of a and b such that

$$
\|\varphi\|_{L^\infty(D^2)} + \|\nabla\varphi\|_{L^2(D^2)} \leq C\ \|\nabla a\|_{L^2(D^2)}\ \|\nabla b\|_{L^2(D^2)}. \tag{2.99}
$$

2.3. Integrability by compensation on Riemann surfaces

Integration by compensation is natural in the framework of surfaces. An example provides the following generalization of Wente's inequality to Riemann surfaces, due to Topping ([**Top97**]) and Ge ([**Ge98**]).

THEOREM 2.27. *Let (Σ, g) be a Riemann surface, $a, b \in W^{1,2}(\Sigma)$ and ϕ the solution in $W^{1,p}(\Sigma)$ - for $1 \leq p < 2$ - of*

$$\left\{ \begin{array}{rcll} *_g \Delta_g \phi & = & da \wedge db & \text{on } \Sigma \\ \phi & = & 0 & \text{on } \partial\Sigma, \text{ if } \partial\Sigma \neq \emptyset \\ \phi(p) & = & 0 & \text{at a certain } p \in \Sigma, \text{ if } \partial\Sigma = \emptyset. \end{array} \right. \tag{2.100}$$

Then ϕ belongs to $W^{1,2} \cap C^0(\Sigma)$ and satisfies

$$\|\phi\|_{L^\infty(\Sigma)} + \|\nabla\phi\|_{L^2(\Sigma)} \leq C \, \|\nabla a\|_{L^2(\Sigma)} \, \|\nabla b\|_{L^2(\Sigma)}, \tag{2.101}$$

where C is a universal constant, not depending on Σ.

Note that the operator $*_g\Delta_g$ in (2.100) as well as the norms in (2.101) are conformally invariant (and thus it makes sense to consider them on a Riemann surface).

2.4. Some more results in integrability by compensation

THEOREM 2.28 ([**Bet92**], [**Ge99**]). *Let a be such that $\nabla a \in L^{2,\infty}(D^2)$. Let $p \in (1, \infty)$ and $b \in W^{1,p}(D^2)$. Then the $W_0^{1,1}(D^2)$-solution ϕ of*[3]

$$\left\{ \begin{array}{rcll} \Delta\phi & = & \partial_{x_1} a \; \partial_{x_2} b - \partial_{x_2} a \; \partial_{x_1} b & \text{in } D^2 \\ \phi & = & 0 & \text{on } \partial D^2 \end{array} \right.$$

is in $W_0^{1,p}(D^2)$ and

$$\|\nabla\phi\|_{L^p(D^2)} \leq C \|\nabla a\|_{L^{2,\infty}(D^2)} \|\nabla b\|_{L^p(D^2)}.$$

3. Existence of isothermal coordinates in the weak framework

Given a Lipschitz immersion $\vec{\Phi} \in \mathcal{E}_\Sigma$ with L^2-bounded second fundamental form, we want to show the existence of isothermal coordinates for $\vec{\Phi}$:

For all $p \in \Sigma$ there exists a neighborhood U of p and a bi-Lipschitz diffeomorphism $\psi \colon D^2 \to U$ such that ψ is a (weakly) isothermal chart for $\vec{\Phi}$, i.e. satisfying

$$\left\{ \begin{array}{ll} \langle \partial_{x_1}(\vec{\Phi} \circ \psi), \partial_{x_2}(\vec{\Phi} \circ \psi) \rangle = 0 & \text{a.e. in } D^2 \\ |\partial_{x_1}(\vec{\Phi} \circ \psi)| = |\partial_{x_2}(\vec{\Phi} \circ \psi)| & \text{a.e. in } D^2. \end{array} \right. \tag{2.102}$$

In the proof of Theorem 1.9 we presented the Chern moving frame method in order to show the existence of isothermal coordinates in the smooth case. We now want to apply this method in the context of Lipschitz immersions with L^2-bounded second

---

[3]If $p \leq 2$, the Jacobian $\partial_{x_1} a \; \partial_{x_2} b - \partial_{x_2} a \; \partial_{x_1} b$ is understood in the weak sense

$$\partial_{x_1} a \; \partial_{x_2} b - \partial_{x_2} a \; \partial_{x_1} b := div \left[a \, \nabla^\perp b \right].$$

Since $\nabla a \in L^{2,\infty}(D^2)$ we have that $a \in L^q(D^2)$ for all $q < \infty$ and hence $a\nabla^\perp b \in L^r(D^2)$ for all $r < p$.

fundamental form and prove the following analogue, where we require an additional energy restriction:

LEMMA 2.29. *Let $\vec{\Phi} \in \mathcal{E}_{D^2}$ such that*

$$\int_{D^2} |\nabla \vec{n}_{\vec{\Phi}}|^2 dx_1 dx_2 < \frac{8\pi}{3}. \tag{2.103}$$

Then there exists a homeomorphism ψ of D^2, locally bi-Lipschitz, such that the map $\vec{\Phi} \circ \psi \colon D^2 \to \mathbb{R}^m$ is conformal, i.e. it satisfies (2.102).

To prove Lemma 2.29 we construct a local Coulomb tangent moving frame with controlled $W^{1,2}$-energy. We shall do so in two steps: In subsection 4 we will explore that under assumption (2.103) it is possible to "lift" the Gauss map and to construct a tangent moving frame with bounded $W^{1,2}$-energy. After that, in subsection 5, we will show how to turn the latter into a Coulomb frame with controlled energy.

Finally, in subsection 6, we will see that, by means of a result on integrability by compensation, we can use the Chern moving frame method also in the weak framework and conclude Lemma 2.29.

4. Hélein's energy controlled lifting theorem

The following **lifting theorem** was proven by F. Hélein in [**Hél02**].

THEOREM 2.30. *Let $\vec{n} \in W^{1,2}(D^2, Gr_{m-2}(\mathbb{R}^m))$.*
Then there exists a constant $C > 0$, such that, if one assumes that

$$\int_{D^2} |\nabla \vec{n}|^2 \, dx_1 dx_2 < \frac{8\pi}{3}, \tag{2.104}$$

then there exist $\vec{e}_1, \vec{e}_2 \in W^{1,2}(D^2, S^{m-1})$ such that

$$\vec{n} = \star(\vec{e}_1 \wedge \vec{e}_2)^4, \tag{2.105}$$

and

$$\int_{D^2} \sum_{i=1}^{2} |\nabla \vec{e}_i|^2 \, dx_1 \, dx_2 \leq C \int_{D^2} |\nabla \vec{n}|^2 \, dx_1, dx_2. \tag{2.106}$$

5. Construction of local Coulomb frames with controlled $W^{1,2}$-energy

PROOF OF THEOREM 2.29. Let $\vec{\Phi} \in \mathcal{E}_{D^2}$ satisfy (2.103). Due to Theorem 2.30, there exists a frame

$$(\vec{f}_1, \vec{f}_2) \in W^{1,2}(D^2, S^{m-1})^2 \tag{2.107}$$

with controlled energy. Consequently, one can proceed as in the proof of Theorem 1.9 and look for the minimizer of

$$\int_{D^2} \left| d\theta + \langle \vec{f}_1, d\vec{f}_2 \rangle \right|_g^2 \, dvol_g,$$

[4]Condition (2.105), together with the facts that the $\vec{e}_i$ are S^{m-1}-valued and $\vec{n}$ has norm one, implies that $\vec{e}_1$ and $\vec{e}_2$ are orthogonal to each other.

among all $\theta \in W^{1,2}(D^2,\mathbb{R})$, in order to produce a Coulomb frame on D^2. This Lagrangian is convex on the Hilbert space $W^{1,2}(D^2,\mathbb{R})$ and goes to $+\infty$ as $\|\theta\|_{W^{1,2}} \to +\infty$. Then there exists a unique minimum satisfying

$$\left\{\begin{array}{rcll} d^{*_g}\left[d\theta + \langle \vec{f}_1, d\vec{f}_2\rangle\right] & = & 0 & \text{in } D^2 \\ \iota^*_{\partial D^2}\left(*_g\left[d\theta + \langle \vec{f}_1, d\vec{f}_2\rangle\right]\right) & = & 0 & \text{on } \partial D^2, \end{array}\right.$$

where $\iota_{\partial D^2}$ is the canonical inclusion of ∂D^2 into $\overline{D^2}$. Then $\vec{e} := \vec{e}_1 + i\vec{e}_2$ given by

$$\vec{e} = e^{i\theta}\vec{f} \tag{2.108}$$

is Coulomb, as desired:

$$\left\{\begin{array}{rcll} d^{*_g}\left[\langle \vec{e}_1, d\vec{e}_2\rangle\right] & = & 0 & \text{in } D^2 \\ \iota^*_{\partial D^2}\left(*_g\left[\langle \vec{e}_1, d\vec{e}_2\rangle\right]\right) & = & 0 & \text{on } \partial D^2. \end{array}\right. \tag{2.109}$$

6. The Chern moving frame method in the weak framework

With a Coulomb frame $(\vec{e}_1, \vec{e}_2)$ satisfying (2.109) at hand, we are now in the position to apply the Chern moving frame method from 5. This however has to be done with the additional difficulty of keeping track of the regularity of the different actors at each step of the construction.

By the weak Poincaré Lemma, there exists $\lambda \in W^{1,2}(D^2)$ satisfying

$$\left\{\begin{array}{rcl} d\lambda & = & *_g\langle \vec{e}_1, d\vec{e}_2\rangle \\ \displaystyle\int_{\partial D^2} \lambda & = & 0. \end{array}\right. \tag{2.110}$$

The second equation of (2.109) implies that the restriction to ∂D^2 of the 1-form $d\lambda$ is equal to zero. Hence this last fact combined with the second equation of (2.110) implies that λ is identically equal to zero on ∂D^2. With the notation introduced in Section 1, observe that

$$d\langle \vec{e}_1, d\vec{e}_2\rangle = \langle d\vec{e}_1, d\vec{e}_2\rangle = \langle d\vec{f}_1, d\vec{f}_2\rangle, \tag{2.111}$$

by (2.108) and the change of gauge formula (1.61). We have then

$$\left\{\begin{array}{rcll} d *_g d\lambda & = & -\langle d\vec{f}_1, d\vec{f}_2\rangle & \text{on } D^2 \\ \lambda & = & 0 & \text{on } \partial D^2. \end{array}\right. \tag{2.112}$$

In the canonical coordinates of D^2, this reads as

$$\left\{\begin{array}{rl} \partial_{x_i}\left[\sqrt{det\,(g_{ij})}\,g^{ij}\partial_{x_j}\lambda\right] & = \langle \partial_{x_1}\vec{f}_1, \partial_{x_2}\vec{f}_2\rangle - \langle \partial_{x_2}\vec{f}_1, \partial_{x_1}\vec{f}_2\rangle \\ & \qquad\qquad \text{on } D^2 \\ \lambda & = 0 \qquad\qquad \text{on } \partial D^2. \end{array}\right.$$

Since

$$\langle \partial_{x_1}\vec{f}_1, \partial_{x_2}\vec{f}_2\rangle - \langle \partial_{x_2}\vec{f}_1, \partial_{x_1}\vec{f}_2\rangle = \sum_{j=1}^m \partial_{x_1}f_{1,j}\cdot\partial_{x_2}f_{2,j} - \partial_{x_2}f_{1,j}\cdot\partial_{x_1}f_{2,j}$$

is a sum of Jacobians of $W^{1,2}$-functions, due to (2.107), we are now in the position to make use of Theorem 2.26, a generalization of Wente's theorem on integrability by compensation. It gives us that

$$\lambda \in L^\infty(D^2, \mathbb{R}). \tag{2.113}$$

Let (f_1, f_2) be the frame on D^2 given by

$$d\vec{\Phi}(f_i) = \vec{f}_i, \qquad i = 1, 2.$$

An easy computation shows that its dual frame satisfies

$$f_i^* = \sum_{k=1}^{2} \langle \vec{e}_i, \partial_{x_k}\vec{\Phi}\rangle \, dx_k \in L^\infty(D^2), \tag{2.114}$$

for $i = 1, 2$. Denote by (e_1, e_2) the frame on D^2 such that

$$d\vec{\Phi}(e_i) = \vec{e}_i.$$

By (2.108), we have

$$e^* = e_1^* + ie_2^* = e^{-i\theta}\left(f_1^* + if_2^*\right)$$

which is in $L^\infty(D^2)$ due to (2.114). From the proof of Theorem 1.9 follows that

$$d(e^{-\lambda} e_i^*) = 0 \qquad \text{in } \mathcal{D}'(D^2).$$

Applying the weak Poincaré Lemma gives a map $\phi = (\phi_1, \phi_2)$ satisfying

$$d\phi_i := e^{-\lambda} e_i^*.$$

Exactly as in the proof of Theorem 1.9, using Corollary 1.16, one shows that there exists a homeomorphism ψ of D^2, locally bi-Lipschitz, realizing an isothermal chart for $\vec{\Phi}$ (i.e. satisfying (2.102)). □

Lemma 2.29 immediately implies the following theorem.

THEOREM 2.31. **[Existence of isothermal coordinates]**
Let $\vec{\Phi} \in \mathcal{E}_\Sigma$. Then for each $p \in \Sigma$ there exists a neighborhood U and a Lipschitz diffeomorphism ψ from D^2 into U such that ψ is an isothermal chart for $\vec{\Phi}$, i.e. satisfying (2.102).

PROOF OF THEOREM 2.31. Let $p \in \Sigma$ be an arbitrary point. Choose any chart $\varphi \colon D^2 \to U$ around $p \in U$. Due to condition iii), we can assume that

$$\int_{D^2} |d\vec{n}_{\vec{\Phi}\circ\varphi}|^2 dx_1 dx_2 < \frac{8\pi}{3},$$

possibly after restricting φ to a sufficiently small neighborhood of $\varphi^{-1}(p) = 0$. Thus, $\vec{\Phi} \circ \varphi \in \mathcal{E}_{D^2}$ satisfies the assumptions of Lemma 2.29, which then gives us a bi-Lipschitz diffeomorphism ψ of $D^2_{1/2}$ such that $\vec{\Phi} \circ \varphi \circ \psi$ is conformal. $\varphi \circ \psi$ is the desired isothermal chart around p. □

COROLLARY 2.32. **[Existence of a smooth conformal structure]** *Let Σ be a closed smooth 2-dimensional manifold. Then any $\vec{\Phi} \in \mathcal{E}_\Sigma$ defines a smooth conformal structure on Σ. In particular there exists a constant curvature metric h of unit volume on Σ and a bilipschitz diffeomorphism Ψ of Σ such that $\vec{\Phi} \circ \Psi$ realizes a conformal immersion of the Riemann surface (Σ, h) and h and $g_{\vec{\Phi}}$ are conformally equivalent.*

DEFINITION 2.33. *If h is a conformal structure on Σ, a map $\vec{\xi}\colon (\Sigma, h) \to \mathbb{R}$ is called* (weakly) conformal *if it satisfies*

$$\begin{cases} \langle \partial_{x_1}\vec{\xi}, \partial_{x_2}\vec{\xi}\rangle = 0 & a.e.\ in\ \Sigma \\ \\ |\partial_{x_1}\vec{\xi}| = |\partial_{x_2}\vec{\xi}| & a.e.\ in\ \Sigma. \end{cases}$$

PROOF OF COROLLARY 2.32. Σ can be covered by finitely many isothermal charts $(U_i, \psi_i)_{i=1,\ldots,n}$, due to Theorem 2.31 and the fact that Σ is compact. For each $i = 1, \ldots, n$, the map $\vec{\Phi} \circ \psi_i$ realizes a conformal Lipschitz immersion of D^2, denoted by

$$\psi_i^* g_{\vec{\Phi}} = e^{2\lambda_i}(dx^1 + dx^2).$$

If $U_i \cap U_j \neq \emptyset$, the map $\psi_j^{-1} \circ \psi_i\colon \psi_i^{-1}(U_i \cap U_j) \to \mathbb{C}$ is conformal and positive (hence holomorphic) almost everywhere:

$$(\psi_j^{-1} \circ \psi_i)^*(dy^1 + dy^2) = \psi_i^*(e^{-2\lambda_j} g_{\vec{\Phi}}) = e^{2(\lambda_i - \lambda_j)}(dx^1 + dx^2).$$

It follows that $\psi_j^{-1} \circ \psi_i$ is harmonic on $U_i \cap U_j$. From standard distribution theory follows that it is holomorphic everywhere on $U_i \cap U_j$. Thus, the system of charts (U_i, ψ_i) defines a smooth conformal structure c on Σ.

The second statement follows from Theorem 1.18. $\square$

LECTURE 3

Sequences of weak immersions

1. Compactness question

Eventually we will be interested in finding a minimizer of the Willmore energy. More specifically, we will look at a minimizing sequence of weak immersions with L^2-bounded second fundamental form and ask if it has a limit with respect to some notion of weak convergence and how this limiting object looks like.

Recall that in Section 3 we showed that the Willmore energy is invariant under conformal transformations in the target. To get an idea what kind of convergence is the best we can hope for, let us study an example: Assume $\vec{\Phi} \in \mathcal{E}_\Sigma$ immerses a torus in $\mathbb{R}^3$. Figures 1,2 and 3 show that composing $\vec{\Phi}$ with sequences of conformal diffeomorphisms of $\mathbb{R}^3$ of different types produces sequences $\vec{\Phi}_k$, whose limit does not immerse a torus anymore.

Here, for $a \in \mathbb{R}^m$, the map $i_a \in \mathrm{Conf}(\mathbb{R}^3 \cup \{\infty\})$ denotes the inversion at a,

$$i_a \colon x \mapsto \frac{x-a}{|x-a|^2}.$$

Note that i_a is a diffeomorphism from $B_R(a) \setminus B_\varepsilon(a)$ to $B_{1/\varepsilon}(a) \setminus B_{1/R}(a)$ for any $0 < \varepsilon < R < \infty$ and thus, for $a \in \mathbb{R}^m \setminus \vec{\Phi}(\Sigma)$,

$$W(i_a \circ \vec{\Phi}) = W(\vec{\Phi}).$$

These examples illustrate that, given a sequence $\vec{\Phi}_k \in \mathcal{E}_\Sigma$, say with $\sup_k \mathbb{I}(\vec{\Phi}_k) < \infty$, we cannot expect that $\vec{\Phi}_k$ has a limit, which still immerses the surface Σ (in a weak sense). In fact, we need to compose with conformal transformations $\vec{\Xi}_k \in \mathrm{Conf}(\mathbb{R}^m \cup \{\infty\})$ (with the center of inversion of $\vec{\Xi}_k$ not being contained in $\vec{\Phi}_k(\Sigma)$) to avoid degeneracies as shown in Figures 1 - 3.

Furthermore, when passing to the limit there might be energy concentration in single points. Figure 4 provides an example of how such a loss of energy in the limit might occur.

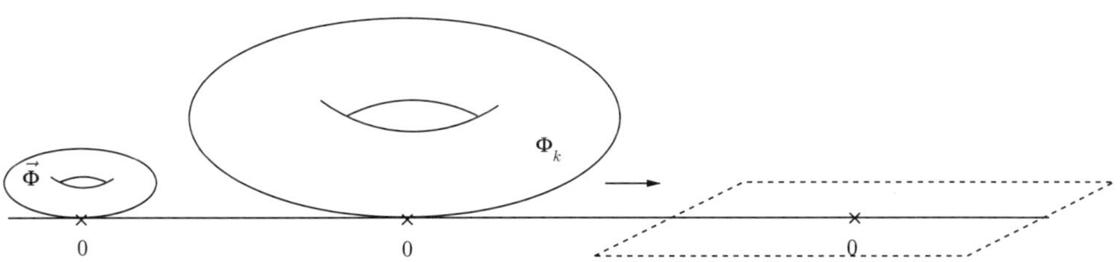

FIGURE 1. Dilation. $\vec{\Phi}_k := k\vec{\Phi}$. (Loss of energy and topology.)

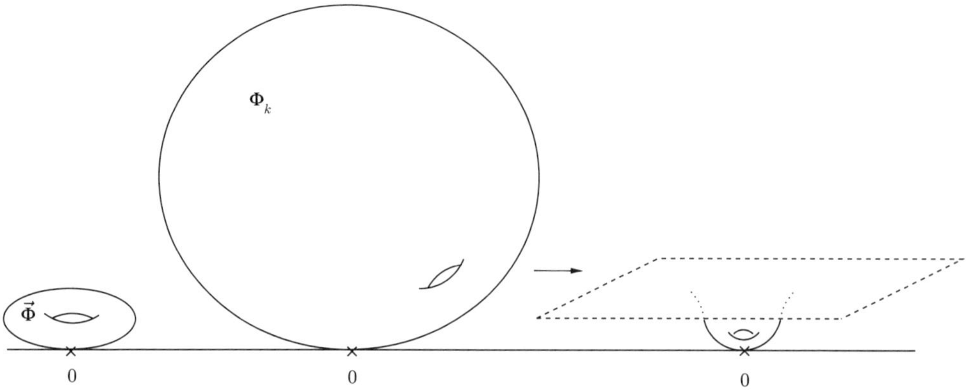

FIGURE 2. Inversion. $\vec{\Phi}_k := i_{a_k} \circ \vec{\Phi}$, where $a_k \to \vec{\Phi}(\Sigma) \setminus \{0\}$. (Loss of energy.)

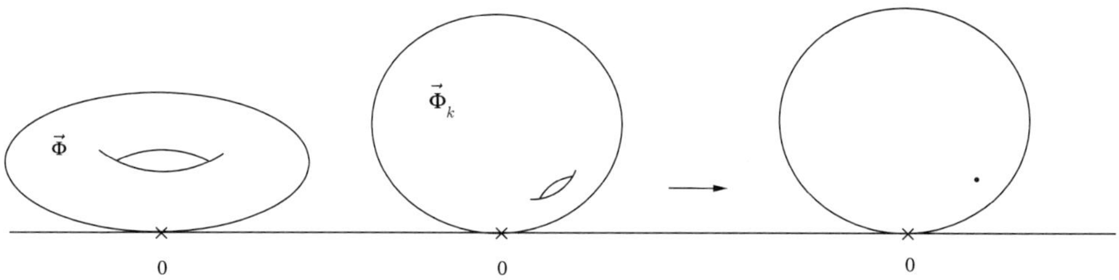

FIGURE 3. Inversion $\circ$ Dilation. $\vec{\Phi}_k := \text{dist}(a_k, \vec{\Phi}(\Sigma))\ i_{a_k} \circ \vec{\Phi}$, where $a_k \to \vec{\Phi}(\Sigma) \setminus \{0\}$. (Loss of energy and topology.)

To allow composing with conformal transformations along the sequence in order to obtain a reasonable (weak) limiting object is not enough: We also need to make a compactness assumption on the conformal classes induced by $\vec{\Phi}_k$. Let $\Psi_k : (\Sigma, h_k) \to \Sigma$ be the Lipschitz diffeomorphisms from Corollary 2.32 such that

$$\vec{\Psi}_k := \vec{\Phi}_k \circ \Psi_k : (\Sigma, h_k) \to \mathbb{R}^m$$

are conformal, where each h_k denotes the reference metric of constant curvature and unit volume of the conformal structure induced by $\vec{\Phi}_k$. Since Ψ_k are elements of the invariance group $\text{Diff}^+(\Sigma)$ of the Willmore functional, we have $W(\vec{\Psi}_k) = W(\vec{\Phi}_k)$. We make the following compactness assumption **(CA)**:

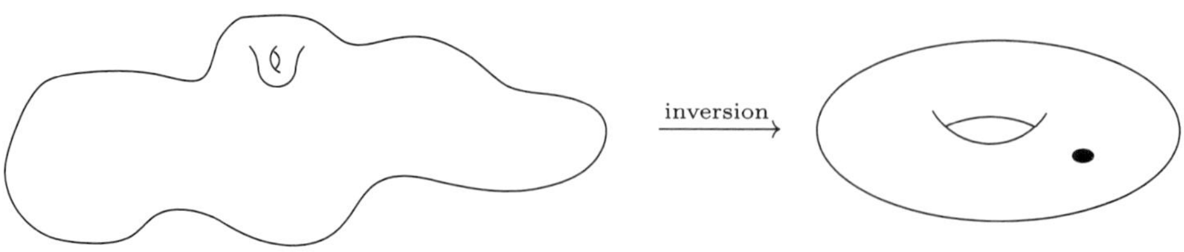

FIGURE 4. Loss of energy, no loss of topology.

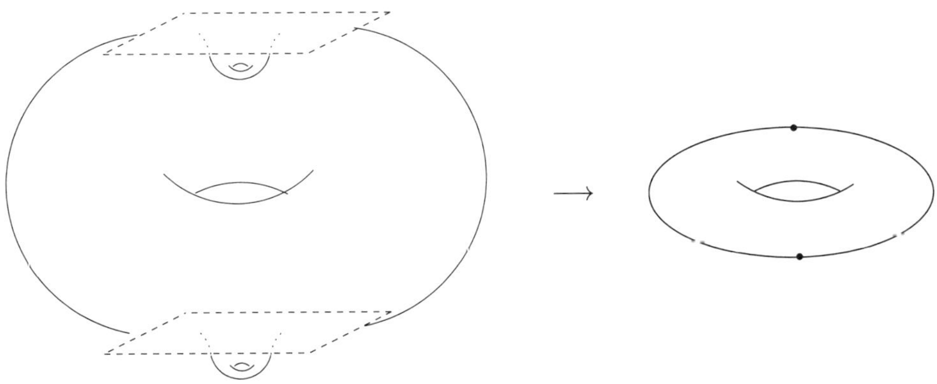

FIGURE 5. Loss of energy, topology and conformal class.

> The conformal classes (Σ, h_k) are contained in a compact subset of $\mathcal{M}_\Sigma$, the moduli space of Σ.

This assumption is necessary for the following reason: If the conformal classes degenerate, we might not only lose energy but also topology in the limit, which is irreversible and has to be avoided. Such a situation is shown in Figure 5. Note that there is no way to preserve the genus in the limit: One can "save" one hole (e.g. by applying dilations and inversions as in Figure 3), but one will lose the other two ones at the same time.

Before using the previous observations to define a notion of *weak convergence*, we note the following fact:

If a sequence of $\vec{\Phi}_k \in \mathcal{E}_\Sigma$ satisfies the compactness assumption (CA), the conformal classes (Σ, h_k) of constant curvature and unit volume, induced by $\vec{\Phi}_k$ resp., satisfy (up to subsequences)

$$h_k \to h_\infty \qquad \text{in } C^l(\Sigma) \quad \forall l \in \mathbb{N},$$

where (Σ, h_∞) is the limiting conformal structure of constant curvature and unit volume.

DEFINITION 3.34. *A sequence $\vec{\Phi}_k \in \mathcal{E}_\Sigma$ satisfying assumption (CA) with $h_k \to h_\infty$ is called* weakly convergent *if there exist Lipschitz diffeomorphisms Ψ_k of Σ, conformal transformations $\vec{\Xi}_k$ of $\mathbb{R}^m \cup \{\infty\}$ with*

$$\vec{\Phi}_k(\Sigma) \cap \{\textit{center of inversion of } \vec{\Xi}_k\} = \emptyset$$

and finitely many points $a_1, \dots, a_N \in \Sigma$, called blow-up points *such that*

$$\vec{\xi}_k := \vec{\Xi}_k \circ \vec{\Phi}_k \circ \Psi_k \colon (\Sigma, h_k) \to \mathbb{R}^m$$

is conformal, and there exists a map $\vec{\xi}_\infty \colon \Sigma \to \mathbb{R}^m$ such that

i)

$$\vec{\xi}_\infty \textit{ is conformal from } (\Sigma, h_\infty) \textit{ into } \mathbb{R}^m;$$

ii)

$$\vec{\xi}_k \rightharpoonup \vec{\xi}_\infty \qquad \textit{weakly in } W^{2,2}_{loc}(\Sigma \setminus \{a_1, \dots, a_N\}), \tag{3.115}$$

iii)

$$\log|d\vec{\xi}_k|^2 \rightharpoonup \log|d\vec{\xi}_\infty|^2 \quad \textit{weakly in } (L^\infty)^*_{loc}(\Sigma\setminus\{a_1,\dots,a_N\}); \tag{3.116}$$

iv)

$$\vec{\xi}_k \rightharpoonup \vec{\xi}_\infty \qquad \textit{weakly in } W^{1,2}\cap(L^\infty)^*(\Sigma). \tag{3.117}$$

The following lemma shows that the Willmore functional W and the energy functional $\mathbb{I}$ are lower semicontinuous under weak convergence.

LEMMA 3.35. *Let (Σ, h_k) be a sequence of conformal structures on Σ, where h_k denotes the associated metric of constant curvature and unit volume. Assume it satisfies assumption (CA), with $h_k \to h_\infty$.*

Let $\vec{\xi}_k\colon (\Sigma, h_k)\to\mathbb{R}^m$ be a sequence of conformal maps in $\mathcal{E}_\Sigma$ with

$$\sup_k \mathbb{I}(\vec{\xi}_k) < \infty \tag{3.118}$$

and $\vec{\xi}_\infty\colon(\Sigma,h_\infty)\to\mathbb{R}^m$ a conformal map such that ii) *and* iii) *from Definition* 3.34 *are satisfied, i.e. there exist $a_1,\dots,a_N\in\Sigma$ such that*

$$\vec{\xi}_k \rightharpoonup \vec{\xi}_\infty \qquad \textit{weakly in } W^{2,2}_{loc}(\Sigma\setminus\{a_1,\dots,a_N\}) \tag{3.119}$$

and

$$\log|d\vec{\xi}_k|^2 \rightharpoonup \log|d\vec{\xi}_\infty|^2 \quad \textit{weakly in } (L^\infty)^*_{loc}(\Sigma\setminus\{a_1,\dots,a_N\}). \tag{3.120}$$

Then for any $K\Subset\Sigma\setminus\{a_1,\dots,a_N\}$, we have

$$\int_K |\vec{H}_{\vec{\xi}_\infty}|^2 dvol_{g_\infty} \le \liminf_k \int_K |\vec{H}_{\vec{\xi}_k}|^2 dvol_{g_k} \tag{3.121}$$

and

$$\int_K |d\vec{n}_{\vec{\xi}_\infty}|^2_{g_\infty} dvol_{g_\infty} \le \liminf_k \int_K |d\vec{n}_{\vec{\xi}_k}|^2_{g_k} dvol_{g_k}. \tag{3.122}$$

PROOF. Since the $\vec{\xi}_k$'s are conformal, the mean curvature vector in isothermal coordinates can be written as

$$\vec{H}_{\vec{\xi}_k} = \frac{1}{2}e^{-2\lambda_k}\Delta\vec{\xi}_k,$$

where $\lambda_k := \log|\partial_{x_1}\vec{\xi}_k|$.

Let $K\Subset\Sigma\setminus\{a_1,\dots,a_N\}$ be compact. Note that, due to (3.119),

$$\Delta\vec{\xi}_k \rightharpoonup \Delta\vec{\xi}_\infty \qquad \text{weakly in } L^2(K). \tag{3.123}$$

Moreover, using again (3.119) and applying Rellich/Kondrachov, we obtain for all $1<p<\infty$ a subsequence such that

$$\partial_{x_1}\vec{\xi}_{k'} \to \partial_{x_1}\vec{\xi}_\infty \qquad \text{strongly in } L^p(K).$$

(3.120) ensures that $|\partial_{x_1}\vec{\xi}_k|\ge C$ a.e. for $C>0$ independent of k. Thus, we have

$$e^{-\lambda_{k'}} = \frac{1}{|\partial_{x_1}\vec{\xi}_{k'}|} \to \frac{1}{|\partial_{x_1}\vec{\xi}_\infty|} = e^{-\lambda_\infty} \qquad \text{strongly in } L^p(K).$$

This, together with (3.123), implies

$$\vec{H}_{\vec{\xi}_{k'}}\sqrt{vol_{g_{k'}}} = \frac{1}{2}e^{-\lambda_{k'}}\Delta\vec{\xi}_{k'} \to \frac{1}{2}e^{-\lambda_\infty}\Delta\vec{\xi}_\infty = \vec{H}_{\vec{\xi}_\infty}\sqrt{vol_{g_\infty}} \quad \text{in } \mathcal{D}'(K). \tag{3.124}$$

But note that $\Delta\vec{\xi}_{k'}$ is uniformly bounded in $L^2(K)$, due to (3.119), and $e^{-\lambda_{k'}}$ is uniformly bounded in $L^\infty(K)$, due to (3.120). It follows that $\vec{H}_{\vec{\xi}_{k'}}\sqrt{vol_{g_{k'}}}$ is uniformly bounded in $L^2(K)$ and consequently, the convergence in (3.124) is a weak convergence in $L^2(K)$. Lower semicontinuity of the L^2-norm under weak L^2-convergence implies the desired result.

To prove (3.122), note that, by $h_k \to h_\infty$, (3.118) implies that

$$\sup_k \int_\Sigma |d\vec{n}_{\vec{\xi}_k}|^2_{h_\infty} dvol_{h_\infty} < \infty.$$

Thus, up to subsequences, there exists $\vec{n}_\infty \in W^{1,2}(\Sigma)$ such that

$$\vec{n}_{\vec{\xi}_{k'}} \rightharpoonup \vec{n}_\infty \qquad \text{weakly in } W^{1,2}(\Sigma).$$

We will show that $\vec{n}_\infty$ equals $\vec{n}_{\vec{\xi}_\infty}$ on any compact set $K \Subset \Sigma \setminus \{a_1, \ldots, a_N\}$, which in turn implies (3.122), by lower semicontinuity of weak $W^{1,2}$-convergence and $h_k \to h_\infty$ again.

Given any $p < \infty$, we have, modulo extraction of a subsequence, strong convergence $\vec{n}_{\vec{\xi}_{k'}} \to \vec{n}_\infty$ in $L^p(\Sigma)$, by Rellich/Kondrachov for compact manifolds (see [**Aub82**], Theorem 2.34). After passing to a further subsequence, we can assume that

$$\vec{n}_{\vec{\xi}_{k'}} \to \vec{n}_\infty \qquad \text{a.e. in } \Sigma. \tag{3.125}$$

Note that in isothermal coordinates, we have

$$\vec{n}_{\vec{\xi}_{k'}} = e^{-2\lambda_{k'}} \star \left(\partial_{x_1}\vec{\xi}_{k'} \wedge \partial_{x_2}\vec{\xi}_{k'}\right). \tag{3.126}$$

Applying Rellich/ Kondrachov, (3.119) implies that for any $p < \infty$, there is a further subsequence such that

$$d\vec{\xi}_{k'} \to d\vec{\xi}_\infty \qquad \text{strongly in } L^p(K).$$

By passing to a further subsequence, we obtain

$$d\vec{\xi}_{k'} \to d\vec{\xi}_\infty \qquad \text{a.e. in } K.$$

This implies in particular that

$$\star\left(\partial_{x_1}\vec{\xi}_{k'} \wedge \partial_{x_2}\vec{\xi}_{k'}\right) \to \quad \star\left(\partial_{x_1}\vec{\xi}_\infty \wedge \partial_{x_2}\vec{\xi}_\infty\right) \quad \text{a.e. in } K. \tag{3.127}$$

Observe that $\sup_{k'} \|\lambda_{k'}\|_{W^{1,2}(K)} \leq C$, due to

$$|\nabla\lambda_{k'}| = \left|\nabla \log\left(\frac{1}{\sqrt{2}}|\nabla\vec{\xi}_{k'}|\right)\right| \leq C + \frac{|\nabla^2\vec{\xi}_{k'}|}{|\nabla\vec{\xi}_{k'}|},$$

(3.119) and (3.120), which implies $\sup_k |\nabla\vec{\xi}_{k'}| \geq c > 0$. Consequently, we can assume that

$$\lambda_{k'} \to \lambda_\infty \qquad \text{in } L^p(K) \text{ and a.e. in } K, \tag{3.128}$$

after extraction of subsequences. (3.126), (3.128) and (3.127) imply that

$$\vec{n}_{\vec{\xi}_{k'}} \to \underbrace{e^{-2\lambda_\infty} \star \left(\partial_{x_1}\vec{\xi}_\infty \wedge \partial_{x_2}\vec{\xi}_\infty\right)}_{=\vec{n}_{\vec{\xi}_\infty}} \quad \text{a.e. in } K.$$

(3.125), together with uniqueness of the limit, gives the desired result that $\vec{n}_{\vec{\xi}_\infty}$ and $\vec{n}_\infty$ coincide on K. □

THEOREM 3.36 (Weak almost-closure theorem). *Let $\vec{\Phi}_k \in \mathcal{E}_\Sigma$ such that*

$$\sup_k \mathbb{I}(\vec{\Phi}_k) < \infty \tag{3.129}$$

and such that assumption (CA) is satisfied.

Then there exists a weakly converging subsequence of $\vec{\Phi}_k$ (in the sense of Definition 3.34).

In the next two subsections we shall prepare the proof of Theorem 3.36, which will be finally given in Subsection 4.

2. Control of the conformal factor

Let $\vec{\Phi}_k \in \mathcal{E}_\Sigma$ be a sequence of weak immersions. Corollary 2.32 tells us how to fix a gauge, namely one can find Lipschitz diffeomorphisms $\Psi_k\colon (\Sigma, h_k) \to \Sigma$ such that

$$\vec{\Psi}_k := \vec{\Phi}_k \circ \Psi_k\colon (\Sigma, h_k) \to \mathbb{R}^m$$

are conformal, where each h_k denotes the reference metric of constant curvature and unit volume of the conformal structure induced by $\vec{\Phi}_k$. Denote

$$g_{\vec{\Phi}_k} = e^{2\alpha_k} h_k.$$

A priori the conformal factors $e^{2\alpha_k}$ could go either to $+\infty$ or 0 as $k \to \infty$. In both cases, the limiting map will not be an element of $\mathcal{E}_\Sigma$, in the first case failing the Lipschitz condition and in the second case failing the non-degeneracy of the metric. The question that we want to investigate now is therefore: Can the logarithms of the conformal factors, that are α_k, be controlled in the L^∞-norm by $\sup_k \mathbb{I}(\vec{\Phi}_k)$, when we let $k \to \infty$?

The first result is a global bound for α_k in the $L^{2,\infty}(\Sigma)$-norm.

THEOREM 3.37. *Let $\vec{\Phi}_k \in \mathcal{E}_\Sigma$ be a sequence of Lipschitz immersions with L^2-bounded second fundamental form such that*

$$\sup_k \mathbb{I}(\vec{\Phi}_k) < \infty.$$

Let Ψ_k be Lipschitz diffeomorphisms of Σ such that

$$\vec{\Psi}_k := \vec{\Phi}_k \circ \Psi_k\colon (\Sigma, h_k) \to \mathbb{R}^m$$

are conformal, where each h_k denotes the reference metric of constant curvature and unit volume of the conformal structure induced by $\vec{\Phi}_k$. Furthermore, we make assumption (CA), that is the conformal classes (Σ, h_k) are contained in a compact subset of $\mathcal{M}_\Sigma$.

Denote $g_k := \vec{\Phi}_k^* g_{\mathbb{R}^m}$ and

$$g_k = e^{2\alpha_k} h_k.$$

Then

$$\sup_k \|d\alpha_k\|_{L^{2,\infty}(\Sigma)} < \infty.$$

PROOF. Since $g_k = e^{2\alpha_k} h_k$, by (1.52), we have for all $k \in \mathbb{N}$

$$-\Delta_{h_k}\alpha_k = e^{2\alpha_k} K_{g_k} - K_{h_k}. \tag{3.130}$$

This identity, together with (1.23), gives us the following estimate for $\Delta_{h_k}\alpha_k$ in $L^1_{h_k}(\Sigma)$:

$$\begin{aligned}\int_\Sigma |\Delta_{h_k}\alpha_k|\, dvol_{h_k} &\leq \int_\Sigma e^{2\alpha_k}|K_{g_k}| dvol_{h_k} + \int_\Sigma |K_{h_k}| dvol_{h_k} \\ &= \int_\Sigma |K_{g_k}| dvol_{g_k} + |K_{h_k}| \leq \frac{1}{2}\int_\Sigma |\mathbb{I}_{\vec{\Phi}_k}|^2_{g_k} dvol_{g_k} + |K_{h_k}| \qquad (3.131) \\ &\leq \frac{1}{2}\sup_k \mathbb{I}(\vec{\Phi}_k) + C =: \tilde{C},\end{aligned}$$

where we used that we chose h_k to be the constant curvature metric of unit volume and $h_k \to h_\infty$. Since $c_k^{-1} g_0 \leq h_k \leq c_k g_0$ and due to $h_k \to h_\infty$, (3.131) gives also the estimate

$$\|\Delta_{h_k}\alpha_k\|_{L^1_{g_0}(\Sigma)} \leq \hat{C}$$

for all $k \in \mathbb{N}$. Applying Theorem 2.24 yields thus the desired uniform bound for $d\alpha_k$ in $L^{2,\infty}(\Sigma)$:

$$\|d\alpha_k\|_{L^{2,\infty}_{g_0}(\Sigma)} \leq C_{g_0}\ \|\Delta_{h_k}\alpha_k\|_{L^1_{g_0}(\Sigma)} \leq C_{g_0} \cdot \hat{C}. \tag{3.132}$$

□

We now shall investigate the evolvement of the logarithms of the conformal factors in the L^∞-norm. This is done locally, wherever the second fundamental form does not concentrate "too much energy".

THEOREM 3.38. *Let $\vec{\Phi}_k \in \mathcal{E}_\Sigma$ be a sequence in $\mathcal{E}_\Sigma$ which satisfies*

$$\sup_k \mathbb{I}(\vec{\Phi}_k) < \infty.$$

Furthermore, we make assumption (CA): the conformal classes (Σ, h_k) are contained in a compact subset of the moduli space of Σ. As before, let Ψ_k be Lipschitz diffeomorphisms of Σ such that

$$\vec{\Psi}_k := \vec{\Phi}_k \circ \Psi_k \colon (\Sigma, h_k) \to \mathbb{R}^m$$

are conformal, where each h_k denotes the reference metric of constant curvature and unit volume of the conformal structure induced by $\vec{\Phi}_k$.

Moreover, let φ_k be a sequence of isothermal charts satisfying

$$\sup_k \int_{D^2} |\nabla \vec{n}_{\vec{\Psi}_k\circ\varphi_k}|^2 dx_1 dx_2 < \frac{8\pi}{3}. \tag{3.133}$$

Denote

$$(\vec{\Psi}_k \circ \varphi_k)^* g_{\mathbb{R}^m} = e^{2\lambda_k}(dx_1^2 + dx_2^2).$$

Then there exist constants $c_k \in \mathbb{R}$ such that

$$\sup_k \|\lambda_k - c_k\|_{L^\infty(\Omega)} \leq C_\Omega, \tag{3.134}$$

for any set $\Omega \Subset D^2$.

The following example shows that in general we cannot control the λ_k in L^∞ when deleting the constants c_k in (3.134): Compose the weak immersions $\vec{\Psi}_k$ with dilations $s_k\cdot$ in $\mathbb{R}^m$, for some $s_k \in \mathbb{R}$. Due to the conformal invariance of the Willmore functional, we have $\mathbb{I}(\vec{\Psi}_k) = \mathbb{I}(s_k\vec{\Psi}_k)$. In contrast, the dilations are reflected in the logarithms of the conformal factors, that is $\lambda_{s_k\vec{\Psi}_k} = \lambda_{\vec{\Psi}_k} + \log s_k$.

Note that this example also shows that the constants c_k are a priori not controlled, when $k \to \infty$.

Proof of Theorem 3.38. Assuming (3.133), for each $k \in \mathbb{N}$, we can apply Hélein's lifting theorem 2.30 to obtain the existence of $\vec{f}_1^k$ and $\vec{f}_2^k$ in $W^{1,2}(D^2, S^{m-1})$ such that

$$\vec{n}_k := \vec{n}_{\vec{\Psi}_k\circ\varphi_k} = \star(\vec{f}_1^k \wedge \vec{f}_2^k),$$

and

$$\int_{D^2} \sum_{i=1}^{2} |\nabla \vec{f}_i^k|^2 \, dx_1\, dx_2 \leq C \int_{D^2} |\nabla \vec{n}_k|^2 \, dx_1 dx_2,$$

where $\vec{n}_k := \vec{n}_{\vec{\Psi}_k\circ\varphi_k}$.

Using the notation introduced in Section 1, the factors λ_k satisfy the following equation:

$$\Delta\lambda_k = -\langle\nabla^\perp \vec{f}_1^k, \nabla \vec{f}_2^k\rangle. \tag{3.135}$$

This can seen by observing that, by (1.45), the latter equation holds for the frame $(\vec{f}_1^k, \vec{f}_2^k)$ being replaced by the frame $(\vec{e}_1^k, \vec{e}_2^k)$ given by

$$\vec{e}_i^k := e^{-\lambda_k} \partial_{x_i}(\vec{\Psi}_k \circ \varphi_k).$$

Furthermore, the change of gauge formula (1.62) leaves the right hand side of (3.135) invariant. Observe that the right hand side of (3.135) is a sum of Jacobians, more specifically

$$\begin{aligned} -\langle\nabla^\perp \vec{f}_1^k, \nabla \vec{f}_2^k\rangle &= \langle\partial_{x_2} \vec{f}_1^k, \partial_{x_1} \vec{f}_2^k\rangle - \langle\partial_{x_1} \vec{f}_1^k, \partial_{x_2} \vec{f}_2^k\rangle \\ &= \sum_{j=1}^{m} \partial_{x_2} f_{1,j}^k \cdot \partial_{x_1} f_{2,j}^k - \partial_{x_1} f_{1,j}^k \cdot \partial_{x_2} f_{2,j}^k. \end{aligned}$$

Let μ_k be the solution to

$$\begin{cases} \Delta\mu_k &= \sum_{j=1}^m \partial_{x_2} f_{1,j}^k \cdot \partial_{x_1} f_{2,j}^k - \partial_{x_1} f_{1,j}^k \cdot \partial_{x_2} f_{2,j}^k \qquad \text{on } D^2 \\ \mu_k &= 0 \qquad \text{on } \partial D^2. \end{cases}$$

Wente's Theorem 2.26 gives us the estimate

$$\begin{aligned} &\|\nabla\mu_k\|_{L^2(D^2)} + \|\mu_k\|_{L^\infty(D^2)} \\ &\leq C\sum_{j=1}^{m}\int_{D^2}\left(|\nabla f_{1,j}^k|^2 + |\nabla f_{2,j}^k|^2\right)dx_1dx_2 \qquad (3.136)\\ &\leq C\int_{D^2}|\nabla\vec{n}_k|^2 dx_1dx_2 \leq C, \end{aligned}$$

where we used (3.133) in the last step such that the constant C is independent of k.

We now consider the harmonic rest $\nu_k := \lambda_k - \mu_k$. Due to Theorem 3.37, which gives a global estimate of $d\alpha_k$ in $L^{2,\infty}(\Sigma)$, and the strong convergence $h_k \to h_\infty$, we have a uniform bound

$$\sup_k \|\nabla\lambda_k\|_{L^{2,\infty}(D^2)} \leq C.$$

Together with (3.136) (and the fact that $L^2 \hookrightarrow L^{2,\infty}$ continuously), this yields $\nu_k \in L^{2,\infty}(D^2)$ and

$$\sup_k \|\nabla\nu_k\|_{L^{2,\infty}(D^2)} \leq C.$$

From (2.90), we know that

$$\|\nabla\nu_k\|_{L^p(D^2)} \leq C_p\ \|\nabla\nu_k\|_{L^{2,\infty}(D^2)}$$

for all $p < 2$. Applying Poincaré's inequality yields

$$\|\nu_k - \bar{\nu}_k\|_{L^p(D^2)} \leq C_p\ \|\nabla\nu_k\|_{L^p(D^2)},$$

where $\bar{\nu}_k$ denotes the average of ν_k on D^2.

Since $W^{1,p}(D^2) \hookrightarrow W^{1-\frac{1}{p},p}(\partial D^2) \hookrightarrow L^1(\partial D^2)$, we obtain

$$\sup_k \|\nu_k - \bar{\nu}_k\|_{L^1(\partial D^2)} \leq C.$$

Using the Poisson representation formula for harmonic functions on D^2 yields

$$\|\nu_k - \bar{\nu}_k\|_{C^1(\Omega)} \leq C_\Omega\ \|\nu_k - \bar{\nu}_k\|_{L^1(\partial D^2)}$$

for any $\Omega \Subset D^2$ and thus

$$\sup_k \|\nu_k - \bar{\nu}_k\|_{L^\infty(\Omega)} \leq \tilde{C}_\Omega.$$

Combining this result with (3.136) gives us

$$\sup_k \|\lambda_k - c_k\|_{L^\infty(\Omega)} \leq \hat{C}_\Omega,$$

for any set $\Omega \Subset D^2$, where $c_k := \bar{\nu}_k$. □

3. The monotonicity formula and consequences

As always, let Σ be a smooth closed oriented surface. The following lemma is a corollary of Simon's monotonicity formula.

LEMMA 3.39 (Simon's Monotonicity formula). *Let $\vec{\Phi} \in \mathcal{E}_\Sigma$ be a Lipschitz immersion with L^2-bounded second fundamental form. Denote by $M := \vec{\Phi}(\Sigma)$ the immersed surface.*

Then for any point $x_0 \in \mathbb{R}^m$ and any $0 < t < T < \infty$, we have

$$\frac{\text{Area}\,(M \cap B_T(x_0))}{T^2} - \frac{\text{Area}\,(M \cap B_t(x_0))}{t^2}$$

$$\geq -\frac{1}{4} \int_{M \cap (B_T(x_0) \setminus B_t(x_0))} |\vec{H}|^2 \, dvol_g$$

$$-\frac{1}{T^2} \int_{M \cap B_T(x_0)} \langle x - x_0, \vec{H} \rangle \, dvol_g + \frac{1}{t^2} \int_{M \cap B_t(x_0)} \langle x - x_0, \vec{H} \rangle \, dvol_g.$$

For a proof see [**Sim84**] or [**Riv13**].

DEFINITION 3.40. *Let $\vec{\Phi} \in \mathcal{E}_\Sigma$. For $x_0 \in \mathbb{R}^m$, define the density $\theta_{x_0} \in \mathbb{N}$ at x_0 as*

(3.137)
$$\theta_{x_0} := \lim_{t \to 0^+} \frac{1}{\pi t^2} \text{Area}\left(\vec{\Phi}(\Sigma) \cap B_t(x_0)\right),$$

whenever the limit exists.

To simplify notation in the proofs of the following corollaries, we introduce for fixed $x_0 \in \mathbb{R}^m$ the quantities

(3.138)
$$W(t) := \int_{M \cap B_t(x_0)} |\vec{H}|^2 \, dvol_g,$$

(3.139)
$$Y(t) := \frac{1}{t^2} \text{Area}\left(\vec{\Phi}(\Sigma) \cap B_t(x_0)\right).$$

COROLLARY 3.41 (Existence of the density). *Let $\vec{\Phi} \in \mathcal{E}_\Sigma$. Then the density θ_{x_0} exists for every point $x \in \mathbb{R}^m$.*

PROOF. Let $x_0 \in \mathbb{R}^m$ be an arbitrary point.

Using the Cauchy-Schwarz inequality, we obtain for any $t > 0$:

$$\frac{1}{t^2} \int_{M \cap B_t(x_0)} \langle x - x_0, \vec{H} \rangle \, dvol_g$$

(3.140)
$$\leq \frac{1}{t} \left(\int_{M \cap B_t(x_0)} \frac{|x - x_0|^2}{t^2} \, dvol_g \right)^{1/2} \left(\int_{M \cap B_t(x_0)} |\vec{H}|^2 \, dvol_g \right)^{1/2}$$

$$\leq Y(t)^{1/2} \, W(t)^{1/2} \leq \frac{1}{2} \left(Y(t) + W(t) \right),$$

where we again denote $M := \vec{\Phi}(\Sigma)$. Since $W(\vec{\Phi}) < \infty$, we have $W(t) \to 0$ for $t \to 0^+$ and thus, as a direct consequence of Lemma 3.39, we obtain for $0 < t < T < \infty$,

$$(1 - o_t(1))\, Y(t) \leq Y(T)\, (1 + o(T)) + o_{t,T}(1).$$

Hence,

$$\lim_{t \to 0^+} Y(t) = \theta_{x_0}$$

exists. □

COROLLARY 3.42 (Li-Yau inequality). *Let $\vec{\Phi} \in \mathcal{E}_\Sigma$. Then for any $x_0 \in \mathbb{R}^m$,*

$$\theta_{x_0} \leq \frac{1}{4\pi} \int_\Sigma |\vec{H}|^2 \, dvol_g. \tag{3.141}$$

PROOF. Let $x_0 \in \mathbb{R}^m$ be an arbitrary point. Note that for $T > 0$,

$$\begin{aligned} &\frac{1}{T^2} \int_{M \cap B_T(x_0)} \langle x - x_0, \vec{H} \rangle \, dvol_g \\ &\leq \frac{1}{T^2} \left(\|\vec{\Phi}\|_{L^\infty(\Sigma)} + |x_0| \right) \mathrm{Area}(\vec{\Phi}(\Sigma))^{1/2}\, W(\vec{\Phi})^{1/2} \qquad (3.142) \\ &\xrightarrow{T \to \infty} 0, \end{aligned}$$

with $M := \vec{\Phi}(\Sigma)$. Moreover,

$$\frac{1}{T^2} \mathrm{Area}(\vec{\Phi}(\Sigma) \cap B_T(x_0)) \leq \frac{1}{T^2} \mathrm{Area}(\vec{\Phi}(\Sigma)) \xrightarrow{T \to \infty} 0. \tag{3.143}$$

Using Lemma 3.39, (3.140), (3.142) and (3.143) gives

$$\frac{1}{4} \int_\Sigma |\vec{H}|^2 \, dvol_g \geq (1 - o_t(1)) Y(t) + o_{1/T}(1).$$

Hence, if we let $T \to \infty$, we obtain

$$\theta_{x_0} = \lim_{t \to 0^+} \frac{1}{\pi} Y(t) \leq \frac{1}{4\pi} \int_\Sigma |\vec{H}|^2 \, dvol_g.$$

□

COROLLARY 3.43. *Let $\vec{\Phi} \in \mathcal{E}_\Sigma$ and*

$$W(\vec{\Phi}) < 8\pi. \tag{3.144}$$

Then the immersion $\vec{\Phi}$ is in fact an embedding.

PROOF. For every point $x_0 \in \mathbb{R}^m$, the density θ_{x_0} exists and is an element of $\mathbb{N}$. If (3.144) holds true, Corollary 3.42 implies that $\theta_{x_0} \in \{0, 1\}$. This means that $\vec{\Phi}$ is an embedding. □

COROLLARY 3.44. *Let $\vec{\Phi} \in \mathcal{E}_\Sigma$. For any $x_0 \in \Sigma$ and $t > 0$,*

$$\mathrm{Area}(\vec{\Phi}(\Sigma) \cap B_t(x_0)) \leq \frac{3}{2} W(\vec{\Phi}) t^2. \tag{3.145}$$

PROOF. By Lemma 3.39, (3.140), (3.142) and (3.143) we have

$$\frac{1}{2}Y(t) \leq \left(\frac{1}{4}+\frac{1}{2}\right) W(\vec{\Phi}) + o_{1/T}(1).$$

For $T \to \infty$, the result follows. □

COROLLARY 3.45. *Let $\vec{\Phi} \in \mathcal{E}_\Sigma$ and $x_0 \in \mathbb{R}^m$. If $\theta_{x_0} \neq 0$, then for any $T > 0$ we have*

$$\text{Area}(\vec{\Phi}(\Sigma) \cap B_T(x_0)) \geq \frac{2\pi}{3}T^2 - \frac{T^2}{2}\int_{M\cap B_T(x_0)} |\vec{H}|^2 dvol_g. \tag{3.146}$$

PROOF. By Lemma 3.39, we obtain for $T > 0$,

$$\begin{aligned}\pi\theta_{x_0} &= \lim_{t\to 0^+} Y(t)\\ &\leq Y(T) + \frac{1}{4}W(T) + \frac{1}{T^2}\int_{M\cap B_T(x_0)} \langle x - x_0, \vec{H}\rangle\; dvol_g\\ &\leq \frac{3}{2}Y(T) + \frac{3}{4}W(T),\end{aligned}$$

where we used (3.140) in the last step. Hence, if $\theta_{x_0} \geq 1$, then

$$\text{Area}\left(\vec{\Phi}(\Sigma) \cap B_T(x_0)\right) \geq \frac{2\pi}{3}T^2 - \frac{T^2}{2}W(T).$$

□

4. Proof of the almost-weak closure theorem

PROOF OF THEOREM 3.36. Let $\vec{\Phi}_k \in \mathcal{E}_\Sigma$ be a sequence satisfying the assumptions of Theorem 3.36. Let Ψ_k be diffeomorphisms of Σ such that $\vec{\Psi}_k := \vec{\Phi}_k \circ \Psi_k \colon (\Sigma_k, h_k) \to \mathbb{R}^m$ is conformal for any $k \in \mathbb{N}$, where h_k is the metric of constant curvature and unit volume of the conformal structure induced by $\vec{\Phi}_k$.

Define for each $k \in \mathbb{N}$ and $x \in \Sigma$ a number $\rho_{k,x} > 0$ by

$$\rho_{k,x} := \inf\left\{\rho > 0 \quad \text{s.t.} \quad \int_{B_\rho^{h_k}(x)} |d\vec{n}_{\vec{\Psi}_k}|^2_{h_k} dvol_{h_k} \geq \frac{8\pi}{3}\right\}. \tag{3.147}$$

and

$$\rho_{\infty,x} := \liminf_{k\to\infty} \rho_{k,x}. \tag{3.148}$$

Step 1: The case when there are no concentration points. We first want to investigate the case in which the second fundamental form does not concentrate anywhere, i.e. we assume that

$$\rho_\infty := \inf_{x\in\Sigma} \rho_{\infty,x} > 0. \tag{3.149}$$

We will prove that under this assumption, no blow-up points occur such that the limit $\vec{\xi}$ is in $\mathcal{E}_\Sigma$.

Step 1a): Translating and dilating $\vec{\Psi}_k$ to control the conformal factors in L^∞ and to bound the image. Assumption (3.149) gives us

$$\int_{B^{h_k}_{\rho_\infty/2}(x)} |d\vec{n}_{\vec{\Psi}_k}|^2_{h_k} dvol_{h_k} < \frac{8\pi}{3} \tag{3.150}$$

for all $x \in \Sigma$ and large enough $k \geq k_x$. By conformal invariance of the Dirichlet energy, (3.150) implies that assumption (3.133) is satisfied for any sequence of conformal charts $\{\varphi^x_k\}_{k \geq k_x}$ around an arbitrary point $x \in \Sigma$ with $\varphi^x_k(D^2) = B^{h_k}_{\rho_\infty/2}(x)$. Since we further suppose (3.129) and condition (CA), we are in the assumptions of Theorem 3.38. Thus for any $\Omega \Subset D^2$, we have

$$\sup_{x, k \geq k_x} \|\lambda^x_k - c^x_k\|_{L^\infty(\Omega)} \leq C_\Omega, \tag{3.151}$$

where λ^x_k (c^x_k resp.) are the logarithms of the conformal factors (the obtained constants resp.) in the charts φ^x_k.

By compactness, we can extract a finite subcovering $\{B^{h_\infty}_{\rho_\infty/8}(x_i)\}_{i=1,\ldots,n}$ from $\{B^{h_\infty}_{\rho_\infty/8}(x)\}_{x\in\Sigma}$. Then, since $h_k \to h_\infty$, we can assume that the n balls

$$\{B^{h_k}_{\rho_\infty/4}(x_i)\}_{i=1,\ldots,n}$$

also cover Σ, for $k \in \mathbb{N}$ large enough.

We denote $\lambda_k = \alpha_k + \sigma_k$ for $h_k = e^{2\sigma_k}(dx_1^2 + dx_2^2)$ and $\vec{\Psi}^*_k g_{\mathbb{R}^m} = e^{2\alpha_k} h_k$. Again due to the strong convergence $h_k \to h_\infty$, there is a uniform bound

$$\max_{i=1,\ldots,n} \|\sigma^{x_i}_k\|_{L^\infty(B^{h_k}_{\rho_\infty/4}(x_i))} \leq M.$$

This and (3.151) imply for all $k \geq k_0 := \max\{k_{x_1}, \ldots, k_{x_n}\}$ and $i = 1, \ldots, n$,

$$\begin{aligned} &\|\alpha_k - c^{x_i}_k\|_{L^\infty(B^{h_k}_{\rho_\infty/4}(x_i))} \\ &\leq \|\lambda^{x_i}_k - c^{x_i}_k\|_{L^\infty(B^{h_k}_{\rho_\infty/4}(x_i))} + \|\sigma^{x_i}_k\|_{L^\infty(B^{h_k}_{\rho_\infty/4}(x_i))} \\ &\leq M + C_\Omega =: \tilde{C}, \end{aligned} \tag{3.152}$$

where $\Omega \Subset D^2$ is chosen in such a way that $\varphi^{x_i}_k(\Omega) \supset B_{\rho_\infty/4}(x_i)$ for all $i = 1, \ldots, n$.

As a result of (3.152), observe that if $B^{h_k}_{\rho_\infty/4}(x_i) \cap B^{h_k}_{\rho_\infty/4}(x_j) \neq \emptyset$, we have

$$|c^{x_i}_k - c^{x_j}_k| \leq 2\tilde{C}.$$

Since Σ is path-connected, this yields

$$\sup_{k \geq k_0, i, j} |c^{x_i}_k - c^{x_j}_k| \leq n\tilde{C}. \tag{3.153}$$

Next, we compose each $\vec{\Psi}_k$ with a translation and dilation, conformal transformations in $\mathbb{R}^m$, in the following way: Define

$$\tilde{\vec{\Psi}}_k := e^{-c_k}\left(\vec{\Psi}_k - \vec{\Psi}_k(x_0)\right) \tag{3.154}$$

for some $x_0 \in \Sigma$ and $c_k := c^{x_1}_k$.

The Willmore energy does not change, i.e. we have

$$W(\vec{\tilde{\Psi}}_k) = W(\vec{\Psi}_k) = W(\vec{\Phi}_k).$$

What we have achieved by dilating, however, is that $\tilde{\alpha}_k$, the logarithms of the conformal factors of the new conformal immersions $\vec{\tilde{\Psi}}_k$, are uniformly bounded in $L^\infty(D^2)$:

We have $\tilde{\alpha}_k = \alpha_k - c_k^{x_1}$, and thus for $i = 1, \ldots, n$ and all $k \geq k_0$,

$$\begin{aligned}\|\tilde{\alpha}_k\|_{L^\infty(B^{h_k}_{\rho_\infty/4}(x_i))} &\leq \|\alpha_k - c_k^{x_i}\|_{L^\infty(B^{h_k}_{\rho_\infty/4}(x_i))} + \|c_k^{x_i} - c_k^{x_1}\|_{L^\infty(B^{h_k}_{\rho_\infty/4}(x_i))} \\ &\leq (1+n)\tilde{C},\end{aligned}$$

by (3.152) and (3.153). Since $\{B^{h_k}_{\rho_\infty/4}(x_i)\}_{i=1,\ldots,n}$ cover Σ, this yields

$$\sup_{k \geq k_0} \|\tilde{\alpha}_k\|_{L^\infty(\Sigma)} \leq (1+n)\tilde{C} =: \hat{C}. \tag{3.155}$$

Further, we performed the translation in order to bound the image of $\vec{\tilde{\Psi}}_k$ uniformly: $\vec{\tilde{\Psi}}_k$ maps $x_0 \in \Sigma$ to $0 \in \mathbb{R}^m$ and for any point $y \in \Sigma$, we have

$$|\vec{\tilde{\Psi}}_k(y) - \vec{\tilde{\Psi}}_k(x_0)| \leq \int_{x_0}^{y} e^{\tilde{\alpha}_k} dl_{h_k} \leq C, \tag{3.156}$$

where the constant is independent of $k \in \mathbb{N}$, since $\sup_k \|\tilde{\alpha}_k\|_{L^\infty(\Sigma)}$ and $\sup_k \operatorname{diam}(\Sigma, h_k)$ are finite. It follows that

$$\vec{\tilde{\Psi}}_k(\Sigma) \subset B_C(0) \quad \text{for all } k \in \mathbb{N}.$$

Step 1b): Weak $W^{2,2}$-convergence of $\vec{\tilde{\Psi}}_k$. Let φ_k be a sequence of conformal charts and denote the logarithms of the conformal factors of $\vec{\tilde{\Psi}}_k \circ \varphi_k$ as $\tilde{\lambda}_k$. Then, by (1.42) the mean curvature vector can be written as

$$\vec{H}_k = \frac{e^{-2\tilde{\lambda}_k}}{2} \Delta\left(\vec{\tilde{\Psi}}_k \circ \varphi_k\right)$$

and for all $k \in \mathbb{N}$,

$$\begin{aligned}\left\|\Delta\left(\vec{\tilde{\Psi}}_k \circ \varphi_k\right)\right\|^2_{L^2(D^2)} &= \frac{1}{4}\int_{D^2} e^{4\tilde{\lambda}_k}|\vec{H}_k|^2 dx_1 dx_2 \\ &\leq \frac{1}{4}\|e^{2\tilde{\lambda}_k}\|_{L^\infty(D^2)} \int_{\varphi_k(D^2)} |\vec{H}_k|^2 dvol_{g_{\vec{\tilde{\Psi}}_k}} \leq \frac{1}{4}\|e^{2\tilde{\lambda}_k}\|_{L^\infty(D^2)}\, W(\vec{\Phi}_k) \leq C.\end{aligned} \tag{3.157}$$

Moreover, for all $k \in \mathbb{N}$ we have

$$\left\|\nabla\left(\vec{\tilde{\Psi}}_k \circ \varphi_k\right)\right\|^2_{L^2(D^2)} = 2\int_{D^2} e^{2\tilde{\lambda}_k} dx_1 dx_2 \leq C. \tag{3.158}$$

(3.157) and (3.158) imply that

$$\sup_k \|\vec{\tilde{\Psi}}_k \circ \varphi_k\|_{W^{2,2}(D^2_{1/2})} < \infty. \tag{3.159}$$

To see this, let μ_k be the solution to

$$\begin{cases} \Delta\mu_k &= \Delta\left(\tilde{\vec{\Psi}}_k \circ \varphi_k\right) \qquad \text{on } D^2 \\ \mu_k &= 0 \qquad \text{on } \partial D^2. \end{cases}$$

Then

$$\|\mu_k\|_{W^{2,2}(D^2)} \le C \left\|\Delta\left(\tilde{\vec{\Psi}}_k \circ \varphi_k\right)\right\|_{L^2(D^2)} \le C \tag{3.160}$$

for all $k \in \mathbb{N}$. For the harmonic rest $\nu_k := \tilde{\vec{\Psi}}_k \circ \varphi_k - \mu_k$, we get for $l = 1, 2$, all $k \in \mathbb{N}$ and $\Omega \Subset D^2$,

$$\|\nabla^l \nu_k\|_{L^2(\Omega)} \le C_{l,\Omega} \cdot \|\nu_k\|_{L^1(\partial D^2)} \le \tilde{C}_{l,\Omega}, \tag{3.161}$$

using $W^{1,2}(D^2) \hookrightarrow L^1(\partial D^2)$, the estimates (3.158) and (3.160) and the Poisson representation formula for harmonic functions on D^2. (3.160) and (3.161) imply the desired result (3.159).

Due to the strong convergence $h_k \to h_\infty$, (3.159) implies

$$\sup_k \|\tilde{\vec{\Psi}}_k\|_{W^{2,2}(\Sigma)} < \infty. \tag{3.162}$$

Thus, we can extract a subsequence $\tilde{\vec{\Psi}}_{k'}$ such that

$$\tilde{\vec{\Psi}}_{k'} \rightharpoonup \vec{\xi}_\infty \qquad \text{weakly in } W^{2,2}(\Sigma) \tag{3.163}$$

for some $\vec{\xi}_\infty \in W^{2,2}(\Sigma)$.

Step 1c): $\vec{\xi}_\infty$ is conformal and $\log|d\tilde{\vec{\Psi}}_k|^2 \overset{*}{\rightharpoonup} \log|d\vec{\xi}_\infty|^2$ in $(L^\infty)^*(\Sigma)$. Applying Rellich/ Kondrachov, (3.163) implies that for any $p < \infty$, there is a further subsequence, also denoted by $\tilde{\vec{\Psi}}_{k'}$, such that

$$d\tilde{\vec{\Psi}}_{k'} \to d\vec{\xi}_\infty \qquad \text{strongly in } L^p(\Sigma).$$

By passing to a further subsequence, again denoted by $\tilde{\vec{\Psi}}_{k'}$, we obtain

$$d\tilde{\vec{\Psi}}_{k'} \to d\vec{\xi}_\infty \qquad \text{a.e. in } \Sigma. \tag{3.164}$$

This implies that in any sequence of conformal charts, we have for $i, j = 1, 2$,

$$e^{2\tilde{\lambda}_{k'}}\delta_{ij} = \partial_{x_i}\tilde{\vec{\Psi}}_{k'}\,\partial_{x_j}\tilde{\vec{\Psi}}_{k'} \to \partial_{x_i}\vec{\xi}_\infty\,\partial_{x_j}\vec{\xi}_\infty \qquad \text{a.e. in } D^2.$$

This yields

$$\partial_{x_i}\vec{\xi}_\infty\,\partial_{x_j}\vec{\xi}_\infty = e^{2\tilde{\lambda}_\infty}\delta_{ij}, \tag{3.165}$$

for $\tilde{\lambda}_\infty := \frac{1}{2}\log\frac{1}{2}|d\vec{\xi}_\infty|^2$. As a result $\vec{\xi}_\infty$ is conformal.

Denoting $g_{\tilde{\vec{\Psi}}_k} = e^{2\tilde{\alpha}_k}h_k$ and $g_{\vec{\xi}_\infty} = e^{2\tilde{\alpha}}h_\infty$, we can use (3.155), (3.164) and the dominated convergence theorem to conclude that

$$\tilde{\alpha}_{k'} \overset{*}{\rightharpoonup} \tilde{\alpha}_\infty \qquad \text{weakly}^* \text{ in } (L^\infty(\Sigma))^*.$$

Step 2: The general case. We consider the general case and drop assumption (3.149), which means that we allow concentration of energy of the second fundamental form. This will imply the occurrence of blow-up points.

Step 2a): Detecting the concentration points. For given $k \in \mathbb{N}$, the collection $\{B^{h_k}_{\rho_{k,x}}(x)\}_{x\in\Sigma}$ forms a Besicovitch covering of Σ, where $\rho_{k,x}$ was defined in (3.147). The Besicovitch covering theorem ([**Mat95**]) gives a subcovering $\{B^{h_k}_{\rho_{k,i}}(x^k_i)\}_{i\in I_k}$ such that any point in Σ is covered by at most $c_\Sigma \in \mathbb{N}$ balls, where c_Σ does not depend on $k \in \mathbb{N}$. In fact, I_k is finite and its cardinality uniformly bounded in k since

$$|i \in I_k| \cdot \frac{8\pi}{3} \leq \sum_{i\in I_k} \int_{B_{\rho_{k,i}}(x^k_i)} |dn_{\vec{\Psi}_k}|^2_{h_k}\, dvol_{h_k}$$

$$= \int_\Sigma \left|i \in I_k :\ x \in B_{\rho_{k,i}}(x^k_i)\right| |dn_{\vec{\Psi}_k}|^2_{h_k}\, dvol_{h_k}$$

$$\leq c_\Sigma \int_\Sigma |dn_{\vec{\Psi}_k}|^2_{h_k}\, dvol_{h_k} = c_\Sigma\ \mathbb{I}(\vec{\Phi}_k) \leq C.$$

Thus, we can extract a subsequence such that I is independent of k (and finite) and such that for all $i \in I$,

$$x^i_k \to x^i_\infty, \tag{3.166}$$

$$\rho_{k,i} \to \rho_{\infty,i} \tag{3.167}$$

as $k \to \infty$, for some $x^i_\infty \in \Sigma$ and $\rho_{\infty,i} \geq 0$. Let

$$J := \{i \in I \quad \text{s.t. } \rho_{\infty,i} = 0\} \quad \text{and} \quad I_0 := I \setminus J.$$

It is clear that $\bigcup_{i\in I_0} \overline{B^{h_\infty}_{\rho_{\infty,i}}(x^i_\infty)}$ covers Σ. Note that the balls $B^{h_\infty}_{\rho_{\infty,i}}(x^i_\infty)$ are strictly convex: this holds if $K_{h_\infty} \leq 0$ and if $K_{h_\infty} = 1$ we assume w.l.o.g. that $\rho_{\infty,i} < \frac{\pi}{2}$. Consequently, the points in Σ which are not contained in the union of the finitely many open balls cannot accumulate and therefore are isolated and hence finite:

$$\{a_1, \ldots, a_N\} := \Sigma \setminus \left(\bigcup_{i\in I_0} B^{h_\infty}_{\rho_{\infty,i}}(x^i_\infty)\right). \tag{3.168}$$

Step 2b): Applying Step 1a away from the concentration points. For arbitrary $i_0 \in I_0$, choose $s^{i_0} < \rho_{\infty,i_0}$. Note that for k large enough, $B^{h_k}_{s_{i_0}}(x^{i_0}_\infty) \subset B^{h_k}_{\rho_{k,i_0}}(x^{i_0}_k)$ because $\rho_{k,i_0} \to \rho_{\infty,i_0}$ and $x^{i_0}_k \to x^{i_0}_\infty$. Therefore we can assume that for all $k \in \mathbb{N}$,

$$\int_{B^{h_k}_{s_{i_0}}(x^{i_0}_\infty)} |d\vec{n}_{\vec{\Psi}_k}|^2_{h_k} dvol_{h_k} < \frac{8\pi}{3}.$$

Theorem 3.38 gives a constant c_k such that

$$\sup_k \|\alpha_k - c_k\|_{L^\infty(B_r(x^{i_0}_\infty))} \leq C_r \tag{3.169}$$

for any $r < s_{i_0}$. Define

$$\vec{\tilde{\Psi}}_k := e^{-c_k}\left(\vec{\Psi}_k - \vec{\Psi}_k(x_0)\right)$$

for some $x_0 \in \Sigma$. Let $K \Subset \Sigma \setminus \{a_1, \ldots, a_N\}$ be any compact set. Note that, due to (3.168),

$$\rho_\infty := \inf_{x\in K} \rho_{\infty,x} > 0, \tag{3.170}$$

where ρ_x was defined in (3.148). Thus, we can apply Step 1a to any such compact K (and since $K \subset \Sigma \setminus (\cup_{i=1,\dots,N} B_\delta(a_i))$ for some $\delta > 0$, we restrict to compact sets of the latter form). For $\delta < \inf_{i \in I_0} \rho_{\infty,i}$, we obtain

$$\sup_k \|\tilde{\alpha}_k\|_{L^\infty(\Sigma \setminus (\cup_{i=1,\dots,n} B_\delta(a_i)))} < C_\delta, \tag{3.171}$$

for a constant C_δ depending on δ. Note that (3.171) implies, similarly as in (3.156), that

$$\vec{\tilde{\Psi}}_k \left(\Sigma \setminus (\cup_{i=1,\dots,n} B_\delta(a_i))\right) \subset B_{C_\delta}(0) \tag{3.172}$$

for $k \in \mathbb{N}$.

Step 2c): Using inversions to control $\vec{\tilde{\Psi}}_k(\Sigma)$. The previous results do not give estimates on what happens in the balls $B_\delta(a_i)$. In order to get an area control on the limiting map, we would like to improve (3.172) and have the entire image $\vec{\tilde{\Psi}}_k(\Sigma)$ contained in a ball.

Since $\vec{\tilde{\Psi}}_k(B_\delta(a_i))$ could degenerate to infinity as $k \to \infty$, the strategy is to bring this back to a ball around 0 by using inversions. If we can find a ball $B_r(p_0)$ such that for all $k \in \mathbb{N}$, $\vec{\tilde{\Psi}}_k(\Sigma) \cap B_r(p_0) = \emptyset$, then composing the maps $\vec{\tilde{\Psi}}_k$ with the inversion $i\colon x \mapsto \frac{x-p_0}{|x-p_0|^2}$ yields $i \circ \vec{\tilde{\Psi}}_k(\Sigma) \subset B_{1/r}(0)$, as desired.

The existence of such $p_0 \in B_1(0) \subset \mathbb{R}^m$, $r > 0$ is given by the following lemma, which follows from the monotonicity formula.

LEMMA 3.46. *Let $\vec{\Phi}_k \in \mathcal{E}_\Sigma$ with*

$$\sup_k W(\vec{\Phi}_k) < \infty. \tag{3.173}$$

Then there exists $p_0 \in \mathbb{R}^m$ and $r < 1 - |p_0|$ such that

$$\vec{\Phi}_k(\Sigma) \cap B_r(p_0) = \emptyset \qquad \textit{for all } k \in \mathbb{N}.$$

PROOF OF LEMMA 3.46. Let $S > 0$ and place disjoint balls $B_S(p_i)$ in the unit ball obtaining a total number of balls proportional to $1/S^m$ (consider for instance a grid of length $2S$ and put a ball $B_S(p_i)$ in each cube).

Fix $k \in \mathbb{N}$. If for a ball we have

$$B_{S/2}(p_i) \cap \vec{\Phi}_k(\Sigma) \neq \emptyset,$$

there exists $q_i \in B_{S/2}(p_i)$ with $\theta_{k,q_i} \geq 1$. Since $B_{S/2}(q_i) \subset B_S(p_i)$, Corollary 3.45 gives

$$\begin{aligned}\text{Area}\left(\vec{\Phi}_k(\Sigma) \cap B_S(p_i)\right) &\geq \text{Area}\left(\vec{\Phi}_k(\Sigma) \cap B_{S/2}(q_i)\right) \\ &\geq \frac{S^2\pi}{6} - \frac{S^2}{8}\int_{B_{S/2}(q_i)} |\vec{H}|^2 dvol_{g_{\vec{\Phi}_k}}.\end{aligned}$$

Since the balls $B_S(p_i)$ are disjoint and all contained in $B_1(0)$,

$$\begin{aligned} &\frac{S^2\pi}{6}\cdot\Big|\{i \quad \text{s.t. } B_{S/2}(p_i)\cap\vec{\Phi}_k(\Sigma)\neq\emptyset\}\Big| \\ &\leq \text{Area}\left(\vec{\Phi}_k(\Sigma)\cap B_1(0)\right)+\frac{S^2}{8}\int_{B_1(0)}|\vec{H}|^2 dvol_{g_{\vec{\Phi}_k}} \qquad (3.174)\\ &\leq\left(\frac{3}{2}+\frac{S^2}{8}\right)W(\vec{\Phi}_k), \end{aligned}$$

where we applied Corollary 3.44 in the last step.

Consequently, due to assumption (3.173), for $S>0$ chosen small enough, there exists for each $k\in\mathbb{N}$ a point p_{i_k} such that

$$B_{S/2}(p_{i_k})\cap\vec{\tilde{\Psi}}_k(\Sigma)=\emptyset.$$

(If $\frac{c}{S^m}$ is the total number of balls $B_{S/2}(p_i)$ in $B_1(0)$, choose $S>0$ in such a way that $\frac{c\pi}{6S^{m-2}}>\left(\frac{3}{2}+\frac{S^2}{8}\right)\sup_k W(\vec{\Phi}_k)$.) Extract a subsequence such that $p_{i_k}=p_0$ is independent of $k\in\mathbb{N}$. $B_{S/2}(p_0)$ is the ball we have been looking for. □

Proof of Theorem 3.36 *continued.* Recall (1.30) and apply Lemma 3.46 to the sequence $\vec{\tilde{\Psi}}_k$ and let $B_r(p_0)$ be the obtained ball free of mass. Consider the inversion

$$i_0\colon x\mapsto\frac{x-p_0}{|x-p_0|^2}, \qquad (3.175)$$

which is a conformal transformation of $\mathbb{R}^m\cup\{\infty\}$ such that

$$\vec{\tilde{\Psi}}_k(\Sigma)\cap\{\text{center of inversion of } i_0\}=\emptyset.$$

Note that i_0 is a diffeomorphism from $B_R(p_0)\setminus B_r(p_0)$ into $B_{1/r}(0)\setminus B_{1/R}(0)$, for any $R\in(0,\infty)$. Thus,

$$\|\nabla i_0\|_{L^\infty(B_R(p_0)\setminus B_r(p_0))}+\|\nabla i_0^{-1}\|_{L^\infty\left(B_{1/r}(0)\setminus B_{1/R}(0)\right)}\leq C_R. \qquad (3.176)$$

Since i_0 is conformal it satisfies the equation

$$di_0(x)=e^{\nu(x)}R$$

for $R\in O(m)$ being some orthogonal matrix. (3.176) implies that for the conformal factor, we have

$$\|\nu\|_{L^\infty(B_R(p_0)\setminus B_r(p_0))}\leq\tilde{C}_R. \qquad (3.177)$$

Define for $k\in\mathbb{N}$,

$$\vec{\hat{\Psi}}_k:=i_0\circ\vec{\tilde{\Psi}}_k,$$

and let $\hat{\alpha}_k=\tilde{\alpha}_k+\nu$ denote its conformal factor satisfying $e^{2\hat{\alpha}_k}h_k=g_{\vec{\hat{\Psi}}_k}$.

From (3.172) and the choice of $B_r(p_0)$, we know that for $\delta>0$ and all $k\in\mathbb{N}$,

$$\vec{\tilde{\Psi}}_k\left(\Sigma\setminus(\cup_{i=1,\ldots,n}B_\delta(a_i))\right)\subset B_{C_\delta}(0)\setminus B_r(p_0).$$

Together with (3.171) and (3.177), this implies that for the conformal factors, we have again

$$\sup_k\|\hat{\alpha}_k\|_{L^\infty(\Sigma\setminus(\cup_{i=1,\ldots,n}B_\delta(a_i)))}\leq\tilde{C}_\delta. \qquad (3.178)$$

What we have gained by inverting is that

$$\vec{\hat{\Psi}}_k(\Sigma) \subset B_{1/r}(0) \qquad \text{for all } k \in \mathbb{N}. \tag{3.179}$$

Corollary 3.44 implies that for all $k \in \mathbb{N}$,

$$\int_\Sigma e^{2\hat{\alpha}_k} dvol_{h_k} = \text{Area}(\vec{\hat{\Psi}}_k(\Sigma)) \leq \frac{3}{2r^2} \sup_k W(\vec{\Phi}_k) \leq C, \tag{3.180}$$

by (1.30).

Step 2d): Weak convergence of $\vec{\hat{\Psi}}_k$ to $\vec{\xi}_\infty$. Since (3.178) holds for any $\delta > 0$, and due to (3.180), we can argue exactly as in Step 1b) and extract a subsequence such that

$$\vec{\hat{\Psi}}_{k'} \rightharpoonup \vec{\xi}_\infty \qquad \text{weakly in } W^{2,2}_{\text{loc}}(\Sigma \setminus \{a_1, \ldots, a_N\}). \tag{3.181}$$

Furthermore, Step 1c) shows that $\vec{\xi}_\infty$ is conformal and we have

$$\log |d\vec{\hat{\Psi}}_k|^2 \overset{*}{\rightharpoonup} \log |d\vec{\xi}_\infty|^2 \qquad \text{in } (L^\infty)^*_{\text{loc}}(\Sigma \setminus \{a_1, \ldots, a_N\}).$$

It remains to prove Condition iv) from Definition 3.34. Since $h_k \to h_\infty$, (3.180) implies that

$$\sup_k \int_\Sigma |d\vec{\hat{\Psi}}_k|^2_{h_\infty} \, dvol_{h_\infty} < \infty.$$

Together with (3.181), this implies that for any $\delta > 0$,

$$\int_{\Sigma \setminus (\cup_{i=1,\ldots,n} B_\delta(a_i))} |d\vec{\xi}_\infty|^2_{h_\infty} dvol_{h_\infty} \leq \liminf_k \int_\Sigma |d\vec{\hat{\Psi}}_k|^2_{h_\infty} \, dvol_{h_\infty} \leq C,$$

where C is independent of δ. Thus, $\vec{\xi}_\infty$ extends to a map in $W^{1,2}(\Sigma)$ and we have

$$\vec{\hat{\Psi}}_k \rightharpoonup \vec{\xi}_\infty \qquad \text{weakly in } W^{1,2}(\Sigma).$$

By (3.179),

$$\sup_k \|\vec{\hat{\Psi}}_k\|_{L^\infty(\Sigma)} < \infty$$

which implies in a similar way

$$\vec{\hat{\Psi}}_k \overset{*}{\rightharpoonup} \vec{\xi}_\infty \qquad \text{weakly in } (L^\infty)^*(\Sigma).$$

This finishes the proof of Theorem 3.36 for $\vec{\xi}_k := \vec{\hat{\Psi}}_k$. □

5. Weak branched immersions

5.1. Expansion at a blow-up point

Motivated by the compactness result of Theorem 3.36, the purpose of this section is to find out more about the limit object $\vec{\xi}$ of a weakly convergent sequence $\vec{\Phi}_k \in \mathcal{E}_\Sigma$ satisfying $\sup_k \mathbb{I}(\vec{\Phi}_k) < \infty$. The first observation is on the Gauss map $\vec{n}_{\vec{\xi}}$.

LEMMA 3.47. *Let $\vec{\xi}$ be the weak limit of a weakly convergent sequence $\vec{\Phi}_k \in \mathcal{E}_\Sigma$ in the sense of Definition 3.34 which satisfies $\sup_k \mathbb{I}(\vec{\Phi}_k) < \infty$. Then*

$$\vec{n}_{\vec{\xi}} \in W^{1,2}(\Sigma).$$

PROOF. Denote the blow-up points of $\vec{\xi}$ by $a_1, \ldots, a_N$. If follows from Lemma 3.35 that for any $\delta > 0$, we have

$$\int_{\Sigma \setminus (\cup_{i=1,\ldots,n} B_\delta(a_i))} |d\vec{n}_{\vec{\xi}}|_g^2 \, dvol_g \leq \liminf_k \int_\Sigma |d\vec{n}_{\vec{\Phi}_k}|_{g_k}^2 \, dvol_{g_k}$$
$$= \liminf_k \mathbb{I}(\vec{\Phi}_k) \leq C.$$

Hence, $\vec{n}_{\vec{\xi}} \in W^{1,2}_{\text{loc}}(\Sigma \setminus \{a_1, \ldots, a_N\})$ and since C is independent of δ, $\vec{n}_{\vec{\xi}}$ extends to a map in $W^{1,2}(\Sigma)$. □

The following lemma helps to understand the behavior of $\vec{\xi}$ at its blow-up points.

LEMMA 3.48. *Let* $\vec{\xi}\colon D^2 \to \mathbb{R}^m$ *be a weakly conformal map such that* $\log |\nabla \vec{\xi}| \in L^\infty_{loc}(D^2 \setminus \{0\})$ *and* $\vec{\xi} \in W^{2,2}_{loc}(D^2 \setminus \{0\})$. *Assume* $\vec{\xi}$ *extends to a map in* $W^{1,2}(D^2)$ *and that the corresponding Gauss map* $\vec{n}_{\vec{\xi}}$ *also extends to a map in* $W^{1,2}(D^2, Gr_{m-2}(\mathbb{R}^m))$.

Then $\vec{\xi} \in W^{1,\infty}(D^2)$ *and there exists* $n \in \mathbb{N} \setminus \{0\}$ *and a constant* C *such that*

$$(C - o(1))\, |z|^{n-1} \leq |\partial_z \vec{\xi}| \leq (C + o(1))\, |z|^{n-1}. \tag{3.182}$$

REMARK 3.49. (3.182) *tells us that the behavior of* $\vec{\xi}$ *at its blow-up point is just the one of a holomorphic curve such as*

$$\mathbb{C} \to \mathbb{C}^2, \quad z \mapsto (z^2, z^3).$$

We thus call a blow-up point branch point *if it has positive branching order* $n - 1 > 0$, *where* $n \in \mathbb{N} \setminus \{0\}$ *is given by Lemma 3.48. (Note that if* $n - 1 = 0$, *there is no branching and we can remove the singularity.)*

PROOF OF LEMMA 3.48. We can localize in order to ensure that

$$\int_{D^2} |\nabla \vec{n}_{\vec{\xi}}|^2 \, dx\, dy < \frac{8\pi}{3}.$$

Exactly as in Subsection 5, using Hélein's lifting theorem, we deduce the existence of a framing

$$\vec{e} := (\vec{e}_1, \vec{e}_2) \in W^{1,2}(D^2, S^{m-1} \times S^{m-1})$$

such that

$$\langle \vec{e}_1, \vec{e}_2 \rangle = 0, \qquad \vec{n}_{\vec{\xi}} = \star(\vec{e}_1 \wedge \vec{e}_2), \tag{3.183}$$

$$\int_{D^2} \left[|\nabla \vec{e}_1|^2 + |\nabla \vec{e}_2|^2 \right] \, dx\, dy \leq C \int_{D^2} |\nabla \vec{n}_{\vec{\xi}}|^2 \, dx\, dy \tag{3.184}$$

and satisfying the Coulomb conditon

$$\begin{cases} div(\vec{e}_1, \nabla \vec{e}_2) = 0 & \text{in } D^2 \\[2ex] \left(\vec{e}_1, \dfrac{\partial \vec{e}_2}{\partial \nu} \right) = 0 & \text{on } \partial D^2. \end{cases} \tag{3.185}$$

We introduce $e_i := d\vec{\xi}^{-1} \vec{e}_i$ and e_i^* to be the dual framing. Denoting $|\partial_x \vec{\xi}|^2 = |\partial_y \vec{\xi}|^2 = e^{2\lambda}$ we have that the metric $g := \vec{\xi}^* g_{\mathbb{R}^m}$ is given by $g = e^{2\lambda} \left[dx^2 + dy^2 \right]$. Hence with respect to the flat metric $g_0 := \left[dx^2 + dy^2 \right]$ one has

$$|e_i|_{g_0}^2 = g_0(e_i, e_i) = e^{-2\lambda} \, g(e_i, e_i) = e^{-2\lambda}.$$

and since $e_j^*(e_i) = \delta_{ij}$ we have that $|e_i^*|^2_{g_0} = e^{2\lambda}$. Since $\vec{\xi}$ is assumed to be in $W^{1,2}(D^2)$, we deduce that

$$e_i^* \in L^2(D^2).$$

Since $\vec{\xi}$ is in $W^{1,\infty} \cap W^{2,2}_{loc}(D^2 \setminus \{0\}, \mathbb{R}^m)$ and $\log|\nabla\vec{\xi}| \in L^\infty_{loc}(D^2 \setminus \{0\})$ we have that the framing given by $\vec{f}_i := e^{-\lambda}\, \partial_{x_i}\vec{\xi}$ is in $L^\infty_{loc} \cap W^{1,2}_{loc}(D^2 \setminus \{0\}, \mathbb{R}^m)$. Since $\vec{\xi}$ is conformal the unit framing $(\vec{f}_1, \vec{f}_2)$ is Coulomb:

$$div(\vec{f}_1, \nabla\vec{f}_2) = 0 \qquad \text{in } D^2 \setminus \{0\}.$$

Denoting by $e^{i\theta}$ the rotation which passes from $(\vec{f}_1, \vec{f}_2)$ to $(\vec{e}_1, \vec{e}_2)$, the Coulomb condition satisfied by the two framings implies that θ is harmonic on $D^2 \setminus \{0\}$ and hence analytic on this domain. This implies that

$$e_i^* \in L^\infty_{loc} \cap W^{1,2}_{loc}(D^2 \setminus \{0\}).$$

As in Subsection 6 we introduce $f \in W^{1,2}(D^2)$ as the solution to

$$\left\{\begin{array}{rcll} df &=& *_g\langle\vec{e}_1, d\vec{e}_2\rangle & \text{on } D^2 \\ \displaystyle\int_{\partial D^2} f &=& 0. & \end{array}\right. \tag{3.186}$$

Then f satisfies

$$\left\{\begin{array}{rcll} \Delta_{g_0} f &=& (\nabla^\perp\vec{e_1}, \nabla\vec{e}_2) & \text{on } D^2 \\ f &=& 0 & \text{on } \partial D^2 \end{array}\right.$$

and Theorem 2.25 implies that $f \in C^0(D^2)$. As in Subsection 6, we obtain for $i = 1, 2$

$$d[e^{-f} e_i^*] = 0 \qquad \text{in } \mathcal{D}'(D^2 \setminus \{0\}).$$

By the Schwartz Lemma the distribution $d[e^{-f} e_i^*]$ is a finite linear combination of successive derivatives of the Dirac Mass at the origin but since $e^{-f} e_i^* \in L^2(D^2)$, this linear combination can only be 0. Hence we have for $i = 1, 2$

$$d[e^{-f} e_i^*] = 0 \qquad \text{in } \mathcal{D}'(D^2).$$

Hence, by Poincaré's Lemma, there exists $(\sigma_1, \sigma_2) \in W^{1,2}(D^2, \mathbb{R}^2)$ such that

$$d\sigma_i = e^{-f} e_i^*.$$

The dual basis $(\partial/\partial\sigma_1, \partial/\partial\sigma_2) = e^f (e_1, e_2)$ is positive and orthogonal on $D^2 \setminus \{0\}$. Hence $\sigma = \sigma_1 + i\sigma_2$ is an holomorphic function on $D^2 \setminus \{0\}$ which extends to a $W^{1,2}$-map on D^2. The classical point removability theorem for holomorphic maps implies that σ extends to an holomorphic function on D^2. Possibly after modifying σ by a constant, we can assume that $\sigma(0) = 0$. The holomorphicity of σ implies in particular that

$$|d\sigma|_{g_0} = \sqrt{2}\, e^{\lambda - f}$$

is uniformly bounded and, since $f \in L^\infty(D^2)$, we deduce that λ is bounded from above on D^2. This fact implies that $\vec{\xi}$ extends to a Lipschitz map on D^2. Though $|d\sigma|_{g_0}$ has no zero on D^2, σ' might have a zero at the origin: there exists an holomorphic function $h(z)$ on D^2 satisfying $h(0) = 0$, a complex number c_0 and an integer n such that

$$\sigma(z) = c_0\, z^n\, (1 + h(z)). \tag{3.187}$$

We have that locally

$$\partial_\sigma\vec{\xi} = \partial_{\sigma_1}\vec{\xi} - i\partial_{\sigma_2}\vec{\xi} = d\vec{\xi}(e^f e_1) - i d\vec{\xi}(e^f e_2) = e^f[\vec{e}_1 - i\vec{e}_2].$$

Hence, since f is continuous, we have that

$$|\partial_\sigma\vec{\xi}| = \sqrt{2}\, e^{f(0)}\, (1+o(1)). \tag{3.188}$$

Combining (3.187) and (3.188) gives

$$|\partial_z\vec{\xi}| = |\partial_\sigma\vec{\xi}|\ |\partial_z\sigma| = c_0\ n\ \sqrt{2}\ e^{f(0)}\ |z|^{n-1}\ (1+o(1)). \tag{3.189}$$

This last identity implies (3.182). □

DEFINITION 3.50. *Let (Σ, h) be a conformal structure on Σ, where h denotes the associated metric of constant curvature and unit volume. The space $\mathcal{F}^{conf}_{(\Sigma,h)}$ denotes the set of measurable maps $\vec{\xi}\colon \Sigma \to \mathbb{R}^m$ that satisfy*

i) $\vec{\xi} \in W^{1,\infty}(\Sigma)$;

ii) $\vec{\xi}\colon (\Sigma, h) \to \mathbb{R}^m$ *is weakly conformal;*

iii) there exist finitely many blow-up points $a_1, \dots a_N \in \Sigma$ s.t.

$$\log|d\vec{\xi}| \in L^\infty_{\text{loc}}(\Sigma \setminus \{a_1, \dots, a_N\});$$

iv) $\vec{n}_{\vec{\xi}} \in W^{1,2}(\Sigma, Gr_{m-2}(\mathbb{R}^m))$.

REMARK 3.51. *Let $\vec{\Phi}_k$ be a sequence in $\mathcal{E}_\Sigma$ with $\sup_k \mathbb{I}(\vec{\Phi}_k) < \infty$. Let h_k denote the respective metrics of constant curvature and unit volume of the induced conformal structures, which are assumed to satisfy condition (CA) with $h_k \to h_\infty$. Suppose $\vec{\Phi}_k$ weakly converges to $\vec{\xi}_\infty$ in the sense of Definition* 3.34. *Then Lemma* 3.47 *and Lemma* 3.48 *imply that $\vec{\xi}_\infty$ is an element of $\mathcal{F}^{conf}_{(\Sigma,h_\infty)}$.*

We are now ready to introduce the space of *weak branched immersions*, which contains the closure of $\mathcal{E}_\Sigma$ under weak convergence.

DEFINITION 3.52. *Define the space $\mathcal{F}_\Sigma$ of* weak branched immersions *as the space of measurable maps $\vec{\Phi}\colon \Sigma \to \mathbb{R}^m$ such that there exists a bi-Lipschitz diffeomorphism Ψ of Σ and a conformal structure on Σ, with h being the associated constant curvature metric of unit volume, such that $\vec{\Phi} \circ \Psi \in \mathcal{F}^{conf}_{(\Sigma,h)}$.*

Let $\vec{\xi} \in \mathcal{F}^{\text{conf}}_{(\Sigma,h)}$ be a weak branched conformal immersion with branch points $\{b_j\}$ and respective branching orders $\{n_j - 1\}$, given by Lemma 3.48. Taking isothermal coordinates around b_j, Lemma 3.48 gives us information on the behavior of the conformal factor

$$\lambda = \log|\partial_{x_1}\vec{\xi}| = \log|\partial_{x_2}\vec{\xi}| = \log|\partial_z\vec{\xi}| - \log\sqrt{2}$$

at $0 = \psi^{-1}(b_j)$. More specifically, we have

$$\begin{cases} -\Delta\lambda = e^{2\lambda}K & \text{in } D^2 \setminus \{0\} \\ \\ \lambda(z) = (n_j - 1)\log|z| + O(1) & \text{in } D^2, \end{cases} \tag{3.190}$$

where we used Lemma 1.5 on the regular part of $\vec{\xi}$.

$$-\Delta\lambda - e^{2\lambda}K \tag{3.191}$$

is a distribution on D^2 and its support is contained in $\{0\}$. The Schwartz Lemma implies that it is a finite linear combination of δ_0 and its derivatives. Since (3.191) is in $\bigcap_{p<\infty} L^p(D^2)$, standard techniques show that no derivatives of δ_0 can occur. The second line in (3.190) yields that

$$-\Delta\lambda = e^{2\lambda}K - 2\pi(n_j - 1)\delta_0. \tag{3.192}$$

In the same way, Lemmas 1.6 and 3.48 imply the following extension of Liouville's equation to maps in $\mathcal{F}_\Sigma^{\text{conf}}$.

LEMMA 3.53. *Let $\vec{\xi} \in \mathcal{F}_{(\Sigma,h)}^{conf}$ have the branch points $b_1, \ldots, b_N$ with respective branching orders $n_1 - 1, \ldots, n_N - 1 \in \mathbb{N} \setminus \{0\}$, given by Lemma 3.48 and let*

$$g = e^{2\alpha}h.$$

Then α satisfies the following PDE in $\mathcal{D}'(\Sigma)$:

$$-\Delta_h\alpha = e^{2\alpha}K_g - K_h - 2\pi\sum_{j=1}^{N}(n_j - 1)\delta_{b_j}. \tag{3.193}$$

The following lemma gives a control of the branch points with multiplicity.

LEMMA 3.54. *Let $\vec{\xi} \in \mathcal{F}_{(\Sigma,h)}^{conf}$ have the branch points $b_1, \ldots, b_N$ with respective branching orders $n_1 - 1, \ldots, n_N - 1 \in \mathbb{N} \setminus \{0\}$, given by Lemma 3.48 and let*

$$g = e^{2\alpha}h.$$

Then

$$\sum_{j=1}^{N}(n_j - 1) \le \frac{1}{4\pi}\mathbb{I}(\vec{\xi}) - \chi(\Sigma). \tag{3.194}$$

PROOF OF LEMMA 3.54. Applying the identity (3.193) of distributions to the constant function 1 on Σ yields

$$\begin{aligned} 2\pi\sum_{j=1}^{N}(n_j - 1) &= \int_\Sigma e^{2\alpha}K_g\, dvol_h - \int_\Sigma K_h\, dvol_h \\ &= \int_\Sigma K_g\, dvol_g - 2\pi\chi(\Sigma) \le \frac{1}{2}\mathbb{I}(\vec{\xi}) - 2\pi\chi(\Sigma), \end{aligned}$$

where we used (1.23) in the last step. □

6. Weak sequentially closedness of $\mathcal{F}_\Sigma$

We called Theorem 3.36 a weak "almost-closure theorem" because starting from a sequence $\vec{\Phi}_k$ in $\mathcal{E}_\Sigma$ the weak limit map $\vec{\xi}_\infty$ is in general not contained in the class $\mathcal{E}_\Sigma$, but only in the strictly larger space $\mathcal{F}_\Sigma$. In the following theorem we show that the space $\mathcal{F}_\Sigma$ of weak branched immersions is in fact closed under weak convergence, i.e. we obtain a *weak closure theorem.*

THEOREM 3.55 (Weak closure theorem). *Let $\vec{\Phi}_k \in \mathcal{F}_\Sigma$ be a sequence such that*

$$\sup_k \mathbb{I}(\vec{\Phi}_k) < \infty. \tag{3.195}$$

Suppose assumption (CA) is satisfied and thus, up to subsequences, for the constant curvature metrics h_k of unit volume of the conformal structures induced by $\vec{\Phi}_k$, we have

$$h_k \to h_\infty \qquad \text{in } C^l(\Sigma), \qquad \text{for all } l \in \mathbb{N},$$

for h_∞ being the constant curvature metric of unit volume of some conformal structure on Σ.

Then there exists a subsequence of $\vec{\Phi}_k$ which, in the sense of Definition 3.34, weakly converges to an element of the space $\mathcal{F}^{conf}_{(\Sigma,h_\infty)} \subset \mathcal{F}_\Sigma$.

PROOF OF THEOREM 3.55. Compose with diffeomorphisms Ψ_k to obtain $\vec{\xi}_k := \vec{\Phi}_k \circ \Psi_k \in \mathcal{F}^{\rm conf}_{(\Sigma,h_k)}$. Denote the branch points of $\vec{\xi}_k$ by $b_1^k, \dots, b_{N_k}^k$ with respective branching orders $n_1^k - 1, \dots, n_{N_k}^k - 1 \in \mathbb{N}\setminus\{0\}$. (3.193) and (3.194) imply that

$$\|\Delta\alpha_k\|_{\mathcal{M}(\Sigma)} \leq 2\left(\frac{1}{2}\mathbb{I}(\vec{\Phi}_k) - 2\pi\chi(\Sigma)\right) \leq C.$$

Consequently, Theorem 2.24 implies that, as in the unbranched case, we get a global bound

$$\sup_k \|d\alpha_k\|_{L^{2,\infty}(\Sigma)} < \infty.$$

Thus, Theorem 3.38 holds true for a sequence of isothermal charts φ_k satisfying

$$\sup_k \int_{D^2} |\nabla \vec{n}_{\vec{\xi}_k \circ \varphi_k}|^2 dx_1 dx_2 < \frac{8\pi}{3} \tag{3.196}$$

and containing none of the branch points $b_1^k, \dots, b_{N_k}^k$.

Lemma 3.54 and condition (3.195) imply that the number of branch points is uniformly bounded:

$$N_k \leq \sum_{j=1}^{N_k}(n_j^k - 1) \leq \frac{1}{4\pi}\mathbb{I}(\vec{\Phi}_k) - \chi(\Sigma) \leq C.$$

Hence, we can extract a subsequence of $\vec{\xi}_k$ such that $N_{k'} := N_0$ is independent of k'.

We perform Step 2a) in the proof of Theorem 3.36 with the only difference that when extracting a further subsequence in order to obtain (3.166) and (3.167), we do this in such a way that additionally, for each $j \in \{1, \dots, N_0\}$ there exists $b_j^\infty \in \Sigma$ with

$$b_j^{k'} \to b_j^\infty, \tag{3.197}$$

as $k' \to \infty$. Define

$$\{d_1, \dots, d_M\} := \{b_1^\infty, \dots, b_{N_0}^\infty\} \cup \{a_1, \dots, a_N\}, \tag{3.198}$$

where the latter set of points is as defined in (3.168).

In Step 2b), if $B^{h_\infty}_{s_{i_0}}(x^{i_0}_\infty)$ contains any of the points $b_1^\infty, \dots, b_{N_0}^\infty$, choose a smaller ball $\hat{B}_t \subset B^{h_\infty}_{s_{i_0}}(x^{i_0}_\infty)$ that is free of these points.

Let $0 < \delta < \min\{\min_{i\in I_0} \rho_{i,\infty}, t\}$ be arbitrary small. Step 1a) can now be applied to $K = \Sigma \setminus \cup_{i=1}^M B_\delta(d_i)$, with

$$\rho_\infty = \min\left\{\inf_{x\in K} \rho_{x,\infty}, \frac{\delta}{2}\right\}.$$

The choice of ρ_∞ makes sure that Theorem 3.38 can be applied, i.e. we have a cover of balls satisfying (3.196) and *containing no branch points.*

The rest of Step 2b) and Steps 2c) and d) imply weak convergence to some $\vec{\xi}_\infty \in \mathcal{F}^{\rm conf}_{(\Sigma,h_\infty)}$ with blow-up points $d_1, \dots, d_M$. □

LECTURE 4

The Willmore surface equation

In this section we shall work mostly in the simplest framework of weak immersions into $\mathbb{R}^3$, though most of the arguments below can be extended to the most general framework of weak immersions into arbitrary Riemannian manifolds ([**Riv08**], [**MR13**]).

As always, Σ denotes a smooth closed oriented 2-dimensional manifold. In this section we will work locally in conformal charts most of the time.

We first introduce the notion of *weak Willmore immersions.*

DEFINITION 4.56. *Let* $\vec{\Phi}\colon \Sigma \to \mathbb{R}^m$ *be a weak immersion in* $\mathcal{E}_\Sigma$. $\vec{\Phi}$ *is a* critical point *for* W *if*

$$\forall \vec{w} \in C^\infty(\Sigma, \mathbb{R}^m) \qquad \frac{d}{dt} W(\vec{\Phi} + t\vec{w})\bigg|_{t=0} = 0. \tag{4.199}$$

Such a weak immersion is called Willmore.

Weak Willmore immersions are characterized by an Euler-Lagrange equation. We want to study it for the case $m = 3$.

THEOREM 4.57 (Willmore surface equation). *Let* $\vec{\Phi}\colon D^2 \to \mathbb{R}^3$ *be a conformal weak immersion in* $\mathcal{E}_{D^2}$. Φ *is Willmore if and only if*

$$div\left[2\nabla\vec{H} - 3H\ \nabla\vec{n} + \vec{H} \times \nabla^\perp\vec{n}\right] = 0 \quad in\ \mathcal{D}'(D^2). \tag{4.200}$$

Since it is a local property for a weak immersion $\vec{\Phi} \in \mathcal{E}_\Sigma$ to be Willmore, one can work locally in a disc-neighborhood of a point and use isothermal coordinates on this disc. Thus Theorem 4.57 implies immediately the following.

COROLLARY 4.58. $\vec{\Phi} \in \mathcal{E}_\Sigma$ *is a weak Willmore immersion if and only if in any conformal chart, the following holds true:*

$$div\left[2\nabla\vec{H} - 3H\ \nabla\vec{n} + \vec{H} \times \nabla^\perp\vec{n}\right] = 0 \quad in\ \mathcal{D}'(D^2). \tag{4.201}$$

PROOF OF THEOREM 4.57. Let $\vec{\Phi}\colon \Sigma \to \mathbb{R}^3$ be conformal and in $\mathcal{E}_{D^2}$, and let

$$\vec{\Phi}_t := \vec{\Phi} + t\vec{w}$$

be any variation, for $\vec{w} \in C_c^\infty(D^2, \mathbb{R}^3)$. For $|t|$ small enough, $\vec{\Phi}_t$ is still in $\mathcal{E}_{D^2}$ and we consider

$$W(\vec{\Phi}_t) = \int_{D^2} |\vec{H}_t|^2\ dvol_{g_t} = \int_{D^2} H_t^2\ dvol_{g_t},$$

where $\vec{H}_t := H_t\ \vec{n}_t$. We want to compute

$$\frac{d}{dt} W(\vec{\Phi}_t)\bigg|_{t=0} = 2\int_{D^2} H\ \frac{d}{dt} H_t\bigg|_{t=0}\ dvol_g + \int_{D^2} H^2\ \frac{d}{dt}\left(dvol_{g_t}\right)\bigg|_{t=0}. \tag{4.202}$$

Recall from (1.16) that $H_t = \frac{1}{2}\sum_{i,j}(g_t)^{ij}(\mathbb{I}_t)_{ij}$ where we denote

$$(\mathbb{I}_t)_{ij} = \langle \vec{\mathbb{I}}_t(\partial_{x_i}, \partial_{x_j}), \vec{n}_t\rangle = -\langle \partial_{x_i}\vec{n}_t, \partial_{x_j}\vec{\Phi}\rangle.$$

Consequently,

$$\begin{aligned}\frac{d}{dt}H_t\Big|_{t=0} &= -\frac{1}{2}\sum_{i,j}\frac{d}{dt}(g_t)^{ij}\Big|_{t=0}\langle\partial_{x_i}\vec{n},\partial_{x_j}\vec{\Phi}\rangle \\ &\quad + g^{ij}\left[\left\langle\partial_{x_i}\frac{d}{dt}\vec{n}_t\Big|_{t=0},\partial_{x_j}\vec{\Phi}\right\rangle + \langle\partial_{x_i}\vec{n},\partial_{x_j}\vec{w}\rangle\right].\end{aligned} \tag{4.203}$$

We have $(g_t)_{ij} = \langle\partial_{x_i}\vec{\Phi}_t, \partial_{x_j}\vec{\Phi}_t\rangle$, thus

$$\frac{d}{dt}(g_t)_{ij}\Big|_{t=0} = \langle\partial_{x_i}\vec{w},\partial_{x_j}\vec{\Phi}\rangle + \langle\partial_{x_i}\vec{\Phi},\partial_{x_j}\vec{w}\rangle. \tag{4.204}$$

Since $\sum_i (g_t)^{ki}(g_t)_{ij} = \delta_{kj}$ and $g_{ij} = e^{2\lambda}\, I_2$, where I_2 is the (2×2)-identity matrix, we have

$$\frac{d}{dt}(g_t)^{kj}\Big|_{t=0} e^{2\lambda} + e^{-2\lambda}\,\frac{d}{dt}(g_t)_{kj}\Big|_{t=0} = 0, \tag{4.205}$$

from which we deduce

$$\frac{d}{dt}(g_t)^{kj}\Big|_{t=0} = -e^{-4\lambda}\,\frac{d}{dt}(g_t)_{kj}\Big|_{t=0} = -e^{-4\lambda}\left(\langle\partial_{x_k}\vec{w},\partial_{x_j}\vec{\Phi}\rangle + \langle\partial_{x_k}\vec{\Phi},\partial_{x_j}\vec{w}\rangle\right). \tag{4.206}$$

Note that we can write

$$\frac{d}{dt}\vec{n}_t\Big|_{t=0} = a\,\vec{e}_1 + b\,\vec{e}_2,$$

for two functions a and b. They can be identified as follows:

$$e^{\lambda}\,a = \left\langle e^{\lambda}\,\vec{e}_1, \frac{d}{dt}\vec{n}_t\Big|_{t=0}\right\rangle = -\left\langle\frac{d}{dt}\left(\partial_{x_1}\vec{\Phi}_t\right)\Big|_{t=0},\vec{n}\right\rangle = -\langle\partial_{x_1}\vec{w},\vec{n}\rangle,$$

and similarly one obtains $b = -e^{-\lambda}\,\langle\partial_{x_2}\vec{w},\vec{n}\rangle$. Thus,

$$\begin{aligned}&\sum_{i,j} g^{ij}\left\langle\partial_{x_i}\frac{d}{dt}\vec{n}_t\Big|_{t=0},\partial_{x_j}\vec{\Phi}\right\rangle \\ &= -e^{-2\lambda}\,\partial_{x_1}\left(e^{-2\lambda}\,\langle\langle\partial_{x_1}\vec{w},\vec{n}\rangle\;\partial_{x_1}\vec{\Phi} + \langle\partial_{x_2}\vec{w},\vec{n}\rangle\;\partial_{x_2}\vec{\Phi},\partial_{x_1}\vec{\Phi}\rangle\right) \\ &\quad - e^{-2\lambda}\,\partial_{x_2}\left(e^{-2\lambda}\,\langle\langle\partial_{x_1}\vec{w},\vec{n}\rangle\;\partial_{x_1}\vec{\Phi} + \langle\partial_{x_2}\vec{w},\vec{n}\rangle\;\partial_{x_2}\vec{\Phi},\partial_{x_2}\vec{\Phi}\rangle\right) \\ &= -e^{-2\lambda}\,\left(\partial_{x_1}\langle\partial_{x_1}\vec{w},\vec{n}\rangle + \partial_{x_2}\langle\partial_{x_2}\vec{w},\vec{n}\rangle\right).\end{aligned} \tag{4.207}$$

Observe

$$
(4.208)\qquad
\begin{aligned}
\frac{d}{dt}\left(dvol_{g_t}\right)\Big|_{t=0} &= \frac{d}{dt}\left(det(g_t)_{ij}\right)^{1/2}\Big|_{t=0} dx_1\wedge dx_2\\
&= \frac{1}{2}e^{-2\lambda}\ \frac{d}{dt}\left((g_t)_{11}\ (g_t)_{22}-(g_t)_{12}^2\right)\Big|_{t=0} dx_1\wedge dx_2\\
&= \frac{1}{2}\left(\frac{d}{dt}(g_t)_{11}\Big|_{t=0}+\frac{d}{dt}(g_t)_{22}\Big|_{t=0}\right) dx_1\wedge dx_2\\
&= \Big(\langle\partial_{x_1}\vec{\Phi},\partial_{x_1}\vec{w}\rangle+\langle\partial_{x_2}\vec{\Phi},\partial_{x_2}\vec{w}\rangle\Big)\, dx_1\wedge dx_2.
\end{aligned}
$$

where the last step is due to (4.204).

Plugging (4.203), (4.205), (4.207) and (4.208) in (4.202) yields

$$
\begin{aligned}
&\frac{d}{dt}W(\vec{\Phi}_t)\Big|_{t=0}\\
&= \int_{D^2} H\, e^{-4\lambda}\sum_{i,j}\Big(\langle\partial_{x_i}\vec{w},\partial_{x_j}\vec{\Phi}\rangle+\langle\partial_{x_i}\vec{\Phi},\partial_{x_j}\vec{w}\rangle\Big)\,\langle\partial_{x_i}\vec{n},\partial_{x_j}\vec{\Phi}\rangle\, e^{2\lambda}\, dx_1dx_2\\
&\quad+\int_{D^2} H\, e^{-2\lambda}\left(\partial_{x_1}\langle\partial_{x_1}\vec{w},\vec{n}\rangle+\partial_{x_2}\langle\partial_{x_2}\vec{w},\vec{n}\rangle\right)\, e^{2\lambda}\, dx_1dx_2\\
&\quad-\int_{D^2} H\, e^{-2\lambda}\left(\langle\partial_{x_1}\vec{n},\partial_{x_1}\vec{w}\rangle+\langle\partial_{x_2}\vec{n},\partial_{x_2}\vec{w}\rangle\right)\, e^{2\lambda}\, dx_1dx_2\\
&\quad+\int_{D^2} H^2\,\Big(\langle\partial_{x_1}\vec{\Phi},\partial_{x_1}\vec{w}\rangle+\langle\partial_{x_2}\vec{\Phi},\partial_{x_2}\vec{w}\rangle\Big)\, dx_1dx_2\\
&= \int_{D^2} H\, e^{-2\lambda}\sum_{i,j}\langle\partial_{x_i}\vec{n},\partial_{x_j}\vec{\Phi}\rangle\langle\partial_{x_j}\vec{\Phi},\partial_{x_i}\vec{w}\rangle\, dx_1dx_2\\
&\quad+\int_{D^2} H\,\left(\partial_{x_1}\langle\partial_{x_1}\vec{w},\vec{n}\rangle+\partial_{x_2}\langle\partial_{x_2}\vec{w},\vec{n}\rangle\right)\, dx_1dx_2\\
&\quad+\int_{D^2} H^2\,\Big(\langle\partial_{x_1}\vec{\Phi},\partial_{x_1}\vec{w}\rangle+\langle\partial_{x_2}\vec{\Phi},\partial_{x_2}\vec{w}\rangle\Big)\, dx_1dx_2,
\end{aligned}
$$

where we used that

$$
\langle\partial_{x_1}\vec{n},\partial_{x_1}\vec{w}\rangle+\langle\partial_{x_2}\vec{n},\partial_{x_2}\vec{w}\rangle=\sum_{i,j}\langle\partial_{x_i}\vec{n},\partial_{x_j}\vec{\Phi}\rangle\langle\partial_{x_j}\vec{\Phi},\partial_{x_i}\vec{w}\rangle.
$$

and

$$
\langle\partial_{x_i}\vec{n},\partial_{x_j}\vec{\Phi}\rangle=\langle\partial_{x_j}\vec{n},\partial_{x_i}\vec{\Phi}\rangle.
$$

Partial integration gives

$$
\begin{aligned}
&\left.\frac{d}{dt}W(\vec{\Phi}_t)\right|_{t=0}\\
&= -\int_{D^2}\left\langle\partial_{x_1}\left(He^{-2\lambda}\langle\partial_{x_1}\vec{n},\partial_{x_1}\vec{\Phi}\rangle\,\partial_{x_1}\vec{\Phi}+He^{-2\lambda}\langle\partial_{x_1}\vec{n},\partial_{x_2}\vec{\Phi}\rangle\,\partial_{x_2}\vec{\Phi}\right),\vec{w}\right\rangle dx_1dx_2\\
&\quad -\int_{D^2}\left\langle\partial_{x_2}\left(He^{-2\lambda}\langle\partial_{x_2}\vec{n},\partial_{x_1}\vec{\Phi}\rangle\,\partial_{x_1}\vec{\Phi}+He^{-2\lambda}\langle\partial_{x_2}\vec{n},\partial_{x_2}\vec{\Phi}\rangle\,\partial_{x_2}\vec{\Phi}\right),\vec{w}\right\rangle dx_1dx_2\\
&\quad +\int_{D^2}\langle\partial_{x_1}\left(\partial_{x_1}H\,\vec{n}\right)+\partial_{x_2}\left(\partial_{x_2}H\,\vec{n}\right),\vec{w}\rangle\, dx_1dx_2\\
&\quad -\int_{D^2}\left\langle\partial_{x_1}\left(H^2\,\partial_{x_1}\vec{\Phi}\right)+\partial_{x_2}\left(H^2\,\partial_{x_2}\vec{\Phi}\right),\vec{w}\right\rangle dx_1dx_2\\
&= \int_{D^2}\left\langle\partial_{x_1}\left(-H\,\pi_T(\partial_{x_1}\vec{n})+\partial_{x_1}H\,\vec{n}-H^2\,\partial_{x_1}\vec{\Phi}\right)\right.\\
&\qquad\quad \left.+\partial_{x_2}\left(-H\,\pi_T(\partial_{x_2}\vec{n})+\partial_{x_2}H\,\vec{n}-H^2\,\partial_{x_2}\vec{\Phi}\right),\vec{w}\right\rangle dx_1dx_2\\
&= \int_{D^2}\left\langle div\left(-H\nabla\vec{n}+\nabla H\,\vec{n}-H^2\nabla\vec{\Phi}\right),\vec{w}\right\rangle dx_1dx_2.
\end{aligned}
$$

Thus, (4.199) holds if and only if

$$div\left(-H\nabla\vec{n}+\nabla H\,\vec{n}-H^2\nabla\vec{\Phi}\right)=0. \tag{4.209}$$

Using Lemma 1.60 gives the desired result.

□

Note that in the proof of Theorem 4.57, the assumption of $\vec{\Phi}$ being an element of $\mathcal{E}_{D^2}$ was enough to make sense of each line. In particular, the quantity in (4.200) is an "honest" distribution in $\mathcal{D}'(D^2)$. Indeed, $\vec{H}\in L^2(D^2)$ and consequently

$$\nabla\vec{H}\in H^{-1}(D^2),\quad H\,\nabla\vec{n}\in L^1(D^2),\quad \vec{H}\times\nabla^{\perp}\vec{n}\in L^1(D^2). \tag{4.210}$$

Note that first an alternative Euler-Lagrange equation for *smooth* Willmore immersions was discovered: by Shadow-Thomsen ([**Tho23**]) in dimension 3, for general $m\geq 3$ by Weiner ([**Wei78**]). In dimension 3, it is

$$\Delta_g H+2H\,(H^2-K)=0. \tag{4.211}$$

In [**Riv08**] the equivalence of (4.200) and (4.211) for smooth conformal immersions is shown. Note that equation (4.211) contains the nonlinearity $2H\,(H^2-K)$, which is *cubic* in the second fundamental form. Thus, it has no meaning for weak immersions with second fundamental form bounded in L^2.

In the next subsection we want to address the question whether weak Willmore immersions are actually *smooth.*

Observe that (4.201) is in *conservative-elliptic form* which is *critical in dimension* 2 under the assumption of L^2-bounded second fundamental form. Indeed, we write the equation as follows:

$$\Delta\vec{H} = div\left[\frac{3}{2}H\,\nabla\vec{n} - \frac{1}{2}\vec{H}\times\nabla^\perp\vec{n}\right]. \tag{4.212}$$

As observed in (4.210), the second fundamental form being in L^2 implies that

$$\frac{3}{2}\,H\,\nabla\vec{n} - \frac{1}{2}\vec{H}\times\nabla^\perp\vec{n} \in L^1(D^2).$$

Theorem 2.23 implies that

$$\frac{1}{2\pi}\log|x| * div\left[\frac{3}{2}H\,\nabla\vec{n} + \frac{1}{2}\nabla^\perp\vec{n}\times\vec{H}\right] \in L^{2,\infty}(D^2).$$

Inserting this information back in (4.212), we obtain $\vec{H} \in L^{2,\infty}_{loc}(D^2)$ which is almost the information we started from. This phenomenon characterizes *critical elliptic systems.*

Conservation laws for weak Willmore immersions

The key for studying the regularity of weak Willmore immersions will be to discover conservation laws for them.

THEOREM 5.59. *Let $\vec{\Phi} \in \mathcal{E}_{D^2}$ be a conformal weak Willmore immersion. Then there exists $\vec{L} \in L^{2,\infty}_{loc}(D^2, \mathbb{R}^3)$ such that*

$$\nabla^\perp \vec{L} = 2\, \nabla \vec{H} - 3H\, \nabla \vec{n} + \vec{H} \times \nabla^\perp \vec{n}. \tag{5.213}$$

Moreover the following conservation laws are satisfied:

$$\begin{cases} div\left[\langle \vec{L}, \nabla^\perp \vec{\Phi}\rangle\right] = 0 & (5.214a) \\ \\ div\left[\vec{L} \times \nabla^\perp \vec{\Phi} + 2H\, \nabla^\perp \vec{\Phi}\right] = 0. & (5.214b) \end{cases}$$

PROOF OF THEOREM 5.59. Since $\vec{\Phi}$ satisfies (4.201), by the weak Poincaré Lemma there exists $\vec{L} \in \mathcal{D}'(D^2)$ such that

$$\nabla^\perp \vec{L} = 2\, \nabla \vec{H} - 3H\, \nabla \vec{n} + \vec{H} \times \nabla^\perp \vec{n}. \tag{5.215}$$

Assuming $\vec{\Phi} \in \mathcal{E}_{D^2}$, the right hand side of (5.215) is in $H^{-1} \cap L^1(D^2)$. We deduce from Theorem 2.23 that $\vec{L}$ is the sum of a harmonic function and a function in $L^{2,\infty}(D^2)$, thus

$$\vec{L} \in L^{2,\infty}_{\text{loc}}(D^2).$$

For proving (5.214a), note that

$$\begin{aligned} div\left[\langle \vec{L}, \nabla^\perp \vec{\Phi}\rangle\right] &= \langle \nabla \vec{L}, \nabla^\perp \vec{\Phi}\rangle = -\langle \nabla^\perp L, \nabla \vec{\Phi}\rangle \\ &= -\langle -H\, \nabla \vec{n} + \vec{H} \times \nabla^\perp \vec{n}, \nabla \vec{\Phi}\rangle. \end{aligned} \tag{5.216}$$

Using (1.60), we obtain

$$\langle H\, \nabla \vec{n} + \vec{H} \times \nabla^\perp \vec{n}, \nabla \vec{\Phi}\rangle = -4\, e^{2\lambda}\, H^2. \tag{5.217}$$

Finally,

$$\langle -2H\, \nabla \vec{n}, \nabla \vec{\Phi}\rangle = 2H\, (\mathbb{I}(\partial_{x_1}, \partial_{x_1}) + \mathbb{I}(\partial_{x_2}, \partial_{x_2})) = 4\, e^{2\lambda}\, H^2, \tag{5.218}$$

which, together with (5.216) and (5.217) gives the first conservation law (5.214a).

For showing the second one, recall from (1.10) that

$$div\left[\vec{L} \times \nabla^\perp \vec{\Phi}\right] = \nabla \vec{\Phi} \times \nabla^\perp \vec{L}. \tag{5.219}$$

Using again (1.60), we have

$$\nabla\vec{\Phi}\times\left(H\ \nabla\vec{n}+\vec{H}\times\nabla^{\perp}\vec{n}\right)=-2H^2\ \nabla\vec{\Phi}\times\nabla\vec{\Phi}=0. \tag{5.220}$$

We compute

$$\begin{aligned}
&\nabla\vec{\Phi}\times(2\ \nabla H\cdot\vec{n}-2H\ \nabla\vec{n})\\
&=2\ \nabla H\cdot\left(\nabla\vec{\Phi}\times\vec{n}\right)-2H\underbrace{\left(\partial_{x_1}\vec{\Phi}\times\partial_{x_1}\vec{n}+\partial_{x_2}\vec{\Phi}\times\partial_{x_2}\vec{n}\right)}_{=0}\\
&=2\ \nabla H\cdot\nabla^{\perp}\vec{\Phi}.
\end{aligned} \tag{5.221}$$

Identities (5.219), (5.220) and (5.220) imply the second conservation law (5.214b). □

Having now found two new conserved quantities, as we did for the first one $2\ \nabla\vec{H}-3H\ \nabla\vec{n}+\vec{H}\times\nabla^{\perp}\vec{n}$, we can apply the weak Poincaré Lemma in order to obtain "primitives" of these quantities.

THEOREM 5.60. *Let $\vec{\Phi}\in\mathcal{E}_{D^2}$ be a conformal weak Willmore immersion and let $\vec{L}\in L^{2,\infty}_{\text{loc}}(D^2,\mathbb{R}^3)$ be as in Theorem* 5.59, *satisfing the conservation laws* (5.214a) *and* (5.214b).

There exist[1] $S\in W^{1,(2,\infty)}_{loc}(D^2,\mathbb{R})$ *and* $\vec{R}\in W^{1,(2,\infty)}_{loc}(D^2,\mathbb{R}^3)$ *such that*

$$\begin{cases}\nabla^{\perp}S=\langle\vec{L},\nabla^{\perp}\vec{\Phi}\rangle & (5.222a)\\[2ex] \nabla^{\perp}\vec{R}=\vec{L}\times\nabla^{\perp}\vec{\Phi}+2H\ \nabla^{\perp}\vec{\Phi} & (5.222b)\end{cases}$$

and the following equations hold:

$$\begin{cases}\nabla S=-\langle\vec{n},\nabla^{\perp}\vec{R}\rangle & (5.223a)\\[2ex] \nabla\vec{R}=\vec{n}\times\nabla^{\perp}\vec{R}+\nabla^{\perp}S\cdot\vec{n}. & (5.223b)\end{cases}$$

PROOF OF THEOREM 5.60. Due to conservations laws (5.214a) and (5.214b), and the Poincaré Lemma, there exists $S\in\mathcal{D}'(D^2,\mathbb{R})$ and $\vec{R}\in\mathcal{D}'(D^2,\mathbb{R}^3)$ satisfying (5.222a) and (5.222b).

Since $\vec{\Phi}\in W^{1,\infty}(D^2)$ and $\vec{L}\in L^{2,\infty}_{\text{loc}}(D^2)$, ∇S and $\nabla\vec{R}$ are in $L^{2,\infty}_{\text{loc}}(D^2)$.

Next, we want to show (5.223b). Note that, by (5.222b),

$$\vec{n}\times\nabla^{\perp}\vec{R}=\vec{n}\times\left(\vec{L}\times\nabla^{\perp}\vec{\Phi}\right)+2H\ \vec{n}\times\nabla^{\perp}\vec{\Phi}. \tag{5.224}$$

We have

$$\vec{n}\times\nabla^{\perp}\vec{\Phi}=\nabla\vec{\Phi} \tag{5.225}$$

[1] We denote by $W^{1,(2,\infty)}$ the space of distributions in L^2 with gradient in $L^{2,\infty}$.

and

$$\begin{aligned}
&\vec{n} \times \left(\vec{L} \times \nabla^\perp \vec{\Phi}\right) \\
&= -\nabla^\perp \vec{\Phi} \times \left(\vec{n} \times \vec{L}\right) - \vec{L} \times \left(\nabla^\perp \vec{\Phi} \times \vec{n}\right) \\
&= -\left(\langle \vec{L}, \nabla^\perp \vec{\Phi}\rangle \cdot \vec{n} - \langle \nabla^\perp \vec{\Phi}, \vec{n}\rangle \cdot \vec{L}\right) + \vec{L} \times \nabla \vec{\Phi} \\
&= -\langle \vec{L}, \nabla^\perp \vec{\Phi}\rangle \cdot \vec{n} + \vec{L} \times \nabla \vec{\Phi}.
\end{aligned} \tag{5.226}$$

From (5.222a) we know that

$$\langle \vec{L}, \nabla^\perp \vec{\Phi}\rangle = \nabla^\perp S. \tag{5.227}$$

Combining the identities (5.224), (5.225), (5.226) and (5.227) yields

$$\vec{n} \times \nabla^\perp \vec{R} = -\nabla^\perp S \cdot \vec{n} + \vec{L} \times \nabla \vec{\Phi} + 2H\ \nabla \vec{\Phi}.$$

On the other hand, from (5.222b) we know that

$$\nabla \vec{R} = \vec{L} \times \nabla \vec{\Phi} + 2H\ \nabla \vec{\Phi}.$$

The two last identities together imply (5.223b).

The first equation (5.223a) now follows easily from the second one: Using (5.223b), we have

$$\langle \vec{n}, \nabla^\perp \vec{R}\rangle = -\underbrace{\langle \vec{n}, \vec{n} \times \nabla \vec{R}\rangle}_{=0} - \langle \vec{n}, \nabla S \cdot \vec{n}\rangle = -\nabla S.$$

□

COROLLARY 5.61. *Let $\vec{\Phi} \in \mathcal{E}_{D^2}$ be a conformal weak Willmore immersion. Let $\vec{L} \in L^{2,\infty}_{loc}(D^2, \mathbb{R}^3)$ be as in Theorem 5.59 and $S \in W^{1,(2,\infty)}_{loc}(D^2, \mathbb{R})$ and $\vec{R} \in W^{1,(2,\infty)}_{loc}(D^2, \mathbb{R}^3)$ as in Theorem 5.60.*

Then the triple $(\vec{\Phi}, S, \vec{R})$ satisfies the following system:

$$\left\{\begin{aligned}
&\Delta S = -\langle \nabla \vec{n}, \nabla^\perp \vec{R}\rangle && (5.228a)\\
&\Delta \vec{R} = \nabla \vec{n} \times \nabla^\perp \vec{R} + \nabla^\perp S \cdot \nabla \vec{n} && (5.228b)\\
&\Delta \vec{\Phi} = \frac{1}{2}\left(\nabla^\perp S \cdot \nabla \vec{\Phi} + \nabla^\perp \vec{R} \times \nabla \vec{\Phi}\right). && (5.228c)
\end{aligned}\right.$$

PROOF OF COROLLARY 5.61. (5.228a) and (5.228b) are obtained by taking the divergence of (5.223a) and (5.223b) respectively, recalling (1.9) and (1.10).

Furthermore, using (5.222b), we have

$$
\begin{aligned}
&\nabla\vec{\Phi}\times\nabla^\perp\vec{R}\\
&= \nabla\vec{\Phi}\times\left(\vec{L}\times\nabla^\perp\vec{\Phi}\right)+2H\;\nabla\vec{\Phi}\times\nabla^\perp\vec{\Phi}\\
&= \langle\nabla\vec{\Phi},\nabla^\perp\vec{\Phi}\rangle\;\vec{L}-\langle\nabla\vec{\Phi},\vec{L}\rangle\cdot\nabla^\perp\vec{\Phi}-2H\;\partial_{x_1}\vec{\Phi}\times\partial_{x_2}\vec{\Phi}\\
&= -\langle\nabla\vec{\Phi},\vec{L}\rangle\cdot\nabla^\perp\vec{\Phi}-4He^{2\lambda}\;\vec{n},
\end{aligned}
\tag{5.229}
$$

where $e^\lambda := |\partial_{x_1}\vec{\Phi}| = |\partial_{x_1}\vec{\Phi}|$. Using (5.222a), we obtain

$$4e^{2\lambda}\;\vec{H} = -\nabla S\cdot\nabla^\perp\vec{\Phi}-\nabla\vec{\Phi}\times\nabla^\perp\vec{R}.$$

This, together with the representation (1.42) of the mean curvature vector, implies the desired identity (5.228c). □

REMARK 5.62. *Recently, Bernard ([**Ber**]) found that the three conservations laws (4.201), (5.214a) and (5.214b) are due to Noether's theorem, i.e. they correspond to particular symmetries of the Willmore functional.*

We recall Noether's theorem for a functional of the form

$$L(u) = \int_{D^2} l(u,\nabla u)\,dxdy, \qquad u\in W^{1,2}(D^2,\mathbb{R}^m),$$

for $l(z,p)$ being C^1 wrt z and C^2 wrt p.

A vector field X on $\mathbb{R}^m$ is called infinitesimal symmetry *of l if for all $u\in W^{1,2}(D^2)$*

$$l(u,\nabla u) = l(F(t,u),\nabla(F(t,u))),$$

where $F(t,z)$ is the flow of X at time t started from $z\in\mathbb{R}^m$ at time 0.

THEOREM 5.63 (Emmy Noether, 1918). *Let X be an infinitesimal symmetry of l. If u is a critical point of L, then*

$$div\left(\frac{\partial l}{\partial p}\cdot X(u)\right) = 0.$$

$J := \frac{\partial l}{\partial p}\cdot X(u)$ is called the Noether Current *associated to the symmetry X.*

*In [**Ber**], Bernard considers variations of a smooth immersion $\vec{\Phi}\colon\Sigma\to\mathbb{R}^m$ of the form*

$$\vec{\Phi}_t := \vec{\Phi}+t(A^j\partial_j\vec{\Phi}+\vec{B}), \tag{5.230}$$

for $\vec{B}=B\vec{n}$. (As before, we shall only consider the case $m=3$ in the sequel.) He derives

$$
\begin{aligned}
&\frac{d}{dt}\left(\int_{\Sigma_0}|\vec{H}_t|^2 dvol_{g_t}\right)\Bigg|_{t=0}\\
&= \int_{\Sigma_0}\Bigg[\langle\vec{B},\vec{\mathcal{W}}\rangle+div\underbrace{\left(\langle\vec{H},\nabla\vec{B}\rangle-\langle\vec{B},\nabla\vec{H}\rangle+H^2\begin{pmatrix}A^1\\A^2\end{pmatrix}\right)}_{:=J}\Bigg]dvol_g,
\end{aligned}
\tag{5.231}
$$

where $\Sigma_0 \subset \Sigma$ is any smooth subsurface and

$$\vec{\mathcal{W}} = (\Delta_g H + 2H(H^2 - K))\vec{n}$$

the Willmore operator. Recall from (4.211) *that $\vec{\mathcal{W}}$ vanishes for a Willmore surface, i.e. a critical point of the Willmore functional. Assuming that $\vec{\Phi}$ is a Willmore surface and choosing $A^j\partial_j\vec{\Phi} + \vec{B}$ in* (5.230) *as a translation, dilation and rotation gives then the following associated Noether currents:*

Translation: *For $\vec{a} \in \mathbb{R}^3$, let*

$$\vec{\Phi}_t = \vec{\Phi} + t\vec{a}, \qquad \text{i.e. } A^j = \langle \vec{a}, g^{ij}\partial_i\vec{\Phi}\rangle, \quad B = \langle \vec{n}, \vec{a}\rangle.$$

Noether current:

$$J_{\vec{a}} = \langle \vec{a}, -\nabla H\vec{n} + H\nabla\vec{n} + H^2\nabla\vec{\Phi}\rangle.$$

Note that

$$div(-\nabla H\vec{n} + H\nabla\vec{n} + H^2\nabla\vec{\Phi}) = 0$$

is equivalent to the Willmore equation (4.201) *in conservative form (cf.* (4.209)*).*

Dilation: *For $\mu \in \mathbb{R} \setminus \{0\}$, let*

$$\vec{\Phi}_t = \vec{\Phi} + t\mu\vec{\Phi}, \qquad \text{i.e. } A^j = \mu\langle g^{ij}\partial_i\vec{\Phi}, \vec{\Phi}\rangle, \quad B = \mu\langle \vec{n}, \vec{\Phi}\rangle.$$

Noether current:

$$J_\mu = \frac{\mu}{2}\langle \vec{L}, \nabla^\perp\vec{\Phi}\rangle,$$

where L is defined as in (5.213).

$$div\langle \vec{L}, \nabla^\perp\vec{\Phi}\rangle = 0$$

is conservation law (5.214a).

Rotation: *For $b \in \mathbb{R}^3$, let*

$$\vec{\Phi}_t = \vec{\Phi} + t\vec{b} \times \vec{\Phi}, \qquad \text{i.e. } A^j = -\langle \vec{b}, g^{ij}\partial_i\vec{\Phi} \times \vec{\Phi}\rangle, \quad B = -\langle \vec{b}, \vec{n} \times \vec{\Phi}\rangle.$$

Noether current:

$$J_{\vec{b}} = \langle \vec{b}, -\frac{1}{2}\vec{L} \times \nabla^\perp\vec{\Phi} + \vec{H} \times \nabla\vec{\Phi}\rangle,$$

Note that

$$div(-\frac{1}{2}\vec{L} \times \nabla^\perp\vec{\Phi} + \vec{H} \times \nabla\vec{\Phi})) = -\frac{1}{2}div(\vec{L} \times \nabla^\perp\vec{\Phi} + 2H\nabla^\perp\vec{\Phi}) = 0$$

is conservation law (5.214b).

Inversion: *Finally, considering a variation corresponding to a translation and an inversion of the form*

$$\vec{\Phi}_t = \vec{\Phi} + t(|\vec{\Phi}|^2\vec{a} - 2\langle \vec{\Phi}, \vec{a}\rangle\vec{\Phi}),$$

for $a \in \mathbb{R}^3$, leads to (5.228c)*, i.e. the equation that establishes a connection between the potentials S and $\vec{R}$, obtained as primitives of the two former conservations laws, and the immersion $\vec{\Phi}$.*

1. The regularity of weak Willmore immersions

We are now ready to prove that weak Willmore immersions are C^∞ in conformal parametrization. The starting point is the elliptic system with quadratic nonlinearities which are made of linear combinations of Jacobians. It is somehow reminiscent to the CMC (constant mean curvature) equation

$$\Delta u = 2H\, \partial_{x_1} u \times \partial_{x_2} u, \tag{5.232}$$

for $H \in \mathbb{R}$ being a constant.

Wente showed that any $W^{1,2}(D^2, \mathbb{R}^3)$-solution of (5.232) is actually smooth. This makes us hope to get the same fact for weak Willmore immersions. To put ourselves in the same starting position as in the case of the CMC equation, we show that ∇S and $\nabla \vec{R}$ are not only in $L^{2,\infty}$, but in fact in L^2. This is an easy consequence of the previous corollary, together with some result on integrability by compensation.

COROLLARY 5.64. *Let $\vec{\Phi} \in \mathcal{E}_{D^2}$ be a conformal weak Willmore immersion and $\vec{L} \in L^{2,\infty}_{loc}(D^2, \mathbb{R}^3)$, $S \in W^{1,(2,\infty)}_{loc}(D^2, \mathbb{R})$ and $\vec{R} \in W^{1,(2,\infty)}_{loc}(D^2, \mathbb{R}^3)$ be as in Theorems 5.59 and 5.60.*

Then $\nabla S \in W^{1,2}_{loc}(D^2, \mathbb{R})$ and $\nabla \vec{R} \in W^{1,2}_{loc}(D^2, \mathbb{R}^3)$.

PROOF. Appying Theorem 2.28 to the equations (5.228a) and (5.228b) gives the result. □

We can now attack the proof of the smoothness of weak Willmore immersions, just as it works in the case of the CMC equation. We will need the following lemma.

LEMMA 5.65. *Let v be a harmonic function on D^2. Then for every point $p \in D^2$, the function*

$$r \mapsto \frac{1}{r^2} \int_{B_r(p)} |\nabla v|^2\, dx_1 dx_2$$

is increasing.

Proof. See [**Riv**], Lemma IV.1.

THEOREM 5.66 (Weak Willmore immersions are smooth.). *Let $\vec{\Phi} \in \mathcal{E}_\Sigma$ be a weak Willmore immersion. Then $\vec{\Phi}$ is C^∞ in conformal paratmetrization.*

PROOF. For any conformal chart, we can apply Theorems 5.59 and 5.60 as well as Corollaries 5.61 and 5.64 and obtain $\nabla S \in W^{1,2}_{loc}(D^2, \mathbb{R})$ and $\nabla \vec{R} \in W^{1,2}_{loc}(D^2, \mathbb{R}^3)$ such that the following system is satisfied:

$$\begin{cases} \Delta S = -\langle \nabla \vec{n}, \nabla^\perp \vec{R} \rangle & (5.233a) \\ \\ \Delta \vec{R} = \nabla \vec{n} \times \nabla^\perp \vec{R} + \nabla^\perp S \cdot \nabla \vec{n} & (5.233b) \\ \\ \Delta \vec{\Phi} = \dfrac{1}{2} \left(\nabla^\perp S \cdot \nabla \vec{\Phi} + \nabla^\perp \vec{R} \times \nabla \vec{\Phi} \right). & (5.233c) \end{cases}$$

Step 1: Morrey decrease. Our first aim is to prove the existence of a positive constant α such that

$$\sup_{r<1/4,\ p\in B_{1/2}(0)} r^{-\alpha} \int_{B_r(p)} \left(|\nabla S|^2 + |\nabla \vec{R}|^2\right)\, dx_1dx_2 < \infty. \tag{5.234}$$

Once this Morrey decrease for $|\nabla S|^2+|\nabla \vec{R}|^2$ is established, by (5.233a) and (5.233b), we get one for $|\Delta S| + |\Delta \vec{R}|$ as well:

$$\sup_{r<1/4,\ p\in B_{1/2}(0)} r^{-\alpha/2} \int_{B_r(p)} \left(|\Delta S| + |\Delta \vec{R}|\right)\, dx_1dx_2 < \infty. \tag{5.235}$$

Then a classical estimate on Riesz potentials ([**Ada75**]) gives that

$$\nabla S, \nabla \vec{R} \in L^p_{\text{loc}}(B_{1/2}(0)) \qquad \text{for some } p > 2, \tag{5.236}$$

from which we will start the bootstrapping in Step 2.

To prove (5.234), we let $\varepsilon_0 > 0$ fixing its value later. There exists some radius $r_0 > 0$ such that

$$\sup_{p\in B_{1/2}(0)} \int_{B_{r_0}(p)} |\nabla \vec{n}|^2\, dx_1dx_2 < \varepsilon_0.$$

Let $p \in B_{1/2}(0)$ be arbitrary. Let Ψ_S and $\vec{\Psi}_{\vec{R}}$ be the solutions of

$$\begin{cases} \Delta\Psi_S = -\langle \nabla\vec{n}, \nabla^\perp \vec{R}\rangle & \text{in } B_{r_0}(p) \\ \Psi_S = 0 & \text{on } \partial B_{r_0}(p) \end{cases} \tag{5.237}$$

and

$$\begin{cases} \Delta\vec{\Psi}_{\vec{R}} = \nabla\vec{n}\times\nabla^\perp\vec{R} + \nabla^\perp S\cdot\nabla\vec{n} & \text{in } B_{r_0}(p) \\ \vec{\Psi}_{\vec{R}} = 0 & \text{on } \partial B_{r_0}(p). \end{cases} \tag{5.238}$$

By Lemma 5.65 and the Dirichlet principle, the harmonic rests

$$v_S := S - \Psi_S \qquad \vec{v}_{\vec{R}} := \vec{R} - \vec{\Psi}_{\vec{R}}$$

satisfy

$$\begin{aligned} &\int_{B_{r_0/2}(p)} \left(|\nabla v_S|^2 + |\nabla \vec{v}_{\vec{R}}|^2\right)\, dx_1dx_2 \\ &\leq \frac{1}{4}\int_{B_{r_0}(p)} \left(|\nabla S|^2 + |\nabla \vec{R}|^2\right)\, dx_1dx_2. \end{aligned} \tag{5.239}$$

Applying Wente's Theorem 2.25 to (5.237) and (5.238) yields

$$
\begin{aligned}
&\int_{B_{r_0}(p)} \left(|\nabla \Psi_S|^2 + |\nabla \vec{\Psi}_{\vec{R}}|^2\right) \, dx_1 dx_2 \\
&\leq C \int_{B_{r_0}(p)} |\nabla \vec{n}|^2 \, dx_1 dx_2 \int_{B_{r_0}(p)} \left(|\nabla S|^2 + |\nabla \vec{R}|^2\right) \, dx_1 dx_2 \\
&\leq C\varepsilon_0 \int_{B_{r_0}(p)} \left(|\nabla S|^2 + |\nabla \vec{R}|^2\right) \, dx_1 dx_2 .
\end{aligned}
\tag{5.240}
$$

Putting (5.239) and (5.240) together yields

$$
\begin{aligned}
&\int_{B_{r_0/2}(p)} \left(|\nabla S|^2 + |\nabla \vec{R}|^2\right) \, dx_1 dx_2 \\
&\leq 2 \int_{B_{r_0}(p)} \left(|\nabla \Psi_S|^2 + |\nabla \vec{\Psi}_{\vec{R}}|^2\right) \, dx_1 dx_2 \\
&\qquad + 2 \int_{B_{r_0/2}(p)} \left(|\nabla v_S|^2 + |\nabla \vec{v}_{\vec{R}}|^2\right) \, dx_1 dx_2 \\
&\leq \left(2C\varepsilon_0 + \frac{1}{2}\right) \int_{B_{r_0}(p)} \left(|\nabla S|^2 + |\nabla \vec{R}|^2\right) \, dx_1 dx_2 .
\end{aligned}
\tag{5.241}
$$

We now choose $\varepsilon_0 := 1/(8C)$. Then iterating (5.241) yields

$$
\begin{aligned}
&\int_{B_{2^{-j} r_0}(p)} \left(|\nabla S|^2 + |\nabla \vec{R}|^2\right) \, dx_1 dx_2 \\
&\leq \left(\frac{3}{4}\right)^j \int_{B_{r_0}(p)} \left(|\nabla S|^2 + |\nabla \vec{R}|^2\right) \, dx_1 dx_2 \\
&\leq C_{r_0} \, \left(2^{-j} r_0\right)^{\alpha} ,
\end{aligned}
\tag{5.242}
$$

where we choose

$$
\alpha := \log_2\left(\frac{4}{3}\right), \qquad C_{r_0} := r_0^{-\alpha} \int_{B_1(0)} \left(|\nabla S|^2 + |\nabla \vec{R}|^2\right) \, dx_1 dx_2 .
$$

Since α and r_0 are independent of $p \in B_{1/2}(0)$, this gives (5.234) for the sup taken over all $r \leq r_0$. Noting that

$$
\begin{aligned}
\sup_{r_0 < r < 1/4, \ p \in B_{1/2}(0)} & r^{-\alpha} \int_{B_r(p)} \left(|\nabla S|^2 + |\nabla \vec{R}|^2\right) \, dx_1 dx_2 \\
& \leq r_0^{-\alpha} \int_{B_1(0)} \left(|\nabla S|^2 + |\nabla \vec{R}|^2\right) \, dx_1 dx_2 < \infty
\end{aligned}
$$

yields (5.234).

Bootstrapping. Note that (5.236) from Step 1 and (5.233c) imply that also

$$\nabla\vec{n} \in L^p_{\mathrm{loc}}(B_{1/2}(0)).$$

This and (5.236) we can use to bootstrap in (5.233a) and (5.233b) and obtain that

$$\nabla S,\ \nabla\vec{R} \in L^q_{\mathrm{loc}}(B_{1/2}(0)) \qquad \text{for all } q < \infty.$$

Bootstrapping this information in (5.233c) yields

$$\vec{\Phi} \in W^{2,q}_{\mathrm{loc}}(B_{1/2}(0)) \qquad \text{for all } q < \infty,$$

from which we deduce that

$$\vec{n} \in W^{1,q}_{\mathrm{loc}}(B_{1/2}(0)) \qquad \text{for all } q < \infty.$$

The latter information injected back in (5.233a) and (5.233b) gives

$$\nabla^2 S,\ \nabla^2\vec{R} \in L^q_{\mathrm{loc}}(B_{1/2}(0)) \qquad \text{for all } q < \infty,$$

and so on and so forth. Iteration gives that

$$\vec{\Phi} \in W^{k,p}_{\mathrm{loc}}(B_{1/2}(0)) \qquad \text{for all } k \in \mathbb{N},\ 1 \leq p \leq \infty.$$

Hence

$$\vec{\Phi} \in C^\infty_{\mathrm{loc}}(B_{1/2}(0)),$$

which finishes the proof. □

Finally, we want to remark that if one performs the steps in the regularity proof more carefully and bootstraps in (4.212), one obtains the following result.

THEOREM 5.67 (ε-regularity for Willmore). *Let $\vec{\Phi} \in \mathcal{E}_{D^2}$ be a conformal weak Willmore immersion with $|\partial_{x_1}\vec{\Phi}| = |\partial_{x_2}\vec{\Phi}| = e^\lambda$. Assume that*

$$\|\nabla\lambda\|_{L^{2,\infty}(D^2)} \leq C_0.$$

Then there exists $\varepsilon_0 > 0$ with the following property:
If

$$\int_{D^2} |\nabla\vec{n}|^2 \, dx_1 dx_2 \leq \varepsilon_0,$$

then

$$\begin{aligned} &\|e^\lambda\, |H|\|^2_{L^\infty(D^2_{1/2})} + \|\nabla\vec{n}\|^2_{L^\infty(D^2_{1/2})} \\ &\leq C\left(\varepsilon_0, \|\nabla\lambda\|_{L^{2,\infty}(D^2)}\right) \int_{D^2_{3/4}} |\nabla\vec{n}|^2 \, dx_1 dx_2. \end{aligned} \tag{5.243}$$

From the ε-regularity (and Theorem 3.37), one sees easily that one can pass to the limit in the Willmore surface equation (4.200), if one considers a sequence of weak Willmore immersions with a uniform L^2-bound on the second fundamental form satisfying (CA). This is possible at all points where there is no energy concentration. More precisely, one obtains the following theorem.

THEOREM 5.68. *Let $\vec{\Phi}_k \in \mathcal{E}_\Sigma$ be a sequence of weak Willmore immersions, which satisfies the compactness assumption (CA) and*

$$\sup_k \mathbb{I}(\vec{\Phi}_k) < \infty.$$

Let $\vec{\xi}_\infty$ be the weak limit of a subsequence of $\vec{\Phi}_k$ in the sense of Definition 3.34, which exists due to Theorem 3.36.

Then $\vec{\xi}$ is weakly Willmore away from its blow-up points.

2. A minimization procedure for the Willmore energy among weak branched immersions

We will now merge the compactness theorem from Section 5 and the regularity theorem from Section 1 in order to present a general approach for minimizing Willmore energy under various constraints.

The first one is the topological constraint that Σ (or in other words the genus) is prescribed. We shall give a new proof of the following classical result which has originally been derived with different techniques ([**Sim86**],[**BK03**]).

THEOREM 5.69. *Let Σ be a closed, orientable 2-dimensional smooth manifold. Then*

$$\inf_{\vec{\Phi}\in\mathcal{E}_\Sigma} W(\vec{\Phi})$$

is achieved by a smooth embedding.

PROOF. We present the proof for $m = 3$.

Because of the closure theorem 3.55, it is natural to work in $\mathcal{F}_\Sigma$, the class of weak branched immersions, instead of $\mathcal{E}_\Sigma$. We thus take a minimizing sequence $\vec{\Phi}_k \in \mathcal{F}_\Sigma$, i.e. satisfying

$$W(\vec{\Phi}_k) \searrow \inf_{\vec{\Phi}\in\mathcal{F}_\Sigma} W(\vec{\Phi}).$$

In order to be able to apply Theorem 3.55 and extract a weakly convergent subsequence, we need that the sequence $\vec{\Phi}_k$ satisfies the compactness assumption (CA). To see this, we note two facts:

i) For every genus g there exists a smooth immersion $\vec{\Phi}\colon \Sigma_g \to \mathbb{R}^3$ such that $W(\vec{\Phi}) < 8\pi$ (for $m = 3$ see e.g. [**Kus89**]).

ii) If a sequence $\vec{\Phi}_k \in \mathcal{E}_\Sigma$ satisfies $W(\vec{\Phi}_k) < \min\{8\pi, \omega_g^3\}$, it meets the compactness assumption (CA) (see [**Riv13**], where also Simon's definition of ω_g^n is recalled). For our purposes, it is sufficient to know that $\inf_{\vec{\Phi}\in\mathcal{F}_\Sigma} W(\vec{\Phi}) < \omega_g^n$. This follows in particular from [**BK03**].

From i) follows that for sufficiently large k, we have

$$W(\vec{\Phi}_k) < 8\pi.$$

The Li-Yau inequality (3.141) and Lemma 3.47 imply that any $\vec{\Phi} \in \mathcal{F}_\Sigma$ with $W(\vec{\Phi}) < 8\pi$ is in fact a weak embedding, i.e. there are neither branch points nor multiple points. Thus we can assume that $\vec{\Phi}_k \in \mathcal{E}_\Sigma$.

By ii) (and again i)) we can assume that $\vec{\Phi}_k$ satisfy (CA). Now we can apply Theorem 3.55, which gives us a weakly convergent subsequence with limit $\vec{\xi}_\infty \in \mathcal{F}_\Sigma^{\text{conf}}$. By lower semicontinuity of the Willmore functional (Lemma 3.35), we get $W(\vec{\xi}_\infty) = \inf_{\vec{\Phi}\in\mathcal{F}_\Sigma} W(\vec{\Phi})$. As above, by Li-Yau and the branch point lemma, we deduce that $\vec{\xi}_\infty$ is actually in $\mathcal{E}_\Sigma$ and does not have multiple points. Furthermore,

$$W(\vec{\xi}_\infty) = \min_{\vec{\Phi}\in\mathcal{F}_\Sigma} W(\vec{\Phi}) = \min_{\vec{\Phi}\in\mathcal{E}_\Sigma} W(\vec{\Phi}). \tag{5.244}$$

To conclude the proof one uses the following lemma, which is proven in [**MR13**] for weak immersions from S^2 into a Riemannian manifold.

LEMMA 5.70 ([**MR13**], Lemma IX.5). *Let $\vec{\xi} \in \mathcal{E}_\Sigma$ be a conformal weak immersion. Then W is Fréchet differentiable at $\vec{\xi}$ with respect to variations $\vec{w} \in W^{1,\infty} \cap W^{2,2}(\Sigma, \mathbb{R}^3)$.*

We apply Lemma 5.70 to the conformal weak immersion $\vec{\xi}_\infty \in \mathcal{E}_\Sigma$. Note that for small δ, variations of the form $\vec{\xi}_\infty + \vec{w}$, for $\|\vec{w}\|_{W^{1,\infty}\cap W^{2,2}(\Sigma)} \leq \delta$, are still elements of $\mathcal{E}_\Sigma$. Thus, (5.244) implies that

$$dW_{\vec{\xi}_\infty} = 0.$$

Corollary 4.58 implies that

$$div\left[-2\nabla H_{\vec{\xi}}\, \vec{n}_{\vec{\xi}} + H_{\vec{\xi}} \nabla \vec{n}_{\vec{\xi}} - \vec{H}_{\vec{\xi}} \times \nabla^\perp \vec{n}_{\vec{\xi}}\right] = 0.$$

The regularity theorem 5.66 implies that $\vec{\xi} \in C^\infty$. The fact that $\vec{\xi}$ is an embedding was noted before. This finishes the proof. □

Bibliography

[Ada75] David R. Adams, *A note on Riesz potentials*, Duke Math. J. **42** (1975), no. 4, 765–778. MR0458158 (56:16361)

[Aub82] Thierry Aubin, *Nonlinear analysis on manifolds. Monge-Ampère equations*, Grundlehren der Mathematischen Wissenschaften [Fundamental Principles of Mathematical Sciences], vol. 252, Springer-Verlag, New York, 1982. MR681859 (85j:58002)

[Ber] Yann Bernard, *Noether's Theorem and the Willmore Functional*, in prep.

[Bet92] Fabrice Bethuel, *Un résultat de régularité pour les solutions de l'équation de surfaces à courbure moyenne prescrite*, C. R. Acad. Sci. Paris Sér. I Math. **314** (1992), no. 13, 1003–1007. MR1168525 (93g:53011)

[BK03] Matthias Bauer and Ernst Kuwert, *Existence of minimizing Willmore surfaces of prescribed genus*, Int. Math. Res. Not. (2003), no. 10, 553–576. MR1941840 (2003j:53086)

[Bla29] Wilhelm Blaschke, *Vorlesungen über Differentialgeometrie und geometrische Grundlagen von Einsteins Relativitätstheorie. III: Differentialgeometrie der Kreise und Kugeln. Bearbeitet von G. Thomsen.*, X + 474 S. Berlin, J. Springer (Grundlehren der mathematischen Wissenschaften in Einzeldarstellungen), 1929 (German).

[Che74] Bang-yen Chen, *Some conformal invariants of submanifolds and their applications*, Boll. Un. Mat. Ital. (4) **10** (1974), 380–385. MR0370436 (51:6663)

[CL92] Sagun Chanillo and Yan Yan Li, *Continuity of solutions of uniformly elliptic equations in* $\mathbf{R}^2$, Manuscripta Math. **77** (1992), no. 4, 415–433. MR1190215 (93k:35033)

[dC76] Manfredo P. do Carmo, *Differential geometry of curves and surfaces*, Prentice-Hall Inc., Englewood Cliffs, N.J., 1976. MR0394451 (52:15253)

[dC92] Manfredo Perdigão do Carmo, *Riemannian geometry*, Mathematics: Theory & Applications, Birkhäuser Boston Inc., Boston, MA, 1992. MR1138207 (92i:53001)

[Ge98] Yuxin Ge, *Estimations of the best constant involving the* L^2 *norm in Wente's inequality and compact* H*-surfaces in Euclidean space*, ESAIM Control Optim. Calc. Var. **3** (1998), 263–300 (electronic). MR1634837 (2000a:58038)

[Ge99] Yuxin Ge, *A remark on generalized harmonic maps into spheres*, Nonlinear Anal. **36** (1999), no. 4, Ser. A: Theory Methods, 495–506. MR1675268 (2000a:58046)

[GHL04] Sylvestre Gallot, Dominique Hulin, and Jacques Lafontaine, *Riemannian geometry*, third ed., Universitext, Springer-Verlag, Berlin, 2004. MR2088027 (2005e:53001)

[Gra08] Loukas Grafakos, *Classical Fourier Analysis*, Graduate Texts in Mathematics, Springer, 2008. MR2445437 (2011c:42001)

[GT01] David Gilbarg and Neil S. Trudinger, *Elliptic partial differential equations of second order*, Classics in Mathematics, Springer-Verlag, Berlin, 2001. MR1814364 (2001k:35004)

[Hél02] Frédéric Hélein, *Harmonic maps, conservation laws and moving frames*, second ed., Cambridge Tracts in Mathematics, vol. 150, Cambridge University Press, Cambridge, 2002. MR1913803 (2003g:58024)

[IT92] Y. Imayoshi and M. Taniguchi, *An introduction to Teichmüller spaces*, Springer-Verlag, Tokyo, 1992. MR1215481 (94b:32031)

[Jos97] Jürgen Jost, *Compact Riemann surfaces*, Universitext, Springer-Verlag, Berlin, 1997. MR1632873 (2000k:32011)

[KaRu] Dan Katzman and Jacob Rubinstein, *Method for the design of multifocal optical elements*, United States Patent No.: US006302540B1, Oct. 16, 2001.

[KMR] Laura Keller, Andrea Mondino, and Tristan Rivière, *Embedded surfaces of arbitrary genus minimizing the Willmore energy under isoperimetric constraint*, Arch. Ration. Mech. Anal. **212** (2014), no. 2, 645–682.

[Ki] Gustav Kirchhoff, *Über das Gleichgewicht und die Bewegung einer elastischen Scheibe*, J. Reine Angew. Math. **40** (1850).

[Kus89] Rob Kusner, *Comparison surfaces for the Willmore problem*, Pacific J. Math. **138** (1989), no. 2, 317–345. MR996204 (90e:53013)

[KS] Ernst Kuwert and Reiner Schätzle, *Minimizers of the Willmore energy under fixed conformal class*, J. Differential Geometry **93** (2013), 471–530.

[LaLi] L. D. Landau and E. M. Lifschitz, *Theory of elasticity* - Course of theoretical physics, volume 7 - third edition revised and enlarged by E. M. Lifshitz, A. M. Kosevich and L. P. Pitaevskii, Butterwoth-Heinemann 1986.

[MN] Fernando C. Marques and André Neves, *Min-max theory and the Willmore conjecture*, Ann. of Math. (2) **179** (2014), no. 2, 683D782.

[Mat95] Pertti Mattila, *Geometry of sets and measures in Euclidean spaces*, Cambridge Studies in Advanced Mathematics, vol. 44, Cambridge University Press, Cambridge, 1995, Fractals and rectifiability. MR1333890 (96h:28006)

[MoN] Andrea Mondino and Huy Nguyen, *The Global Conformal Invariants of Sub-manifolds*, arXiv:1501.07527 (2015).

[MR13] Andrea Mondino and Tristan Rivière, *Willmore spheres in compact Riemannian manifolds*, Adv. Math. **232** (2013), 608–676. MR2989995

[Riv] Tristan Rivière, *Conformally invariant variational problems*, http://www.math.ethz.ch/ riviere/pub.html.

[Riv06] Tristan Rivière, *Analysis aspects of Willmore surfaces*, arXiv:math/0612526 (2006).

[Riv08] Tristan Rivière, *Analysis aspects of Willmore surfaces*, Invent. Math. **174** (2008), no. 1, 1–45. MR2430975 (2009k:53154)

[Riv10] Tristan Rivière, *Variational Principles for immersed Surfaces with L^2-bounded Second Fundamental Form*, arXiv:1007.2997 (2010). Published in J. Reine und Angewandte Math. **695**, (2014).

[Riv13] Tristan Rivière, *Lipschitz conformal immersions from degenerating Riemann surfaces with L^2-bounded second fundamental forms*, Adv. Calc. Var. **6** (2013), no. 1, 1–31. MR3008339

[Riv15] Tristan Rivière, *Critical weak immersed surfaces within sub-manifolds of the Teichmüller space*, Adv. Math. **283** (2015), 232–274.

[Riv16] Tristan Rivière, *Willmore Minmax Surfaces and the Cost of the Sphere Eversion*, arXiv:1512.08918

[Rud87] Walter Rudin, *Real and complex analysis*, third ed., McGraw-Hill Book Co., New York, 1987. MR924157 (88k:00002)

[Sim93] Leon Simon, *Existence of surfaces minimizing the Willmore functional*, Comm. Anal. Geom., **1** (1993), no 2, 281–325.

[Sim84] Leon Simon, *Lectures on geometric measure theory*, Proc. Centre Math. Anal. Austral. Nat. Univ., Australian National University, 1984. MR756417 (87a:49001)

[Sim86] Leon Simon, *Existence of Willmore surfaces*, Miniconference on geometry and partial differential equations (Canberra, 1985), Proc. Centre Math. Anal. Austral. Nat. Univ., vol. 10, Australian National University, Canberra, 1986, pp. 187–216. MR857667 (87m:53006)

[Tho23] G. Thomsen, *Über konforme geometrie 1: Grundlagen der konformen flächentheorie*, Lütcke & Wulff, 1923.

[Top97] Peter Topping, *The optimal constant in Wente's L^∞ estimate*, Comment. Math. Helv. **72** (1997), no. 2, 316–328. MR1470094 (98k:35075)

[Wei78] Joel L. Weiner, *On a problem of Chen, Willmore, et al*, Indiana Univ. Math. J. **27** (1978), no. 1, 19–35. MR0467610 (57:7466)

[Wen69] Henry C. Wente, *An existence theorem for surfaces of constant mean curvature*, J. Math. Anal. Appl. **26** (1969), 318–344. MR0243467 (39:4788)

[Wi] T. J. Willmore, *Note on embedded surfaces*, An. Sti. Univ. "Al. I. Cuza" Iasi Sect. I a Mat. (N.S.) 11B (1965) 493D496.

Introduction to Minimal Surface Theory

Brian White

IAS/Park City Mathematics Series
Volume 22, 2013

Introduction to Minimal Surface Theory

Brian White

Minimal surfaces have been studied by mathematicians for several centuries, and they continue to be a very active area of research. Minimal surfaces have also proved to be a useful tool in other areas, such as in the Willmore problem and in general relativity. Furthermore, various general techniques in geometric analysis were first discovered in the study of minimal surfaces.

The following notes are slightly expanded versions of four lectures presented at the 2013 summer program of the Institute for Advanced Study and the Park City Mathematics Institute. The goal was to give beginning graduate students an introduction to some of the most important basic facts and ideas in minimal surface theory. I have kept prerequisites to a minimum: the reader should know basic complex analysis and elementary differential geometry (in particular, the second fundamental form and the Gauss-Bonnet Theorem).

For readers who wish to pursue the subject further, there are a number of good books available, such as [**CM11**], [**Law80**], and [**DHS10**]. Readers may also enjoy exploring Matthias Weber's extensive online collection of minimal surface images: `http://www.indiana.edu/~minimal/archive/`. All but one of the illustrations in these notes are taken from that collection.

If I had a little more time, I would have talked more about the maximum principle and about the structure of the intersection set for pairs of minimal surfaces. See for example [**CM11**, 1.§7, 6.§1, 6.§2].

If I had a lot more time, I would have talked about geometric measure theory, which has had and continues to have an enormous impact on minimal surface theory. Almgren's 1969 expository article [**Alm69**] remains an excellent introduction. Morgan's book [**Mor09**] is a very readable account of the main concepts and results. In many cases, he describes the key ideas of proofs without giving any details. For complete proofs, I recommend [**Sim83**].

In the last few years, there have been a number of spectacular breakthroughs in minimal surface theory that are not mentioned here. See [**MP11**] for a survey of many of the recent results.

I would like to thank Alessandro Carlotto for running problem sessions for the graduate students attending my lectures and for carefully reading an early version of these notes and making a great many suggestions. I would also like to thank David Hoffman for additional suggestions. The notes are much improved as a result of their input.

Mathemtics Department, Stanford University, Stanford, CA 94305 USA
E-mail address: bcwhite@stanford.edu
Preparation of these notes was partially funded by NSF grants DMS–1105330 and DMS–1404282.

LECTURE 1

The First Variation Formula and Consequences

Let M be an m-dimensional surface in $\mathbf{R}^n$ with boundary Γ. We say that M is a **least-area surface** if its area is less than or equal to the area of any other surface having the same boundary. To make this definition precise, one has to specify the exact meaning of the words "surface", "area", and "boundary". For example, do we require the surfaces to be oriented? But for the moment we will be vague about such matters, since the topics we consider now are independent of such specifications.

In the Plateau problem, one asks: given a boundary Γ in Euclidean space (or, more generally, in a Riemannian manifold), does there exist a least-area surface M with boundary Γ? If so, how smooth is M? For example, if $m = 1$ and Γ consists of a pair of points in $\mathbf{R}^n$, then the solution M of the Plateau problem is the straight line segment joining them.

In general, however, even proving existence is very nontrivial. Indeed, in 1936 Jesse Douglas won one of the first two[1] Fields Medals for his existence and regularity theorems for the $m = 2$ case of the Plateau problem. We will discuss those results in lecture 4.

Now we consider a related question: given a surface M, how do we tell if it a least-area surface? In general, it is very hard to tell, but the "first-derivative test" provides a necessary condition for M to be a least-area surface: If M_t is a one-parameter family of surfaces each with boundary Γ, and if $M_0 = M$, then

$$\left(\frac{d}{dt}\right)_{t=0} \text{area}(M_t)$$

should be 0.

For the test to be useful, we need a way of calculating the first derivative:

THEOREM 1 (The first variational formula). *Let M be a compact m-dimensional manifold in $\mathbf{R}^n$. Let $\phi_t : M \to \mathbf{R}^n$ be a smooth one-parameter family of smooth maps such that*

$$\phi_0(p) \equiv p.$$

Let $X(p) = \left(\frac{d}{dt}\right)_{t=0} \phi_t(p)$ be the initial velocity vectorfield. Then

$$\begin{aligned}\left(\frac{d}{dt}\right)_{t=0} \text{area}(\phi_t M) &= \int_M \text{div}_M(X)\, dS \\ &= \int_{\partial M} X \cdot \nu_{\partial M}\, ds - \int_M H \cdot X\, dS,\end{aligned}$$

[1] Ahlfors won the other one.

where

$$\operatorname{div}_M X := \sum_{i=1}^{m} \mathbf{e}_i \cdot \nabla_{\mathbf{e}_i} X$$

for any orthonormal basis $\mathbf{e}_1(p), \dots, \mathbf{e}_m(p)$ *of* $\operatorname{Tan}_p M$, *where* $H(p)$ *is mean curvature vector of* M *at* p, *and where* $\nu_{\partial M}(x)$ *is the unit vector in the tangent plane to* M *at* x *that is normal to* ∂M *and that points away from* M.[2]

It is perhaps better to refer to the first equality

$$\left(\frac{d}{dt}\right)_{t=0} \operatorname{area}(\phi_t M) = \int_M \operatorname{div}_M(X)\, dS,$$

as the "first variation formula" and to the second equality

$$\int_M \operatorname{div}_M X\, dS = \int_{\partial M} X \cdot \nu_{\partial M}\, ds - \int_M H \cdot X\, dS$$

as the "generalized divergence theorem". Note that when the vectorfield X is tangent to M, the generalized divergence theorem is just the ordinary divergence theorem.

Proof of the first variation formula. Note that

$$\operatorname{area}(\phi_t M) = \int_M J_m(D\phi_t)\, dS$$

where $J_m(D\phi_t)$ is the Jacobian determinant $\sqrt{\det(D\phi^T D\phi)}$, so

$$\left(\frac{d}{dt}\right)_{t=0} \operatorname{area}(\phi_t M) = \int_M \left(\frac{d}{dt}\right)_{t=0} J_m(D\phi_t)\, dS.$$

Thus it suffices to calculate that

$$\left(\frac{d}{dt}\right)_{t=0} J_m(D\phi_t) = \operatorname{div}_M X.$$

By definition of X,

$$\phi_t(p) = p + t\, X(p) + o(t),$$

so

$$D\phi_t(\mathbf{e}_i) = \nabla_{\mathbf{e}_i}(\phi_t) \cong \nabla_{\mathbf{e}_i}(p + tX(p)) = \mathbf{e}_i + t\nabla_{\mathbf{e}_i} X,$$

where $a \cong b$ means $a - b = o(t)$. Thus

$$\begin{aligned} J_m(D\phi_t) &= \sqrt{\det(D\phi_t(\mathbf{e}_i) \cdot D\phi_t(\mathbf{e}_j))} \\ &\cong \sqrt{\det((\mathbf{e}_i + t\nabla_{\mathbf{e}_i} X) \cdot (\mathbf{e}_j + t\nabla_{\mathbf{e}_j} X))} \\ &\cong \sqrt{\det(\delta_{ij} + t(\mathbf{e}_i \cdot \nabla_{\mathbf{e}_j} X + \mathbf{e}_j \cdot \nabla_{\mathbf{e}_i} X))} \end{aligned}$$

Recall that for a square matrix A,

$$\det(I + tA) = 1 + t \operatorname{trace}(A) + o(t).$$

[2] The notation $\nu_{\partial M}$ suggests that the vector depends only on ∂M. However, it depends on M and thus arguably should be written in some other way such as $\nu_{\partial M; M}$.

Thus

$$\begin{aligned} J_m(D\phi_t) &\cong \sqrt{1 + 2t \sum_i (\mathbf{e}_i \cdot \nabla_{\mathbf{e}_i} X)} \\ &\cong \sqrt{1 + 2t \operatorname{div}_M X} \cong 1 + t \operatorname{div}_M X. \end{aligned}$$

□

PROOF OF THE GENERALIZED DIVERGENCE THEOREM. Splitting X into the portion X^T tangent to M and the portion X^N normal to M, we have, by the ordinary divergence theorem,

$$\begin{aligned} \int_M \operatorname{div}_M X \, dS &= \int_M \operatorname{div}_M(X^T) \, dS + \int_M \operatorname{div}_M(X^N) \, dS \\ &= \int_{\partial M} X^T \cdot \nu_{\partial M} \, ds + \int_M \operatorname{div}_M(X^N) \, dS \\ &= \int_{\partial M} X \cdot \nu_{\partial M} \, ds + \int_M \operatorname{div}_M(X^N) \, dS. \end{aligned}$$

Thus we need only show that $\operatorname{div}_M(X^N) = -H \cdot X$. Using the summation convention,

$$\begin{aligned} \operatorname{div}_M(X^N) &= \mathbf{e}_i \cdot \nabla_{\mathbf{e}_i}(X^N) \\ &= \nabla_{\mathbf{e}_i}(\mathbf{e}_i \cdot X^N) - (\nabla_{\mathbf{e}_i} \mathbf{e}_i) \cdot X^N \\ &= \nabla_{\mathbf{e}_i}(0) - (\nabla_{\mathbf{e}_i} \mathbf{e}_i)^N \cdot X \\ &= -H \cdot X. \end{aligned}$$

□

REMARK. The first variation formula and the generalized divergence theorem are also true for submanifolds of a Riemannian manifold N. They can be deduced from the special case by isometrically embedding[3] N in a Euclidean space $\mathbf{R}^n$. The first variation formula in N then becomes a special case of the first variation formula in $\mathbf{R}^n$. To see that the generalized divergence theorem is true in N, note that the submanifold $M \subset N \subset \mathbf{R}^n$ has two mean curvatures vectorfields: the mean curvature H_N of M as a submanifold of N and the mean curvature H of M as a submanifold of $\mathbf{R}^n$. Note that $H_N(p)$ is the orthogonal projection of H to $\operatorname{Tan}_p N$. If X is a vectorfield on M that is tangent to N (as it will be if $\phi_t(M) \subset N$ for all t), then $H \cdot X = H_N \cdot X$, and therefore

$$\begin{aligned} \int_M \operatorname{div}_M X \, dS &= \int_{\partial M} X \cdot \nu_{\partial M} \, ds - \int_M X \cdot H \, dS \\ &= \int_{\partial M} X \cdot \nu_{\partial M} \, ds - \int_M X \cdot H_N \, dS, \end{aligned}$$

which is the generalized divergence theorem for M as a submanifold of N.

[3]Such an isometric embedding exists by the Nash Embedding Theorem [**Nas56**]. A more elementary proof was discovered by Günther [**Gün89**]. See [**Yan98**] for a very readable account of Günther's proof.

Exercise: Let $u : \mathbf{R}^n \to \mathbf{R}$ be a C^1 function. Prove (under the assumptions of theorem 1) that

$$\left(\frac{d}{dt}\right)_{t=0} \int_{\phi_t M} u\,dS = \int_M \nabla u \cdot X\,dS + \int_M u \operatorname{div}_M X\,dS$$
$$= \int_{\partial M} u X \cdot \nu_{\partial M}\,ds + \int_M ((\nabla u)^\perp - uH) \cdot X\,dS.$$

where $(\nabla u)^\perp$ is the component of ∇u normal to M.

DEFINITION. An m-dimensional submanifold $M \subset \mathbf{R}^n$ (or of a Riemannian manifold) is called **minimal** (or **stationary**) provided its mean curvature is everywhere 0, i.e., provided it is a critical point for the area functional.

THEOREM 2. *Let M be a compact m-dimensional minimal submanifold of $\mathbf{R}^n$. Then*

$$m \operatorname{area}(M) = \int_{x \in \partial M} x \cdot \nu_{\partial M}\,ds. \tag{1}$$

PROOF. We apply

$$\int_M \operatorname{div}_M X\,dS = \int_{\partial M} X \cdot \nu_{\partial M} - \int_M X \cdot H\,dS$$

to the vectorfield $X(x) \equiv x$. Now $\operatorname{div}_M X \equiv m$ and $H \equiv 0$, so we get (1). (To see that $\operatorname{div}_M X = m$, note that $\nabla_{\mathbf{v}} X = \mathbf{v}$ for every vector $\mathbf{v}$, and thus that $\sum_i (\mathbf{e}_i \cdot \nabla_{\mathbf{e}_i} X) = m$ for any orthonormal vectors $\mathbf{e}_1, \dots, \mathbf{e}_m$.) □

REMARK. Even if M is not minimal, the same proof shows that

$$m \operatorname{area}(M) = \int_{x \in \partial M} x \cdot \nu_{\partial M}\,ds - \int_{x \in M} x \cdot H\,dS. \tag{2}$$

Monotonicity

THEOREM 3 (Monotonicity Theorem). *Let M be a minimal submanifold of $\mathbf{R}^n$ and let $p \in \mathbf{R}^n$. Then*

$$\Theta(M, p, r) := \frac{\operatorname{area}(M \cap \mathbf{B}(p,r))}{\omega_m r^m}$$

is an increasing function of r for $0 < r \le R := \operatorname{dist}(p, \partial M)$.
Indeed,

$$\left(\frac{d}{dr}\right) \Theta(M, p, r) \ge 0$$

with equality if and only if M intersects $\partial \mathbf{B}(p,r)$ orthogonally.

Here ω_m is the m-dimensional area (i.e., Lebesgue measure) of the unit ball in $\mathbf{R}^m$. Thus $\Theta(M,p,r)$ (which is called the **density ratio** of M in $\mathbf{B}(p,r)$) is the area of $M \cap \mathbf{B}(p,r)$ divided by the area of the cross-sectional m-disk in $\mathbf{B}(p,r)$. Equivalently, it is the area of $M \cap \mathbf{B}(p,r)$ after dilating by $1/r$.

PROOF OF MONOTONICITY. We may assume that $p = 0$. Let $M_r = M \cap \mathbf{B}(0,r)$, so

$$\partial M_r = M \cap \partial \mathbf{B}(0,r).$$

Let $A(r)$ be the m-dimensional area of M_r and $L(r)$ be the $(m-1)$-dimensional measure of ∂M_r. (When $m = 2$, $A(r)$ is an area and $L(r)$ is a length.)

Then

$$A'(r) \geq L(r).$$

This follows from the coarea formula applied to the function $x \in M \mapsto |x|$. But intuitively (in the case $m = 2$ for simplicity) $A(r + dr) \setminus A(r)$ is a thin ribbon of surface: the length of the ribbon is $L(r)$ and the width is $\geq dr$. (The width is equal to dr at a point $p \in \partial M_r$ if and only if M is orthogonal to $\partial \mathbf{B}(0, r)$ at p.) Hence $A(r + dr) - A(r) \geq L(r)\, dr$.

By theorem 2,

$$mA(r) = \int_{\partial M_r} x \cdot \nu_{\partial M_r}\, ds \leq rL(r).$$

Combining these last two inequalities gives:

$$A' - mr^{-1}A \geq 0,$$

so

$$r^{-m}A' - mr^{-m-1}A \geq 0,$$

and therefore

$$(r^{-m}A)' \geq 0.$$

□

REMARK. The monotonicity theorem follows from the more general monotonicity identity [**Sim83**, 17.4]:

$$\begin{aligned}\Theta(M, p, b) - \Theta(M, p, a) = &\int_{M(a,b)} \frac{|(\nabla r)^{\perp}|^2}{r^m}\, dS \\ &+ \frac{1}{m}\int_{x \in M(a,b)} \left(\frac{1}{r^m} - \frac{1}{b^m}\right)(x - p) \cdot H\, dS \\ &+ \frac{1}{m}\int_{x \in M(0,a)} \left(\frac{1}{a^m} - \frac{1}{b^m}\right)(x - p) \cdot H\, dS,\end{aligned}$$

where $0 < a < b \leq \operatorname{dist}(p, \partial M)$, where $r(\cdot) = \operatorname{dist}(\cdot, p) = |(\cdot) - p|$, and where $M(a, b) = M \cap (\mathbf{B}(p, b) \setminus \mathbf{B}(p, a))$. Note that if M is minimal, then the last two integrals vanish. One can use the monotonicity identity to prove a version of the monotonicity theorem in a general Riemannian manifold N by embedding N isometrically in $\mathbf{R}^n$. (If M is a minimal submanifold of $N \subset \mathbf{R}^n$, then M has locally bounded mean curvature as a submanifold of $\mathbf{R}^n$.) See [**Sim83**, 17.6], for example.

We define the **density** of M at a point $p \in M \setminus \partial M$ to be

$$\Theta(M, p) := \lim_{r \to 0} \Theta(M, p, r).$$

For a smooth, immersed surface, the density of M at a point $p \in M \setminus \partial M$ is equal to the number of sheets of M that pass through p. In particular, $\Theta(M, p) \geq 1$.

Density at infinity

Let M be a properly immersed minimal surface without boundary in $\mathbf{R}^n$. Then $\Theta(M, p, r)$ is increasing for $0 < r < \infty$. Thus $\lim_{r \to \infty} \Theta(M, p, r)$ exists. (It may be infinite.) Note that

$$\mathbf{B}(p, r) \subset \mathbf{B}(q, r + |p - q|)$$

FIGURE 1. The catenoid (left) and Scherks's surface (right) each have density 2 at infinity.

from which it easily follows that $\lim_{r\to\infty} \Theta(M, p, r)$ is independent of p and therefore can be written without ambiguity as $\Theta(M)$. We call $\Theta(M)$ **the density of M at infinity**.

For example, the density at infinity of a plane is 1, and the density at infinity of a union of k planes is k. Near infinity, a catenoid (figure 1) looks like a multiplicity 2 plane. (To be precise, if we dilate the catenoid by $1/n$ about its center and let $n \to \infty$, then the resulting surfaces converge smoothly (away from the center) to a plane with multiplicity 2.) It follows that the catenoid has density 2 at infinity.

Similarly, Scherk's surface (figure 1) resembles two orthogonal planes near infinity, so its density at infinity is also 2.

The following theorem characterizes the plane by its density at infinity:

THEOREM 4. *Let M be a properly immersed minimal m-manifold without boundary in $\mathbf{R}^n$. Then $\Theta(M) \geq 1$, with equality if and only if M is a multiplicity 1 plane.*

PROOF. Let $p \in M$. Then by monotonicity,

$$1 \leq \Theta(M, p, r) \leq \Theta(M). \tag{3}$$

This proves the inequality. If $1 = \Theta(M)$, then we would have equality in (3), so M would intersect $\partial\mathbf{B}(p, r)$ orthogonally for every r. That implies that M is invariant under dilations about p, i.e., that M is a cone with vertex p. Since we are assuming that M is smooth, M must in fact be a union of planes (with multiplicity) passing through p. Since $\Theta(M) = 1$, M is a single plane with multiplicity 1. □

Extended monotonicity

According to the monotonicity theorem (theorem 3), if M is minimal, then the density ratio

$$\Theta(M, p, r) = \frac{\operatorname{area}(M \cap \mathbf{B}(p, r))}{\omega_m r^m} \tag{4}$$

is an increasing function of r for $0 < r < R = \operatorname{dist}(p, \partial M)$. The theorem is false without the restriction $r < \operatorname{dist}(p, \partial M)$. For example, if $M \subset \mathbf{B}(p, \hat{R})$, then the density ratio is strictly *decreasing* for $r \geq \hat{R}$, because the numerator of the fraction (4) is constant for $r \geq \hat{R}$.

However, there is an extension of the monotonicity theorem that gives information for all r:

THEOREM 5 (Extended Monotonicity Theorem [**EWW02**]). *Suppose that $M \subset \mathbf{R}^n$ is a compact, minimal m-manifold with boundary Γ, and that $p \in \mathbf{R}^n \setminus \Gamma$. Let $E = E(p, \Gamma)$ denote the* **exterior cone** *with vertex p over Γ:*

$$E = \cup_{q \in \Gamma} \{p + t(q - p) : t \geq 1\}.$$

Let $\tilde{M} = M \cup E$. Then the density ratio

$$\Theta(\tilde{M}, p, r) := \frac{\text{area}(\tilde{M} \cap \mathbf{B}(p, r))}{\omega_m r^m}$$

is an increasing function of r for all $r > 0$. Indeed,

$$\left(\frac{d}{dr}\right) \Theta(\tilde{M}, p, r) \geq 0,$$

with equality if and only if: (i) $\nu_{\partial M} + \nu_{\partial E} \equiv 0$ on $\Gamma \cap \mathbf{B}(p, r)$ and (ii) $\tilde{M}$ intersects $\partial \mathbf{B}(p, r)$ orthogonally.

REMARK. In the definition of $\Theta(\tilde{M}, p, r)$, we count area with multiplicity. For example, if exactly two portions of E overlap in a region, we count the area of that region twice. Of course if M is embedded and if p is in general position, then such overlaps do not occur.

In proving the extended monotonicity theorem, we may assume that $p = 0$. As before, we will apply the first variation formula (or, more precisely, the generalized divergence theorem) to the vectorfield $X(x) = x$.

LEMMA 6. *Let E be the exterior cone over Γ with vertex 0. Among all unit vectors $\mathbf{v}$ that are normal to Γ at $x \in \Gamma$, the maximum value of $x \cdot \mathbf{v}$ is attained by $\mathbf{v} = -\nu_{\partial E}(x)$.*

Consequently, $x \cdot \mathbf{v} \leq -x \cdot \nu_{\partial E}(x)$ and therefore $x \cdot (\mathbf{v} + \nu_{\partial E}(x)) \leq 0$ for every such vector $\mathbf{v}$. The proof of the lemma is left as an exercise.

PROOF OF EXTENDED MONOTONICITY. Let M_r, E_r, $\tilde{M}_r$, and Γ_r be the portions of M, E, $\tilde{M}$, and Γ inside the ball $\mathbf{B}_r = \mathbf{B}(0, r)$. By the generalized divergence theorem,

$$\begin{aligned} \int_{M_r} \operatorname{div}_M X \, dS &= \int_{\partial M_r} X \cdot \nu_{\partial M_r} \, ds - \int_{M_r} H \cdot X \, dS \\ &= \int_{\partial M_r} X \cdot \nu_{\partial M_r} \, ds \end{aligned} \tag{5}$$

since $H \equiv 0$ on M. Similarly,

$$\begin{aligned} \int_{E_r} \operatorname{div}_E X \, dS &= \int_{\partial E_r} X \cdot \nu_{\partial E_r} \, ds - \int_{E_r} H \cdot X \, dS \\ &= \int_{\partial E_r} X \cdot \nu_{\partial E_r} \, ds \end{aligned} \tag{6}$$

because $H \cdot X \equiv 0$ on E, since H is perpendicular to E and X is tangent to E. Also, $\operatorname{div}_M X \equiv \operatorname{div}_E X \equiv m$, so the left sides of these equations are $m \operatorname{area}(M_r)$ and $m \operatorname{area}(E_r)$. Adding equations (5) and (6) gives

$$m \operatorname{area}(\tilde{M}_r) \le \int_{\partial M_r} x \cdot \nu_{\partial M_r}\, dS + \int_{\partial E_r} x \cdot \nu_{\partial E_r}\, dS. \tag{7}$$

Note that ∂M_r consists of two parts: $M \cap \partial \mathbf{B}_r$ and Γ_r. Likewise ∂E_r consists of $E \cap \partial \mathbf{B}_r$ and Γ_r. By combining the two integrals over $M \cap \partial \mathbf{B}_r$ and $E \cap \partial \mathbf{B}_r$, and by also combining the two integrals over Γ_r, we can rewrite (7) as

$$m \operatorname{area}(\tilde{M}_r) \le \int_{\partial \tilde{M}_r} x \cdot \nu_{\partial \tilde{M}_r}\, ds + \int_{\Gamma_r} x \cdot (\nu_{\partial M_r} + \nu_{\partial E_r})\, ds.$$

By lemma 6, the second integrand is everywhere nonpositive. Thus

$$m \operatorname{area}(\tilde{M}_r) \le \int_{x \in \partial \tilde{M}_r} x \cdot \nu_{\partial \tilde{M}_r}\, ds.$$

The rest of the proof is exactly the same as the proof of the monotonicity theorem. □

COROLLARY 7. *Let $p \in M$, Γ, and $E = E(p,\Gamma)$ be as in the extended monotonicity theorem. If $p \in M \setminus \Gamma$, then $1 \le \Theta(M \cup E)$, with equality if and only if $M \cup E$ is a multiplicity 1 plane, i.e., if only if M is a star-shaped region (with respect to p) in an m-dimensional plane.*

PROOF. The proof is almost identical to the proof of theorem 4. □

As a consequence of the extended monotonicity theorem, we can show that a minimal surface must stay reasonably close to its boundary:

THEOREM 8. *Let M be a compact m-dimensional minimal submanifold of $\mathbf{R}^n$. Then for $p \in M$,*

$$m\, \omega_m \operatorname{dist}(p, \partial M)^{m-1} \le |\partial M|.$$

where $|\partial M|$ is the $(m-1)$-dimensional measure of ∂M. Furthermore, equality holds if and only if M is a flat m-disk centered at p with multiplicity 1.

PROOF. We may assume that $p = 0$. Let $\Gamma = \partial M$, let C be the (entire) cone over Γ:

$$\{tq : t \ge 0,\ q \in \Gamma\}$$

and let $E = \{tq : t \ge 1,\ q \in \Gamma\}$ be the exterior cone. Let Γ^* be the result of radially projecting Γ to $\partial \mathbf{B}(0,R)$, where $R = \operatorname{dist}(p, \partial M)$. Then by extended monotonicity,

$$1 \le \Theta(M \cup E) = \Theta(C) = \Theta(C, 0, R) = \frac{1}{\omega R^m} \frac{R}{m} |\Gamma^*| \le \frac{|\Gamma|}{m\, \omega_m R^{m-1}}, \tag{8}$$

which is the asserted inequality. Equality of the first two terms in (8) implies that $M \cup E$ is a plane with multiplicity 1, and equality of the last two terms implies that $\Gamma^* = \Gamma$, which implies that the function $\operatorname{dist}(\cdot, 0)$ is constant on Γ. □

The following corollary implies (for example) that two short curves bounding a connected minimal surface cannot be too far apart:

COROLLARY 9. *If $M \subset \mathbf{R}^n$ is a compact, connected m-dimensional minimal submanifold such that Γ is the union of two (not necessarily connected) components Γ_1 and Γ_2, then*

$$\operatorname{dist}(\Gamma_1, \Gamma_2) \leq 2 \left(\frac{|\partial M|}{m\,\omega_m} \right)^{1/(m-1)}.$$

PROOF. Since the function $\operatorname{dist}(\cdot, \Gamma_1) - \operatorname{dist}(\cdot, \Gamma_2)$ is negative on Γ_1 and positive on Γ_2, there must be a point $p \in M$ at which it vanishes. (Here dist denotes the straight line distance in $\mathbf{R}^n$.) Let

$$R = \operatorname{dist}(p, \Gamma_1) = \operatorname{dist}(p, \Gamma_2) = \operatorname{dist}(p, \Gamma).$$

By the triangle inequality, $\operatorname{dist}(\Gamma_1, \Gamma_2) \leq 2R$, which is at most

$$2 \left(\frac{|\partial M|}{m\,\omega_m} \right)^{1/(m-1)}$$

by theorem 8. □

In [**EWW02**], the extended monotonicity theorem was used to solve a long-open problem in minimal surface theory: if $\Gamma \subset \mathbf{R}^3$ is a smooth, simple closed curve with total curvature at most 4π, must the unique[4] minimal immersed disk bounded by Γ be embedded? (The total curvature of a smooth curve is the integral of the norm of the curvature vector with respect to arclength.)

THEOREM 10. [**EWW02**] *Let M be an immersed minimal surface (possibly with branch points[5]) in $\mathbf{R}^n$ bounded by a smooth embedded curve Γ whose total curvature is at most 4π. Then M is smoothly embedded (without branch points).*

PROOF. For simplicity, we give the proof for unbranched surfaces and for curves of total curvature strictly less than 4π, and we prove only that $M \setminus \Gamma$ has no points of self-intersection. Let p be a point in $M \setminus \Gamma$. Let C and E be the cone and the exterior cone over Γ with vertex p, as in the extended monotonicity theorem 5. It is left as an exercise to the reader to show that

$$\Theta(C) \leq \frac{1}{2\pi} \,(\text{the total curvature of } \Gamma)\,.$$

(This is just a fact about the geometry of curves.) Thus by hypothesis, $\Theta(C) < 2$. If we let $\tilde{M} = M \cup E$ as in the Extended Monotonicty Theorem 5, then $\Theta(\tilde{M}) = \Theta(C) < 2$, so

$$\Theta(M, p) = \Theta(\tilde{M}, p) \leq \Theta(\tilde{M}) < 2.$$

Since $\Theta(M, p)$ is the number of sheets of M passing through p, we see that only one sheet passes through p. Since p is an arbitrary point in $M \setminus \Gamma$, we are done. □

We have not yet discussed branch points, but exactly the same argument rules out interior branch points (i.e., branch points not in Γ): one only needs to know the fact that the density of a minimal surface at a branch point is at least 2. That fact is easily proved using the Weierstrass representation (for example), which will

[4]Nitsche [**Nit73**] had proved that a curve of total curvature less than 4π bounds a unique minimal disk, and that the disk is smoothly immersed. Whether such a curve can bound a minimal surface of nonzero genus is an interesting open question. Such curves can bound minimal Möbius strips [**EWW02**]. Nitsche's uniqueness theorem was extended to curves of total curvature at most 4π by X. Li and Jost [**LJ94**].

[5]Branch points will be discussed in lecture 2.

be discussed in Lecture 2. A similar argument rules out branch points and self-intersections at the boundary.

COROLLARY 11 (Farey-Milnor Theorem). *If Γ is a smooth, simple closed curve in $\mathbf{R}^3$ with total curvature at most 4π, then Γ is unknotted.*

PROOF. Let M be a least-area disk bounded by Γ. (The disk exists by the Douglas-Rado Theorem, which will be discussed in Lecture 4.) By theorem 10, M is a smooth embedded disk. But for any smoothly embedded curve, bounding a smooth embedded disk implies (indeed, is equivalent to) being unknotted. (To see the implication, let $F : \overline{D} \to M$ be a smooth, conformal parametrization of M by the unit disk in $\mathbf{R}^2$. Then

$$h : (0,1] \times \partial D \to \mathbf{R}^3,$$
$$h : (t,p) = F(tp)$$

provides an isotopy from Γ to small, nearly circular curves, which are clearly unknotted.) □

Theorem 10 is sharp: there exist smooth embedded curves, including unknotted ones, that have total curvature slightly larger than 4π and that bound many immersed minimal surfaces (see theorem 46). The Farey-Milnor Theorem is also sharp: consider, for example, a trefoil knot that is a slight perturbation of a twice-traversed circle.

One can also define total curvature for arbitrary continuous curves. Theorem 10 and corollary 11 remain true for continuous simple closed curves with total curvature at most 4π. (The surface M will be embedded, though of course it will in general be smoothly embedded only away from its boundary.) See [**EWW02**].

The isoperimetric inequality

The following fundamental theorem was proved by Allard [**All72**] (with a constant that was allowed to depend on dimension n of the ambient space) and by Michael and Simon [**MS73**]:

THEOREM 12 (Isoperimetric inequality). *Let M be a smooth, compact m-dimensional submanifold of $\mathbf{R}^n$. Then*

$$|M| \le c_m \left(|\partial M| + \int_M |H|\, dS \right)^{m/(m-1)}$$

where M is the m-dimensional measure of M and $|\partial M|$ is the $(m-1)$-dimensional measure of ∂M.

See [**CM11**, 3§2], [**Sim83**, §18], or the original papers for the proof.[6]

> **Exercise**: Prove the isoperimetric inequality for a two-dimensional surface with connected boundary. (Use theorem 2.)

The value of the best constant in the isoperimetric inequality, even in the case $H \equiv 0$ of minimal surfaces, is an interesting open problem. For minimal surfaces, it is conjectured that the best constant is attained by a ball in an m-dimensional plane.

[6]In some of the references, the inequality is stated as a Sobolev inequality for a function u that is compactly supported on $M \setminus \partial M$. The isoperimetric inequality follows by taking a suitable sequence of such u's that converge to 1 on $M \setminus \partial M$.

Almgren [**Alm86**] proved the conjecture assuming that M is area-minimizing. For two-dimensional minimal surfaces, the conjecture has been proved in some cases, such as when ∂M has at most two connected components [**LSY84**]. See [**CS09**] for some more recent developments.

LECTURE 2

Two-Dimensional Minimal Surfaces

The theory of two-dimensional surfaces has many features that do not generalize to higher dimensional manifolds. For example, every two-dimensional surface with a smooth Riemannian metric admits local isothermal coordinates, i.e., can be parametrized locally by conformal maps from domains in the plane.[1]

Relation to harmonic maps

THEOREM 13. *Let $F : \Omega \subset \mathbf{R}^2 \to \mathbf{R}^n$ be a conformal immersion. Then $F(\Omega)$ is minimal if and only if F is harmonic.*

PROOF. One way to show this is to calculate that the mean curvature H is equal to the Laplacian $\Delta_g F$ of F with respect to the pullback by F of the metric on $\mathbf{R}^n$. (This is also true for immersions of m-dimensional manifolds M into general Riemannian manifolds.) Thus M is minimal if and only if F is harmonic with respect to the metric g. The theorem follows immediately because, for two-dimensional surfaces, harmonic functions remain harmonic under conformal change of metric on the domain. □

COROLLARY 14. *Every two-dimensional C^2 minimal surface in $\mathbf{R}^n$ is real-analytic.*

This is also true for m-dimensional minimal submanifolds of $\mathbf{R}^n$, but by a different proof.

THEOREM 15 (Convex hull property). *Let M be a two-dimensional minimal surface in $\mathbf{R}^n$.*

(1) *If $\phi : \mathbf{R}^n \to \mathbf{R}$ is a C^2 function, then $\phi|M$ cannot have a local maximum at any point of $M \setminus \partial M$ where $D^2\phi$ is positive definite.*
(2) *If M is compact, then it lies in the convex hull of its boundary.*

PROOF. Let $p \in M \setminus \partial M$ be a point at which $D^2\phi$ is positive definite. Let $F : D \subset \mathbf{R}^2 \to M$ with $F(0) = p$ be a conformal (and therefore harmonic) parametrization of a neighborhood of M. Then (using $y_1, \dots, y_n$ and x_1, x_2 as coordinates for $\mathbf{R}^n$ and $\mathbf{R}^2$ and summing over repeated indices) one readily calculates by the chain rule that

$$\Delta(\phi \circ F) = \frac{\partial \phi}{\partial y_i} \cdot \Delta F_i + \frac{\partial^2 \phi}{\partial y_i \, \partial y_j} \frac{\partial F_i}{\partial x_k} \frac{\partial F_j}{\partial x_k} \geq \lambda |DF|^2 \tag{9}$$

[1]Existence of isothermal coordinates was proved by Gauss for analytic metrics in 1822, and by Korn and by Lichtenstein for $C^{1,\alpha}$ metrics in 1916. A number of other proofs have been given subsequently. See, for example, lemma 1 and the paragraph following its proof in [**DK81**].

since F is harmonic, where λ is the lowest eigenvalue of $D^2\phi$. This is strictly positive at a point where $D^2\phi$ is positive definite. Consequently $(\partial/\partial x_k)^2(\phi \circ F)$ must be positive for $k = 1$ and/or $k = 2$, which proves (1).

To prove (2), it suffices to show that that if ∂M lies in a closed ball, then so does M, since the convex hull of ∂M is the intersection of all such balls. If this failed for some ball $\mathbf{B}(p, r)$, then the function $x \in M \mapsto |x - p|^2$ would attain its maximum at an interior point of M, contradicting (1). □

Theorem 15 is also true for m-dimensional minimal submanifolds of $\mathbf{R}^n$ by essentially the same proof. (In particular, (9) is true at a point if Δ denotes the Laplacian with respect to the metric on M induced from $\mathbf{R}^n$ and if $x_1, \dots, x_m$ are normal coordinates at that point.) Theorem 15 is a special case of much more general maximum principles for (possibly singular) minimal varieties. See for example [**Whi10**].

Exercises:

(1) Suppose that M is a compact, simply connected, two-dimensional minimal surface in $\mathbf{R}^n$ and that $\mathbf{B} \subset \mathbf{R}^n$ is a ball. Prove that $M \cap \mathbf{B}$ is also simply connected.

(2) In theorem 15, show that $\phi|M$ cannot have a local maximum at any point of $M \setminus \partial M$ at which the average of the two smallest eigenvalues of $D^2\phi$ is strictly positive. (Hint: show that the inequality (9) for harmonic, conformal maps F remains true if we let λ denote the average of the smallest two eigenvalues of $D^2\phi$.)

(3) Show that if $M \subset \mathbf{R}^3$ is a connected minimal surface contained in a region N with smooth boundary, then $M \setminus \partial M$ cannot touch ∂N at any point where the mean curvature of ∂N is nonzero and points into N. (Hint: Let ϕ be the signed distance function to ∂N such that $\phi < 0$ in the interior of N.)

(4) (For readers familiar with the strong maximum principle in partial differential equations.) Show in exercise (3) that if the mean curvature vector of ∂N is a nonnegative multiple of the inward unit normal at all points and that if $M \setminus \partial M$ touches ∂N, then M is contained in ∂N.

All four exercises generalize to m-dimensional minimal submanifolds. (In exercise (1), replace "simply connected" by "having trivial $(m-1)^{\text{th}}$ homology". In exercise (2), replace "smallest two eigenvalues" by "smallest m eigenvalues".) Also, theorem 15 and the exercises remain true for branched minimal surfaces (which will be discussed in lecture 2).

Conformality of the Gauss map

Recall that if M is an oriented surface in $\mathbf{R}^3$, then its Gauss map $\mathbf{n} : M \to \mathbf{S}^2$ is the map that maps each point in M to the unit normal to M at that point.

THEOREM 16. *Let M be a minimal surface in $\mathbf{R}^3$. Then M is minimal if and only if the Gauss map $\mathbf{n} : M \to \mathbf{S}^2$ is almost conformal and orientation-reversing.*

Recall that a map $F : M \to N$ between Riemannian manifolds is almost conformal provided

$$DF(p)\mathbf{u} \cdot DF(p)\mathbf{v} \equiv \lambda(p)\mathbf{u} \cdot \mathbf{v}$$

for all $\mathbf{u}, \mathbf{v} \in \operatorname{Tan}_p M$, where λ is a non-negative function. The map is conformal if λ is everywhere positive.

In theorem 16, the orientation on $\mathbf{S}^2$ is the standard orientation, i.e., the orientation given by the outward unit normal.

PROOF. Let $\mathbf{e}_1$ and $\mathbf{e}_2$ be the principal directions of M at $p \in M$. Then $\{\mathbf{e}_1, \mathbf{e}_2\}$ is an orthonormal basis for $\operatorname{Tan}_p M$ and also for $\operatorname{Tan}_{\mathbf{n}(p)} \mathbf{S}^2$. With respect to this basis, the matrix for $D\mathbf{n}(p)$ is

$$\begin{bmatrix} \kappa_1 & 0 \\ 0 & \kappa_2 \end{bmatrix}$$

where κ_1 and κ_2 are the principal curvatures. □

Total curvature

For any surface $M \subset \mathbf{R}^3$, the Gauss curvature $K = \kappa_1 \kappa_2$ is the signed Jacobian of the Gauss map. The **total absolute curvature** (which we will call the total curvature, for short) of M is

$$TC(M) = \int_M |K| \, dS,$$

which is equal to the area (counting multiplicity) of the image of M under the Gauss map:

$$TC(M) = \operatorname{area}(\mathbf{n}, M) = \int_{p \in \mathbf{S}^2} \#\mathbf{n}^{-1}(p) \, dp.$$

If M is minimal, $K \leq 0$, so

$$TC(M) = \int_M |K| \, dS = -\int_M K \, dS.$$

THEOREM 17 (Osserman [**Oss63**]). *Let $M \subset \mathbf{R}^3$ be a complete, connected, orientable minimal surface of finite total curvature:*

$$TC(M) = \int_M |K| \, dS = \int_M (-K) \, dS < \infty.$$

Then

1. *M is conformally equivalent to a compact Riemann surface minus finitely many points:*
$$M \cong \Sigma \setminus \{p_1, \ldots, p_k\}.$$
2. *The Gauss map extends analytically to the punctures.*
3. *There is a nonnegative integer m such that for almost every $v \in \mathbf{S}^2$, exactly m points in M have unit normal $\mathbf{n} = v$.*
4. *The total curvature of M is a equal to $4\pi m$.*
5. *M is proper.*

Properness means that if we go off to infinity in M, we also go off to infinity in $\mathbf{R}^n$. More precisely, we say that a sequence $p_i \in M$ diverges in M if no subsequence converges with respect to the induced arclength metric on M to a point in M. We say that a sequence diverges in $\mathbf{R}^n$ if no subsequence converges with respect to the metric on $\mathbf{R}^n$. We say that M is a proper in $\mathbf{R}^n$ if every sequence $p_i \in M$ that diverges in M also diverges in $\mathbf{R}^n$.

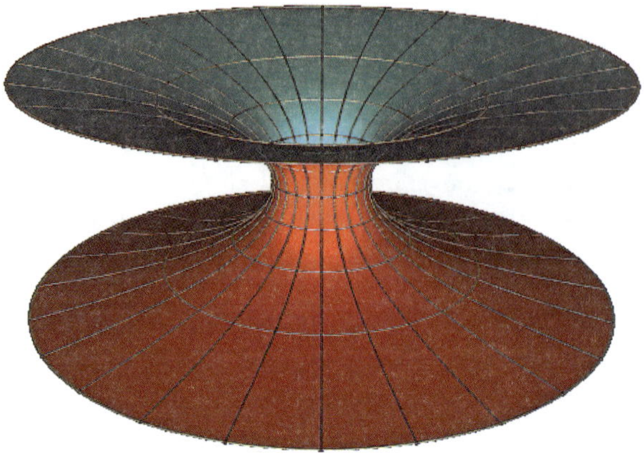

FIGURE 1. The catenoid has total curvature 4π.

For example, the curve $C = \{(x, \sin(\pi/x) : x > 0\}$ is not proper in $\mathbf{R}^2$: the sequence $(1/n, 0) \in C$ diverges in C but converges in $\mathbf{R}^2$. (The disconnected curve $C \cup Y$, where Y is the y-axis, also is not a proper submanifold of $\mathbf{R}^2$ for the same reason.)

The catenoid M with a vertical axis of rotational symmetry (figure 1) provides a good example of Osserman's theorem. The Gauss map is a conformal diffeomorphism from M to $\mathbf{S}^2 \setminus \{NP, SP\}$, where $NP = (0,0,1)$ and $SP = (0,0,-1)$ are the north and south poles on $\mathbf{S}^2$. In particular, M is conformally diffeomorphic (in this case by the map $\mathbf{n}$) to a twice-punctured sphere. The Gauss map extends continuously to the punctures. The total curvature is the area of the Gaussian image, namely 4π.

PROOF OF OSSERMAN'S THEOREM. The first assertion is a special case of an intrinsic theorem due to Huber [**Hub57**]: if M is a complete, connected surface such that

$$\int_M K^- \, dS < \infty$$

then M is conformally a punctured Riemann surface. Here

$$K^- = \begin{cases} |K| & \text{if } K \le 0 \\ 0 & \text{if } K > 0. \end{cases}$$

To prove the second assertion, let us (for the moment) orient $\mathbf{S}^2$ by the inward-pointing unit normal, so that the Gauss map becomes orientation preserving. Let $U \subset \Sigma$ be a neighborhood of one of the punctures, p. By Picard's Theorem, either $\mathbf{n} : U \setminus \{p\} \to \mathbf{S}^2 \cong \mathbf{C} \cup \{\infty\}$ is meromorphic at p (and therefore extends to p), or $\mathbf{n} : U \setminus \{p\} \to \mathbf{S}^2$ takes all but two values in $\mathbf{S}^2$ infinitely many times. The latter implies that $\int_U |K| \, dS = \infty$, a contradiction. Thus $\mathbf{n}$ extends continuously (indeed analytically) to U.

We have shown that the Gauss map $\mathbf{n} : \Sigma \to \mathbf{S}^2$ is holomorphic (with respect to orientation on $\mathbf{S}^2$ induced by the inward unit normal.) Let m be its mapping degree. Then assertion (3) holds by standard complex analysis or differential topology.

The fourth assertion (that the total curvature is $4\pi m$) follows immediately, since

$$TC(M) = \int_{\mathbf{S}^2} \#\mathbf{n}^{-1}(\cdot)\, dS.$$

The last assertion (properness) can be proved using the Weierstrass representation (discussed below). Alternatively, one can show that if $S \subset \mathbf{R}^3$ is a complete surface diffeomorphic to a closed disk minus its center and if the slope of $\mathrm{Tan}(S,p)$ is uniformly bounded, then S is proper in $\mathbf{R}^3$; see [**Whi87a**]. One applies this result to a small neighborhood in Σ of a puncture, on which one can assume (by rotating) that the unit normal is very nearly vertical. □

REMARKS. (1). Every multiple of 4π does occur as the total curvature of such a surface. (2). For a proof of generalization of Osserman's Theorem that does not use Huber's Theorem, see [**Whi87a**].

Osserman's Theorem has an extension, due to Chern and Osserman [**CO67**], to two-dimensional surfaces in $\mathbf{R}^n$. All the conclusions remain true, except that the total curvature[2] is a multiple of 2π rather than of 4π. And every multiple of 2π does occur as the total curvature of such a surface. For example,

$$\{(z,w) \in \mathbf{C}^2 \cong \mathbf{R}^4 : w = z^n\}.$$

is a complete, embedded minimal surface with total curvature $2\pi(n-1)$. The surface is minimal and indeed area minimizing by the Federer-Wirtinger theorem (theorem 39). To see that it has total curvature $2\pi(n-1)$, let M_r be the portion of the surface with $\{|w| \le r\}$. Note that if r is large, then ∂M_r is very nearly a circle of radius r traversed n times. Thus by the Gauss Bonnet theorem,

$$2\pi n \cong \int_{\partial M_r} k\, ds = 2\pi - \int_{M_r} K\, dS.$$

Letting $r \to \infty$ gives $2\pi n = 2\pi + TC(M)$.

Theorem 4 characterized the plane by its density at infinity. Using Osserman's Theorem, we can give another characterization of the plane:

COROLLARY 18. *If $M \subset \mathbf{R}^3$ is a complete, orientable minimal surface of total curvature $< 4\pi$, then M is a plane.*

PROOF. If the total curvature is less than 4π, it must be 0, so $K \equiv 0$. But for a minimal surface in $\mathbf{R}^3$, $K(p) = 0$ implies that the principal curvatures at p are 0. □

Similarly, using the Chern-Osserman Theorem, one sees that a complete, orientable minimal surface in $\mathbf{R}^n$ with total curvature $< 2\pi$ must be a plane.

As will be explained in lecture 3, corollary 18 implies a useful curvature estimate (theorem 23) for minimal surfaces.

---

[2]Here we can take the total curvature to be the integral of the absolute value of the scalar curvature. Since the scalar curvature is everywhere nonpositive, this is equal to minus the integral of the scalar curvature. Since M is minimal, the scalar curvature is equal to $-\frac{1}{2}|A|^2$, so we could also define the total curvature to be the integral of $\frac{1}{2}|A|^2$,where $|A|$ is the norm of the second fundamental form, so we could also define the total curvature to be the integral of $\frac{1}{2}|A|^2$.

The Weierstrass Representation

We have seen that immersed minimal surfaces in $\mathbf{R}^3$ are precisely those that can be parametrized locally by conformal, harmonic maps $F : \Omega \subset \mathbf{R}^2 \to \mathbf{R}^3$. Following work of Riemann, Weierstrass and Enneper[3] independently found a nice way to generate all such F.

Write $z = x + iy$ in $\mathbf{R}^2$, and

$$\frac{\partial}{\partial z} = \frac{1}{2}\left(\frac{\partial}{\partial x} - i\frac{\partial}{\partial y}\right) \quad , \quad \frac{\partial}{\partial \bar{z}} = \frac{1}{2}\left(\frac{\partial}{\partial x} + i\frac{\partial}{\partial y}\right).$$

Note that if u is a map from $\Omega \subset \mathbf{R}^2 \cong \mathbf{C}$ to $\mathbf{C}$ (or more generally to $\mathbf{C}^n$), then $u_{\bar{z}} = 0$ if and only if u is holomorphic. (The real and imaginary parts of the equation $u_{\bar{z}} = 0$ are precisely the Cauchy-Riemann equations for u.)

Note also that

$$\frac{\partial}{\partial \bar{z}}\frac{\partial}{\partial z} = \frac{1}{4}\left(\frac{\partial}{\partial x} + i\frac{\partial}{\partial y}\right)\left(\frac{\partial}{\partial x} - i\frac{\partial}{\partial y}\right) = \frac{1}{4}\left(\frac{\partial}{\partial x}^2 + \frac{\partial}{\partial y}^2\right) = \frac{1}{4}\Delta.$$

Thus

$$F \text{ is harmonic} \iff F_{z\bar{z}} = 0 \iff F_z \text{ is holomorphic.}$$

Concerning conformality, note that if we extend the Euclidean inner product from $\mathbf{R}^n$ to $\mathbf{C}^n$ by making it complex linear in both arguments, then

$$\begin{aligned} F_z \cdot F_z &= \frac{1}{4}(F_x - iF_y) \cdot (F_x - iF_y) \\ &= \frac{1}{4}\left(|F_x|^2 - |F_y|^2 - 2iF_x \cdot F_y\right), \end{aligned}$$

so the real and imaginary parts of $F_z \cdot F_z$ are

$$\Re(F_z \cdot F_z) = |F_x|^2 - |F_y|^2 \quad , \quad \Im(F_z \cdot F_z) = 2F_x \cdot F_y.$$

Consequently, F is almost conformal if and only if $F_z \cdot F_z \equiv 0$.

We have proved:

THEOREM 19. *Let $F : \Omega \subset \mathbf{C} \to \mathbf{R}^n$ be a C^2 map. Then F is harmonic if and only if F_z is holomorphic, and F is almost conformal if and only if $F_z \cdot F_z \equiv 0$.*

Let $\phi = 2F_z = (\phi_1, \phi_2, \phi_3) \in \mathbf{C}^3$. Our goal is to find holomorphic ϕ such that $\phi \cdot \phi \equiv 0$. We then recover F by

$$F = \Re\left(\int 2F_z\, dz\right) = \Re\left(\int \phi\, dz\right).$$

Note that $\phi(z) = 2F_z = F_x - iF_y$ determines the oriented tangent plane to M at $F(z)$, and therefore the image $\mathbf{n}(z)$ of $F(z)$ under the Gauss map, and thus the image $g(z) \in \mathbf{R}^2 \cong \mathbf{C}$ of $\mathbf{n}(z)$ under stereographic projection. Indeed, one can check that

$$g(z) = \frac{\phi_3}{\phi_1 - i\phi_2}.$$

[3]In the interests of brevity, I use the conventional name "Weierstrass Representation" rather than the more accurate "Enneper-Weierstrass Representation".

FIGURE 2. The catenoid has Weierstrass data $\eta(z) = 1/z\,dz$ and $g(z) = z$ on $\mathbf{C} \setminus \{0\}$. The helicoid has Weierstrass data $\eta(z) = dz$ and $g(z) = e^{iz}$ on $\mathbf{C}$.

One can solve for ϕ_1 and ϕ_2 in terms of g and ϕ_3:

$$\phi_1 = \frac{1}{2}(g^{-1} - g)\phi_3,$$
$$\phi_2 = \frac{i}{2}(g^{-1} + g)\phi_3.$$

THEOREM 20 (Weierstrass Representation). *Let $\Omega \subset \mathbf{C}$ be simply connected, let ϕ_3 and g be a holomorphic function and a meromorphic function on Ω, and let η be the one form $\phi_3(z)\,dz$. Suppose also that wherever g has a pole or zero of order m, the function ϕ_3 has a zero of order $\geq 2m$. Then*

$$F(z) = \Re \int^z \left(\frac{1}{2}(g^{-1} - g), \frac{i}{2}(g^{-1} + g), 1\right)\eta \tag{10}$$

is a harmonic, almost conformal mapping of Ω into $\mathbf{R}^3$.

Furthermore, every harmonic, almost conformal map $F : \Omega \to \mathbf{R}^3$ arises in this way, unless the image of F is a horizontal planar region.

The function g may take the value ∞ because the surface may have points where the unit normal is $(0,0,1)$. Thus g may have poles. At the poles and zeroes of g, F would have poles unless ϕ_3 has zeroes to counteract the poles of $g^{-1} \pm g$. Hence the condition about orders of poles and zeroes.

The "unless" proviso is needed because if $F(\Omega)$ is a horizontal planar region, then $g \equiv 0$ or $g \equiv \infty$ and thus (10) does not make sense.

More generally, η and g can be any holomorphic differential and any meromorphic function on any Riemann surface Ω. But if Ω is not simply connected, the expression (10) may be well-defined only on the universal cover of Ω. To be precise, the Weierstrass representation (10) gives a mapping of Ω (rather than of its universal cover) if and only if the closed one forms

$$\frac{1}{2}(g^{-1} - g)\eta, \ \frac{i}{2}(g^{-1} + g)\eta, \text{ and } \eta$$

have no real periods.

Since g and ϕ_3 (or η) determine the surface, all geometric quantities can be expressed in terms of them. For example, the conformal factor λ is given by

$$\lambda = |F_x| = |F_y|,$$

so

$$|\phi| = |F_x - iF_y| = \sqrt{2}\lambda,$$

and thus (calculating $|\phi|$ from (10)) we have

$$\lambda = \frac{1}{\sqrt{2}}|\phi| = \frac{(|g|^{-1} + |g|)}{2}|\phi_3|. \tag{11}$$

That is,

$$ds^2 = \left[\frac{(|g|^{-1} + |g|)}{2}\right]^2 |\phi_3|^2 \, |dz|^2 = \left[\frac{(|g|^{-1} + |g|)}{2}\right]^2 |\eta|^2$$

One can also calculate the Gauss curvature:

$$K = -\left[\frac{4|g'|}{|\phi_3 g|(|g|^{-2} + |g|^2)^2}\right]^2 = -\left[\frac{4}{(|g|^{-2} + |g|^2)^2}\left|\frac{dg}{g\eta}\right|\right]^2$$

(Note: dg and $g\eta$ are both meromorphic 1-forms, so their ratio is a meromorphic function.) The points where F fails to be conformal are called **branch points**. Using the expression (11) for the conformal factor, we can identify the branch points:

PROPOSITION 21. *Suppose in theorem 20 that g has a pole or zero of order $m \geq 0$ at p and that ϕ_3 has a zero of order $k \geq 2m$ at p. Then F is an immersion at p if and only $k = 2m$. Thus F has a branch point at p if and only if $k > 2m$.*

The difference $k - 2m$ is called the **order** of the branching at p. It is not hard to show that for sufficiently small $\epsilon > 0$, the density of $F(\mathbf{B}(p, \epsilon))$ at $F(p)$ is 1 plus the order of branching at p.

The geometric meaning of the Weierstrass data

As explained above, the function g in the Weierstrass representation has a simple geometric meaning: it is the Gauss map (or, more precisely, the Gauss map followed by stereographic projection to $\mathbf{R}^2 \cup \{\infty\} \cong \mathbf{C} \cup \{\infty\}$).

As for ϕ_3, note by theorem 20 and the discussion preceding it that

$$\phi_3 = 2\frac{\partial F_3}{\partial z} = \frac{\partial F_3}{\partial x} - i\frac{\partial F_3}{\partial y},$$

where $F = (F_1, F_2, F_3)$. In other words, ϕ_3 encodes the derivative of the height function F_3 with respect to the parametrization. More generally, as mentioned above, if the domain of F is a Riemann surface, we should think of the Weierstrass data as being g together with a holomorphic one form η (corresponding to $\phi_3(z)\, dz$.) In this case,

$$\eta = 2\,\partial F_3,$$

where ∂ is the Dolbeault operator which (in any local holomorphic chart) is given by $\partial(\cdot) = \frac{\partial}{\partial z}(\cdot)\, dz$.

It can be difficult to determine for a particular g and ϕ_3 (or g and η) whether the real periods vanish and (if they do vanish) whether the resulting surface is embedded. For those reasons, great ingenuity is often required to prove the existence of specific kinds of embedded, genus g surfaces using the Weierstrass representation.

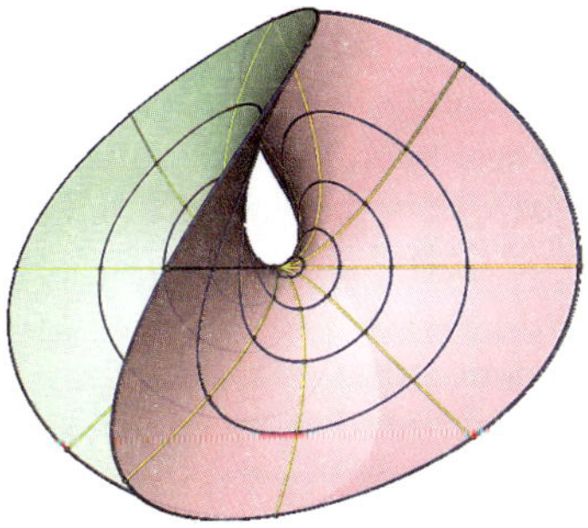

FIGURE 3. Enneper's surface has Weierstrass data $\eta(z) = z\,dz$ and $g(z) = z$. The surface is intrinsically rotationally symmetric about the point $z = 0$.

Part of the discussion above carries over without change to two-dimensional minimal surfaces in $\mathbf{R}^n$. In particular, $F : \Omega \subset \mathbf{R}^n$ is harmonic and almost conformal if and only F can be written as

$$\Re\left(\int \phi\, dz\right)$$

where $\phi = (\phi_1, \ldots, \phi_n) : \Omega \to \mathbf{C}^n$ is a holomorphic map such that $\phi \cdot \phi \equiv 0$. Of course we can use the equation $\phi \cdot \phi \equiv 0$ to express ϕ in terms of $(n-1)$ holomorphic functions; those $(n-1)$ functions can then be chosen more-or-less arbitrarily. See [**Oss86**, §12] for more details.

As an example of how the Weierstrass representation gives nontrivial information about a surface, consider Enneper's surface M (figure 3), i.e., the surface with Weierstrass data $\eta(z) = \phi_3(z)\,dz = z\,dz$ and $g(z) = z$ on the entire plane. As suggested by the picture, there is a finite group of congruences of M, i.e., of isometries of $\mathbf{R}^3$ that map M to itself. Though it is not evident from the picture, in fact M has an infinite group of intrinsic isometries: M is intrinsically rotationally symmetric about $z = 0$, since the metric depends only on $|\phi_3|$ and $|g|$, which in this case are both equal to $|z|$.

Rigidity and Flexibility

A surface $M \subset \mathbf{R}^3$ is **flexible** if it can be deformed (non-trivially) through a one-parameter family of smooth isometric immersions. Otherwise it is **rigid**. (Here "trivially" means "by rigid motions of $\mathbf{R}^3$".) For example, a flat rectangle (or, more generally, any planar domain) is flexible, as can be demonstrated by rolling up a piece of paper. According to a classical theorem, every smooth, closed, convex surface in $\mathbf{R}^3$ is rigid. Whether there exists a flexible smooth, closed, non-convex surface M is a long-open problem.

Is a round hemisphere rigid? Perhaps there is no intuitive reason why it should or should not be rigid. However, it (or, more generally, any proper closed subset of any closed, uniformly convex surface) is flexible, as explained (if the closed surface is a sphere) in [**HCV52**, §32.10]. On the other hand, in the surface $z = (x^2 + y^2)^3$ (for example), arbitrarily small neighborhoods of the origin are rigid [**Usm96**]. So whether particular surfaces are rigid or flexible can be rather subtle.

Now we impose an extra condition that makes flexing a surface much more difficult: is there a non-trivial one-parameter family $F_t : M \to \mathbf{R}^3$ of smooth isometric immersions of a surface such that the unit normal at each point remains constant? In other words, if $\mathbf{n}(p,t)$ is the unit normal to $F_t(M)$ at $F_t(p)$, we require that $\mathbf{n}(p,t)$ be independent of t. If $M_0 = F_0(M)$ is a planar region, for example, the answer is "no", since in that case the only allowed deformations are translations and rotations about axes perpendicular to M_0.

Intuitively, non-trivial isometric deformations keeping the normals constant should be impossible. However, such deformation do exist for *every* simply connected, nonplanar minimal surface! In the Weierstrass representation, replace η by $e^{i\theta}\eta$ and let θ vary. The metric depends only on $|g|$ and $|\eta|$, so it does not change, and the Gauss map g also does not change. (The proof that the resulting deformations are nontrivial for nonplanar M is left as an exercise.)

For example, if we start with the helicoid, this gives an isometric deformation of the helicoid to the catenoid (covered infinitely many times). One may see animations of the deformation online, for example at

`http://www.foundalis.com/mat/helicoid.htm`,

`http://www.indiana.edu/~minimal/archive/Classical/Classical/AssociateCatenenoid/web/qt.mov`,

and

`http://virtualmathmuseum.org/Surface/helicoid-catenoid/helicoid-catenoid.mov`.

Incidentally, one can show that minimal surfaces are the only surfaces in $\mathbf{R}^3$ that can be isometrically deformed keeping the normals fixed, and that the one-parameter family obtained by replacing η by $e^{i\theta}\eta$ is the only such deformation of a minimal surface.

See [**HK97**] or [**Web05**] for more information about the Weierstrass Representation. (Note: in those works, the authors use dh to denote the holomorphic one-form η. If one thinks of h as the height function on the surface, then their dh is *not* the exterior derivative of h, but rather $2\,\partial h$. The exterior derivative of the height function is the real part of their dh.)

LECTURE 3

Curvature Estimates and Compactness Theorems

In many situations, one wants to take limits of minimal surfaces. For example, David Hoffman, Martin Traizet, and I needed to do this in our recent work on genus-g helicoids. For several centuries, the plane and the helicoid were the only known complete, properly embedded minimal surfaces in $\mathbf{R}^3$ with finite genus and with exactly one end.[1] Jacob Bernstein and Christine Breiner [**BB11**] proved (using work of Colding and Minicozzi) that any such surface other than a plane must be asymptotic to a helicoid at infinity. (Later Meeks and Peréz [**MP**] gave a different proof, also based on the work of Colding and Minicozzi.) Hence such a surface of genus g is called an embedded **genus-g helicoid**.

But do embedded genus-g helicoids exist for $g \neq 0$? In 1992, Hoffman, Karcher, and Wei [**KWH93**] used the Weierstrass Representation to prove the existence but not the embeddedness of a genus-one helicoid. In 2004, Hoffman, Weber, and Wolf [**WHW09**] proved existence of an embedded genus-1 helicoid as shown in figure 1. (See [**HW08**] for a different, somewhat shorter proof.) Although genus-2 examples were found numerically as early as 1993 (see figure 2), existence was not known rigorously until 2013, when Hoffman, Traizet, and I [**Hof05a**, **Hof05b**] proved existence of embedded genus-g helicoids for every positive integer g. In our proof, first we construct analogous surfaces in $\mathbf{S}^2 \times \mathbf{R}$ (which, oddly enough, turns out to be easier), and then we get examples in $\mathbf{R}^3$ by letting the radius of the $\mathbf{S}^2$ tend to infinity. Of course we need to know that the surfaces in $\mathbf{S}^2 \times \mathbf{R}$ converge smoothly (after passing to subsequences) to limit surfaces in $\mathbf{R}^3$.

In general, it is very useful to have compactness theorems: conditions on a sequence of minimal surfaces that guarantee existence of a smoothly converging subsequence.

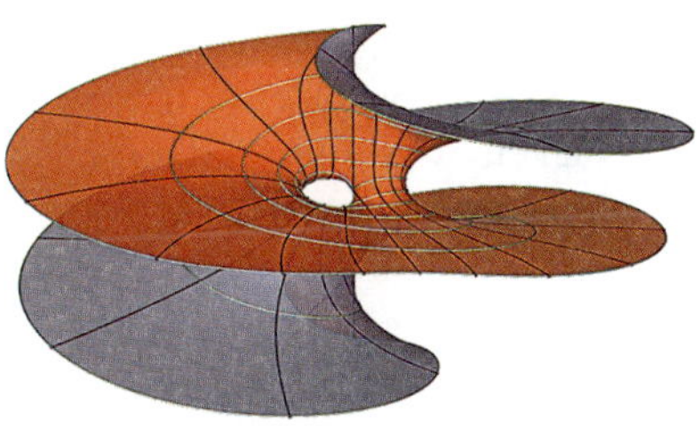

FIGURE 1. A genus-1 helicoid.

[1]For a complete minimal surface M properly immersed in $\mathbf{R}^n$, the number of ends is the limit as $r \to \infty$ of the number of connected components of $M \setminus \mathbf{B}(0,r)$. It follows from the convex hull property (theorem 15) that the number of those components cannot decrease as r increases, and thus that the limit exists.

FIGURE 2. A genus-2 helicoid. (Picture by Martin Traizet (1993)).

THEOREM 22 (basic compactness theorem). *Let M_i be a sequence of minimal submanifolds of $\mathbf{R}^n$ (or of a Riemannian manifold) such that second fundamental forms are uniformly bounded.*

Then locally there exist smoothly converging subsequences.

In particular, if $p_i \in M_i$ is a sequence bounded in $\mathbf{R}^N$ and if $\text{dist}(p_i, \partial M_i) \geq R > 0$, then (after passing to a subsequence),

$$M_i \cap \mathbf{B}(p_i, R)$$

converges smoothly to a limit minimal surface M^*. Here dist is **intrinsic** distance in M_i, and $\mathbf{B}(p_i, r)$ is the open geodesic ball of radius r in M_i.

The theorem (if interpreted properly) is also true when dist is exterior distance and $\mathbf{B}(p, r)$ is the extrinsic open ball. In this case the conclusion is that there is an r with $0 < r < R$ such that the connected component of $M_i \cap \mathbf{B}(p_i, r)$ containing p_i converges smoothly (after passing to a subsequence) to a limit surface M^*. Furthermore, under rather mild hypotheses, the various M^* will fit together to form a nice surface. For example, if the M_i are properly immersed in an open set Ω and if the areas of the M_i in compact subsets of Ω are uniformly bounded, then a subsequence of the M_i converges smoothly in Ω to a limit surface that is properly immersed in Ω. Even without area bounds, if the M_i are properly embedded hypersurfaces in an open set Ω, then the M^* fit together to form a lamination[2] of Ω in which the leaves are smooth minimal hypersurfaces.

PROOF OF THE BASIC COMPACTNESS THEOREM IN $\mathbf{R}^n$. By scaling, we can assume that the principle curvatures are bounded by 1, and that $\text{dist}(p_i, \partial M_i) \geq \pi/2$.

We can assume the p_i converge to a limit p and that $\text{Tan}(M_i, p_i)$ converge to a limit plane. Indeed, in $\mathbf{R}^n$, we can assume by translating and rotating that $p_i \equiv 0$ and that $\text{Tan}(M_i, p_i)$ is the horizontal plane through 0.

[2]A lamination is like a foliation except that gaps are allowed. For example, if $S \subset \mathbf{R}$ is an arbitrary closed set, then the set of planes $\mathbf{R}^2 \times \{z\}$ where $z \in S$ forms a lamination of $\mathbf{R}^3$ by planes.

For each i, let S_i be the connected component of

$$M_i \cap (\mathbf{B}^m(0,1/2) \times \mathbf{R}^{N-m})$$

containing 0. The hypotheses imply that S_i is the graph of a function

$$F_i : \mathbf{B}^m(0,1/2) \to \mathbf{R}^{N-m}$$

where the C^2 norm of the F_i is uniformly bounded. Hence by Arzelá-Ascoli, we may assume (by passing to a subsequence) that the F_i converge in $C^{1,\alpha}$ to a limit function F.

So far we have not used minimality. Since the surfaces M_i are minimal, the F_i are solutions to an elliptic partial differential equation (or system of equations if $n > m+1$), the minimal surface equation (or system). According to the theory of such equations, convergence in $C^{1,\alpha}$ on $\mathbf{B}^m(0,1/2)$ implies convergence in $C^{k,\alpha}$ on $\mathbf{B}^m(0,1/2-\epsilon)$ (for any k and ϵ).

We have proved convergence in sufficiently small geodesic balls. We leave it to the reader to piece such balls together to get the convergence of the $M_i \cap \mathbf{B}(p_i,R)$. □

This theorem indicates the importance of curvature estimates: curvature estimates for a class of minimal surfaces imply smooth subsequential convergence for sequences of such surfaces.

The 4π curvature estimate

THEOREM 23 (4π curvature estimate [**Whi87b**]). *For every $\lambda < 4\pi$, there is a $C < \infty$ with the following property. If $M \subset \mathbf{R}^3$ is an orientable minimal surface with total curvature $\leq \lambda$, then*

$$|A(p)| \operatorname{dist}_M(p, \partial M) \leq C.$$

Here $|A(p)|$ is the norm of the second fundamental form of M at p, i.e, the square root of the sum of the squares of the principal curvatures at p, and dist_M denotes intrinsic distance in M.

The theorem is false for $\lambda = 4\pi$, since the catenoid has total curvature 4π and is not flat. (Earlier, Choi and Schoen [**CS85**] proved that there exists a $\lambda > 0$ and a $C < \infty$ for which the conclusion holds.)

PROOF. It suffices to prove the theorem when M is a smooth, compact manifold with boundary, since a general surface can be exhausted by such M.

Suppose the theorem is false. Then there is a sequence $p_i \in M_i$ of examples with total curvature $TC(M_i) \leq \lambda$ and with

$$|A_i(p_i)| \operatorname{dist}(p_i, \partial M_i) \to \infty. \tag{12}$$

(Throughout the proof, all distances are intrinsic, so we write dist rather than $\operatorname{dist}_{M_i}$.) We may assume that each p_i has been chosen in M_i to maximize the left side of (12). By translating and scaling, we may also assume that $p_i = 0$ and that $|A_i(p_i)| = 1$, and therefore that $\operatorname{dist}(0, \partial M_i) \to \infty$. We may also replace M_i by the geodesic ball of radius $R_i := \operatorname{dist}(0, \partial M_i)$ about 0.

Thus we have

$$|A_i(0)| = 1,$$
$$R_i = \operatorname{dist}(0, \partial M_i) \to \infty, \text{ and}$$
$$|A_i(x)| \operatorname{dist}(x, \partial M_i) \le \operatorname{dist}(0, \partial M_i).$$

Now $\operatorname{dist}(0,x) + \operatorname{dist}(x, \partial M_i) = \operatorname{dist}(0, \partial M_i) = R_i$, so

$$|A_i(x)| \le \frac{R_i}{\operatorname{dist}(x, \partial M_i)} = \frac{R_i}{R_i - \operatorname{dist}(0,x)} \le \frac{R_i}{R_i - r}$$

if $\operatorname{dist}(x, \partial M_i) \le r$.

We have shown for each r that

$$\sup_{\operatorname{dist}(x,0) \le r} |A_i(x)| \le \frac{R_i}{R_i - r} \to 1.$$

Hence the M_i converge smoothly by theorem 22 (the basic compactness theorem) to a complete minimal surface M with $|A_M(0)| = 1$.

Thus by corollary 18 (the corollary to Osserman's Theorem), $TC(M) \ge 4\pi$. However, $TC(M) \le \liminf_i TC(M_i) \le \lambda < 4\pi$, a contradiction. □

REMARK. The theorem is also true (with the same proof) in $\mathbf{R}^n$, but with 4π replaced by 2π. This is because Osserman's theorem is also true in $\mathbf{R}^n$, but with 4π replaced by 2π.

Theorem 23 can be generalized to manifolds in various ways. For example:

THEOREM 24. *Suppose that N is a 3-dimensional submanifold of Euclidean space $\mathbf{R}^n$ with the induced metric. Let*

$$\rho(N) = (\sup |A_N| + \sup |\nabla A_N|^{1/2})^{-1}.$$

For every $\lambda < 4\pi$, there is a $C = C_{\lambda,n} < \infty$ with the following property. If M is an orientable, immersed minimal surface in N and if the total absolute curvature of M is at most λ, then

$$|A_M(p)| \min\{\operatorname{dist}_M(p, \partial M), \rho(N)\} \le C$$

for all $p \in M$.

The proof is almost exactly the same as the proof of theorem 23. In particular, we get a sequence $p_i \in M_i \subset N_i$ with $|A_{M_i}(p_i)| = 1$ and with

$$\min\{\operatorname{dist}_{M_i}(p_i, \partial M_i), \rho(N_i)\} \to \infty.$$

The fact that $\rho_{N_i}(p_i) \to \infty$ means that the N_i are converging (in a suitable sense) to $\mathbf{R}^3$, so in the limit we get a complete minimal immersed surface M in $\mathbf{R}^3$, exactly as in the proof of theorem 23.

A general principle about curvature estimates

Recall that we have proved:

(1) A complete, nonflat minimal surface in $\mathbf{R}^3$ has total curvature $\ge 4\pi$.
(2) For any minimal $M \subset \mathbf{R}^3$ with $TC(M) \le \lambda < 4\pi$,

$$|A(p)| \operatorname{dist}_M(p, \partial M) \le C_\lambda.$$

We deduced (2) from (1). But conversely, (2) implies (1): if the M in (1) is complete, then $\text{dist}(p, \partial M) = \infty$, so $|A(p)| = 0$ according to (2). Thus (1) and (2) may be regarded as global and local versions of the same fact.

The equivalence of statements (1) and (2) is an example of general principle: any "Bernstein-type" theorem (i.e., a theorem asserting that certain complete minimal surfaces must be flat) should be equivalent to a local curvature estimate. Indeed, the Bernstein-type theorem in Euclidean space should imply a local curvature estimate in arbitrary ambient manifolds (as in theorem 24).

An easy version of Allard's Regularity Theorem

As an example of the general principle discussed above, consider the following:

(1) Global theorem: If $M \subset \mathbf{R}^n$ is a proper minimal submanifold without boundary and if $\Theta(M) \leq 1$, then M is a plane.

(2) Local estimate: there exist $\lambda > 1$, $\epsilon > 0$, and $C < \infty$ with the following property. If $M \subset \mathbf{R}^N$ is minimal, $\text{dist}(p, \partial M) \geq R$, and $\Theta(M, p, R) \leq \lambda$, then

$$\sup_{x \in \mathbf{B}(p, \epsilon R)} |A(q)| \, \text{dist}(q, \partial \mathbf{B}(p, \epsilon R)) \leq C,$$

where dist denotes Euclidean distance in $\mathbf{R}^n$. (Hence $|A(q)| \leq 2C/(R\epsilon)$ for $q \in \mathbf{B}(p, \epsilon R/2)$.)

We have already proved statement (1) (see theorem 4). Statement (2) is a special case of Allard's Regularity Theorem.

Clearly (2) implies (1), and proof that (1) implies (2) is very similar to the proof of the 4π curvature estimate (theorem 23). Furthermore, as suggested in the discussion of the general principle above, statement (1) implies a version of statement (2) in Riemannian manifolds.

Allard's theorem (see [**All72**, §8] or [**Sim83**, §23–§24]) is much more powerful than statement (2) because Allard does not assume that M is smooth: it can be any minimal variety ("stationary integral varifold"). He *concludes* that $M \cap \mathbf{B}(p, \epsilon R)$ is smooth (with estimates).[3]

Exercise: Provide the details of the proof that (1) implies (2).

Bounded total curvatures

For total curvatures that are bounded, but not bounded by some number $\lambda < 4\pi$, we have the following theorem, which says that for a sequence of minimal surfaces with uniformly bounded total curvatures, we get smooth subsequential convergence except at a finite set of points where curvature concentrates:

THEOREM 25 (Concentration Theorem [**Whi87b**]). *Suppose that $M_i \in \Omega \subset \mathbf{R}^n$ are two-dimensional, orientable, minimal surfaces, that $\partial M_i \subset \partial\Omega$, and that $TC(M_i) \leq \Lambda < \infty$.*

---

[3] I am describing Allard's theorem specialized to minimal varieties. His theorem is stated more generally for varieties with mean curvature in $\mathcal{L}^p$ where p can be any number larger than the dimension of the variety. In this generality, the conclusion is not that $M \cap \mathbf{B}(p, \epsilon)$ is smooth, but rather that it is $C^{1,\alpha}$ for suitable α. If the variety is minimal, smoothness then follows by standard PDE arguments. The easy version of Allard's Regularity Theorem first appeared in [**Whi05**].

Then (after passing to a subsequence) there is a set $S \subset \Omega$ of at most $\frac{\Lambda}{2\pi}$ points such that M_i converges smoothly in $\Omega \backslash S$ to a limit minimal surface M. The surface $M \cup S$ is a smooth, possibly branched minimal surface.

Now suppose that $\Omega \subset \mathbf{R}^3$. Then S contains at most $\frac{\Lambda}{4\pi}$ points. Also, if the M_i are embedded, then $M \cup S$ is a smooth embedded surface (with multiplicity, but without branch points.)

The theorem remains true (with essentially the same proof) in Riemannian manifolds.

To illustrate the concentration theorem, let M_k be obtained by dilating the catenoid by $1/k$. Then M_k converges to a plane with multiplicity 2, and the convergence is smooth except at the origin.

Of course, the concentration theorem is only useful if the hypothesis (uniformly bounded total curvatures of the M_i) holds in situations of interest. Fortunately, there are many situations in which the hypothesis does hold. For example, suppose the $M_i \subset \mathbf{R}^n$ all have the same finite topological type. Suppose also that the boundary curves ∂M_i are reasonably well-behaved:

$$\sup_i \int_{\partial M_i} |\kappa_{\partial M_i}|\, ds < \infty,$$

where $\kappa_{\partial M_i}$ denotes the curvature vector of the curve ∂M_i. (In other words, suppose that the boundary curves have uniformly bounded total curvatures.) Then the hypothesis $\sup_i TC(M_i) < \infty$ holds by the Gauss-Bonnet Theorem.

Proof of part of the concentration theorem in $\mathbf{R}^3$. Define measures μ_i on Ω by

$$\mu_i(U) = TC(M_i \cap U).$$

By passing to a subsequence, we can assume that the μ_i converge weakly to a limit measure μ with $\mu(\Omega) \leq \Lambda$.

Let S be the set of points p such that $\mu\{p\} \geq 4\pi$. Then $|S| \leq \frac{\Lambda}{4\pi}$, where $|S|$ is the number of points in S.

Suppose $x \in \Omega \setminus S$. Then $\mu\{x\} < \lambda < 4\pi$ for some λ. Thus there is a closed ball $\mathbf{B} = \mathbf{B}(x,r) \subset \Omega$ with $\mu(\mathbf{B}) < \lambda$. Hence

$$TC(M_i \cap \mathbf{B}) = \mu_i(\mathbf{B}) < \lambda$$

for all sufficiently large i. Consequently, $|A_i(\cdot)|$ is uniformly bounded on $\mathbf{B}(x, r/2)$ by the 4π curvature estimate (theorem 23).

Since $|A_i(\cdot)|$ is locally uniformly bounded in $\Omega \setminus S$, we get subsequential smooth convergence on $\Omega \setminus S$ by the basic compactness theorem 22.

Let $p \in S$. By translation, we may assume that p is the origin. Note that we can find $\mathbf{B}(0,\epsilon)$ for which $\mu(\mathbf{B}(0,\epsilon) \setminus \{0\})$ is arbitrarily small. It follows (by the 4π curvature estimate 23 and the basic compactness theorem 22) that if we dilate M about 0 by a sequence of numbers tending to infinity, a subsequence of the dilated surfaces converges smoothly on $\mathbf{R}^3 \setminus \{0\}$ to a limit minimal surface with total curvature 0, i.e., to a union of planes. By monotonicity, the number of those planes is finite. It follows that (for small r), the surface $M \cap (\mathbf{B}(0,r) \setminus \{0\})$ is topologically a finite union of punctured disks. In fact, it is not hard to show that the smooth subsequential convergence of the dilated surfaces to planes implies that the components of $M \cap (\mathbf{B}(0,r) \setminus \{0\})$ are not just topologically punctured disks, but actually conformally punctured disks.

Let $F : D\backslash\{0\} \to \mathbf{R}^3$ be a conformal (and therefore harmonic) parametrization of one of those punctured disks. Since isolated singularities of bounded harmonic functions are removable, in fact F extends smoothly to D. This proves that $M \cup S$ is a smooth (possibly branched) minimal surface.

Finally, one can show that if $M \subset \mathbf{R}^3$ is not an embedded surface (possibly with multiplicity) then portions of it intersect each other transversely. (This is true for any minimal surface in a 3-manifold.) But then the smooth convergence $M_i \to M$ away from S implies that the M_i would also have self intersections. In other words, if the M_i are embedded, then $M \cup S$ is also embedded (possibly with multiplicity). □

If the surfaces M_i in theorem 25 are simply connected, one can say more:

THEOREM 26. *Suppose in the concentration theorem that $M_i \subset \Omega \subset \mathbf{R}^3$ and that the M_i are embedded and simply connected. Then $S = \emptyset$.*

PROOF. Suppose to the contrary that S contains a point p. Near p, the surface $M \cup \{p\}$ is a smooth embedded surface with some multiplicity Q. Thus we can choose a small closed ball $\mathbf{B}$ around p such that $M \cap \partial\mathbf{B}$ is very nearly circular and so that $\mathbf{B} \cap S = \{p\}$. The smooth convergence $M_i \to M$ away from S implies that (for large i) $M_i \cap \partial\mathbf{B}$ is the union of Q very nearly circular curves. (This is where we use embeddedness of the M_i: if the M_i were not embedded, $M_i \cap \partial\mathbf{B}$ might contain a component that is perturbation of a circle transversed multiple times.) By the convex hull property (see exercise (1) after theorem 15), $M_i \cap \mathbf{B}$ is a union of simply connected components. Since (for large i) each such component has very nearly circular boundary, its total curvature is close to 0 by the Gauss-Bonnet Theorem. But then the curvatures of the M_i are uniformly bounded on compact subsets of the interior of $\mathbf{B}$ by the 4π curvature estimate (theorem 23). □

Theorem 26 and its proof generalize to Riemannian 3-manifolds, but one has to be careful because in some 3-manifolds, simple connectivity of a minimal surface M does not imply that the components of its intersection with a small ball (say a geodesic ball) are simply connected. Thus one needs to make some additional hypothesis. For example, one could assume that the ambient space is simply connected and has non-positive sectional curvatures or, more generally, that for each $p \in \Omega$ and $r > 0$, the set $\{x \in \Omega : \text{dist}(x,p) \le r\}$ has smooth boundary and that the mean curvature vector of that boundary points into the set. (See exercise 3 after theorem 15).

To apply the concentration theorem, we need uniform local bounds on total curvature. Such bounds are implied by uniform local bounds on genus and area:

THEOREM 27. [**Ilm95**, Theorem 3] *Let Ω be an open subset of a smooth Riemannian manifold. Suppose that M_i are minimal surfaces in Ω with $\partial M_i \subset \partial\Omega$. Suppose also that*

$$\sup_i \text{genus}(M_i) < \infty$$

and that

$$\sup_i \text{area}(M_i \cap U) < \infty \qquad \textit{for } U \subset\subset \Omega.$$

Then

$$\sup_i TC(M_i \cap U) < \infty \qquad \textit{for } U \subset\subset \Omega.$$

Thus under the hypotheses of this theorem, we get the conclusion of the concentration theorem: smooth convergence (after passing to a subsequence) away from a discrete set S.

Note: Ilmanen's Theorem 3 is about surfaces, not necessarily minimal, in Euclidean space; it gives local bounds on total curvature (integral of the norm of the second fundamental form squared) in terms of genus, area, and integral of the square of the mean curvature. To deduce Theorem 27 from that result, isometrically embed Ω into a Euclidean space.

Stability

Let M be a compact minimal submanifold of a Riemannian manifold. We say that M is **stable** provided

$$\left(\frac{d}{dt}\right)^2_{t=0} \text{area}(\phi_t M) \geq 0$$

for all deformations ϕ_t with $\phi_0(x) \equiv x$ and $\phi_t(y) \equiv y$ for $y \in \partial M$. For noncompact M, we say that M is stable provided each compact portion of M is stable.

If $M \subset \mathbf{R}^n$ is an oriented minimal hypersurface and if $X(x) = \left(\frac{d}{dt}\right)_{t=0} \phi_t(x)$ is a normal vectorfield, we can write $X = u\nu$ where $u : M \to \mathbf{R}$ and ν is the unit normal vectorfield. Note that $\phi_t(x) = x$ for $x \in \partial M$ implies that $u \equiv 0$ on ∂M.

THEOREM 28 (The second variation formula). *Under the hypotheses above,*

$$\begin{aligned}\left(\frac{d}{dt}\right)^2_{t=0} \text{area}(\phi_t M) &= \frac{1}{2}\int_M (|\nabla u|^2 - |A|^2 u^2)\, dS \\ &= \frac{1}{2}\int_M (-\Delta u - |A|^2) u\, dS.\end{aligned}$$

To prove the theorem, one observes that

$$\left(\frac{d}{dt}\right)^2 \text{area}(\phi_t M) = \int \left(\frac{d}{dt}\right)^2 J_m(D\phi_t)\, dS,$$

and calculates $\left(\frac{d}{dt}\right)^2 J_m(D\phi_t)$ as in the proof of the first variation formula. (Integrate by parts to get the second expression from the first.) The formula remains true in an oriented ambient manifold N, except that one replaces $|A|^2$ by $|A|^2 + \text{Ricci}_N(\nu, \nu)$ where ν is the unit normal vectorfield to M. See [**Sim83**, §9] or [**CM11**, 1.§8], for example, for details.

The following theorem is one of the most important and useful facts about stable surfaces. It was discovered independently by Do Carmo and Peng [**dCP79**] and by Fischer-Colbrie and Schoen [**FCS80**]. A few years later Pogorelov gave another proof [**Pog81**].

THEOREM 29. *(1) A complete, stable, orientable minimal surface in $\mathbf{R}^3$ must be a plane.*

(2) If M is a stable, orientable minimal surface in $\mathbf{R}^3$, then

$$|A(p)| \operatorname{dist}(p, \partial M) \leq C$$

for some $C < \infty$.

As usual, (1) and (2) are equivalent. Also, a version of (2) holds in Riemannian 3-manifolds. In some ways, Fischer-Colbrie and Schoen get the best results, because they get results in 3-manifolds of nonnegative scalar curvature that include (1) as a special case. However, the proof below is a slight modification[4] of Pogorelov's. First we prove some preliminary results.

THEOREM 30 (Fischer-Colbrie/Schoen). *Suppose M is an oriented minimal hypersurface in $\mathbf{R}^n$. Then M is stable if and only if there is a positive solution of*

$$\Delta u + |A|^2 u = 0$$

on $M \setminus \partial M$.

This is actually a very general fact about the lowest eigenvalue of self-adjoint, second-order elliptic operators (first proved by Barta for the Laplace operator). In particular, theorem 30 is true in Riemannian manifolds with $|A|^2$ replaced by $|A|^2 + \mathrm{Ricci}(\nu, \nu)$. See [**FCS80**] or [**CM11**, 1.§8, proposition 1.39] for the proof.

COROLLARY 31. *Let M be as in theorem 30. If M is stable, then so is its universal cover.*

PROOF. Lift the function u from M to its universal cover. □

PROPOSITION 32. *Let M be a complete, simply connected surface with $K \leq 0$. Let $A(r) = A_p(r)$ be the area of the geodesic ball $\mathbf{B}_r$ of radius r about some point p. Let*

$$\theta(M) = \lim_{r \to \infty} \frac{A(r)}{\pi r^2}.$$

Then

$$\theta(M) = 1 - \frac{1}{2\pi} \int_M K \, dS = 1 + \frac{TC(M)}{2\pi}.$$

Note that $\theta(M)$ is an intrinsic analog of $\Theta(M)$, the density at infinity of a properly immersed minimal surface (without boundary) in Euclidean space discussed in lecture 1.

PROOF. Let $L(r)$ be the length of ∂B_r. Then $A' = L$, so

$$A'' = L' = \int_{\partial B_r} k \, ds = 2\pi - \int_{B_r} K \, dS.$$

(The formula for L' is a special case of the first variation formula.) Thus

$$\lim_{r \to \infty} A''(r) = 2\pi - \int_M K \, dS = 2\pi + TC(M).$$

The result follows easily. □

COROLLARY 33. *If M (as above) is a minimal surface in $\mathbf{R}^3$ and if $\theta(M) < 3$, then M is a plane.*

PROOF. If $\theta(M) < 3$, then $TC(M) < 4\pi$ (by the proposition), and therefore M is a plane (by corollary 18). □

[4]Here corollary 33 is used in place of one of the lemmas that Pogorelov proves.

LEMMA 34 (Pogorelov). *Let $M \subset \mathbf{R}^3$ be a simply connected, minimal immersed surface. Suppose B_R is a geodesic ball in M of radius R about some point $p \in M$ such that the interior of B_R contains no points of ∂M, i.e., such that* $\operatorname{dist}(p, \partial M) \geq R$.

If $A(R) := \operatorname{area}(B_R) > \frac{4}{3}\pi R^2$, then B_R is unstable.

PROOF. We may assume that $M = B_R$. To prove instability, it suffices (by the second variation formula) to find a function u in B_R with $u|\partial B_R = 0$ such that $Q(u) < 0$, where

$$Q(u) = \int_M (|\nabla u|^2 - |A|^2 u^2)\, dS = \int_M |\nabla u|^2\, dS + 2\int_M Ku^2\, dS \tag{13}$$

(The second equality holds because $|H|^2 = |A|^2 + 2K$ for any surface.) Let r and θ be geodesic polar coordinates in M centered at the point p. Thus the metric has the form

$$ds^2 = dr^2 + g^2\, d\theta^2$$

for some nonnegative function $g(r, \theta)$ such that $g(0, 0) = 0$ and $g_r(0, 0) = 1$. Recall that the Gauss curvature is given by

$$K = -\frac{g_{rr}}{g}.$$

Thus the second integral in (13) becomes

$$\begin{aligned} Q_2(u) &=: 2\int_M u^2 K\, dS \\ &= 2\int_0^{2\pi}\int_0^R u^2 K g\, dr\, d\theta \\ &= -2\int_0^{2\pi}\int_0^R u^2 g_{rr}\, dr\, d\theta. \end{aligned}$$

Integrating by parts twice gives

$$\begin{aligned} Q_2(u) &= 4\pi u(0)^2 - 2\int_0^{2\pi}\int_0^R (u^2)_{rr} g\, dr\, d\theta \\ &= 4\pi u(0)^2 - 4\int_0^{2\pi}\int_0^R (u_r)^2 g\, dr\, d\theta - 4\int_0^{2\pi}\int_0^R u u_{rr} g\, dr\, d\theta \\ &= 4\pi u(0)^2 - 4\int_M (u_r)^2\, dS - 4\int_0^{2\pi}\int_0^R u u_{rr} g\, dr\, d\theta. \end{aligned} \tag{14}$$

Now let $u(r, \theta) = u(r) = (R - r)/R$, so that $u(r)$ decreases linearly from $u(0) = 1$ to $u(R) = 0$. Then the last integral in (14) vanishes, and $(u_r)^2 = |\nabla u|^2 = 1/R^2$, so combining (13) and (14) gives

$$Q(u) = 4\pi - \frac{3}{R^2}A(R),$$

which is negative if $A(R) > \frac{4}{3}\pi R^2$. □

Now we can give the proof of theorem 29:

PROOF. By corollary 31, we may assume that M is simply connected. Suppose that M is not a plane. Then by corollary 33,

$$\theta(M) \geq 3 > \frac{4}{3},$$

so $\frac{A(r)}{\pi r^2} > \frac{4}{3}$ for large r. But then M is unstable by lemma 34. □

LECTURE 4

Existence and Regularity of Least-Area Surfaces

Our goal today is existence and regularity of a surface of least area bounded by a smooth, simple closed curve Γ in $\mathbf{R}^N$. As mentioned in lecture 1, the nature of such surfaces depends (in an interesting way!) on what we mean by "**surface**", "**area**", and "**bounded by**". There are different possible definitions of these terms, and they lead to different versions of the Plateau problem.

In the most classical version of the Plateau problem: "surface" means "continuous mapping $F : D \to \mathbf{R}^n$ of a disk", "bounded by Γ" means "such that $F : \partial D \to \Gamma$ is a monotonic parametrization", and "area" means "mapping area" (as in multivariable calculus):

$$A(F) := \int J(DF)\, dS$$

where

$$J(DF) = \sqrt{|F_x|^2|F_y|^2 - (F_x \cdot F_y)^2}.$$

($F_x = \partial F/\partial y$ and $F_y = \partial F/\partial y$.)

THEOREM 35 (The Douglas-Rado Theorem). *Let Γ be a smooth, simple closed curve in $\mathbf{R}^N$. Let $\mathcal{C}$ be the class of continuous maps $F : \overline{D} \to \mathbf{R}^N$ such that $F|D$ is locally Lipschitz and such that $F : \partial D \to \Gamma$ is a monotonic parametrization. Then there exists a map $F \in \mathcal{C}$ that minimizes the mapping area $A(F)$. Indeed, there exists such a map that is harmonic and almost conformal, and that is a smooth immersion on D except (possibly) at isolated points ("branch points").*

REMARK. The theorem remains true even if Γ is just a continuous simple closed curve, provided one assumes that the class $\mathcal{C}$ contains a map of finite area. (If Γ is smooth, or, more generally, if it has finite arclength, then $\mathcal{C}$ does contain a finite-area map.) With fairly minor modifications, the proof presented below establishes the more general result. See [**Law80**], for example, for details. Even more generally, Douglas proved that $\mathcal{C}$ contains a harmonic, almost conformal map without the assumption that $\mathcal{C}$ contains a finite-area map. Morrey [**Mor48**] generalized the Douglas-Rado theorem by replacing $\mathbf{R}^n$ by a general Riemannian n-manifold N under a rather mild hypothesis ("homogeneous regularity") on the behavior of N at infinity.

We say that a continuous map $\phi : \partial D \to \Gamma$ is a monotonic parametrization provided it is continuous, surjective, and has the following property: the inverse image of each point in Γ is a connected subset of ∂D. Roughly speaking, this means that if a point p goes once around ∂D, always moving in one direction (e.g., counterclockwise), then $\phi(p)$ goes once around Γ, always in one direction. (Note that ϕ is allowed to map arcs of ∂D to a single points in Γ.)

Note that we need some condition on F to guarantee that $A(F)$ makes sense. Requiring that F be locally Lipschitz on D is such a condition, since such an F is differentiable almost everywhere by Rademacher's Theorem. (Alternatively, one could work in the Sobolev space of mappings whose first derivatives are in L^2.)

The most natural approach for proving existence for this (or any other minimization problem) is the "direct method", which we describe now. Let α be the infimum of $A(F)$ among all $F \in \mathcal{C}$. Then there exists a **minimizing sequence** F_i, i.e., a sequence $F_i \in \mathcal{C}$ such that $A(F_i) \to \alpha$. Now one hopes that there exists a subsequence $F_{i(j)}$ that converges to a limit $F \in \mathcal{C}$ with $A(F) = \alpha$.

For the direct method to work, one needs two ingredients: a compactness theorem (to guarantee existence of a subsequential limit $F \in \mathcal{C}$), and lowersemicontinuity of the functional $A(\cdot)$ (to guarantee that $A(F) = \alpha$.)

For the Plateau problem, a minimizing sequence need not have a convergent subsequence[1]. For example, there exists a minimizing sequence F_i such that the images $F_i(D)$ converge as sets to all of $\mathbf{R}^N$:

$$\text{dist}(p, F_i(D)) \to 0 \text{ for every } p \in \mathbf{R}^N. \tag{15}$$

(Think of $F_i(D)$ as a flat disk with a long, thin tentacle attached near the center. Even if the tentacle is very long, its area can be made arbitrarily small by making it sufficiently thin. By making the tentacle meander more and more as $i \to \infty$, we can arrange for (15) to hold, even though $A(F_i)$ converges to the area of the flat disk.)

One can also find a minimizing sequence F_i such that $F_i \mid D$ converges pointwise to a constant map. For example, suppose that Γ is the unit circle $x^2 + y^2 = 1$ in the plane $z = 0$, and let

$$F_k : \overline{D} \subset \mathbf{R}^2 \to \mathbf{R}^3$$
$$F_k(x, y) = (x^2 + y^2)^k (x, y, 0).$$

To avoid such pathologies, instead of using an arbitrary minimizing sequence, we choose a well-behaved minimizing sequence. For that, we make use of the energy functional. The **energy** of a map $F : \Omega \subset \mathbf{R}^2 \to \mathbf{R}^N$ is

$$E(F) = \frac{1}{2} \int_M |DF|^2 \, dS$$

where $|DF|^2 = |F_x|^2 + |F_y|^2$.

We need several facts about energy:

LEMMA 36 (Area-Energy Inequality). *For $F \in \mathcal{C}$,*

$$A(F) \le E(F),$$

with equality if and only if F is almost conformal.

PROOF. For any two vectors $\mathbf{u}$ and $\mathbf{v}$ in $\mathbf{R}^n$,

$$\sqrt{|\mathbf{u}|^2|\mathbf{v}|^2 - (\mathbf{u} \cdot \mathbf{v})^2} \le \sqrt{|\mathbf{u}|^2|\mathbf{v}|^2} = |\mathbf{u}|\,|\mathbf{v}| \le \frac{1}{2}(|\mathbf{u}|^2 + |\mathbf{v}|^2),$$

[1]In the geometric measure theory approach to Plateau's problem, one works with a class of surfaces and a suitable notion of convergence for which minimizing sequences do have convergent subsequences. One disadvantage (compared to the classical approach described here) is that a limit of simply connected surfaces need not not be simply connected.

with equality if and only if **u** are **v** orthogonal and have the same length. Apply that fact to F_x and F_y and integrate. □

LEMMA 37. *Suppose $F : \overline{D} \to \mathbf{R}^N$ is smooth and harmonic. Then*

$$E(F) \le E(G)$$

for all smooth $G : \overline{D} \to \mathbf{R}^N$ with $G|\partial D = F|\partial D$, with equality if and only if $G = F$.

PROOF. Let $V = G - F$. Then

$$\begin{aligned} E(G) &= E(F+V) \\ &= E(F) + E(V) + \int DF \cdot DV \, dS \\ &= E(F) + E(V) - \int \Delta F \cdot V \, dS \\ &= E(F) + E(V). \end{aligned}$$

□

(The proof actually shows that the lemma holds for domains in arbitrary Riemannian manifolds, and in the Sobolev space of mappings whose first derivatives are in L^2.)

PROOF OF THE DOUGLAS-RADO THEOREM. Let $\alpha = \inf\{A(F) : F \in \mathcal{C}\}$. We begin with four claims, each of which implies that we can find a minimizing sequence consisting of functions in $\mathcal{C}$ with some additional desirable properties.

CLAIM 1. *For every $\beta > \alpha$, there is a smooth map $F \in \mathcal{C}$ with $A(F) < \beta$.*

PROOF OF CLAIM 1. We will show that there is an $F \in \mathcal{C}$ such that F is Lipschitz on $\overline{D}$, such that F is smooth near ∂D, and such that $A(F) < \beta$. The assertion of claim 1 then readily follows by standard approximation theorems.

By definition of α, there is an $G \in \mathcal{C}$ with $A(G) < \beta$. Let $R > 0$ be the **reach** of the curve Γ, i.e., the supremum of numbers ρ such that every point p with $\operatorname{dist}(p, \Gamma) < \rho$ has a unique nearest point $\Pi(p)$ in Γ. For $\delta < R/2$, let $\Phi_\delta : \mathbf{R}^n \to \mathbf{R}^n$ be the map

$$\Phi_\delta(p) = \begin{cases} p & \text{if } \operatorname{dist}(p,\Gamma) \ge 2\delta, \\ \Pi(p) & \text{if } \operatorname{dist}(p,\Gamma) \le \delta, \text{ and} \\ \Pi(p) + \left(\frac{\operatorname{dist}(p,\Gamma)}{\delta} - 1\right)(p - \Pi(p)) & \text{if } \delta \le \operatorname{dist}(p,\Gamma) \le 2\delta. \end{cases}$$

Then for every $\delta \in (0, R)$, the map $\Phi_\delta \circ G$ is in the class $\mathcal{C}$. Furthermore, $A(\Phi_\delta \circ G) \to A(G)$ as $\delta \to 0$. Now let $F = \Phi_\delta \circ G$ for a $\delta > 0$ small enough that $A(F) < \beta$.

Note that there is an r with $0 < r < 1$ such that F maps the annular region $A := \{z : r \le |z| \le 1\}$ to Γ: $F(A) = \Gamma$. Now it is straightforward to modify the definition of F on A so that $F(A)$ remains Γ, so that F is Lipschitz, and so that F is smooth near ∂D and maps ∂D diffeomorphically to ∂D. (Note that this modification does not change $A(F)$.) □

CLAIM 2. *If $\beta > \alpha$, then there exists a smooth map $G \in \mathcal{C}$ with $E(G) \le \beta$.*

Since $A(G) \le E(G)$ for every map G, claim 2 is stronger than claim 1.

PROOF OF CLAIM 2. By claim 1, there is a smooth map $F \in \mathcal{C}$ with $A(F) < \beta$. Although F is smooth, its image need not be a smooth surface. That is, F need not be an immersion. To get around this, for $\delta > 0$, we define a new map

$$F_\delta : \overline{D} \to \mathbf{R}^n \times \mathbf{R}^2 \cong \mathbf{R}^{n+2},$$
$$F_\delta(z) = (F(z), \delta z).$$

By choosing δ small, we can assume that $A(F_\delta) < \beta$.

Now $F_\delta(\overline{D})$ is a smooth, embedded disk. Hence (by existence of conformal coordinates and the Riemann mapping theorem), we can parametrize $F_\delta(\overline{D})$ by a smooth **conformal** map $\Phi : \overline{D} \to \mathbf{R}^N$. Let $G = \Pi \circ \Phi$, where $\Pi : \mathbf{R}^n \times \mathbf{R}^2 \to \mathbf{R}^n$ is the projection map. Then

$$E(G) \le E(\Phi) = A(\Phi) = A(F_\delta) < \beta,$$

where $E(\Phi) = A(\Phi)$ by conformality of Φ and where $A(\Phi) = A(F_\delta)$ because Φ and F_δ parametrize the same surface. □

CLAIM 3. *For every $\beta > \alpha$, there is a smooth harmonic map $F \in \mathcal{C}$ such that $E(F) \le \beta$.*

PROOF OF CLAIM 3. By claim 2, there is a smooth map $G \in \mathcal{C}$ with $E(G) < \beta$. Now let $F : \overline{D} \to \mathbf{R}^n$ be the harmonic map with the same boundary values as G. By lemma 37, $E(F) \le E(G) < \beta$. □

CLAIM 4. *Let a, b, and c be three distinct points in ∂D, and let $\hat{a}$, $\hat{b}$, and $\hat{c}$ be three distinct points in Γ. For every $\beta > \alpha$, there is a smooth harmonic map $F \in \mathcal{C}$ such that $E(F) < \beta$ and such that F maps a, b, and c to $\hat{a}$, $\hat{b}$, and $\hat{c}$.*

PROOF. By claim 3, there is a smooth harmonic map $F \in \mathcal{C}$ such that $E(F) < \beta$. Let a', b', and c' be points in ∂D that are mapped by F to $\hat{a}$, $\hat{b}$, and $\hat{c}$. Let $u : \overline{D} \to \overline{D}$ be the unique conformal diffeomorphism that maps a, b, and c to a', b', and c'. Then $F \circ u$ has the desired properties. (For any map F with a two-dimensional domain and for any conformal diffeomorphism u of the domain, note that $E(F) = E(F \circ u)$, and that if F is harmonic, then so is $F \circ u$.) □

By claim 4, we can find a sequence of smooth, harmonic maps $F_i \in \mathcal{C}$ such that

$$E(F_i) \to \alpha = \inf_{F \in \mathcal{C}} A(F).$$

Furthermore, we can choose the F_i so that they map a, b, and c in ∂D to $\hat{a}$, $\hat{b}$, and $\hat{c}$ in Γ.

By the maximum principle for harmonic functions (applied to $L \circ F_i$, for each linear function $L : \mathbf{R}^n \to \mathbf{R}$), the F_i are uniformly bounded:

$$\max_{\overline{D}} |F_i(\cdot)| = \max_{\partial D} |F_i(\cdot)| = \max_{p \in \Gamma} |p|. \tag{16}$$

Thus by passing to a subsequence, we can assume that the F_i converge smoothly on the interior of the disk[2] to a harmonic map F. However, we need uniform convergence on the closed disk.

[2]For readers not familiar with this fact about harmonic maps (which holds more generally for solutions of second-order linear elliptic partial differential equations under mild conditions on the coefficients), note that each coordinate of F_i is the real part of a holomorphic function. By (16), those holomorphic functions take values in a strip in the complex plane, and hence form a normal family.

CLAIM (Equicontinuity). *The maps F_i are equicontinuous.*

PROOF OF EQUICONTINUITY. Suppose not. Then (by the smooth convergence on the interior) there exist point $p_i \in \partial D$ and $q_i \in \overline{D}$ such that

$$\delta_i := |p_i - q_i| \to 0$$

and such that $|F_i(p_i) - F_i(q_i)| \not\to 0$. By passing to a subsquence (and by relabeling, if necessary) we may assume that the p_i converge to a point $p \in \partial D$ that does not lie on the closed arc joining b to c (and disjoint from a.) Let $E = \sup_i E(F_i)$. By the Courant-Lebesgue Lemma (lemma 38 below), there exist arcs

$$C_i = D \cap \partial \mathbf{B}(p_i, r_i)$$

with $r_i \in [\delta_i, \sqrt{\delta_i}]$ such that the arclength L_i of $F|C_i$ satisfies

$$L_i \leq \sqrt{\frac{8\pi E}{|\ln(\delta_i)|}}$$

which tends to 0 as $i \to \infty$.

Let $D_i = D \cap \mathbf{B}(p_i, r_i)$. The boundary of D_i consists of two arcs, C_i and an arc C_i' in ∂D, namely $\mathbf{B}(p_i, r_i) \cap \partial D$. The two arcs have the same endpoints. Since the length of $F(C_i)$ tends to 0, the distance between the endpoints tends to 0.

Thus $F(C_i')$ is an arc in Γ, and the distance between the endpoints tends to 0. Thus, for large i, $F(C_i)$ is either a (i) very short arc in Γ or (ii) all of Γ except for a very short arc. Since $F(C_i')$ contains $F(p)$ and (for large i) is disjoint from the arc in Γ joining $\hat{b}$ to $\hat{c}$, in fact $F(C_i)$ must be very short arc in Γ: its length tends to 0 as $i \to \infty$.

We have shown that the arclength and therefore the diameter[3] of $F(\partial D_i)$ tends to 0. By the maximum principle for harmonic functions, $F_i(\overline{D_i})$ is contains in the convex hull of $F_i(\partial D_i)$, so the diameter of $F_i(\overline{D_i})$ tends to 0. Therefore $|F(p_i) - F(q_i)| \to 0$. This completes the proof of equicontinuity. □

By equicontinuity, we can (by passing to a subsequence) assume that the F_i converge uniformly on $\overline{D}$ to a limit map F. As already mentioned, F is harmonic on the interior. The uniform convergence implies that $F \in \mathcal{C}$, so

$$\alpha \leq A(F) \leq E(F) \leq \liminf E(F_i) \leq \alpha.$$

Since $A(F) = E(F)$, the map is almost conformal. □

LEMMA 38 (Courant-Lebesgue Lemma). *Let $\Omega \subset \mathbf{R}^2$ and $F : \Omega \to \mathbf{R}^n$ be a map with energy E. Let p be a point in $\mathbf{R}^2$ and let $L(r)$ be the arclength of $F|\partial\mathbf{B}(p, r)$. Then*

$$\int_0^\infty \frac{L(r)^2}{r}\, dr \leq 4\pi E.$$

Consequently,

$$\min_{a \leq r \leq b} L(r)^2 \leq \frac{4\pi E}{\ln(b/a)}.$$

[3]The diameter of a subset of a metric space is the supremum of the distance between pairs of points in the subset.

PROOF. It suffices to consider the case $p = 0$. Using polar coordinates,

$$|DF|^2 = |F_r|^2 + \frac{1}{r^2}|F_\theta|^2 \geq \frac{1}{r^2}|F_\theta|^2.$$

Thus

$$\begin{aligned} L(r)^2 &= \left(\int_0^{2\pi} F_\theta \, d\theta\right)^2 \\ &\leq 2\pi \int_0^{2\pi} |F_\theta|^2 \, d\theta. \\ &\leq 2\pi r^2 \int_0^{2\pi} |DF|^2 \, d\theta. \end{aligned}$$

Therefore

$$\int \frac{L(r)^2}{r} \, dr \leq 2\pi \int_{r=0}^{\infty} \int_{\theta=0}^{2\pi} |DF|^2 \, r \, d\theta \, dr = 4\pi E.$$

□

Boundary regularity

The Douglas-Rado Theorem produces an almost conformal, harmonic map F that is continuous on the closed disk and is such that $F|\partial D$ gives a monotonic parametrization of the curve Γ. It is not hard to show that any such map (whether or not it minimizes area) cannot be constant on any arc of ∂D. (See for example [**Oss86**, lemma 7.4] or [**Law80**, proposition 11].) It follows from the monotonicity of $F|\partial D$ that $F : \partial D \to \Gamma$ is a homeomorphism. Later, every such map was proved to be smooth on the closed disk provided Γ is smooth. Roughly speaking, such a map $F : \overline{D} \to \mathbf{R}^n$ turns out to be as regular as Γ. For example, if Γ is $C^{k,\alpha}$ for some $k \geq 1$ and $\alpha \in (0,1)$, then so is F, and if Γ is analytic, then so is F. (Lewy first proved that minimal surfaces in $\mathbf{R}^n$ with analytic boundary curves are analytic up to the boundary. The fundamental breakthrough was due to Hildebrandt [**Hil69**], who, in the case of area-minimizing surfaces, extended Lewy's result to arbitrary ambient manifolds and who also proved the corresponding result for C^4 boundaries. Later Heinz and Hildebrandt [**HH70**] proved such results for surfaces that are minimal but not necessarily area minimizing. See also [**Kin69**].)

Branch points

Let $F : D \to \mathbf{R}^N$ be a non-constant, harmonic, almost conformal map (such as given by the Douglas-Rado theorem).

Recall that harmonicity of F means that the map $F_z = \frac{1}{2}(F_x - iF_y)$ from D to $\mathbf{C}^n$ is holomorphic. Thus F_z can vanish only at isolated points. Those points are called "branch points". Away from the branch points, the map is a smooth, conformal immersion.

Using the Weierstrass representation, it is easy to give examples of minimal surfaces with branch points. (The branch points are the points where g has a pole of order m (possibly 0) and where ν has a zero of order strictly greater than $2m$.) But are there area-minimizing examples? The following theorem implies that there are such examples in $\mathbf{R}^n$ for $n \geq 4$:

THEOREM 39 (Federer, following Wirtinger). *Let M be a complex variety in $\mathbf{C}^n$. Then (as a real variety in $\mathbf{R}^{2n}$) M is absolutely area minimizing in the following sense: if S is a compact portion of M, and if S' is an oriented variety with the same oriented boundary as S, then* $\text{area}(S) \le \text{area}(S')$.

Here "with the same oriented boundary" means that $\partial S' = \partial S$ and that S' and S induce the same orientation on the boundary. For the proof, see [**Fed65**] or [**Law80**, pp. 37–40].

Using the Federer-Wirtinger Theorem, we can give many examples of branched, area-minimizing surfaces. For example, the map

$$F : D \subset \mathbf{R}^2 \cong \mathbf{C}^2 \to \mathbf{R}^4 \cong \mathbf{C}^2$$
$$F(z) = (z^2, z^3)$$

has a branch point at the origin and is area-minimizing by the Federer-Wirtinger Theorem.

Whether there exist any examples other than the ones provided by the Federer-Wirtinger Theorem is a very interesting open question. In other words, must a connected least-area surface with a true[4] branch point in $\mathbf{R}^{2n} \cong \mathbf{C}^n$ be holomorphic after a suitable rotation of $\mathbf{R}^{2n}$? The paper [**MW95**] is suggestive in this regard.

The theorems of Gulliver and Osserman

Osserman and Gulliver proved in $\mathbf{R}^3$ (or more generally in any Riemannian 3-manifold) that the Douglas-Rado solution cannot have any **interior** branch points.[5] Thus (away from the boundary), the map F is a smooth immersion.

Whether the map F in the Douglas-Rado Theorem can have boundary branch points (for a 3-dimensional manifold) is one of longest open questions in minimal surface theory. Using the Federer-Wirtinger Theorem, one can give examples in $\mathbf{R}^n$ for $n \ge 4$, such as

$$F : \{x + iy : x \ge 0\} \to \mathbf{C}^2 \cong \mathbf{R}^4$$
$$F(z) = (z^3, e^{-1/\sqrt{z}}).$$

There are some situations in which boundary branch points are known not to occur:

(1) If Γ lies on the boundary of a compact, strictly convex region in $\mathbf{R}^n$. In this case, one need not assume area minimizing: minimality suffices. (The proof is a slight modification of the proof of theorem 15, together with the Hopf boundary point theorem.)
(2) If Γ is a real analytic curve in $\mathbf{R}^n$ or more generally in an analytic Riemannian manifold [**Whi97**].

[4]A branch point $p \in D$ of F is called **false** if there is a neighborhood U of p such that the image $F(U)$ is a smooth, embedded surface. Otherwise the branch point is **true**. For example, if $F : D \subset \mathbf{C} \to \mathbf{R}^n$ is a smooth immersion, the $z \mapsto F(z^2)$ has a false branch point at $z = 0$.

[5]Osserman ruled out true branch points in $\mathbf{R}^3$, and Gulliver extended Osserman's result to 3-manifolds and also ruled out false branch points. Alt [**Alt72**] independently proved some of Gulliver's results. See [**CM11**], [**Law80**], [**DHS10**], or the original papers for details.

Higher genus surfaces

Let Γ be a simple closed curve in $\mathbf{R}^n$. Does Γ bound a least-area surface of genus one? Not necessarily. Consider a planar circle Γ in $\mathbf{R}^3$. By the convex hull principle (theorem 15), Γ bounds only one minimal surface: the flat disk M bounded by Γ. We can take a minimizing sequence of genus one surfaces, but (for this example) in the limit, the handle shrinks to point, and we end up with the disk.

Technically speaking, the planar circle does bound a least area genus 1 suface in the sense of mappings. Let Σ be a smooth genus 1 surface consisting a a disk with a handle attached. There is a smooth map $F : \Sigma \to M$ that collapses the handle to the center p of the disk M bounded by Γ, and that maps the rest of Σ diffeomorphically to $M \setminus \{p\}$. However, there is no "nice" area-minimizing map $F : \Sigma \to \mathbf{R}^3$ with boundary Γ. For example, there is no such map that is an immersion except at isolated points.

DEFINITION. Let Γ be a smooth, simple closed curve in $\mathbf{R}^n$. Let $\alpha(g)$ be the infimum of the area of genus g surfaces bounded by Γ.

PROPOSITION 40. $\alpha(g) \le \alpha(g-1)$.

PROOF. Take a surface of genus $g-1$ whose area is close to $\alpha(g-1)$, and then attach a very small handle. □

THEOREM 41 (Douglas[6]). *If $\alpha(g) < \alpha(g-1)$, then there exists a domain Σ consisting of a genus g Riemanan surface with an open disk removed, and a continuous map*

$$F : \Sigma \to \mathbf{R}^n$$

that is harmonic and almost conformal in the interior of F, that maps $\partial\Sigma$ monotonically onto Γ, and that has area $A(F)$ equal to $\alpha(g)$.

The proof is similar to the proof of the Douglas-Rado Theorem, but more complicated because not all genus g domains are conformally equivalent. For example, up to conformal equivalence, there is a 3-parameter family of genus-one domains with one boundary component. As a result, we have to vary the domain as well as the map.

The Douglas theorem can be restated slightly informally as follows:

THEOREM 42. *Let g be a nonnegative integer. The least area among all surfaces of genus $\le g$ bounded by Γ is attained by a harmonic, almost conformal map.*

PROOF (USING THE DOUGLAS THEOREM). Let k be the smallest integer such that $\alpha(k) = \alpha(g)$. Then $0 \le k \le g$. If $k = 0$, then the Douglas-Rado solution is a disk that attains the desired infimum $\alpha(g) = \alpha(0)$. If $k > 0$, then $\alpha(k) < \alpha(k-1)$, so the genus k surface given by the Douglas Theorem attains the desired infimum $\alpha(g) = \alpha(k)$. □

The theorems of Gulliver and Osserman also hold for these higher genus surfaces: in $\mathbf{R}^3$ (and in 3-manifolds) they must be smooth immersions except possibly at the boundary.

[6]It seems that Douglas never gave a complete proof of "Douglas's Theorem". A result very similar to Douglas's Theorem, but for minimal surfaces without boundary in Riemannian manifolds, was proved by Schoen and Yau [**SY79**]. Later, Jost [**Jos85**] gave a complete proof of Douglas's original theorem.

Summary: For a fixed genus g and curve Γ, we cannot in general minimize area among surfaces of genus equal to g and get a nice surface: minimizing sequences may converge to surfaces of lower genus. However, we can always minimize area among surfaces of genus $\leq g$: the minimum will be attained by a harmonic, almost conformal map. Intuitively, the Douglas theorem is true because when we take the limit of a minimizing sequence of genus g surfaces, we can loose handles but we cannot gain them.

What happens as the genus increases?

Fix a smooth, simple closed curve Γ in $\mathbf{R}^3$. As above, we let $\alpha(g)$ denote the least area among genus g surfaces bounded by Γ. According to proposition 40, $\alpha(g)$ is a decreasing function of g. The following provides a sufficient condition for $\alpha(g)$ to be strictly less than $\alpha(g-1)$. (Recall that by the Douglas Theorem, the strict inequality implies existence of a least-area genus g surface bounded by Γ.)

THEOREM 43. *Suppose $M \subset \mathbf{R}^3$ is a minimal surface of genus $(g-1)$ bounded by Γ. Suppose also that $M \setminus \Gamma$ is not embedded. Then Γ bounds a genus g surface whose area is strictly less than the area of M. In particular, if* $\text{area}(M) = \alpha(g-1)$, *then $\alpha(g) < \alpha(g-1)$.*

PROOF. One can show that if $M \setminus \Gamma$ is not embedded, then there is a curve C along which two portions of M cross transversely. (There may be many such curves.) We will use that fact without proof here. Note that we can cut and paste M along an arc of C to get a new surface M^*. There are two ways to do the surgery: one produces an orientable surface and the other a non-orientable surface. We do the surgery that makes M^* orientable. The new surface is piecewise smooth but not smooth. It has the same area as M and has genus g. By rounding the corners of M^*, we can make a new genus g surface whose area is strictly less than $\text{area}(M^*) = \text{area}(M)$. □

THEOREM 44. *For each g, there exists a smooth, simple closed curve Γ in $\mathbf{R}^3$ such that $\alpha(0) > \alpha(1) > \cdots > \alpha(g)$, and such that for every $k < g$, each genus-k least area surface is non-embedded.*

PROOF. The **genus** of a simple closed curve in $\mathbf{R}^3$ is defined to be the smallest genus of any embedded minimal surface bounded by the curve. Using elementary knot theory, one can show that there are smooth curves of every genus. Let Γ be such a curve of genus g. For $k = 0, 2, \dots, g-1$, let M_k be a least-area surface of genus $\leq k$ bounded by Γ, so that $\text{area}(M_k) = \alpha(k)$. Since Γ has genus $g > k$, the surface M_k cannot be embedded. Therefore $\alpha(k+1) < \alpha(k)$ by theorem 43. □

Actually, the relevant notion is not the genus of Γ, but rather the "convex hull genus" of Γ: the smallest possible genus of an embedded surface bounded by Γ and lying in the convex hull of Γ.

THEOREM 45 (Almgren-Thurston [**AT77**]). *For every $\epsilon > 0$ and for every positive integer g, there exists a smooth, unknotted, simple closed curve Γ in $\mathbf{R}^3$ whose convex hull genus is g and whose total curvature is less than $4\pi + \epsilon$.*

(Recall that the total curvature of a smooth curve is the integral with respect to arclength of the norm of the curvature vector.)

Later Hubbard [**Hub80**] gave a beautiful, very simple proof of this theorem and gave an explicit formula for calculating the convex hull genus of a large, interesting family of curves.

THEOREM 46. *For every $\epsilon > 0$ and for every positive integer g, there exists a smooth, unknotted simple closed curve Γ in $\mathbf{R}^3$ with total curvature $\le 4\pi + \epsilon$ such that $\alpha(0) > \alpha(1) > \cdots > \alpha(g)$, and such that for every $k < g$, each genus-k least area surface is non-embedded.*

PROOF. Let Γ be a curve satisfying the conclusion of theorem 45. By the convex hull property (theorem 15), any embedded minimal surface bounded by Γ has genus $\ge g$. The rest of the proof is exactly the same as the proof of theorem 44. □

However, for a smooth curve, eventually the function $\alpha(\cdot)$ must stabilize according to the following theorem of Hardt and Simon [**HS79**]:

THEOREM 47. *Let Γ be a smooth simple closed curve in $\mathbf{R}^3$. Let $\alpha = \inf \alpha(\cdot)$ be the infimum of the areas of all orientable surfaces bounded by Γ. Then*

(1) The infimum is attained, and any surface that attains the infimum is smoothly embedded (including at the boundary).
(2) The set of surfaces that attain the infimum is finite.

In particular, if g is the genus of a surface that attains the infimum, then the $\alpha(g) \equiv \alpha(k)$ for all $k \ge g$.

On the other hand, one can construct a simple closed curve Γ that is smooth except at one point such that $\alpha(g) > \alpha(g+1)$ for all g. For example, take such a curve of infinite genus or or even just of infinite convex hull genus, or see [**Alm69**] for an example (due to Fleming [**Fle56**]) for which $\alpha(\cdot)$ is strictly decreasing and for which the Douglas solutions are all embedded. Indeed, all kinds of pathologies can happen once one allows a point at which the curve is not smooth:

THEOREM 48. [**Whi94**, 1.3] *There exists a simple closed curve Γ in $\partial\mathbf{B}(0,1) \subset \mathbf{R}^3$ and a number $A < \infty$ such that Γ is smooth except at one point p and such that the following holds: for every area $a \in [A, \infty]$, for every genus g with $0 \le g \le \infty$, and for every index I with $0 \le I \le \infty$, the curve Γ bounds uncountably many embedded minimal surfaces that are smooth except at p and that have area a, genus g, and index[7] of instability I.*

This is in sharp contrast to the case of an everywhere smooth, simple closed curve Γ in the boundary of a convex set in $\mathbf{R}^3$. For such a curve, one can show that for each genus $g < \infty$, the set of embedded genus-g minimal surfaces bounded by Γ is compact with respect to smooth convergence [**Whi87b**]. It follows that (for each g) the set of possibly indices of instability is finite. With a little more work, one can show that the set of areas of such surfaces (for each g) is a finite set. Of course, if Γ is smooth, then the areas of all the minimal surfaces (regardless of genus) are bounded above according to theorem 2.

Embeddedness: The Meeks-Yau Theorem

THEOREM 49 (Meeks-Yau [**MY82**]). *Let N be a Riemannian 3-manifold and let $F : \overline{D} \to N$ be a least-area disk (parametrized almost conformally) with a smooth boundary curve Γ. Suppose $F(D)$ is disjoint from Γ. Then F is a smooth embedding.*

[7] See, for example, [**CM11**, 1.8] for the definition and the basic properties of the index.

The disjointness hypothesis holds in many situations of interest. In particular, it holds if Γ lies on the boundary of a compact, convex subset of $\mathbf{R}^3$. (This follows from the strong maximum principle.) More generally, it holds if N is a a compact, mean convex 3-manifold and if Γ lies in ∂N. (Mean convexity of N means that the mean curvature vector at each point of the boundary is a nonnegative multiple of the inward unit normal.)

Idea of the proof. Suppose M is immersed but not embedded. One can show that it contains an arc along which it intersects itself transversely. One can cut and paste M along such arcs to get a new piecewise smooth (but not smooth) surface $\tilde{M}$. Such surgery is likely to produce a surface of higher genus (as in the proof of theorem 43). However, Meeks and Yau show that it is possible to do the surgery (simultaneously on many arcs) in such a way that $\tilde{M}$ is still a disk. Thus

$$\text{area}(\tilde{M}) = \text{area}(M),$$

so $\tilde{M}$ is also area minimizing. However, where $\tilde{M}$ has corners, one can round the corners to get a disk with less area than $\tilde{M}$, a contradiction. □

Bibliography

[All72] William K. Allard, *On the first variation of a varifold*, Ann. of Math. (2) **95** (1972), 417–491. MR0307015 (46 #6136)

[Alm69] F. J. Almgren Jr., *Measure theoretic geometry and elliptic variational problems*, Bull. Amer. Math. Soc. **75** (1969), 285–304. MR0240689 (39 #2034)

[Alm86] F. Almgren, *Optimal isoperimetric inequalities*, Indiana Univ. Math. J. **35** (1986), no. 3, 451–547, DOI 10.1512/iumj.1986.35.35028. MR855173 (88c:49032)

[AT77] Frederick J. Almgren Jr. and William P. Thurston, *Examples of unknotted curves which bound only surfaces of high genus within their convex hulls*, Ann. of Math. (2) **105** (1977), no. 3, 527–538. MR0442833 (56 #1209)

[Alt72] Hans Wilhelm Alt, *Verzweigungspunkte von H-Flächen. I* (German), Math. Z. **127** (1972), 333–362. MR0312404 (47 #965b)

[BB11] Jacob Bernstein and Christine Breiner, *Conformal structure of minimal surfaces with finite topology*, Comment. Math. Helv. **86** (2011), no. 2, 353–381, DOI 10.4171/CMH/226. MR2775132 (2012c:53009)

[CO67] Shiing-shen Chern and Robert Osserman, *Complete minimal surfaces in euclidean n-space*, J. Analyse Math. **19** (1967), 15–34. MR0226514 (37 #2103)

[CS09] Jaigyoung Choe and Richard Schoen, *Isoperimetric inequality for flat surfaces*, Proceedings of the 13th International Workshop on Differential Geometry and Related Fields [Vol. 13], Natl. Inst. Math. Sci. (NIMS), Taejŏn, 2009, pp. 103–109. MR2641128 (2011e:53003)

[CS85] Hyeong In Choi and Richard Schoen, *The space of minimal embeddings of a surface into a three-dimensional manifold of positive Ricci curvature*, Invent. Math. **81** (1985), no. 3, 387–394, DOI 10.1007/BF01388577. MR807063 (87a:58040)

[CM11] Tobias Holck Colding and William P. Minicozzi II, *A course in minimal surfaces*, Graduate Studies in Mathematics, vol. 121, American Mathematical Society, Providence, RI, 2011. MR2780140

[DK81] Dennis M. DeTurck and Jerry L. Kazdan, *Some regularity theorems in Riemannian geometry*, Ann. Sci. École Norm. Sup. (4) **14** (1981), no. 3, 249–260. MR644518 (83f:53018)

[DHS10] Ulrich Dierkes, Stefan Hildebrandt, and Friedrich Sauvigny, *Minimal surfaces*, 2nd ed., Grundlehren der Mathematischen Wissenschaften [Fundamental Principles of Mathematical Sciences], vol. 339, Springer, Heidelberg, 2010. With assistance and contributions by A. Küster and R. Jakob. MR2566897 (2012b:49002)

[dCP79] M. do Carmo and C. K. Peng, *Stable complete minimal surfaces in* $\mathbf{R}^3$ *are planes*, Bull. Amer. Math. Soc. (N.S.) **1** (1979), no. 6, 903–906, DOI 10.1090/S0273-0979-1979-14689-5. MR546314 (80j:53012)

[EWW02] Tobias Ekholm, Brian White, and Daniel Wienholtz, *Embeddedness of minimal surfaces with total boundary curvature at most* 4π, Ann. of Math. (2) **155** (2002), no. 1, 209–234, DOI 10.2307/3062155. MR1888799 (2003f:53010)

[Fed65] Herbert Federer, *Some theorems on integral currents*, Trans. Amer. Math. Soc. **117** (1965), 43–67. MR0168727 (29 #5984)

[FCS80] Doris Fischer-Colbrie and Richard Schoen, *The structure of complete stable minimal surfaces in 3-manifolds of nonnegative scalar curvature*, Comm. Pure Appl. Math. **33** (1980), no. 2, 199–211, DOI 10.1002/cpa.3160330206. MR562550 (81i:53044)

[Fle56] Wendell H. Fleming, *An example in the problem of least area*, Proc. Amer. Math. Soc. **7** (1956), 1063–1074. MR0082046 (18,489c)

[Gul73] Robert D. Gulliver II, *Regularity of minimizing surfaces of prescribed mean curvature*, Ann. of Math. (2) **97** (1973), 275–305. MR0317188 (47 #5736)

[Gün89] Matthias Günther, *On the perturbation problem associated to isometric embeddings of Riemannian manifolds*, Ann. Global Anal. Geom. **7** (1989), no. 1, 69–77, DOI 10.1007/BF00137403. MR1029846 (91a:58023)

[HS79] Robert Hardt and Leon Simon, *Boundary regularity and embedded solutions for the oriented Plateau problem*, Ann. of Math. (2) **110** (1979), no. 3, 439–486, DOI 10.2307/1971233. MR554379 (81i:49031)

[HH70] Erhard Heinz and Stefan Hildebrandt, *Some remarks on minimal surfaces in Riemannian manifolds*, Comm. Pure Appl. Math. **23** (1970), 371–377. MR0259765 (41 #4398)

[HCV52] D. Hilbert and S. Cohn-Vossen, *Geometry and the imagination*, Chelsea Publishing Company, New York, N. Y., 1952. Translated by P. Neményi. MR0046650 (13,766c)

[Hil69] Stefan Hildebrandt, *Boundary behavior of minimal surfaces*, Arch. Rational Mech. Anal. **35** (1969), 47–82. MR0248650 (40 #1901)

[HK97] David Hoffman and Hermann Karcher, *Complete embedded minimal surfaces of finite total curvature*, Geometry, V, Encyclopaedia Math. Sci., vol. 90, Springer, Berlin, 1997, pp. 5–93, DOI 10.1007/978-3-662-03484-2_2. MR1490038 (98m:53012)

[KWH93] Hermann Karcher, Fu Sheng Wei, and David Hoffman, *The genus one helicoid and the minimal surfaces that led to its discovery*, Global analysis in modern mathematics (Orono, ME, 1991; Waltham, MA, 1992), Publish or Perish, Houston, TX, 1993, pp. 119–170. MR1278754 (95k:53011)

[Hof05a] David Hoffman, *Computing minimal surfaces*, Global theory of minimal surfaces, Clay Math. Proc., vol. 2, Amer. Math. Soc., Providence, RI, 2005, pp. 259–282. MR2167263 (2006g:53008)

[Hof05b] David Hoffman, *Computing minimal surfaces*, Global theory of minimal surfaces, Clay Math. Proc., vol. 2, Amer. Math. Soc., Providence, RI, 2005, pp. 259–282. MR2167263 (2006g:53008)

[WHW09] Matthias Weber, David Hoffman, and Michael Wolf, *An embedded genus-one helicoid*, Ann. of Math. (2) **169** (2009), no. 2, 347–448, DOI 10.4007/annals.2009.169.347. MR2480608 (2010d:53011)

[HW08] David Hoffman and Brian White, *Genus-one helicoids from a variational point of view*, Comment. Math. Helv. **83** (2008), no. 4, 767–813. MR2442963 (2010b:53013)

[Hub80] J. H. Hubbard, *On the convex hull genus of space curves*, Topology **19** (1980), no. 2, 203–208, DOI 10.1016/0040-9383(80)90007-5. MR572584 (81h:57002)

[Hub57] Alfred Huber, *On subharmonic functions and differential geometry in the large*, Comment. Math. Helv. **32** (1957), 13–72. MR0094452 (20 #970)

[Ilm95] Tom Ilmanen, *Singularities of mean curvature flow of surfaces* (1995), available at `http://www.math.ethz.ch/~ilmanen/papers/pub.html`. Preprint.

[Jos85] Jürgen Jost, *Conformal mappings and the Plateau-Douglas problem in Riemannian manifolds*, J. Reine Angew. Math. **359** (1985), 37–54, DOI 10.1515/crll.1985.359.37. MR794798 (87d:58049)

[Kin69] David Kinderlehrer, *The boundary regularity of minimal surfaces*, Ann. Scuola Norm. Sup. Pisa (3) **23** (1969), 711–744. MR0262943 (41 #7548)

[Law80] H. Blaine Lawson Jr., *Lectures on minimal submanifolds. Vol. I*, 2nd ed., Mathematics Lecture Series, vol. 9, Publish or Perish, Inc., Wilmington, Del., 1980. MR576752 (82d:53035b)

[Lew51] Hans Lewy, *On the boundary behavior of minimal surfaces*, Proc. Nat. Acad. Sci. U. S. A. **37** (1951), 103–110. MR0049398 (14,168b)

[LSY84] Peter Li, Richard Schoen, and Shing-Tung Yau, *On the isoperimetric inequality for minimal surfaces*, Ann. Scuola Norm. Sup. Pisa Cl. Sci. (4) **11** (1984), no. 2, 237–244. MR764944 (86e:53004)

[LJ94] Xianqing Li-Jost, *Uniqueness of minimal surfaces in Euclidean and hyperbolic 3-space*, Math. Z. **217** (1994), no. 2, 275–285, DOI 10.1007/BF02571945. MR1296397 (95j:53015)

[MP11] William H. Meeks III and Joaquín Pérez, *The classical theory of minimal surfaces*, Bull. Amer. Math. Soc. (N.S.) **48** (2011), no. 3, 325–407, DOI 10.1090/S0273-0979-2011-01334-9. MR2801776 (2012m:53008)

[MP] William H. Meeks III and Joaquín Pérez, *Embedded minimal surfaces of finite topology*, available at `http://www.ugr.es/~jperez/papers/finite-top-sept29.pdf`.

[MY82] William H. Meeks III and Shing Tung Yau, *The classical Plateau problem and the topology of three-dimensional manifolds. The embedding of the solution given by Douglas-Morrey and an analytic proof of Dehn's lemma*, Topology **21** (1982), no. 4, 409–442, DOI 10.1016/0040-9383(82)90021-0. MR670745 (84g:53016)

[MW95] Mario J. Micallef and Brian White, *The structure of branch points in minimal surfaces and in pseudoholomorphic curves*, Ann. of Math. (2) **141** (1995), no. 1, 35–85, DOI 10.2307/2118627. MR1314031 (96a:58063)

[MS73] J. H. Michael and L. M. Simon, *Sobolev and mean-value inequalities on generalized submanifolds of R^n*, Comm. Pure Appl. Math. **26** (1973), 361–379. MR0344978 (49 #9717)

[Mor09] Frank Morgan, *Geometric measure theory*, 4th ed., Elsevier/Academic Press, Amsterdam, 2009. A beginner's guide. MR2455580 (2009i:49001)

[Mor48] Charles B. Morrey Jr., *The problem of Plateau on a Riemannian manifold*, Ann. of Math. (2) **49** (1948), 807–851. MR0027137 (10,259f)

[Nas56] John Nash, *The imbedding problem for Riemannian manifolds*, Ann. of Math. (2) **63** (1956), 20–63. MR0075639 (17,782b)

[Nit73] Johannes C. C. Nitsche, *A new uniqueness theorem for minimal surfaces*, Arch. Rational Mech. Anal. **52** (1973), 319–329. MR0341258 (49 #6008)

[Oss63] Robert Osserman, *On complete minimal surfaces*, Arch. Rational Mech. Anal. **13** (1963), 392–404. MR0151907 (27 #1888)

[Oss70] Robert Osserman, *A proof of the regularity everywhere of the classical solution to Plateau's problem*, Ann. of Math. (2) **91** (1970), 550–569. MR0266070 (42 #979)

[Oss86] Robert Osserman, *A survey of minimal surfaces*, 2nd ed., Dover Publications, Inc., New York, 1986. MR852409 (87j:53012)

[Pog81] A. V. Pogorelov, *On the stability of minimal surfaces* (Russian), Dokl. Akad. Nauk SSSR **260** (1981), no. 2, 293–295. MR630142 (83b:49043)

[SY79] R. Schoen and Shing Tung Yau, *Existence of incompressible minimal surfaces and the topology of three-dimensional manifolds with nonnegative scalar curvature*, Ann. of Math. (2) **110** (1979), no. 1, 127–142, DOI 10.2307/1971247. MR541332 (81k:58029)

[Sim83] Leon Simon, *Lectures on geometric measure theory*, Proceedings of the Centre for Mathematical Analysis, Australian National University, vol. 3, Australian National University, Centre for Mathematical Analysis, Canberra, 1983. MR756417 (87a:49001)

[Usm96] Z. D. Usmanov, *On Efimov surfaces that are rigid "in the small"* (Russian, with Russian summary), Mat. Sb. **187** (1996), no. 6, 119–130, DOI 10.1070/SM1996v187n06ABEH000140; English transl., Sb. Math. **187** (1996), no. 6, 903–915. MR1407683 (97d:53004)

[Web05] Matthias Weber, *Classical minimal surfaces in Euclidean space by examples: geometric and computational aspects of the Weierstrass representation*, Global theory of minimal surfaces, Clay Math. Proc., vol. 2, Amer. Math. Soc., Providence, RI, 2005, pp. 19–63. MR2167255 (2006e:53025)

[Whi87a] Brian White, *Complete surfaces of finite total curvature*, J. Differential Geom. **26** (1987), no. 2, 315–326. Correction in J. Differential Geom. **28** (1988), no. 2, 359–360 MR961520 (89j:53009). MR906393 (88m:53020)

[Whi87b] B. White, *Curvature estimates and compactness theorems in 3-manifolds for surfaces that are stationary for parametric elliptic functionals*, Invent. Math. **88** (1987), no. 2, 243–256, DOI 10.1007/BF01388908. MR880951 (88g:58037)

[Whi94] Brian White, *The bridge principle for unstable and for singular minimal surfaces*, Comm. Anal. Geom. **2** (1994), no. 4, 513–532. MR1336893 (96k:49063)

[Whi97] Brian White, *Classical area minimizing surfaces with real-analytic boundaries*, Acta Math. **179** (1997), no. 2, 295–305, DOI 10.1007/BF02392746. MR1607558 (98m:53018)

[Whi05] Brian White, *A local regularity theorem for mean curvature flow*, Ann. of Math. (2) **161** (2005), no. 3, 1487–1519, DOI 10.4007/annals.2005.161.1487. MR2180405 (2006i:53100)

[Whi10] Brian White, *The maximum principle for minimal varieties of arbitrary codimension*, Comm. Anal. Geom. **18** (2010), no. 3, 421–432, DOI 10.4310/CAG.2010.v18.n3.a1. MR2747434 (2012a:53117)

[Yan98] Deane Yang, *Gunther's proof of Nash's isometric embedding theorem* (1998), available at `arXiv:math/9807169[math.DG]`.

Published Titles in This Series

22 **Hubert L. Bray, Greg Galloway, Rafe Mazzeo, and Natasa Sesum, Editors,** Geometric Analysis, 2016

21 **Mladen Bestvina, Michah Sageev, and Karen Vogtmann, Editors,** Geometric Group Theory, 2014

20 **Benson Farb, Richard Hain, and Eduard Looijenga, Editors,** Moduli Spaces of Riemann Surfaces, 2013

19 **Hongkai Zhao, Editor,** Mathematics in Image Processing, 2013

18 **Cristian Popescu, Karl Rubin, and Alice Silverberg, Editors,** Arithmetic of L-functions, 2011

17 **Jeffery McNeal and Mircea Mustaţă, Editors,** Analytic and Algebraic Geometry, 2010

16 **Scott Sheffield and Thomas Spencer, Editors,** Statistical Mechanics, 2009

15 **Tomasz S. Mrowka and Peter S. Ozsváth, Editors,** Low Dimensional Topology, 2009

14 **Mark A. Lewis, Mark A. J. Chaplain, James P. Keener, and Philip K. Maini, Editors,** Mathematical Biology, 2009

13 **Ezra Miller, Victor Reiner, and Bernd Sturmfels, Editors,** Geometric Combinatorics, 2007

12 **Peter Sarnak and Freydoon Shahidi, Editors,** Automorphic Forms and Applications, 2007

11 **Daniel S. Freed, David R. Morrison, and Isadore Singer, Editors,** Quantum Field Theory, Supersymmetry, and Enumerative Geometry, 2006

10 **Steven Rudich and Avi Wigderson, Editors,** Computational Complexity Theory, 2004

9 **Brian Conrad and Karl Rubin, Editors,** Arithmetic Algebraic Geometry, 2001

8 **Jeffrey Adams and David Vogan, Editors,** Representation Theory of Lie Groups, 2000

7 **Yakov Eliashberg and Lisa Traynor, Editors,** Symplectic Geometry and Topology, 1999

6 **Elton P. Hsu and S. R. S. Varadhan, Editors,** Probability Theory and Applications, 1999

5 **Luis Caffarelli and Weinan E, Editors,** Hyperbolic Equations and Frequency Interactions, 1999

4 **Robert Friedman and John W. Morgan, Editors,** Gauge Theory and the Topology of Four-Manifolds, 1998

3 **János Kollár, Editor,** Complex Algebraic Geometry, 1997

2 **Robert Hardt and Michael Wolf, Editors,** Nonlinear partial differential equations in differential geometry, 1996

1 **Daniel S. Freed and Karen K. Uhlenbeck, Editors,** Geometry and Quantum Field Theory, 1995